Selected Works of
Professor Herbert Kroemer

Selected Works of Professor Herbert Kroemer

Editor

C K Maiti

Indian Institute of Technology, Kharagpur, India

World Scientific

NEW JERSEY · LONDON · SINGAPORE · BEIJING · SHANGHAI · HONG KONG · TAIPEI · CHENNAI

Published by

World Scientific Publishing Co. Pte. Ltd.

5 Toh Tuck Link, Singapore 596224

USA office: 27 Warren Street, Suite 401-402, Hackensack, NJ 07601

UK office: 57 Shelton Street, Covent Garden, London WC2H 9HE

British Library Cataloguing-in-Publication Data
A catalogue record for this book is available from the British Library.

SELECTED WORKS OF PROFESSOR HERBERT KROEMER

ISBN-13 978-981-270-901-1
ISBN-10 981-270-901-0

Printed in Singapore by World Scientific Printers

Contents

Introduction

The **2000** Nobel Prize in Physics honored Jack Kilby, Zhores Alferov and Herbert Kroemer ``for basic work on information and communication technology,'' which is a major force in the transition from an industrial society to an information and knowledge based society. Kilby was honored ``for his part in the invention of the integrated circuit.'' Zhores Alferov and Herbert Kroemer were awarded the Nobel Prize in Physics and the citation read: ``for developing semiconductor heterostructures used in high-speed- and opto-electronics.'' Along with his contributions to the field of semiconductor lasers and optoelectronic devices, concepts developed by Kroemer beginning in the 1950s have been remarkably enduring that helped shape the evolution of bipolar transistor technology. This introductory article describes some of the key innovations made by Kroemer and provides a historical context to his contributions.

Herbert Kroemer receiving the Nobel Prize from Sweden's King Carl XVI Gustaf.
Copyright The Nobel Foundation 2000.

Kroemer was born in Germany in 1928 in Weimar, Germany, and studied at the Universities of Jena and Goettingen. Kroemer received his doctorate in theoretical physics in 1952 from the University of Goettingen, Niedersachen, Germany with a dissertation on hot-electron effects in transistors, setting the stage for a career in research on the physics of semiconductors and semiconductor devices. He began wondering why the emerging junction transistors were so slow compared to the earlier point contact transistors. This led him to ask the key question: ``how can an electric field be built into the base region of a junction transistor?''

After his doctorate, he joined the Central Bureau of Telecommunications Technology of the German Postal Service, where he made a pioneering contribution to the then emerging field of solid-state transistors by inventing the drift transistor. He worked in a number of research laboratories in Germany before coming to the U.S. in 1954 where he continued working in heterostructures. Kroemer figured one possible way to speed up transistors, not using a single semiconductor, but a graded base region that started with one material and ended up in another material with a continuous transition between them. Associated with this gradient should then be a force that pushes the charge carriers from the emitter to the collector. Kroemer referred to these new kinds of forces as ``quasi-electric'' fields. His first paper on drift transistor introduced the concept of a doping-engineered electric field in the base to reduce the base transit time. The paper predicted a 8-fold increase in the theoretical frequency limit as compared to Shockley's ``diffusion'' bipolar transistors. The concept of aiding base transport with a built-in electric field resulting from the variation of base doping density is in use in virtually all bipolar transistors fabricated today, and is one of the central concepts in BJT design.

Kroemer expanded his concepts into a new general device design principle going far beyond the starting point of speeding up the bipolar transistors: composition gradient to act as a force on the electrons, that might make possible new devices fundamentally impossible without the new force. These concepts became the backbone of heterostructure bipolar transistor (HBT) technology, even invading silicon technology. An extension of the concept is the basis for much of the speed advantage of present Silicon-Germanium (SiGe) HBTs (see Figure 2). The rf telecommunication industry has built up around the devices Kroemer envisioned in the early years. This 1954 paper introduced a host of far-reaching ideas and an English translation of a portion of this paper is included in this Volume. His landmark 1957 paper in the RCA Review, entitled ``Quasi-Electric and Quasi-Magnetic Fields in Nonuniform Semiconductors'' explained how transistor performance could be improved by the incorporation of quasi-electric fields.

The fact that semiconductors can emit light in response to electric currents has been recognized since the early 1930s. However, it took two decades before this phenomenon could be understood to be due to electron-hole recombination at the interfaces. Also the advent of III-V compound semiconductors altered this situation during the early 1960s. In 1959, Kroemer moved to Central Research Labs at Varian Associates, where he invented the semiconductor heterostructure laser in 1963, which was a straightforward application of the same principle of heterojunctions he worked with and could operate continuously at room temperature. Kroemer was the first to realize the possibilities for carrier and photon confinement offered by a double heterostructure, at a time when the intense research on homojunction semiconductor light emitters seemed to be making little progress. Kroemer suggested that a vastly improved laser could be designed by sandwiching a layer of a narrow band gap semiconductor between two wide band gap semiconductors in another landmark paper in the Proceedings of the IEEE in 1963, a paper that drew little attention at the time.

Sonderdruck

aus

ARCHIV DER ELEKTRISCHEN ÜBERTRAGUNG

Zur Theorie des Diffusions- und des Drifttransistors
III. Dimensionierungsfragen

Von HERBERT KRÖMER*

Mitteilung aus dem Fernmeldetechnischen Zentralamt Darmstadt

(A. E. Ü. 8 [1954], 499—504; eingegangen am 24. Juli 1954)

....

Außer durch inhomogene Dotierung eines homogenen Halbleiters läßt sich ein Driftfeld auch dadurch erzeugen, daß man die Breite des verbotenen Bandes selbst ändert, indem man die Basiszone aus einem nichtstöchiometrischen $-V)\Big]^{-1/2} \cdot (6\,a)$ Mischkristall verschiedener Halbleiter mit verschiedenen Bandabständen (z. B. Ge-Si) herstellt, dessen Zusammensetzung sich innerhalb der Basis stetig ändert. Bei nicht zu hoher Dotierung bleiben dann die Emitterkapazitäten klein, obwohl selbst dann, wenn diese Dotierung konstant ist, ein Driftpotential von

$$\Delta V \approx E_{B,E} - E_{B,C} \qquad (1\,\text{b})$$

erzeugbar wäre. Mit Ge-Si gäbe das etwa 0,4 eV = 16 kT.

Eine Variante dieses Verfahrens besteht darin, zwar in der Basiszone den homogenen Halbleiter mit inhomogener Dotierung beizubehalten, für die Emitterzone jedoch einen Halbleiter mit wesentlich größerem Bandabstand zu wählen[2]. Dann ist es nämlich möglich, die Störstellenkonzentration P_e im Emitter weit unter N_a zu senken, ohne daß der Wirkungsgrad des Emitters abnimmt. Dadurch nehmen aber gemäß Gl. (6a) auch die echten Kapazitäten ab, und unter Umständen kann ΔV noch über die

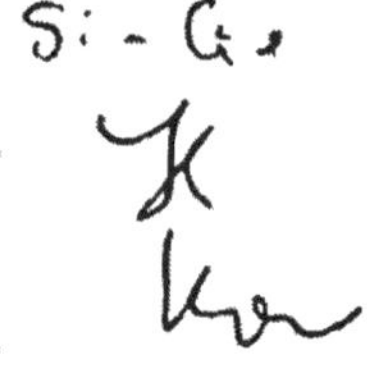

---

[2] Den Hinweis hierauf verdanke ich Herrn A. HÄHNLEIN; siehe hierzu auch LEHOVEC [3].

Although he proposed the idea of the double-heterostructure laser (DHL), he was refused resources to develop the necessary technology, on the grounds that this device could not possibly compete with existing lasers. Interestingly, this novel idea is the basis for the entire modern optoelectronics industry. In fact, these ideas were far ahead of their time, and required the development of modern epitaxial growth technology before they could become mainstream technologies. Not only lasers, but also light emitting diodes (LEDs), such as the blue and green and white LEDs, use the double heterostructure design principle Kroemer envisioned back in 1963. The DHL created its own applications, from the CD player to fiber communications, without which there could never have been an Internet.

Gunn effect was discovered in the early 1960s. Not being able to work on the laser, Kroemer pursued the problem of high-field electron transport, especially negative-resistance effects such as the Gunn effect, and the crucial enabling role of non-trivial energy band structure in such devices. He became the first researcher to explain Gunn effect fully in 1964. He joined the faculty of the University of Colorado in 1968 and moved to UCSB in 1976 where he turned to experimental work and became one of the early pioneers in molecular beam epitaxy, concentrating on applying the technology to new materials systems, such as GaP and GaAs on silicon. 1980s became a decade of ``Heterostructures for Everything'' — a topic that still continues to dominate not only the III-V compound semiconductors but integrated circuits involving the mainstream silicon technology. Kroemer has addressed the understanding of heterojunctions and heterointerfaces with his theoretical prediction of band line-ups and the problems associated with connecting electronic wave functions across heterointerfaces. Kroemer realized that a pure InAs channel may offer the possibility of a very fast electron with the added advantage of excellent confinement offered by AlSb barriers. Thus, in the mid-eighties, his work shifted towards the ``6.1 A group'' of materials including InAs, GaSb, and AlSb.

Kroemer is best described by his favorite talk: ``Heterostructures for Everything.'' Kroemer's career is a fine example of deep, fundamental scientific work having a profound effect on technology and society. He is credited with being the pioneer in the field of heterostructure electronics, which now includes quantum well heterostructures and superlattices. This area involves electron transport in semiconductor superlattices under sufficiently strong electric fields that the electrons undergo oscillations within the tilted energy bands. Such structures might be capable of serving as oscillators — commonly called Bloch oscillators — up to frequencies in the terahertz regime.

Today Kroemer continues his interest with new research areas, like electromagnetic wave propagation in photonic crystals, especially negative-refraction effects, as well as dissimilar materials through his investigations of so-called broken-bandgap combinations of arsenides and antimonides having mid-infrared device applications, and the induced superconducting behavior of semiconductors sandwiched between superconductors. His recent work involves superconductor semiconductor hybrid structures where InAs-AlSb quantum wells are contacted by superconducting niobium electrodes, which induce superconductivity in the semiconductor.

Alan J. Heeger, Alan G. MacDiarmid, and Hideki Shirakawa were awarded 2000 Nobel prize for Chemistry ``for the discovery and development of conductive polymers,'' a revolutionary discovery that plastics can have the properties of metals and semiconductors, a finding that created an important new field of research. There is a great deal of commonality between Heeger and Kroemer, particularly in the conducting and semiconducting materials field and between chemistry and physics, as is evidenced from two joint publications of Heeger and Kroemer.

Kroemer is a Fellow of the IEEE and the American Physical Society, and a Foreign Associate of the U.S. National Academy of Engineering. He has received numerous awards, including the IEEE Medal of Honor, EDS J. J. Ebers and Jack A. Morton Awards, the Heinrich Welker Medal, and the Alexander von Humboldt Research award. He holds honorary doctorates from the Technical University of Aachen, Germany; the University of Lund, Sweden; and from the University of Colorado. He received Germany's Bundesverdienstkreuz (Order of Merit), the highest award given by the Federal Republic of Germany. A German citizen, Kroemer was elected as a foreign associate of the NAS. He even has an asteroid named for him when the German astronomer who discovered it learned of Kroemer's distinguished career.

Kroemer's impact in education has also been significant, with two unique and widely used text books; Thermal Physics by Kittel and Kroemer and Quantum Mechanics. Kroemer has always preferred to work on problems that are one or two generation ahead of established mainstream technology. It will take years to fully exploit his more recent innovations and ideas. A theoretician, Kroemer is not bothered that he has not profited personally from the practical applications of his lifetime research.

His publications may be classified, in general, in the following areas: general principles of heterostructures and HBTs, hot-electron negative resistance effects, GaAs and GaP on Si and related topics, and superconductor-semiconductor hybrids. Some of his important publications reprinted in this volume are listed below. The Volume starts with Kroemer's autobiography and followed by the Nobel lecture. The next two articles on Herbert Kroemer are the ``Oral History from the IEEE'' and ``Not Just Blue Sky'' from IEEE Spectrum - focus on various aspects, such as his early days and Kroemer 'himself' to prevailing political and social conditions in postwar Germany. These two articles are expected to be of special interest to Science historians. The next 35 articles have been selected mainly from the considerations of technical importance and are arranged area wise chronologically.

Technical Articles Reprinted in this Volume

H. Kroemer, ``Zur Theorie des Germaniumgleichrichters und des Transistors,'' Zeitschr. f. Phys., Vol. 134, pp. 435-450, 1953.

H. Kroemer, ``Theory of a Wide-Gap Emitter for Transistors,'' Proc. IRE, Vol. 45(11), pp. 1535-1537, 1957.

H. Kroemer, ``A Proposed Class of Heterojunction Injection Lasers,'' Proc. IEEE, Vol. 51(12), pp. 1782-1783, Dec. 1963. [Discussion ibid., Vol. 52(4), pp. 426-427, 1964].

H. Kroemer, ``Heterostructures for Everything: Device Principle of the 1980's?'' Japan. J. Appl. Phys., Vol. 20 (Suppl. 1), pp. 9-13, 1981.

H. Kroemer, ``Heterostructure Bipolar Transistors and Integrated Circuits,'' Proc. IEEE, Vol. 70(1), pp. 13-25, 1982.

H. Kroemer and G. Griffiths, ``Staggered-Lineup Heterojunctions as Sources of Tunable Below-Gap Radiation: Operating Principle and Semiconductor Selection,'' IEEE Electron Dev. Lett., Vol. EDL-4(1), pp. 20-22, 1983. Rebuttal to Response to ``Critique to Two Recent Theories of Heterojunction Lineups,'' IEEE Electron Dev. Lett., Vol. EDL-4(10), p. 365, 1983.

H. Kroemer, ``Heterostructure Bipolar Transistors: What Should We Build?'' J. Vac. Sci. Technol. B, Vol. 1(2), pp. 126-130, 1983.

H. Kroemer, ``Heterostructure Devices: A Device Physicist Looks at Interfaces,'' Surf. Sci., Vol. 132, pp. 543-576, 1983.

H. Kroemer, ``Barrier Control and Measurements: Abrupt Semiconductor Heterojunctions,'' J. Vac. Sci. Technol. B, Vol. 2(3), pp. 433-439, 1984.

H. Kroemer and H. Okamoto, ``Some Design Considerations for Multi-Quantum-Well Lasers,'' Japan. J. Appl. Phys., Vol. 23, pp. 970-974, 1984.

E. J. Caine, S. Subbanna, H. Kroemer, J. L. Merz, and A. Y. Cho, ``Staggered-Lineup Heterojunctions as Sources of Tunable Below-Gap Radiation: Experimental Verification,'' Appl. Phys. Lett., Vol. 45(10), pp. 1123-1125, 1984.

M. J. Mondry and H. Kroemer, ``Heterojunction Bipolar Transistor Using a (Ga,In)P Emitter on a GaAs Base, Grown by Molecular Beam Epitaxy,'' IEEE Elect. Dev. Lett, Vol. EDL-6(4), pp. 175-177, 1985.

D. I. Babic and H. Kroemer, ``The Role of Nonuniform Dielectric Permittivity in the Determination of Heterojunction Band Offsets by C-V Profiling Through Isotype Heterojunctions,'' Solid-State Electron., Vol. 28(10), pp. 1015-1017, 1985.

H. Kroemer, ``Band Offsets at Heterointerfaces: Theoretical Basis, and Review of Recent Experimental Work,'' Surf. Sci., Vol. 174, pp. 299-306, 1986.

M. A. Rao, E. J. Caine, S. I. Long, and H. Kroemer, ``An (Al,Ga)As/GaAs heterostructure bipolar transistor with non-alloyed graded-gap contacts to the base and emitter,'' IEEE Electron Dev. Lett. EDL-8(1), pp. 30-32, 1987.

G. Tuttle, H. Kroemer, and J. H. English, ``Electron concentrations and mobilities in AlSb/InAs/AlSb quantum wells,'' J. Appl. Phys., Vol. 65(12), pp. 5239-5242, 1989.

P. F. Hopkins, A. J. Rimberg, R. M. Westervelt, G. Tuttle, and H. Kroemer, ``Quantum Hall effect in InAs/AlSb quantum wells,'' Appl. Phys. Lett., Vol. 58(13), pp. 1428-1430, 1991.

I. Sela, D. E. Watkins, B. K. Laurich, D. L. Smith, S. Subbanna, and H. Kroemer, ``Modulated photoabsorption in strained Ga1-xInxAs/GaAs multiple quantum wells,'' Phys. Rev. B, Vol. 43(14), pp. 11884-11892, 1991.

S. A. Chalmers, H. Kroemer, and A. C. Gossard, ``The growth of (Al,Ga)Sb tilted superlattices and their heteroepitaxy with InAs to form corrugated-barrier quantum wells,'' J. Cryst. Growth, Vol. 111, pp. 647-650, 1991.

H. Kroemer, C. Nguyen, and B. Brar, ``Are there Tamm-state donors at the InAs-AlSb quantum well interface?'' J. Vac. Sci. Technol. B, Vol. 10(4), pp. 1769-1772, 1992.

B. Brar, H. Kroemer, and J. H. English, ``'Quasi-direct' narrow GaSb/AlSb (100) quantum wells,'' J. Cryst. Growth, Vol. 127, pp. 752-754, 1993.

H. Kroemer, ``Semiconductor Heterojunctions at the Conference on the Physics and Chemistry of Semiconductor Interfaces: A Device Physicists Perspective,'' J. Vac. Sci. Technol. B, Vol. 11(4), pp. 1354-1361, 1993.

H. Kroemer, ``Proposed Negative-Mass Microwave Amplifier,'' Phys. Rev., Vol. 109(5), p. 1856, 1958.

H. Kroemer, ``Theory of the Gunn Effect,'' Proc. IEEE, Vol. 52(12), p. 1736, 1964.

H. Kroemer, ``Negative Conductance in Semiconductors,'' IEEE Spectrum, Vol. 5(1), pp. 47-56, 1968.

H. Kroemer, ``Generalized Proof of Shockley's Positive Conductance Theorem,'' Proc. IEEE, Vol. 58(11), pp. 1844-1845, Nov. 1970. [Comments on ``Generalized Proof of Shockleys Positive Conductance Theorem'', Proc. IEEE, Vol. 59(8), pp. 1282-1283, 1971].

H. Kroemer, ``Hot-Electron Relaxation Effects in Devices,'' Solid-State Electron., Vol. 21(1), pp. 61-67, 1978.

H. Kroemer, ``Polar-on-Nonpolar Epitaxy,'' J. Cryst. Growth, Vol. 81, pp. 193-204, 1987.

T.-Y. Liu, P. M. Petroff, and H. Kroemer, ``Luminescence of GaAs/(Al,Ga)As superlattices grown on Si substrates, containing a high density of threading dislocations: Strong effect of the superlattice period,'' J. Appl. Phys., Vol. 64(12), pp. 6810-6814, 1988.

H. Kroemer, T.-Y. Liu, and P. M. Petroff, ``GaAs on Si, and Related Systems: Problems and Prospects,'' J. Cryst. Growth, Vol. 95, pp. 96-102, 1989.

H. Kroemer, C. Nguyen, and E. L. Hu, ``Electronic Interactions at Superconductor-Semiconductor Interfaces,'' Solid-State Electron., Vol. 37(4-6), pp. 1021-1025, 1994.

H. Kroemer, ``Superconductor-Semiconductor Devices,'' NATO Adv. Res. Workshop Future Trends in Microelectronics: Reflections on the Road to Nanotechnology, Ile de Bendor, France, S. Luryi, J. Xu, and A. Zaslavsky, Eds., NATO ASI Series; Series E: Applied Sciences, Vol. 323, Kluwer Academic Publishers, pp. 237-250, 1996.

P. M Petroff, K. Ensslin, M. S. Miller, S. A. Chalmers, H. Weman, J. L. Merz, H. Kroemer, and A. C. Gossard, ``Novel Approaches in 2 and 3 Dimensional Confinement Structures: Processing and Properties,'' Superlattices and Microstructures, Vol. 8(1), pp. 35-39, 1990.

H. Kroemer, ``Heterostructures Tomorrow: From Physics to Moore's Law,'' Inst. Phys. Conf. Ser., Vol. 166, pp. 1-11, 1999.

H. Kroemer, ``Speculations about Future Directions,'' J. Cryst. Growth, Vol. 251, pp. 17-22, 2003.

The start of SiGe heterostructures: portion of a page from reference: H. Kroemer, ``Zur Theorie des Diffusions- und des Drifttransistors: III. Dimensionierungs-fragen,'' Archiv d. Elektrischen Ubertragung, Vol. 8, pp. 499-504, 1954.

Außer durch inhomogene Dotierung eines homogenen Halbleiters läßt sich ein Driftfeld auch dadurch erzeugen, daß man die Breite des verbotenen Bandes selbst ändert, indem man die Basiszone aus einem nichtstöchiometrischen Mischkristall verscheidener Halbleiter mit verschiedenen Bandabständen (z. B. Ge-Si) herstellt, dessen Zusammensetzung sich innerhalb der Basis stetig ändert. Bei nicht zu hoher Dotierung bleiben dann die Emitterkapazitäten klein, obwohl selbst dann, wenn diese Dotierung konstant ist, ein Driftpotential von

$$\Delta V \approx E_{B,E} - E_{B,C} \qquad (1b)$$

erreichbar wäre. Mit Ge-Si gäbe das etwa 0.4 eV = 16 kT.

Besides by inhomogeneous doping of a homogeneous semiconductor, a drift field may also be generated through varying the energy gap itself, by making the base region from a non-stoichiometric mixed crystal of different semiconductors with different energy gaps (for example, Ge-Si), the composition of which varies continuously through the base. If the doping is not too high, the emitter capacitances then remain small, although even if the doping is constant, a drift potential of

$$\Delta V \approx E_{B,E} - E_{B,C} \qquad (1b)$$

could be obtained. With Ge-Si this would yield about 0.4 eV = 16 kT.

Eine Variante dieses Verfahrens besteht darin, zwar in der Basiszone den homogenen Halbleiter mit inhomogener Dotierung beizubehalten, für die Emitterzone jedoch einen Halbleiter mit wesentlich größerem Bandabstand zu wählen.[2] Denn ist es nämlich möglich, die Störstellenkonzentration P_e im Emitter weit unter N_a zu senken, ohne daß der Wirkungsgrad des Emitters abnimmt. Dadurch nehmen aber gemäß Gl. (6a) auch die echten Kapazitäten ab, und unter Umständen kann ΔV noch über die durch Gl. (1a) gegebene Grenze erhöht werden. ....

---

[2] Den Hinweis hierauf verdanke ich Herrn A. Hähnlein; siehe hierzu auch Lehoved [3].

A variation of this procedure would consist of retaining, within the basis zone, the homogeneous semiconductor with an inhomogeneous doping, but to select for the emitter zone a semiconductor with a significantly larger energy gap.[2] For then it becomes possible to lower the doping concentration P_e far below N_a [i.e, the donor concentration in the base on the emitter side] without a decrease in the emitter efficiency. But in this way, acording to (6a), the true capacitances also decrease, and ΔV may potentially be increased even beyond the limit given by (1a). ....

---

[2] I owe this suggestion to Mr. A. Hähnlein; see also Lehovec [3] on this matter.

Wir behandeln diese Möglichkeiten in vorliegender Arbeit nicht näher, da über die physikalischen und technologischen Eigenschaften von Halbleiter-Mischphasen außer einer Arbeit von Busch und Winkler [5] hierzu noch keine brauchbaren Untersuchungen vorliegen.

We do not treat these possibilities in the present paper any further, because there are no usable investigations about the physical and technological properties of semiconductor mixed phases, besides a paper by Busch and Winkler [5].

Comment: B&W studied the properties of the semiconductor alloy system $Mg_2(Ge_xSi_{1-x})$.

H. Kroemer,
``Quasi-Electric Fields and Band Offsets: Teaching Electrons New Tricks,''
Les Prix Nobel, The Nobel Prizes 2000, The Nobel Foundation,
Stockholm, pp. 101-121 (Biography on pp. 95-100), 2001.

Herbert Kroemer

Autobiography

I was born on August 25, 1928 in Weimar, Germany. My father was a civil servant working for the city administration of my home town; my mother was a classical German "Hausfrau." Both came from simple skilled-craftsmen families. Neither had a high-school education, but there was never any doubt that they wanted to have their children obtain the best education they could afford. My mother, in particular, pushed relentlessly for top performance in school: simply doing well was not enough. Fortunately, I breezed through 12 years of school almost effortlessly, not once requiring help with homework from my parents.

Despite their insistence on excellence, my parents never pushed me in any particular academic direction; I was completely free to follow my inclinations, which ran towards math, physics, and chemistry. When I finally told my parents that I wanted to study physics, my father merely wondered what that is, and whether I could make a living with it. I certainly could become a physics teacher at a High School, or "Gymnasium," a thoroughly respectable profession.

I did have one major problem in school, though: Discipline! I was often bored, and entertained myself in various disruptive ways. A frequent punishment was an entry into the "Klassenbuch," the daily class ledger. These entries were considered a very serious matter, and if I had not been excellent academically, I would have risked being expelled. Once, after I had again been entered as having disturbed the class, the teacher who had overall responsibility for the class - Dr. Edith Richter, whom I adored - asked me in great exasperation: "Why again?" I told her that I had been bored, whereupon she exploded: "Mr. Kroemer, one of the purposes of a higher education is that you learn to be bored gracefully." I will never forget that outburst - nor have I ever really learned to be bored gracefully.

Another teacher - Willibald Wimmer - had his own clever way of handling me. Before the end of the war, he had been an instructor at a local engineering college, ending up teaching math and physics at our high school. He was used to dealing with more mature students, and he treated us as adults. I was way ahead of the curriculum in math, and kept showing off. Worse, I taught some of my classmates math "thricks," that were not part of the curriculum. So, Mr. Wimmer made a "treaty" with me: While he could not excuse me from attending class, I was guaranteed a top grade without being required to turn in the homework assignments, and was permitted to do whatever I wanted to do during the hour, provided I kept absolutely quiet - except when explicitly asked to speak up. Both of us kept that treaty.

Mr. Wimmer also became our physics teacher, a subject about which he clearly knew little more than what was in the textbook. Realizing that I was deeply into physics, he simply enlisted me and one other student to help him in lecture preparations, like setting up what apparatus had survived the war. Once I even was asked to present the lecture myself, with him sitting in the front row and enjoying the show. It was a wonderful experience.

Having graduated from the gymnasium in 1947, 1 was accepted as a physics student at the University of Jena, where I fell under the spell of the great Friedrich Hund, the most brilliant lecturer I ever encountered. The joy did not last long. In early 1948 the political suppression in East Germany became very severe, especially at rebellious universities like Jena. Every week, some of my fellow students had suddenly disappeared, and you never knew whether they had fled to the West, or had ended up in the German branch of Stalin's Gulag, like the uranium mines near the Czech border. During the Berlin airlift, I was in Berlin as a summer student at the Siemens company, and I decided to go West via one of the empty airlift return flights.

From Berlin, I had written to several west German universities for admission, including Gottingen, but did not receive a reply before leaving Berlin (they had turned me down). I followed the advice of one of my Jena professors "why don't you give my greetings to Professor Konig in Gottingen." Konig told me that physics admissions were closed, but he passed me on for what was ostensibly just a friendly chat to Professor Richard Becker and his alter-ego assistant, Dr. Gnther Leibfried. They in turn passed me on to Wolfgang Paul (Nobel 1989), and I think also to Robert Pohl. It soon dawned on me that this was not just a friendly social chat with people who had nothing better to do, but a thorough examination. I remember one of the questions Paul asked me: "You know that a mirror interchanges left and right? - Then why doesn't it interchange top and bottom?" In the end, I was returned to Becker, who told me that two of the students who had been admitted were not coming, and a meeting was scheduled for the next day to select who would get the two openings. A few days later I received a postcard that I had been accepted.

Post-war Gottingen. was - intellectually - a wonderfully stimulating place. I was attracted to one of the younger instructors - "Privatdozent" Dr. Hellwege - who offered a so-called Proseminar, where pre-research students would present papers assigned to them, and I participated in this for several semesters in a row. Once, the famous Fritz Houtermans visited Hellwege, and sat in on several of the presentations, including mine. I presented someone's data that yielded a reasonable straight line on a double-log plot, and proudly claimed a power law for the data. Houtermans was not impressed: "On a double-log plot, my grandmother fits on a straight line." I keep quoting Houtermans' grandmother to my own students. Eventually, I signed up with Hellwege for a Diploma Thesis, which would probably have led to an experimental study of the optical spectra of some rare-earth salts. But Hellwege had a long waiting list, and in the meantime, Professor Fritz Sauter - a refugee who had found a temporary home as a guest in Becker's Institute for Theoretical Physics - offered me a theoretical Diploma Thesis, based on a talk that I had given in one of his seminars. Hellwege suggested that I accept Sauter's offer: "You will be finished with him before you can start with me." So I became a theorist.

The diploma thesis was an extension of a 1939 paper by Shockley on the nature of surface states in one-dimensional potentials. As one of the elaborations, I looked at the interface between two different periodic potentials, which confronted me for the first time with what we would today call the band offsets at heterojunctions.

There was another early encounter with heterojunctions while working under Sauter. We made a field trip to the AEG research laboratories in Belecke, a small town in Westphalia. There, a Dr. Poganski gave a beautiful demonstration that the selenium rectifier was not a Schottky barrier, but a p-n junction between p-type selenium and n-type CdSe, a true heterojunction - although that term did not exist yet. This must have had an at least sub-conscious influence on me: when I later started thinking about heterojunctions in earnest, the question whether such things could actually exist as real devices had an obvious answer: Of course!

While working on my diploma, I gave another colloquium talk under Sauter, reporting on the famous Bardeen/Brattain paper "Physical Principles Involved in Transistor Action" (or some title like that). At the end I made some suggestion about some open questions raised by the authors. Sauter was intrigued and suggested that as a possible Ph.D. topic. Sometime later, he came into my office and told me to stop further work on my Diploma thesis, and to simply write up what I had done so far. When I protested, he insisted that it was time to move on to the real thing, the Ph.D. dissertation.

I had thus come into contact with one of Sauter's strong beliefs, apparently dating back to the tradition of the 20s: that degrees should not be awarded on the basis of having "served time," but were basically certificates that the recipient had proven capable of executing creative work independently, and no longer required supervision. In fact, he clearly preferred quick dissertations. As a result, I received my Ph.D. before my 24th birthday, fast even for a theorist: Wonderful!

The Ph.D. dissertation involved what we would today call hot-electron effects, in the collector space-charge layer of the then-new transistor. The idea was simple. Almost nothing was known about the energy band structure of Ge, but someone's theoretical estimates suggested - quite incorrectly - very narrow bands, especially for the valence band. In this case, if the field was strong enough, any holes in the valence might undergo what we now call Bloch oscillations. A few lines of algebra suggested that, for a given current density, the traveling hole concentration would increase with increasing field ("Staueffekt"), leading to strong space charge effects. The influence of these space charges on the current-voltage characteristics of point contact diodes and transistors formed the main body of the dissertation.

My algebra also implied a decrease of electron drift velocity with increasing field, implying a negative differential conductivity. Knowing nothing about electrical circuit theory, I was unaware how useful such a phenomenon could be, until Shockley pointed it out to me in a personal discussion two years later.

But it became clear soon that my dissertation was unrelated to reality. My assumptions about the band structure and about an energy-independent mean free path had been invalid, and after the discovery of avalanche breakdown it became obvious that the huge fields required for Bloch oscillations in a bulk semiconductor could never be reached. Twenty years later, after the pathbreaking work of Esaki and Tsu on negative differential conductivity in superlattices, I realized that I had in fact anticipated their basic physics, albeit in a more primitive form: What was not possible in bulk semiconductors, appeared to become possible in superlattices with their much longer period.

Back to Sauter. He was not interested in closely supervising his students; he simply watched what they were doing on their own initiative. Still, he had a tremendous influence on me in matters of methodology. Whenever I came to him with a pure physics idea, he would invariably say, with slight sarcasm: "But Mr. Kroemer, you ought to be able to formulate this mathematically! " If I came to him with a math formulation, I would get, in a similar tone: "But Mr. Kroemer, that is just math, what is the physics?" After a few encounters of this kind, you got the idea: You had to be able to go back and forth with ease. Yet, in the last analysis, concepts took priority over formalism, the latter was simply an (indispensable) means to an end.

This set of priorities clearly showed, and it had a profound influence on me. As a student of Sommerfeld, Sauter was a superb mathematician himself. But he detested it when people were showing off their math skills by using math that was more advanced than necessary for the problem at hand. To the contrary: You were expected to show how simple you could make it. Because he was a great expert on Bessel functions, I once felt compelled to put, into the draft of my dissertation, an ad-hoc problem that required Bessel functions. He was not amused: "This has no business here; you just put it in to impress me. Take it out!"

Richard Becker had exactly the same attitude (the two were close friends), and I later encountered it again in Shockley. Under influences such as these, I never developed into a "hard-core Theorist with a capital T," but became basically a conceptualist who remained acutely aware of his limitations as a formalist, and whose personal role model was Niels Bohr more than anybody else amongst the Greats of Physics.

The German 1952 job market for theoretical physicists was all but nonexistent. New university positions were not created, and there were plenty of more senior people waiting to occupy any vacancies that might open up. So I never even considered a university career. The situation in industry was hardly any better. As luck would have it, the small semiconductor research group at the Central Telecommunications Laboratory (FTZ) of the German postal service was looking for a "house theorist" who knew semiconductor theory, and I got the job. My duties were simple. I had to be available for whatever theoretical questions anybody had, and also take an active role by poking my nose into the work of my experimentalist and technologist colleagues, to look on my own for topics to which I could contribute - provided I would never touch any equipment. Every week or two, I had to give a talk of 1 to 2 hours to the group, on any subject of my choosing of which I thought that the group should be taught about it. Other than that, I was left completely free to pick whatever problems I felt were worth tackling. So I had become a "professor" of sorts after all, teaching a small but highly motivated "class." From day-1 I was forced to learn to communicate, not with other theorists, but with experimentalists and technologists. It was a fascinating challenge, with a range of topics far beyond what I myself had learned in Gottingen, very often going beyond physics, into metallurgy, chemistry, and electrical engineering.

Of course I ceased to be a "real" theoretical physicist - if I ever was one. Call me an Applied Theorist if you want. However, the awareness of doing something truly useful helped overcome the uneasy feelings over ending a theorist career as soon as it had begun. By hindsight, maybe it wasn't such a bad career move after all!

As my research topic at the FTZ, I picked the problem of the severe frequency limitations of the new transistors - and what one might be able to do about them. It was this problem that led directly to heterostructure ideas. In a 1954 publication of mine there are a couple of paragraphs outlining in a rudimentary form the first ideas for what was later to be called the heterostructure bipolar transistor, or HBT. I proposed both a transistor with a graded gap throughout the base, and the simpler form of just a wide-gap emitter. The rest is history. This history is described in some detail in my Nobel Lecture, so I will give here only the highlights.

Some time after joining RCA Laboratories in Princeton, NJ, in 1954, I returned to heterojunctions. I actually tried - unsuccessfully - to build some HBTs with a Ge/Si alloy emitter on a Ge base. But my principal contributions to the field were two theoretical papers. One of these, in the RCA Review, is essentially unknown to this day, but it clearly spelled out the concept of quasielectric fields, which I considered the fundamental design principle for all heterostructures.

The final step came in 1963, while I worked at Varian Associates in Palo Alto, CA. A colleague - Dr. Sol Miller - gave a research colloquium on the new semiconductor diode laser. He reported that experts had concluded that it was fundamentally impossible to achieve a steady-state population inversion at room temperature, because the injected carriers would diffuse out at the opposite side of the junction too rapidly. I immediately protested: "But that's a pile of ... ; all you have to do is give the outer regions a wider energy gap." I wrote up the idea and submitted the paper to Applied Physics Letters, where it was rejected. I was talked into not fighting the rejection, but to submit it to the Proceedings of the IEEE, where it was published, but ignored. I also wrote a patent, which is probably a better paper than the one in Proc. IEEE.

Then came the final irony: I was refused resources to work on the new kind of laser, on the grounds that there could not possibly be any applications for it. By a coincidence, the Gunn effect had just been discovered, and having a long-standing interest in hot-electron negative-resistance effects, I worked on the Gunn effect for the next ten years, and did not participate in the final technological realization of the laser.

I left Varian in 1966, and in 1968 joined the University of Colorado. There I eventually returned to heterostructures, and in the early-70s tackled the theory of band offsets together with my student Bill Frensley - now at UT Dallas - who worked out the first ab-initio theory of the band offsets. Shortly afterwards - now at UCSB - I developed a powerful method to determine band offsets experimentally, by capacitance-voltage profiling through the hetero-interface.

In the late-70s, I returned to the device that had started it all, the HBT. The technology developments that had made possible the DH laser offered great promise also for the HBT, and I became a strong advocate of developing the full potential of that device.

In addition to heterostructures, I have worked on numerous other semiconductor topics, be it in physics, materials, devices, or technology. Second only to heterostructures has been a continuing interest in hot-electron negative-resistance effects, dating back to my Ph.D. dissertation. I already mentioned the work on the Gunn effect, but there was more. During my RCA years, I had come up with a crazy scheme to obtain a negative resistance perpendicular to a strong bias field, by drawing on the fact that some of the heavy holes in Ge have negative transverse effective masses - that is, perpendicular to their velocity. Experimentally, it was another failure, but conceptually, I found it extraordinarily stimulating. So did others, and it earned me a great deal of early notoriety. Today, I am back to one of the sins of my youth: to the superlattice Bloch oscillator, an exciting combination of heterostructures and hot electron physics.

At the opposite end from hot electrons has been recent work on superconducting weak links in which a degenerately modulation-doped InAs/AlSb quantum well acts as a ballistic coupling medium between superconducting Nb electrodes. They exhibit some utterly delightful large discrepancies between experiment and accepted theory.

There are numerous additional topics scattered throughout my career. I have basically been an opportunist - and not at all ashamed of it.

QUASI-ELECTRIC FIELDS AND BAND OFFSETS: TEACHING ELECTRONS NEW TRICKS

Nobel Lecture, December 8, 2000

by

HERBERT KROEMER

ECE Department, University of California, Santa Barbara, CA 93106, USA.

I. INTRODUCTION

Heterostructures, as I use the word here, may be defined as heterogeneous semiconductor structures built from two or more different semiconductors, in such a way that the transition region or interface between the different materials plays an essential role in any device action. Often, it may be said that *the interface is the device.*

The participating semiconductors all involve elements from the central portion of the periodic table of the elements (Table I). In the center is silicon, the backbone of modern electronics. Below Si is germanium. Although Ge is rarely used by itself, Ge-Si alloys with a composition-dependent position play an increasingly important role in today's heterostructure technology. In fact, historically this was the first heterostructure device system proposed, although it was also the system that took longest to bring to practical maturity, largely because of the 4 % mismatch between the lattice constants of Si and Ge.

Table I. Central portion of the periodic table of the elements, showing the element from columns II through VI actively used in current heterostructure technology.

II	III	IV	V	VI
	Al	Si	P	S
Zn	Ga	Ge	As	Se
Cd	In		Sb	Te
Hg				

Silicon plays the same central role in electronic metallurgy that steel plays in structural metallurgy. But just as modern structural metallurgy draws on metals other than steel, electronics draws on semiconductors other than silicon, namely, the compound semiconductors. Every element in column III may be combined with every element in column V to form a so-called III–V compound. From the elements shown, twelve different discrete III–V compounds may be formed. The most widely used compound is GaAs – gallium arsenide – but all of them are used in heterostructures, the specific choice depending on the application. In fact, today the III-V compounds are almost always used in heterostructures, rather than in isolation.

Two or more discrete compounds may be used to form alloys. A common example is aluminum-gallium arsenide, $Al_xGa_{1-x}As$, where x is the fraction of column-III sites in the crystal occupied by Al atoms, $1-x$ is occupied by Ga atoms. Hence we have not just 12 discrete compounds, but a continuous range of materials. As a result, it becomes possible to make compositionally graded heterostructures, in which the composition varies continuously rather than abruptly throughout the device structure.

Similar to the III–V·compounds, every element shown in column II may be used together with every element in column VI to create II–VI compounds, and again alloying is possible to create a continuous range of the latter.

II. BAND DIAGRAMS AND QUASI-ELECTRIC FORCES

Whenever I teach my semiconductor device physics course, one of the central messages I try to get across early is the importance of energy band diagrams. I often put this in the form of "Kroemer's Lemma of Proven Ignorance":

> If, in discussing a semiconductor problem, you cannot draw an **Energy Band Diagram**, this shows that you don't know what you are talking about,

with the corollary

> If you can draw one, but don't, then your audience won't know what you are talking about.

Nowhere is this more true than in the discussion of heterostructures, and much of the understanding of the latter is based on one's ability to draw their band diagrams – and knowing what they mean.

To illustrate the idea, consider first a homogenous piece of semiconductor, say, a piece of uniformly doped silicon, but with an electric field applied. The band diagram then looks like the top diagram in Fig. 1, consisting simply of two parallel tilted lines representing the conduction and valence band edges. The separation between the two lines is the energy gap of the semiconductor; the slope of the two band edges is the elementary charge e multiplied by the electric field E. When an electron or a hole is placed into this structure, a force $-eE$ is acting on the electron, $+eE$ on the hole; the two forces are equal in magnitude and opposite in direction, their magnitude is the slope of the bands, just the signs differ.

In a heterostructure, the energy gap becomes position-dependent, and the two band edge slopes are no longer equal, hence the two forces are no longer equal in magnitude. It would, for example, be possible to have a force acting only upon one kind of the carriers (Fig. 1b), or to have forces that act in the same direction for both types of carriers (Fig. 1c). Purely electrical forces in homogeneous crystals can never do this. This is why I call these forces "quasi-electric." *They present a new degree of freedom for the device designer to enable him to obtain effects that are basically impossible to obtain using only "real" electric fields.*

This is the underlying **general design principle** of all heterostructure de-

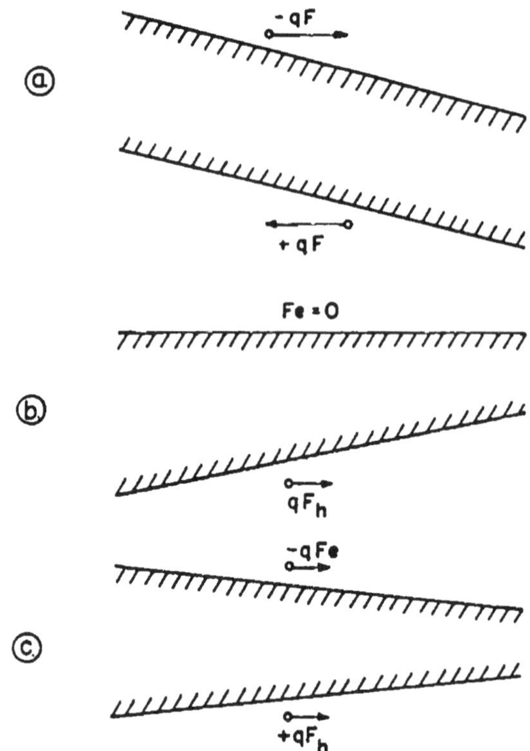

Figure 1. Quasi-Electric Fields: (a) A true electric field simply tilts the bands; (b) quasi-electric fields, with no force on electrons, but a force on holes; (c) quasi-electric fields forcing electrons and holes in the same direction. From Kroemer (1957a).

vices, first spelled out in a 1957 paper of mine (Kroemer, 1957a). In fact, the preceding paragraph is an only slightly edited version of a key paragraph in that paper.

When I wrote those lines, I did not know about Shockley's famous 1951 patent (Shockley), where the possibility of a bipolar transistor with an emitter of wider energy gap is explicitly mentioned. However, the wide-gap emitter idea appears to have been presented principally to cover alternative design possibilities, a procedure typical in patents. The patent gives no indication why such a design would have distinct advantages over a homostructure design, much less a general design principle extending to other kinds of devices. My own formulation might be viewed as a broad generalization of the idea in Shockley's patent. But my point of departure was different: not an *abrupt* energy gap change with accompanying band offset steps, but explicitly a *continuous* energy gap variation of "designable" width, of which the abrupt gap change is simply a limiting case.

Returning to Fig. 1b, it should be emphasized that the zero conduction band slope shown there does not imply a zero electric field. A true electric field is of course present, and it can in principle be determined by the integration of Poisson's equation, provided the local space charge densities are known, often a non-trivial task. But this true field is not part of the band diagram. Nor do the electrons care: The band edge slopes are what matters, not the true electric field. The difference between the two becomes even more drastic in Fig. 1c, where we could not guess even the *direction* of the true field, much less its magnitude.

III. HETEROSTRUCTURE BIPOLAR TRANSISTORS

A. Graded-gap transistor

I had been led to the 1957 principle by a very practical question dating back to 1953/54, when I was working at the telecommunications research laboratory (Fernmeldetechnisches Zentralamt; FTZ) of the German Postal Service: The early bipolar junction transistors were far too slow for practical applications in telecommunications, and I set myself the task of understanding the frequency limitations theoretically – and what to do about them. One approach – not the only one – was to speed up the flow of the minority carriers from the emitter to the collector by incorporating an electric field into the base region. This could be done by using, not a uniform doping in the base, but one that decreased exponentially from the emitter end to the collector end – the so-called *drift transistor* (Krömer, 1953). While working out the details, I realized that

> "... a drift field may also be generated through a variation of the energy gap itself, by making the base region from a non-stoichiometric mixed crystal of different semiconductors with different energy gaps (for example, Ge-Si), with a composition that varies continuously through the base." [Translated from Krömer (1954)]

This was not yet the full general design principle, but it constituted the original conception of what has become known as the heterostructure bipolar transistor (HBT), and ultimately of the heterostructure device field in general.

The appropriate band diagram (Fig. 2) followed in the 1957 paper mentioned earlier, where I gave the 1954 idea as one example of the general design principle. Note that Fig. 2 shows a flat conduction band, as would be the case for a sufficiently heavy uniform doping; the band diagram of Fig. 1b represents essentially the base region of that early concept. The case of Fig. 1c illustrates the generality of the design principle.

Note that the original proposal explicitly gave the Ge-Si system as an example, rather than a III/V compound system. It was to take some four decades until Ge-Si HBTs were finally becoming commercially available, long after devices based on III/V compounds had done so.

B. Wide-gap emitter

The proposed graded-gap base structure was far beyond the technologies then available, a situation that was to remain unchanged for decades. The only possibility one of my colleagues – Mr. Alfons Hähnlein – could envisage

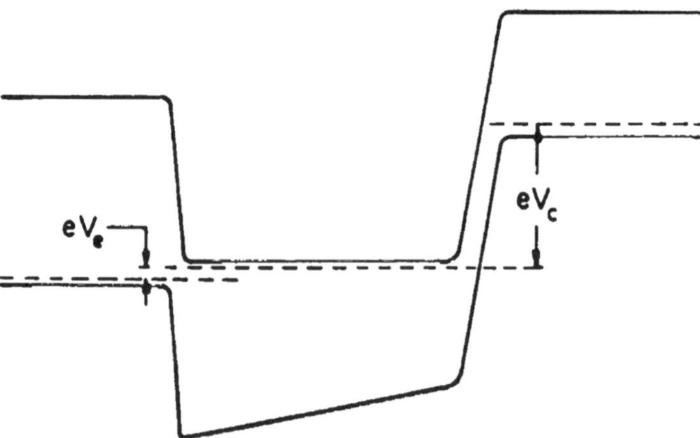

Figure 2. P-n-p transistor with a base region with a graded gap, to speed up minority carrier flow from emitter to collector [from Kroemer (1957a)]. P-n-p transistors were the preferred design for the Ge-based transistors of the mid-50's.

was a design in which the emitter was made from a wider-gap semiconductor than the base, with a quasi-abrupt transition at the interface between the two, leading to a band diagram as in Fig. 3, in essence – but unknowingly – re-inventing Shockley's design.

It was of course obvious that the objective of putting a drift field into the base of the transistor could not be achieved in this way. But on reflecting about what exactly might be the properties of such a structure, I realized that a wide-gap emitter has advantages of its own (Kroemer, 1957b; 1982): One of the problems with all bipolar transistors is minimizing the highly undesirable back-injection of majority carriers from the base (electrons in a p-n-p transis-

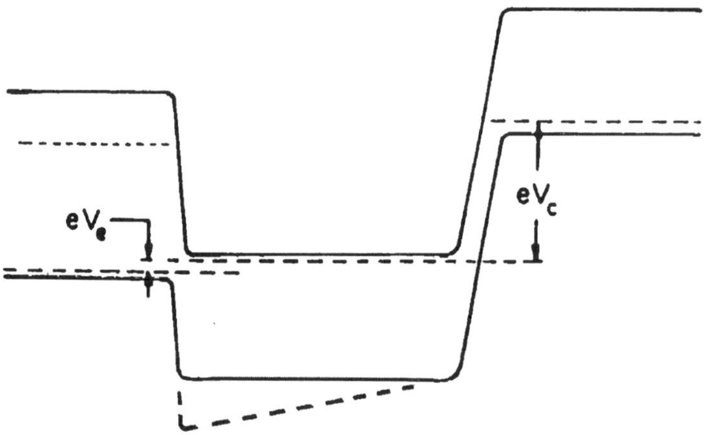

Figure 3. Wide-gap emitter. The energy gap variation has been compressed into a quasi-abrupt transition at the emitter-to-base interface. The base region still has a uniform energy gap without the transport-aiding quasi-field, but there is now a potential barrier for the escape of electrons from the base into the emitter that is larger than the barrier for holes entering the base from the emitter.

tor) into the emitter. In a homojunction transistor, this requirement sharply limits the base doping, which has other undesirable consequences, like a large base access resistance. A wide-gap emitter greatly suppresses this back-injection current: Expressed in terms, not of the quasi-electric forces, but of the associated potentials, any electrons escaping from the base into the emitter must overcome a higher potential barrier than the holes entering the base from the emitter. As a result, the electron escape current density is reduced roughly by a factor $\exp(-\Delta F_G/kT)$, where ΔF_G is the difference in energy gaps. This is very effective: An easily achieved energy gap difference of 0.2eV ($\approx 8kT$) implies a reduction by a factor $e^{-8} \approx 1/3000$.

Given this reduction, it now becomes possible to dope the base much more heavily, to reduce the base resistance. But in the presence of the inevitable junction capacitances, a reduction of base resistance reduces the RC time constants of the device, and thereby enhances its speed .

Because of the much greater technological simplicity of the wide-gap emitter design over the graded-base design, it was the wide-gap emitter design that dominated HBT technology until recently, but the highest-performance HBTs now use both approaches (Kroemer, 1983).

C. Follow-up

Because of the absence of any credible technology, I did not follow up the above 1954 ideas until three years later, after I had joined RCA Laboratories in Princeton, NJ. I realized the generality of the design principle outlined above, and wrote the *RCA Review* paper referred to earlier (Kroemer, 1957a). The paper was almost totally ignored, not only because the *RCA Review* was a somewhat obscure journal, but probably even more because I myself somehow never explicitly referred to the paper (nor to its 1954 precursor) in my own subsequent work until about 40 years later (Kroemer, 1996). The general design principle itself was extensively discussed in a 1982 HBT review (Kroemer, 1982), but without reference to the 1954 paper and the 1957 *RCA Review* paper.

The 1957 paper of mine that *is* widely cited was a second paper in that year, which gives a detailed analysis of the wide-gap emitter version of the HBT (Kroemer, 1957b). Having been published in a more visible journal, it drew considerable attention, and stimulated several attempts by others to realize the wide-gap emitter version of the HBT during the '60s. Unfortunately, technology was still not ready, and none of these early attempts led to anything useful. By 1970, people seemed to have largely given up.

While at RCA, I also made an unsuccessful attempt to build a Ge transistor with a Ge-Si alloy emitter, which might be sufficiently amusing (and characteristic of the primitive state of 1957 technology) to be told here (Kroemer, 1957c). The idea was to utilize the fact that the Au-Si phase diagram exhibits a low-melting (370 °C) eutectic. I prepared such a eutectic, smashed the fairly brittle material with a hammer into a coarse powder, placed small grains of the powder onto a Ge chip, and alloyed the combination at a temperature somewhere between 500 °C and 600 °C. The Au-Si alloy would then melt and

106

penetrate into the Ge chip, dissolving some Ge. Upon cooling, a Ge-Si alloy emitter would re-crystallize (Fig. 4). I actually got one or two transistors to work, but as a rule, the large thermal strains generated during the solidification of the eutectic caused the Ge chip to crack. The attempt was sufficiently unsuccessful that I never published the work. It was followed up by Diedrich and Jötten (1961), who knew about my work, but the technology clearly was unpromising, and Si-Ge HBTs had to wait several decades for their practical realization.

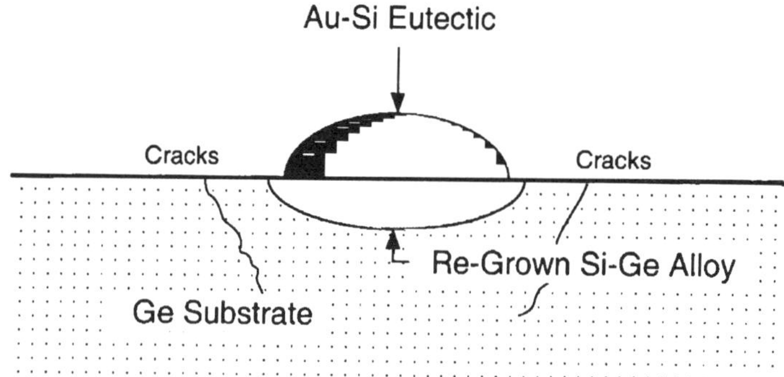

Figure 4. Attempt to realize a Ge transistor with a Ge-Si alloy emitter. A piece of Au-Si eutectic was alloyed into a Ge base, forming a Si-Ge alloy emitter upon cooling. From Kroemer (1957c).

IV. DOUBLE-HETEROSTRUCTURE LASER

Neither the graded-gap HBT nor the wide-gap emitter HBT draw on the full power of the idea expressed in the general design principle that the quasi-electric fields '*enable the device designer to obtain effects that are basically impossible to obtain using only "real" electric fields.*' They represent major improvements, alright, but do they represent something *basically impossible* otherwise?

An example of something that was indeed truly impossible to achieve otherwise emerged abruptly in March 1963. I was working at Varian Associates in Palo Alto at the time, and a colleague of mine – Dr. Sol Miller – had taken a strong interest in the new semiconductor junction lasers that had emerged in 1962, a topic then outside my own range of interests. In a colloquium on the topic he gave a beautiful review of what had been achieved, not failing to point out that successful laser action required either low temperatures or short low-duty-cycle pulses, usually both. Asked what the chances were to achieve continuous operation at room temperature, Miller replied that certain experts had concluded that this was fundamentally impossible.

It is instructive to review this argument here. Consider the (highly over-simplified) energy band diagram of a GaAs p-n junction, heavily doped on both sides, and forward-biased to the point that flatband conditions were reached (Fig. 5). Electrons then diffuse from the n-type side to the p-type side, and holes diffuse in the opposite direction, creating a certain concentration of electron-hole pairs in the junction region proper; their recombination would cause light emission. But in order to obtain *laser* action, a popula-

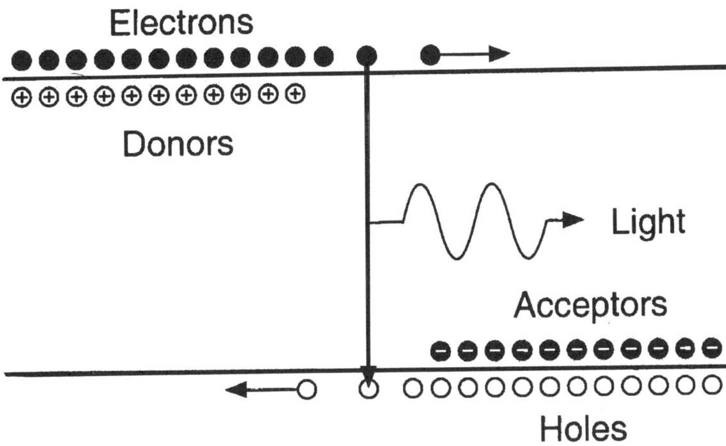

Figure 5. Schematic energy band diagram of a p-n homojunction forward-biased to flatband conditions, creating a high concentration of electron-hole pairs in the vicinity of the junction plane, leading to emission of recombination radiation.

tion *inversion* has to be achieved, which means that, in the active region, the occupation probability of the lowest states in the conduction band has to be higher than that of the highest states in the valence band. A *necessary* condition for such a population inversion is a forward bias larger than the energy gap. But even then, a population inversion is hard to achieve in an ordinary p-n junction. First of all, the electron concentration in the active region will always be lower than in the n-type doped region, with an analogous limitation for the holes. Inversion, therefore, requires degenerate doping on both sides. But even with degenerate doping, both the electrons and holes would diffuse out of the active region immediately into the adjacent oppositely doped region, preventing a population inversion from building up. Increasing the forward bias would not help much, because it would increase the rate of outflow just as much as the rate of injection.

I immediately protested against this argument with words somewhat like "but that is a pile of ..., all one has to do is give the injector regions a wider energy gap ." As is shown in Fig. 6, such a change would cause an electron-repelling quasi-electric field to be present on the p^+ side, and a similar hole-repelling barrier on the n^+ side. Carrier confinement would thus be achieved.

By increasing the forward bias further, potential wells develop for both the electrons and the holes (Fig. 7), with quasi-electric forces on *both* sides pushing *both* electrons and holes towards the active region. As a result, electron and hole concentrations can become much larger than the doping levels in the contact regions, and it becomes readily possible to create the population inversion necessary for laser action. This double-heterostructure (DH) laser finally represented a device truly impossible with only the real electric fields available in homostructures; note that the idea for it arose essentially at the instant I had been made aware that there was a problem.

I wrote up a paper describing the DH idea, along with a patent application.

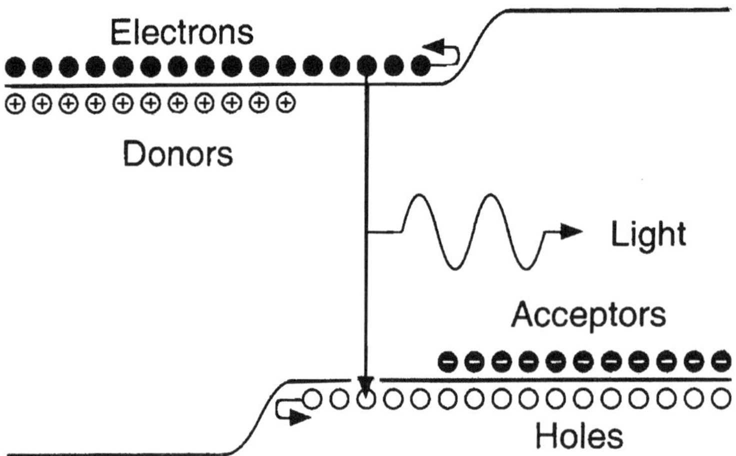

Figure 6. Carrier confinement in a double heterostructure, due to the presence of quasi-electric potential barriers at the ends of the light-emitting active region, preventing the outflow of injected electrons and holes, without interfering with the flow of majority carriers from the injector regions.

The paper was submitted to *Applied Physics Letters*, where it was rejected. I was persuaded not to fight the rejection, but to submit the paper to the *Proceedings of the IEEE* instead, where it was published (Kroemer, 1963) – but largely ignored. Fig. 8 shows the band diagram actually published.

The patent was issued in 1967 (Kroemer, 1967). It is probably a better paper than the *Proc. IEEE* letter. It expired in 1985.

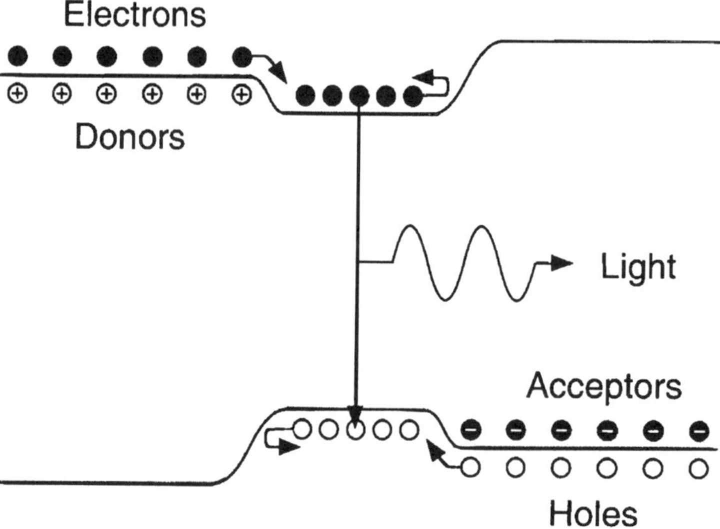

Figure 7. With a further increase of the forward bias, potential wells form for both electrons and holes, which permit the accumulation of the injected carriers to degenerate concentrations much higher than the values in the injector regions.

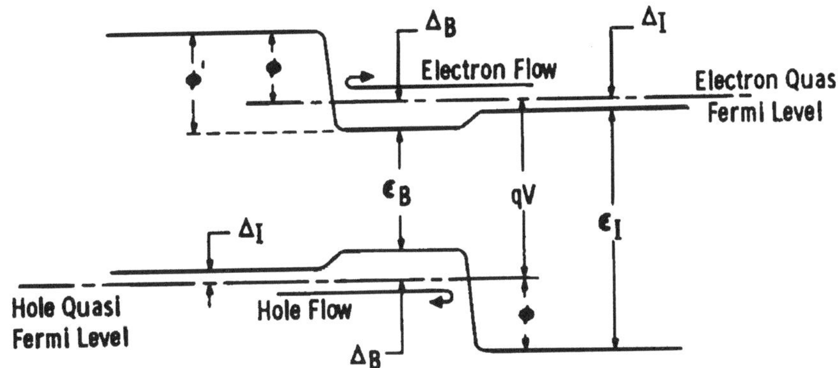

Figure 8. Band diagram of the double-heterostructure laser, as originally published (Kroemer, 1963).

Once again, here was an idea far ahead of any technology to realize it. DH lasers operating continuously at room temperature were finally demonstrated in 1970, first by Alferov et al. (1970), and shortly afterwards by Hayashi *et al.* (1970). For the history of the experimental work, see Alferov (2001); Alferov (1996); Casey and Panish (1978).

For reasons discussed below, I myself was not able to be a participant in the technological realization of the idea. For the next 10 years I worked on research on the Gunn effect, to return to heterostructures in the mid-70s.

V. ON HOW NOT TO JUDGE NEW TECHNOLOGY

When I proposed to develop the technology for the DH laser, I was refused the resources to do so, on the grounds that "this device could not possibly have any practical applications," or words to that effect. By hindsight, it is of course obvious just how wrong this assessment was.

It was really a classical case of judging a fundamentally new technology, not by what new applications it might *create*, but merely by what it might do for already-existing applications. This is extraordinarily short-sighted, but the problem is pervasive, as old as technology itself. The DH laser was simply another example in a long chain of similar examples. Nor will it be the last. I therefore believe it is worthwhile to say a few words about this kind of argument here.

Any detailed look at history provides staggering evidence for what I have called, on another occasion (Kroemer, 1995), the *Lemma of New Technology*:

The principal applications of any
sufficiently new and innovative technology always have been
– and will continue to be –
applications *created* by that technology.

As a rule, such applications have indeed arisen – the DH laser is just a good recent example – although usually not immediately.

110

But this means that we must take a long-term look when judging the applications potential of any new technology: It must *not* be judged simply by how it might fit into already existing applications, where the new discovery may have little chance to be used in the face of competition with already-entrenched technology. Dismissing it on the grounds that it has no known applications will only stifle progress towards those applications that *will* grow out of that technology.

I do not think we can realistically predict which new devices and applications may emerge, but I believe we can create an environment encouraging progress, by not always asking immediately what any new science might be good for (and cutting off the funds if no answer full of fanciful promises is forthcoming). In particular, we must educate our funding agencies about this historical fact. This may not be easy, but it is necessary. We must make it an acceptable answer to the quest for applications to defer that answer, and that at the very least a search for applications should be considered a part of the research itself, rather than a result to be promised in advance. Nobody has expressed this last point better than David Mermin in his recent put-down of so-called "strategic research" (Mermin, 1999):

"I am awaiting the day when people remember the fact that discovery does not work by deciding what you want and then discovering it."

What is *never* acceptable – and what we must refrain from doing – is an attempt to justify the research by promising credibility-stretching mythical improvements in *existing* applications. Most such claims are not likely to be realistic and are easily refuted; they only trigger criticism of just how unrealistic the promises are, thereby discrediting the whole work.

Ultimately, progress in applications is not *deterministic*, but *opportunistic*, exploiting for new applications whatever new science and technology happen to be coming along.

VI. CONSTRAINTS

1. Lattice Matching

Let me now turn to some of the problems in implementing heterostructures.

When two materials with significantly different lattice parameters are grown upon each other, whether graded or not, huge strains rapidly build up with increasing thickness, and eventually misfit dislocations will form, a defect without any redeeming features. As a result, the need for lattice matching is all but obvious. The problem is somewhat less severe in modern structures calling for very thin layers (see below); but even there, the lattice-matched case serves as the conceptual point of departure.

Historically, the importance of lattice matching was recognized almost from the beginning, especially for bipolar devices such as lasers. In my 1967 DH laser patent (Kroemer, 1967), I gave a table listing numerous semiconductors in the order of increasing lattice parameter (see Table II); the accompanying text in the patent called for semiconductor pairs with a lattice mismatch below 0.01Å ($\approx$ 0.2 %) as the most promising ones, indicating a recognition of the stringency of the lattice matching demand. The possibility to achieve lattice matching by alloying was explicitly recognized, though.

Table II. Partial copy of the 1963 table of semiconductors ordered by lattice constant (second column) from ref. (Kroemer, 1967). The third column gives the increase in lattice constant relative to the preceding material. Note that no distinction is made between column-IV elements, the III-V compounds, and the II-VI compounds. Also, the 1963 lattice constant of AlAs was significantly in error: The correct room-temperature value (5.661Å) is actually 0.02Å *larger* than the GaAs value, and the difference is much less at typical crystal growth temperatures. [Only the semiconductors up to ZnSe are shown here; the complete 1963 table can be found in Kroemer (1996)].

Semiconductor	a [Å]	Δa [Å]
ZnS	5.406	
Si	5.428	.022
GaP	5.450	.022
AlP	5.46	.01
AlAs	5.63	.17
GaAs	5.653	.02
Ge	5.658	.005
ZnSe	5.667	.009
...	...	...

Ironically, the 1963 literature value for the lattice constant of AlAs was incorrect. As a result, the GaAs-AlAs pair initially did not seem to meet the proposed stringent criterion, and the known poor stability of (binary) AlAs against oxygen did not help. It took some time to recognize its promise, not so much as a binary material, but as an alloy with GaAs, which greatly reduced the oxidation problems, and reduced the lattice mismatch to a completely negligible level.

A more instructive way to represent the information of Table II, including energy gaps as well, is in terms of what some of us call *The Map of the World*, a display of the energy gaps of semiconductors of interest vs. their lattice constants (Fig. 9), with interconnect lines shown to represent binary alloys.

Much of the reason for the continued dominance of the (Al,Ga)As alloy system in heterostructure studies is precisely the "Great Crystallographic Accident" that AlAs and GaAs have essentially the same lattice parameter. This natural lattice matching means, in particular, that an ideal substrate is readily available for the growth of such heterostructures, namely bulk GaAs, obtainable as high-quality single crystals with low dislocation densities, especially in semi-insulating form. If there remains *one* bad aspect to the (Al,Ga)As system, it is the obnoxious chemical affinity of aluminum to oxygen, the source of many residual defects in (Al,Ga)As. Following a 1983 suggestion by myself (Kroemer, 1983), the use of (Ga,In)P lattice-matched to GaAs has recently drawn some attention as an alternative to (Al,Ga)As, especially in HBTs, for which the band lineups at the (Ga,In)P-GaAs interface are more favorable than those of (Al,Ga)As-GaAs.

A second natural substrate is InP, widely used for both optoelectronic and

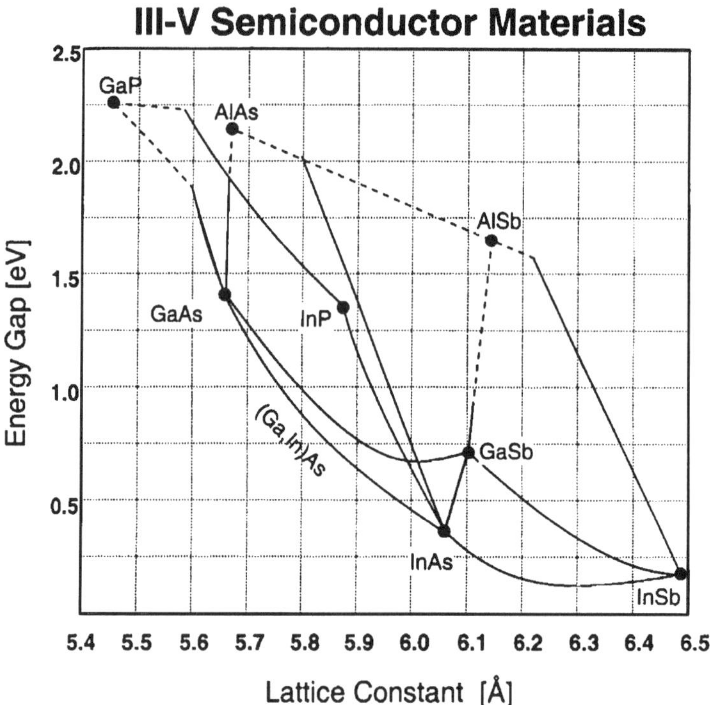

Figure 9. Partial "Map of the World," plotting the energy gap of various III-V compounds vs. lattice constant. The map omits the "Old-World Continents" of the column-IV and the II-VI semiconductors, and the "New World" of the nitrides.

high-speed device applications that call for energy gaps less than that of GaAs. There is no binary III-V compound lattice-matched to InP, but InP is widely used in devices, combined with a wide variety of alloys ranging from (Ga,In)As to Al(As,Sb).

With the emergence of quantum wells, superlattices, and other structures calling for very thin layers, the issue of strain induced by lattice mismatch has lost some of its tyrannical dominance. In sufficiently thin structures, remarkably large strains can be accommodated without dislocation formation, to the point that the modification of the energy band structure of a heterostructure by *deliberate* introduction of strain has become an important device design principle in its own right. The recent evolution of successful Si-Ge HBTs is perhaps the most dramatic triumph of this idea (see, for example, Abstreiter (1996); König (1996), but other examples are close behind, both in field-effect transistors (FETs) and in photonic devices. Some of the recent developments in self-assembling quantum dots are explicitly based on utilizing strain already during the crystal growth process.

2. Valence Matching

If lattice matching were the only constraint, the Ge-GaAs system would be the ideal hetero-system, as was in fact believed by some of us – including myself – in the early-'60s. At that time, the most successful heterojunctions that had been demonstrated were the Ge-on-GaAs heterojunctions studied by Anderson (1960), suggesting a bright future for this system (the term *heterojunction* seems to have appeared first in Anderson's papers). Table II reflects this idea, in the form of combining III–V compounds, II–VI compounds, and group-IV semiconductors into a common table, making the GaAs-Ge system appear to be the most promising candidate It took a few years to realize that this was a blind alley – and why.

It is not a questions of chemical incompatibility, or even of cross-doping effects. Covalent bonds between Ge on the one hand, and Ga or As on the other are readily formed, but they are what I would like to call *valence-mismatched*, meaning that the number of electrons provided by the atoms is not equal to the canonical number of exactly two electrons per covalent bond. Hence the bonds themselves are not electrically neutral, as first pointed out in a 1978 "must-read paper" by Harrison *et al.* (1978).

Consider a hypothetical idealized (001)-oriented interface between Ge and GaAs, with Ge to the left of a mathematical plane, and GaAs to the right (Fig. 10). In GaAs, an As atom brings along 5 electrons (= 5/4 electrons per bond), and expects to be surrounded by 4 Ga atoms, each of which brings along 3 electrons (3/4 per bond), adding up to the correct number of 8/4 = 2 electrons per Ga-As covalent bond. But when, at a (001) interface, an As atom has two Ge atoms as bonding partners, each Ge atom brings along 1 electron per bond, which is one-half electron too many. Loosely speaking, the As atom "does not know" whether it is a constituent of GaAs, or a donor in Ge.

As a result, each Ge-As bond acts as a donor with a fractional charge, and each Ge-Ga bond as an acceptor with the opposite fractional charge. To be electrically neutral, a Ge-GaAs interface would have to have equal numbers of both charges, not only averaged over large distances, but locally. Given chemical bonding preferences, such an arrangement will not occur naturally during epitaxial growth. If only one kind of bonds were present, as in Fig. 10, the interface charge would support an electric field of 4×10^7 V/cm. Such a huge field would force atomic re-arrangements during growth, trying to equalize the number of Ge-As and Ge-Ga bonds. However, these re-arrangements will never go to completion, but will leave behind ill-defined locally fluctuating residual charges, with deleterious consequences for any device application. Interfaces with perfect bond charge cancellation are readily drawn on paper; but in practice there are always going to remain some local deviations from the perfect charge compensation, leading to performance-degrading random potential fluctuations along the interface.

Although Harrison *et al.* discuss only the GaAs-Ge interface, their argument applies to other interfaces combining semiconductors from different columns of the periodic table. In the specific case of compound semiconductor growth on a column-IV elemental semiconductor, the additional problem of antiphase domains on the compound side arises (see, for example, Kroemer (1987)).

The above discussion pertained to the most-widely used (001)-oriented interface. The interface charge at a valence-mismatched interface actually de-

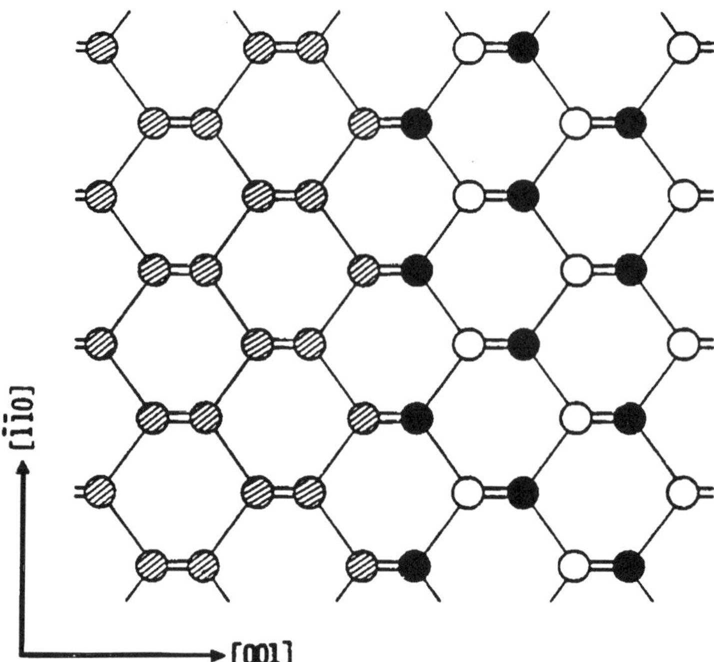

Figure 10. Departure from electrical neutrality at a "mathematically planar" (001)-oriented Ge/GaAs interface. The different atomic species – Ga or As atoms (white and black circles) and Ge atoms (shaded circles) – do not bring along the correct number of electrons to form electrically neutral Ga-Ge or As-Ge covalent bonds of 2 electrons per bond. From Harrison *et al.* (1978).

pends on the crystallographic orientation. It has been shown by Wright *et al.* that an ideal (112) interface exhibits neither an interface charge, nor antiphase domains, and it was in fact possible to demonstrate GaP-on-Si interfaces that had a sufficiently low defect density that they operated as emitters in a GaP-on-Si HBT (Wright *et al.*, 1982; 1984). However, the performance was still sufficiently poor that the approach was not pursued further.

VII. MOLECULAR BEAM EPITAXY AND ABRUPT HETEROSTRUCTURES

The 1970 DH laser demonstration was accomplished by liquid-phase epitaxy (LPE), a beautifully simple technology, but with severe limitations. The big technological breakthrough for heterostructures came only with the emergence of molecular beam epitaxy (MBE) as a practical crystal growth technology, largely pioneered by Al Cho (followed later by organometallic vapor phase epitaxy). In contrast to LPE, MBE permitted combining a wide range of semiconductors, even such hetero-valent combinations as GaP and GaAs on Si. Moreover, it offered a very high degree of control over the local composition, almost on an atomic layer scale. Suddenly, we could realize experimentally almost any band diagram we could draw, at least in the growth direction (lateral control on a similar scale remains an elusive goal to this day). By 1980, the progress in heterostructures had been so large, that I was able to give an invited paper the provocative title "Heterostructures for Everything: Device Principle of the 1980's?" (Kroemer, 1981). It turned out to be an accurate prediction.

In particular, it had become possible to grow almost atomically abrupt heterojunctions. This also meant that two heterojunctions could be placed sufficiently closely together that quantum effects in the space between them became important, and could be utilized for new kinds of devices. The most obvious development was that of quantum wells (QWs), especially for laser applications, which soon became dominated by QW lasers. But we also saw an increasing use of heterostructures in non-bipolar applications, in effect applying the general quasi-electric field design principle outside its range of origin.

One such example is the use of pairs of tunneling barriers in resonant-tunneling diodes, for application as high-frequency sources up into the sub-terahertz frequency range. Another is the idea of Esaki and Tsu to use a periodic heterostructure superlattice as a quasi-bulk negative-resistance medium with an even higher frequency limit (Esaki and Tsu, 1970). It has so far remained an elusive goal, but it continues to be a very active field of research (including by myself).

I would like to single out here a less obvious new concept, that of *modulation doping*, due to Dingle *et al.* (1978). Consider a heterojunction in which only the side with the higher conduction band is doped (Fig. 11). The downward quasi-electric potential step at the interface will cause electrons to drain into the lower conduction band on the other side. Once they are past the range of the quasi-electric potential step associated with the abrupt hetero-interface itself, the electrons still see the ordinary electric field associate with

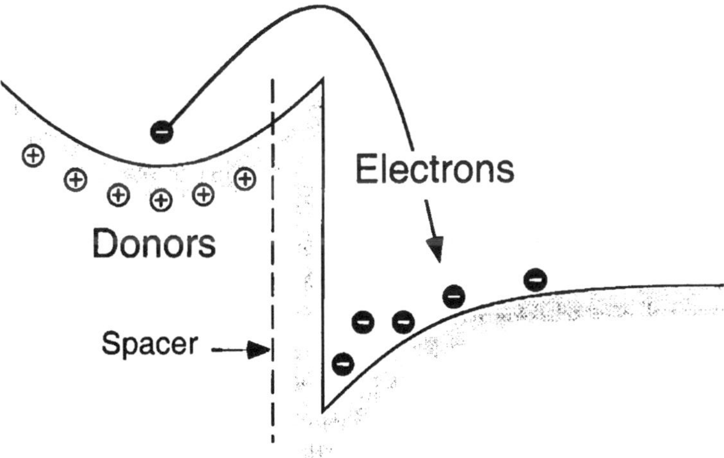

Figure 11. Modulation doping. At an abrupt heterojunction, electrons contributed by donors on the higher-energy side drain onto the lower-energy side, creating a quasi-two-dimensional electron gas there. Because the electrons are now spatially separated from the donors, impurity scattering is reduced, especially if an undoped spacer is inserted on the higher-energy side. The band curvature shown is due to the space charges on the two sides of the interface.

116

the Coulomb attraction by the donors left behind on the other side. It pulls
the electrons towards the interface, creating a 2-dimensional electron gas
(2DEG) inside a roughly triangular quantum well. Moreover – and most importantly – because the electrons have been spatially separated from "their"
donors, impurity scattering is reduced, and the electron mobility is enhanced. To maximize these benefits, an undoped spacer region is left adjacent to the interface.

The idea had extremely far-reaching consequences, both for devices, and
in basic solid-state physics. In devices, it formed the basis of a new class of
field effect transistors (FETs), commonly referred to as HEMTs, meaning
High-Electron-Mobility Transistors (Mimura *et al.*, 1980; Delagebeaudeuf *et
al.*, 1980). Their properties are superior to those of earlier classes of FETs.
Because of their low noise, they are now used as the sensitive input stage in
cellular phones, and thus have contributed to the explosive growth of this aspect of modern information technology.

In basic physics, the suppression of impurity scattering by modulation doping with optimized spacers has permitted the achievement of huge low-temperature mobilities. There is a direct path from the idea of modulation doping to the discovery of the fractional quantum Hall effect, by Tsui, Störmer,
and Gossard (Tsui *et al.*, 1982; Stormer, 1999), in 2DEG samples of unprecedented structural perfection grown by Gossard. The subsequent theoretical
interpretation of the effect by Laughlin (1999) revealed it as a true fundamental breakthrough in solid-state physics, for which Tsui, Störmer, and
Laughlin received the 1998 Nobel Prize in Physics. Unfortunately, the Nobel
statute prohibition against dividing the prize amongst more than three individuals excluded Gossard from sharing in the award.

VIII. BAND OFFSETS

In wake of the emergence of MBE technology in the early-70s, my own research returned to heterostructure problems, especially to the problem of
band offsets at abrupt heterojunctions. In that limit, the energy band structure makes a discontinuous transition, and exactly how the bands on the two
sides are lined up becomes a central question, both experimentally and
theoretically. One of the reasons all my early device band diagrams show
graded transitions was to sidestep this question of band lineups, of which I
was actually well aware.

A. Offset Types

Given two semiconductors, there are evidently three different band lineups
possible (Fig. 12)

1. Straddling Lineups

The most common lineup is the straddling one, with conduction and valence
band offsets of opposite sign. It is, in essence, the abrupt limit of the graded
band structure of Fig. 1c. In quantum wells and superlattices made from such

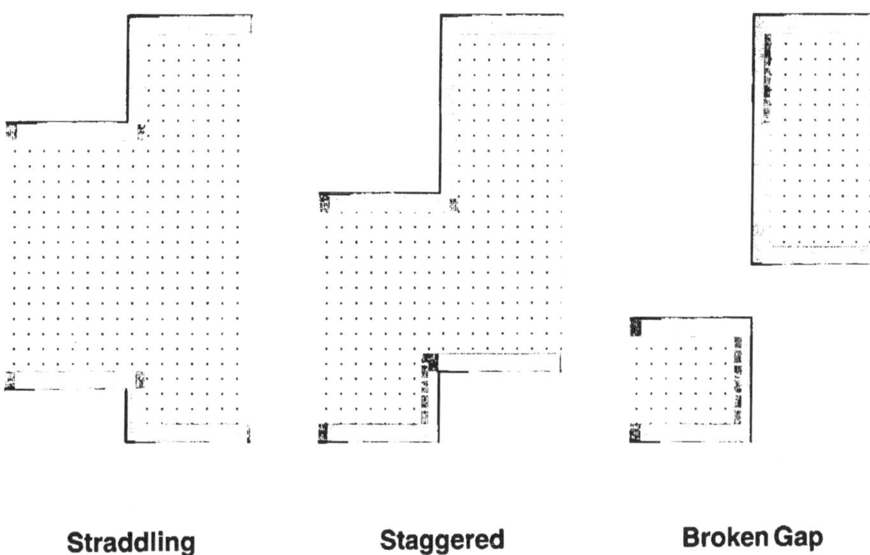

Straddling   **Staggered**   **Broken Gap**

Figure 12. Straddling, staggered, and broken-gap band lineups.

pairs, the lowest conduction band states occur in the same part of the structure as the highest valence band states, which makes these pairs of particular interest for opto-electronic applications, like lasers, which are bipolar kinds of devices, with both electrons and holes involved in the device operation. The two kinds of carriers then occur in the same layers; hence such structures are sometimes referred to as *spatially direct*. Many of today's opto-electronic devices, such as quantum well lasers, are based on such a lineup. The most-widely studied heterojunction system, GaAs-(Al,Ga)As, is of this kind, as are a number of other systems, for example, (Ga,In)As lattice-matched to InP, and (Ga,In)P lattice-matched to GaAs.

2. Staggered Lineups

For some materials pairs, the two bands are shifted in the same direction, leading to a band structure in which the lowest conduction band minimum occurs on one of the sides, the highest valence band maximum on the other, with an energy separation between the two less than the lower of the two bulk gaps. The combination of AlAs-Al$_x$Ga$_{1-x}$As for $x > 0.3$ is of this kind, as is (Al,In)As lattice-matched to InP; there are several others. In bipolar structures with this lineup, the electrons and holes are confined to *different* layers, hence these structures are *spatially indirect*. Nevertheless, the wave functions overlap at the interface, making radiative recombination possible, with a photon energy less than the narrower of the two gaps (Kroemer and Griffiths, 1983; Caine *et al.*, 1984).

Staggered lineups imply large band offsets in either the conduction or the valence band, and for some applications this property is more important than the spatial indirectness. For example, the conduction band lineup at the InAs-AlSb interface, 1.35eV (Nakagawa *et al.*, 1989), is the highest that has been reported for any III-V system, and several applications are based on this property, along with the low electron effective mass in InAs. The fastest resonant tunneling diode reported in the literature (Brown *et al.*, 1991), oscillating up to 712 GHz, was based on this system.

The high barriers also offer superb electron confinement in FETs, and the possibility of achieving extremely high levels of electron concentration (approaching $10^{13}cm^{-2}$) by modulation doping (i. e., putting the donors into the barriers rather than into the wells), while retaining high mobilities. This combination makes the InAs-AlSb system ideal for investigating the properties of quantum wells in the metallic limit, for example as coupling medium in a new class of superconducting weak links (Kroemer *et al.*, 1994).

3. Broken-Gap Lineup

If a staggered lineup is carried to its extreme, the result is a broken-gap lineup, in which the bottom of the conduction band on one side drops below the top of the valence band on the other. There exists at least one nearly-lattice-matched pair of this kind, InAs-GaSb, with a break in the forbidden gap at the interface on the order of 150 meV (Sakaki *et al.*, 1977).

The broken-gap InAs-GaSb lineup by itself is an exotic lineup, of interest especially to research physicist. To the theorist interested in understanding band offsets, the ability to predict such an offset, at least approximately, is one of the litmus tests of any lineup theory, and recent lineup theories pass this test with flying colors.

B. Theory

It should be self-evident from the above that the question as to the exact values of the band offsets at the various semiconductor pairs of interest is a central one, both theoretically and experimentally. I tried to contribute to both.

At the end of the '60s, the only rule for estimating band offsets theoretically was the *electron affinity rule* (Anderson, 1960), according to which the conduction band offset should be equal to the difference in electron affinity at the two free semiconductor surfaces. In a 1975 paper (Kroemer, 1975), I pointed out that this is an extraordinarily unsatisfactory rule. Even if good electron affinity data were available, the validity of the rule depended on hidden assumptions about the relations between the properties of the interface between two semiconductors, and those of the much more drastic vacuum-to-semiconductor interfaces, assumptions that almost certainly were invalid. Harrison aptly characterized the rule by saying that it "replaces one simple problem by two very difficult problems." (Harrison, 1977)

I called for a theory that would determine the band offsets from the *bulk* properties of the participating semiconductors, and I suggested it as a Ph. D. topic to Bill Frensley (now at the University of Texas in Dallas). One of the specific question I asked Bill to look into was whether broken-gap lineups might in fact occur. The resulting theory (Frensley and Kroemer, 1976; 1977), based on pseudopotentials, was the first to give a semi-quantitative derivation, from bulk properties, not only of band offsets that were already known, like

GaAs/AlAs; it also had a considerable predictive value. In particular, the theory predicted that the InAs/GaSb heterojunction either had a broken-gap lineup, or came very close to it.

The Frensley-Kroemer theory has since then been followed by the work of others based on different principles; see Harrison (1977) and Christensen (1988).

C. Band Offsets by C-V Profiling

Sometime in 1979, Jim Harris (then at the Rockwell Science Center, now at Stanford) showed me some capacitance-voltage (C-V) profiling data on an LPE-grown (Al,Ga)As/GaAs heterojunction. C-V profiling is a common technique to determine electron concentrations in semiconductors by measuring the capacitance of a reverse-biased Schottky barrier placed upon the surface of the semiconductor. By varying the bias, one can explore the depth distribution of the electrons over some distance. Near the hetero-interface, Harris' data showed a clear indication of an electron accumulation on the GaAs side, and an electron depletion on the (Al,Ga)As side, as one would expect from an appropriate band diagram. However, the apparent electron concentration was strongly smeared out by averaging over a Debye length. When I tried to understand the averaging process quantitatively, I realized that the dipole moment associated with the accumulation/depletion pair should be preserved during the averaging, and that its measurement should permit a determination of the conduction band offset (Kroemer *et al.*, 1980; Kroemer and Chien, 1981; Kroemer, 1985). The analysis yielded a band offset of approximately 66 % of the energy gap difference (Kroemer *et al.*, 1980), not far from today's generally accepted value of 62%.

The C-V technique has since then been used by many others and has provided some of the best data for band offsets for many heterojunction pairs.

IX. EPILOGUE

Throughout this paper, I have concentrated on my own work towards heterostructures, especially on the early parts of it, through 1963, which were dominated by bipolar device concepts. But today's heterostructure field would not be what it is without the subsequent contributions – technological or conceptual – by numerous others, especially on non-bipolar structures. It was only through this work of numerous others, on topics that went beyond my own contributions, that the significance of the latter eventually emerged. For this I owe all of them my thanks.

X. REFERENCES

Abstreiter, G., 1996, Physica Scripta T68, 68.
Alferov, Z. I., V. M. Andreev, D. Z. Garbuzov, Y. V. Zhilyaev, E. P. Morozov, E. L. Portnoi and
    V. G. Trofim, 1970, Fiz. Tekh. Poluprovodn. 4, 1826. [Sov. Phys. - Semicond. 4, 1573-1575
    (1971)].
Alferov, Z. I., 1996, Physica Scripta **T68**, 32.

Alferov, Z. I., 2001, this volume.

Anderson, R. L., 1960, IBM J. Res. Dev. **4**, 283.

Brown, E. R., J. R. Söderström, C. D. Parker, L. J. Mahoney, K. M. Molvar and T. C. McGill, 1991, Appl. Phys. Lett. **58**, 2291.

Caine, E. J., S. Subbanna, H. Kroemer, J. L. Merz and A. Y. Cho, 1984, Appl. Phys. Lett. **45**, 1123.

Casey, C. and M. Panish, 1978, *Heterostructure Lasers – Part A: Fundamental Principles* (Academic Press, New York). See Sec. 1.2.

Christensen, N. E., 1988, Phys. Rev. B **38**, 12687.

Delagebeaudeuf, D., P. Delescluse, P. Etienne, M. Laviron, J. Chaplart and N. T. Linh, 1980, Electron. Lett. **16**, 667.

Diedrich, H. and K. Jötten, 1961, Procs. of *Colloque international sur les dispositifs à semiconducteurs*, Paris (Editions Chiron, Paris) p. 330.

Dingle, R., H. L. Störmer, A. C. Gossard and W. Wiegmann, 1978, Appl. Phys. Lett. **33**, 665.

Esaki, L. and R. Tsu, 1970, IBM J. Res. Dev. **14**, 61.

Frensley, W. R. and H. Kroemer, 1976, J. Vac. Sci. Technol. **13**, 810.

Frensley, W. R. and H. Kroemer, 1977, Phys. Rev. B **16**, 2642.

Harrison, W. A., 1977, J. Vac. Sci. Technol. **14**, 1016.

Harrison, W. A., E. A. Kraut, J. R. Waldrop and R. W. Grant, 1978, Phys. Rev. B **18**, 4402.

Hayashi, I., M. B. Panish, P. W. Foyt and S. Sumski, 1970, Appl. Phys. Lett. **17**, 109.

König, U., 1996, Physica Scripta **T68**, 90.

Kroemer, H., 1957a, RCA Review **18**, 332. (Re-printed from the Proceedings of the Symposium *"The Role of Solid State Phenomena in Electric Circuits,"* Polytechnic Institute of Brooklyn, April 1957, p. 143)

Kroemer, H., 1957b, Proc. IRE **45**, 1535.

Kroemer, H., 1957c, unpublished.

Kroemer, H., 1963, Proc. IEEE **51**, 1782.

Kroemer, H., 1967, US patent 3,309,553 (filed Aug. 16, 1963).

Kroemer, H., 1975, Crit. Revs. Solid State Sci. **5**, 555.

Kroemer, H., W.-Y. Chien, J. S. Harris and D. D. Edwall, 1980, Appl. Phys. Lett. **36**, 295.

Kroemer, H., 1981, Jpn. J. Appl. Phys. Supplem. **20-1**, 9.

Kroemer, H. and W.-Y. Chien, 1981, Solid-State Electron. **24**, 655.

Kroemer, H., 1982, Proc. IEEE **70**, 13.

Kroemer, H., 1983, J. Vac. Sci. Technol. B **1**, 126.

Kroemer, H. and G. Griffiths, 1983, IEEE Elect. Dev. Lett. **4**, 20.

Kroemer, H., 1985, Appl. Phys. Lett. **46**, 494.

Kroemer, H., 1987, J. Cryst. Growth **81**, 193.

Kroemer, H., C. Nguyen, E. L. Hu, E. L. Yuh, M. Thomas and K. C. Wong, 1994, Physica B **203**, 298.

Kroemer, H., 1995, Procs. of *NATO Adv. Res. Wkshp. on Future Trends in Microelectronics*, Ile de Bendor, France, edited by S. Luryi et al., NATO ASI Series E **323** (Kluwer, Dordrecht) p. 1.

Kroemer, H., 1996, Physica Scripta **T68**, 10.

Krömer, H., 1953, Naturwissensch. **40**, 578.

Krömer, H., 1954, Archiv d. Elekt. Übertragung **8**, 499.

Laughlin, R. B., 1999, Revs. Mod. Phys. **71**, 863.

Mermin, D., 1999 (Aug.), Physics Today **52** (8), 11.

Mimura, T., S. Hiyamizu, T. Fujii and K. Nanbu, 1980, Jpn. J. Appl. Phys. **19**, L225.

Nakagawa, A., H. Kroemer and J. H. English, 1989, Appl. Phys. Lett. **54**, 1893.

Sakaki, H., L. L. Chang, R. Ludeke, C. A. Chang, G. A. Sai-Halasz and L. Esaki, 1977, Appl. Phys. Lett. **31**, 211.

Shockley, W., 1951, US patent 2,569,347 (filed 26 June 1948).

Stormer, H. L., 1999, Revs. Mod. Phys. **71**, 875.

Tsui, D. C., H. L. Störmer and A. C. Gossard, 1982, Phys. Rev. Lett. **48**, 1559.

Wright, S. L., M. Inada and H. Kroemer, 1982, J. Vac. Sci. Technol. **21**, 534.

Wright, S. L., H. Kroemer and M. Inada, 1984, J. Appl. Phys. **55**, 2916.

Publications by H. Kroemer
Journal and Conference Publications

Abbreviations Used:

[AEU: Archiv d. Elektrischen Ubertragung; AJP: American Journal of Physics; APL : Applied Physics Letters; CPC: ChemPhysChem; CRSSS: Crit. Rev. on Solid-State Science; EL: Electronics Letters; FZ: Fernmeldetechn. Zeitschr; IEDM: International Electron Device Meeting; IJMPB: International J. Modern Physics B; IOP: Institute of Physics; IRE: Institution of Radio Engineers; JAP: Journal of Applied Physics; JCG: J. Cryst. Growth; JEM: J. Electronic Materials; JJAP: Japanese Journal of Applied Physics; JVST: Journal of Vac. Sci. and Technology; MGWL: IEEE Microwave and Guided Wave Letters; ME: Microelectronic Engineering; ML: Materials Letter; MSE: Materials Science and Engineering; NAT: Naturwissensch; NTF: Nachrichtentechnische Fachberichte; PhyB: Physica B; PhyE: Physica E; PIEEE: Proceedings of the IEEE; PIRE: Proceedings of the Institution of Radio Engineers; PMRS: Proceedings Materials Research Society; PR: Physical Review; PRL: Physical Review Letters; PS: Physica Scripta; PRB: Physical Review B; QE: IEEE J. Quantum Electronics; RCA: RCA Review; RMP: Review Modern Physics; SM: Superlattices and Microstructures; SS: Surface Science; SSC: Solid-State Communications; SSE: Solid-State Electronics; SSCC: International Solid-state Circuits Conference; SST: Semiconductor Science and Technology; TED: IEEE Trans. on Electron Devices; TNS: IEEE Trans. on Nuclear Science; UFN: Uspekhi Fizicheskikh Nauk; ZP: Zeitschrift fur Physics]

1953-01-NATv40-p163

K. H. Paetzold and H. Kroemer, "Uber die Temperaturabhangigkeit der Elektronenterme in Kristallen," Naturwissensch., Vol. 40, pp. 163-164, March 1953.

1953-02-ZPv134-p435

H. Kroemer, "Zur Theorie des Germaniumgleichrichters und des Transistors," Zeitschr. f. Phys., Vol. 134, pp. 435-450, 1953. (Ph.D. Dissertation)

1953-03-PRv90-p515

R. E. Burgess, H. Kroemer and J. M. Houston, "Corrected Values of Fowler-Nordheim Emission Function v(y) and s(y)," Phys. Rev., Vol. 90, p. 515, 1953.

1953-04-NATv40-p578

H. Kroemer, "Der Drifttransistor," Naturwissensch., Vol. 40, p. 578-579, 1953.

1953-05-FZv6-p438

H. Kroemer, "Leitungsmechanismen in Halbleitern, - P-N Ubergange und Transistoren," Fernmeldetechn. Zeitschr., Vol. 6, pp. 438-443, 1953.

1954-01-AEUv8-p223

H. Kroemer, "Zur Theorie des Diffusions- und des Drifttransistors: I. Die Vierpolmatrix und ihr Niederfrequenzverhalten," Archiv d. Elektrischen Ubertragung, Vol. 8, pp. 223-228, 1954.

1954-02-AEUv8-p363

H. Kroemer, "Zur Theorie des Diffusions- und des Drifttransistors: II. Frequenz-abhangigkeit," Archiv d. Elektrischen Ubertragung, Vol. 8, pp. 363-369, 1954.

1954-03-AEUv8-p499

H. Kroemer, "Zur Theorie des Diffusions- und des Drifttransistors: III. Dimensionierungs-fragen," Archiv d. Elektrischen Ubertragung, Vol. 8, pp. 499-504, 1954.

1954-04-FZv7-p86

H. Kroemer, "Leitungsmechanismen in Halbleitern, - P-N Ubergange und Transistoren," Fernmeldetechn. Zeitschr., Vol. 7, pp. 86-92, 1954.

1955-01-NTFv1-p19

H. Kroemer, "Uber die Entwicklung von Schichttransistoren mit Hoher Frequenzgrenze," Nachrichtentechnische Fachberichte, Vol. 1, pp. 19-24, 1955.

1956-01-RCA-p132

H. Kroemer, "Some Aspects of Thermal Conversion in Germanium," Transistors I, pp. 132-135, RCA Laboratories, Princeton, 1956.

1956-02-RCAv17-p515

H. Kroemer, "The Apparent Contact Potential of a Pseudo-abrupt P-N Junction," RCA Rev., Vol. 17, pp. 515-521, 1956.

1956-03-RCA-p202

H. Kroemer, "The Drift Transistor," Transistors I, pp. 202-220, RCA Laboratories, Princeton, 1956.

1957-01-PIREv45-p1535

H. Kroemer, "Theory of a Wide-Gap Emitter for Transistors," Proc. IRE, Vol. 45, pp. 1535-1537, 1957.

1957-02-RCAv18-p332

H. Kroemer, "Quasi-Electric and Quasi Magnetic Fields in Non-Uniform Semiconductors," RCA Review, Vol. 18, pp. 332-342, 1957.

1958-01-PRv109-p1856

H. Kroemer, "Proposed Negative-Mass Microwave Amplifier," Phys. Rev., Vol. 109, p. 1856, 1958.

1959-01-PIREv47-p397

H. Kroemer, "The Physical Principles of a Negative-Mass Amplifier," Proc. IRE, Vol. 47, pp. 397-406, 1959.

1960-01-SSCC-p86

H. Kroemer, "Microwave interactions in bulk semiconductors: A survey," Dig. Tech. Solid-State Circuits Conf., Vol. III, pp. 86-87, 1960.

1963-01-PIEEEv51-p1782

H. Kroemer, "A Proposed Class of Heterojunction Injection Lasers," Proc. IEEE, Vol. 51, pp. 1782-1783, 1963.

1964-01-PIEEEv52-p426

H. Kroemer, "Considerations Regarding the Use of Semiconductor Heterojunctions for Laser Operation," Proc. IEEE, Vol. 52, pp. 426-427, 1964.

1964-02-PIEEEv52-p1736
H. Kroemer, "Theory of the Gunn Effect," Proc. IEEE, Vol. 52, p. 1736, 1964.

1964-03-SSEv7-p291
H. Kroemer, "On the Theory of Hall Effect Isolators for Tunnel Diode Amplifiers," Solid-State Electron., Vol. 7, pp. 291-310, 1964.

1965-01-JAPv36-p2461
H. Kroemer, G. F. Day, R. D. Fairman, and J. Kinoshita, "Preparation and Some Properties of Mg_2Ge Single Crystals and of Mg_2Ge p-n Junctions," J. Appl. Phys., Vol. 36, p. 2461-2470, 1965.

1965-02-PIEEEv53-p1246
H. Kroemer, "External Negative Conductance of a Semiconductor with Negative Differential Mobility," Proc. IEEE, Vol. 53, p. 1246, 1965.

1965-03-SSEv8-p79
H. Kroemer, "A Possible Tunnel Diode Power Oscillator using a Hall Effect Isolator," Solid-State Electron., Vol. 8, pp. 79-81, 1965.

1966-01-PIEEEv54-p1980
H. Kroemer, "The Effect of a Parasitic Series Resistance on the Performance of Bulk Negative Conductivity Amplifiers," Proc. IEEE, Vol. 54, pp. 1980-1981, 1966.

1966-02-TEDv13-p27
H. Kroemer, "Non-Linear Space-Charge Domain Dynamics in a Semiconductor with Negative Differential Mobility," IEEE Trans. Electron Dev., Vol. ED-13, pp. 27-40, 1966.

1967-01-TEDv14-p476
H. Kroemer, "Detailed Theory of the Negative Conductance of Bulk Negative Mobility Amplifiers, in the Limit of Zero Ion Density," IEEE Trans. Electron Dev.," Vol. ED-14, pp. 476-492, 1967.

1967-02-ConfProc-p264
H. Kroemer, "Negative Conductance in Semiconductors," Festkorperprobleme, Vol. 7, pp. 264-286, 1967.

1968-01-APLv12-p283
M. Shyam and H. Kroemer, "Transverse Negative Differential Mobilities for Hot Electrons and Domain Formation in Germanium," Appl. Phys. Lett., Vol. 12, pp. 283-285, 1968.

1968-02-IEEESpectrumv5-p47
H. Kroemer, "Negative Conductance in Semiconductors," IEEE Spectrum, Vol. 5, pp. 47-56, 1968.

1968-03-TEDv15-p819
H. Kroemer, "The Gunn Effect under Imperfect Cathode Boundary Conditions," IEEE Trans. Electron Dev., Vol. 15, pp. 819-837, 1968.

1970-01-PIEEEv58-p1844
H. Kroemer, "Generalized Proof of Shockley's Positive Conductance Theorem," Proc. IEEE, Vol. 58, pp. 1844-1845, 1970.

1971-01-PIEEEv59-p1282

H. Kroemer, "Comments on Generalized Proof of Shockleys Positive Conductance Theorem," Proc. IEEE, Vol. 59, pp. 1282-1283, 1971.

1972-01-JAPv43-p5124
H. Kroemer, "Slow Gunn Domains with Field-Independent Trapping," J. Appl. Phys., Vol. 43, pp. 5124-5130, 1972.

1974-01-IEDM-p3
H. Kroemer, "Negative bulk mobility devices - what next?," IEDM Tech. Dig., pp. 3-4, 1974.

1975-01-AJPv43-p514
H. Kroemer, "WKB Connection Rules from the Harmonic Oscillator," Am. J. Phys., Vol. 43, pp. 514-517, 1975.

1975-02-PIEEEv63-p988
H. Kroemer, "On the Group Velocity of Bloch Waves," Proc. IEEE, Vol. 63, p. 988, 1975.

1975-03-CRSSSv5-p555
H. Kroemer, "Problems in the Theory of Heterojunction Discontinuities," Crit. Rev. Solid State Science, Vol. 5, pp. 555-564, 1975.

1976-01-JVSTv13-p810
W. R. Frensley and H. Kroemer, "Prediction of Semiconductor Heterojunction Discontinuities," J. Vac. Sci. Technol., Vol. 13, pp. 810-815, 1976.

1976-02-PRBv14-p3335
R. Yeats and H. Kroemer, "The Peierls instability in a nearly-free electron model, including nonlinear screening," Phys. Rev. B, Vol. 14, pp. 3335-3345, 1976.

1976-03-SSCv20-p889
H. Kroemer, "Regular Periodic Explosions of Electron-Hole Drops under Steady-State Illumination and Microwave Heating," Solid State Comm., Vol. 20, pp. 889-891, 1976.

1977-01-PRBv1S-p880
H. Kroemer, "Band-Structure instability in long-period one-dimensional super-lattices under strong population inversion," Phys. Rev. B, Vol. 15, pp. 880-884, 1977.

1977-02-TEDv24-p658
R. H. Hayes, R. M. Raymond, and H. Kroemer, "Computer Simulation of InP Transferred-Electron Amplifiers for Ka-Band," IEEE Trans. Electron Dev., Vol. ED-24, pp. 658-661, 1977.

1977-03-APLv31-p48
W. R. Frensley and H. Kroemer, "Interstitial potential differences, electronegativity differences, and effective ionic charges in zincblende-type semiconductors," Appl. Phys. Lett., Vol. 31, pp. 48-50, 1977.

1977-04-PRBv16-p2642
W. R. Frensley and H. Kroemer, "Theory of the energy-band lineup at an abrupt semiconductor heterojunction," Phys. Rev. B., Vol. 16, pp. 2642-2652, 1977.

1977-05-TEDv24-p192
R. M. Raymond, H. Kroemer, and R. E. Hayes, "Design of Cathode Doping

Notches to Achieve Uniform Fields in Transferred-Electron Devices," IEEE Trans. Electron Dev., Vol. ED-24, pp. 192-195, 1977.

1978-01-SSEv21-p61

H. Kroemer, "Hot-Electron Relaxation Effects in Devices," Solid-State Electron., Vol. 21, pp. 61-67, 1978.

1978-02-TEDv25-p850

H. Kroemer, "The Einstein Relation for Degenerate Carrier Concentrations," IEEE Trans. Electron Dev., Vol. ED-25, p. 850, 1978.

1978-03-APLv33-p749

H. Kroemer, Wu-Yi Chien, H. C. Casey, Jr., and A. Y. Cho, "Photocollection Efficiency and Interface Charges of MBE-grown Abrupt p(GaAs)-N($Al_{0.33}Ga_{0.67}$)As Heterojunctions," Appl. Phys. Lett., Vol. 33, pp. 749-751, 1978.

1980-01-APLv36-p210

S. Wright and H. Kroemer, "Reduction of Oxides on Silicon by Heating in a Gallium Molecular Beam at 800°C," Appl. Phys. Lett., Vol. 36, pp. 210-211, 1980.

1980-02-APLv36-p295

H. Kroemer, Wu-Yi Chien, J. S. Harris, and D. D. Edwall, "Measurement of Isotype Heterojunction Barriers by C-V Profiling," Appl. Phys. Lett., Vol. 36, pp. 295-297, 1980.

1980-03-APLv36-p763

H. Kroemer, K. J. Polasko, and S. C. Wright, "On the (110) Orientation as the Preferred Orientation for the Molecular Beam Epitaxial Growth of GaAs on Ge, GaP on Si, and Similar Zincblende-on-Diamond Systems," Appl. Phys. Lett., Vol. 36, pp. 763-765, 1980.

1980-04-AJPv48-p962

H. Kroemer, "How Incorrect is the Classical Partition Function for the Ideal Gas?," Am. J. Phys., Vol. 48, pp. 962-963, 1980.

1981-01-APLv38-p959

H. Kroemer, "Simple Rate Equation Model for Hypothetical Doubly Stimulated Emission of Both Photons and Phonons in Quantum-Well Lasers," Appl. Phys. Lett., Vol. 38, pp. 959-961, 1981.

1981-02-JAPvS2-p873

H. Kroemer, "Analytic Approximations for Degenerate Accumulation Layers in Semiconductors, with Applications to Barrier Lowering in Isotype Heterojunctions," J. Appl. Phys., Vol. 52, pp. 873-878, 1981.

1981-03-JJAPv20-p9

H. Kroemer, "Heterostructures for Everything: Device Principle of the 1980's?," Japan. J. Appl. Phys., Vol. 20 (Supplement No. 1), pp. 9-13, 1981.

1981-04-PRBv23-p1887

S. M. Latif and H. Kroemer, "Order-disorder Phase Transition of the Iodine Sublattice in tetrathiofulvalene iodide (n≈0.7) in a Strong Electric Field," Phys. Rev. B., Vol. 23, pp. 1887-1895, 1981.

1981-05-SSEv24-p655
H. Kroemer and W.-Y. Chien, "On the Theory of Debye Averaging in the C-V Profiling of Semiconductors," Solid-State Electron., Vol. 24, pp. 655-660, 1981.

1982-01-PIEEEv70-p13
H. Kroemer, "Heterostructure Bipolar Transistors and Integrated Circuits," Proc. IEEE, Vol. 70, pp. 13-25, 1982.

1982-02-JVSTv19-p143
S. L. Wright and H. Kroemer, "Operational Aspects of a Gallium Phosphide Source of P_2 Vapor in Molecular Beam Epitaxy," J. Vac. Sci. Technol., Vol. 19, pp. 143-148, 1982.

1982-03-JVSTv21-p534
S. L. Wright, M. Inada, and H. Kroemer, "Polar-on-Nonpolar Epitaxy: Sublattice Ordering in the Nucleation and Growth of GaP on Si (211) Surfaces," J. Vac. Sci. Technol., Vol. 21, pp. 534-539, 1982.

1982-04-JVSTv21-p551
H. Kroemer and Q.-G. Zhu, "On the Interface Connection Rules for Effective-Mass Wave Functions at an Abrupt Heterojunction Between Two Semiconductors with Different Effective Mass," J. Vac. Sci. Technol., Vol. 21, pp. 551-553, 1982.

1983-01-EDLv4-p20
H. Kroemer and G. Griffiths, "Staggered-Lineup Heterojunctions as Sources of Tunable Below-Gap Radiation: Operating Principle and Semiconductor Selection," IEEE Electron Dev. Lett., Vol. EDL-4, pp. 20-22, 1983.

1983-02-EDLv4-p25
H. Kroemer, "Critique of Two Recent Theories of Heterojunction Lineups," IEEE Electron Dev. Lett., Vol. EDL-4, pp. 25-26, 1983.

1983-03-PRBv27-p3519
Q.-G. Zhu and H. Kroemer, "Interface Connection Rules for Effective-Mass Wave Functions at an Abrupt Heterojunction Between Two Different Semiconductors," Phys. Rev. B, Vol. 27, pp. 3519-3527, 1983.

1983-04-JVSTBv1-p126
H. Kroemer, "Heterostructure Bipolar Transistors: What Should We Build?," J. Vac. Sci. Technol. B, Vol. 1, pp. 126-130, 1983.

1983-05-SSv132-p543
H. Kroemer, "Heterostructure Devices: A Device Physicist Looks at Interfaces," Surf. Sci., Vol. 132, pp. 543-576, 1983.

1983-06-EDLv4-p365
H. Kroemer, "Rebuttal to 'Response to 'Critique to Two Recent Theories of Heterojunction Lineups," IEEE Electron Dev. Lett., Vol. EDL-4, p. 365, 1983.

1983-07-APLv43-p1059
G. Griffiths, K. Mohammed, S. Subbanna, H. Kroemer, and J. L. Merz, "GaSb/AlSb Multiquantum Well Structures: Molecular Beam Epitaxial Growth and Narrow-Well Photoluminescence," Appl. Phys. Lett., Vol. 43, pp. 1059-1061, 1983.

1984-01-MLv2-p189

I. Banerjee, H. Kroemer, and D. W. Chung, "Observation of Phase Separation in $(Ge_2)_x(GaAs)_{1-x}$ Alloys Grown by Molecular Beam Epitaxy," Matls. Lett., Vol. 2, pp. 189-193, 1984.

1984-02-JAPv55-p2916

S. L. Wright, H. Kroemer, and M. Inada, "Molecular-Beam Epitaxial Growth of GaP on Si," J. Appl. Phys., Vol. 55, pp. 2916-2927, 1984.

1984-03-JAPv55-p4239

B. R. Hancock and H. Kroemer, "Relation Between Growth Conditions and Reconstruction on InAs During Molecular Beam Epitaxy Using an As_2 Source," J. Appl. Phys., Vol. 55, pp. 4239-4243, 1984.

1984-04-JVSTBv2-p433

H. Kroemer, "Barrier Control and Measurements: Abrupt Semiconductor Heterojunctions," J. Vac. Sci. Technol. B, Vol. 2, pp. 433-439, 1984.

1984-05-JJAPv23-p970

H. Kroemer and H. Okamoto, "Some Design Considerations for Multi-Quantum-Well Lasers," Japan. J. Appl. Phys., Vol. 23, pp. 970-974, 1984.

1984-06-APLv45-p449

Y.-J. Chang and H. Kroemer, "Protection of an Interrupted Molecular-Beam Epitaxially Grown Surface by a Thin Epitaxial Layer of InAs," Appl. Phys. Lett., Vol. 45, pp. 449-451, 1984.

1984-07-APLv4S-p1123

E. J. Caine, S. Subbanna, H. Kroemer, J. L. Merz, and A. Y. Cho, "Staggered-Lineup Heterojunctions as Sources of Tunable Below-Gap Radiation: Experimental Verification," Appl. Phys. Lett., Vol. 45, pp. 1123-1125, 1984.

1985-01-APLv46-p504

H. Kroemer, "Determination of Heterojunction Band Offsets by Capacitance-Voltage Profiling through Nonabrupt Isotype Heterojunctions," Appl. Phys. Lett., Vol. 46, pp. 504-505, 1985.

1985-02-APLv46-p494

I. Banerjee, D. W. Chung, and H. Kroemer, "Properties of $(Ge_2)_x(GaAs)_{1-x}$ Alloys Grown by Molecular-Beam Epitaxy," Appl. Phys. Lett., Vol. 46, pp. 494-496, 1985.

1985-03-JVSTAv3-p316

M. J. Mondry, E. J. Caine, and H. Kroemer, "A GaP Decomposition Source for Producing a Dimer Phosphorus Molecular Beam Free of Gallium and Tetramer Phosphorus," J. Vac. Sci. Technol. A, Vol. 3, pp. 316-318, 1985.

1985-04-JVSTBv3-p518

Y.-J. Chang and H. Kroemer, "Summary Abstract: Protection of an Interrupted Molecular-Beam Epitaxially Grown Surface by a Thin Epitaxial Layer of InAs," J. Vac. Sci. Technol. B, Vol. 3, pp. 518-519, 1985.

1985-05-JVSTBv3-p538

I. Banerjee, H. Kroemer, and D. W. Chung, "Summary Abstract: Properties of $(Ge_2)_x(GaAs)_{1-x}$ Alloys Grown by Molecular Beam Epitaxy," J. Vac. Sci.

Technol. B, Vol. 3, pp. 538-539, 1985.

1985-06-JVSTBv3-p603
P. N. Uppal and H. Kroemer, "Summary Abstract: MBE Growth of GaAs and GaP on Si (211)," J. Vac. Sci. Technol. B, Vol. 3, p. 603, 1985.

1985-07-EDLv6-p175
M. J. Mondry and H. Kroemer, "Heterojunction Bipolar Transistor Using a (Ga,In)P Emitter on a GaAs Base, Grown by Molecular Beam Epitaxy," IEEE Elect. Dev. Lett, Vol. EDL-6, pp. 175-177, 1985.

1985-08-JAPv58-p2195
P. N. Uppal and H. Kroemer, "Molecular Beam Epitaxial Growth of GaAs on Si (211)," J. Appl. Phys., Vol. 58, pp. 2195-2203, 1985.

1985-09-SSEv28-p1015
D. I. Babic and H. Kroemer, "The Role of Nonuniform Dielectric Permittivity in the Determination of Heterojunction Band Offsets by C-V Profiling Through Isotype Heterojunctions," Solid-State Electron., Vol. 28, pp. 1015-1017, 1985.

1985-10-SSEv28-p1101
H. Kroemer, "Two Integral Relations Pertaining to the Electron Transport Through a Bipolar Transistor with a Non-Uniform Gap in the Base Region," Solid-State Electron., Vol. 28, pp. 1101-1103, 1985.

1985-11-MRS-E-p161
U. Cebulla, A. Forchel, G. Trankle, H. Kroemer, G. Griffiths, S. Subbanna, and J. Wagner, "Indirect and Direct Gap Recombination in GaSb/AlSb Multi Quantum Well Structures," Proc. MRS - Europe 1985, pp. 161-166, 1985.

1985-12-MRS-E-p41
B. Maile, E. Zielinski, H. Schweizer, G. Griffiths, K. Mohammed, S. Subbanna, H. Kroemer, and J. L. Merz, "Optical Gain in GaSb/AlSb Multi Quantum Well Heterostructures," Proc. MRS - Europe 1985, pp. 41-46, 1985.

1986-01-JAPv59-p488
S. Subbanna, H. Kroemer, and J. L. Merz, "Molecular-beam-epitaxial growth and selected properties of GaAs layers and GaAs/(Al,Ga)As superlattices with the (211) orientation," J. Appl. Phys., Vol. 59, pp. 488-494, 1986.

1986-02-AJPv54-p177
H. Kroemer, "On the derivation of h.dk/dt = F, the k-space form of Newton's Law for Bloch waves," Am. J. Phys., Vol. 54, pp. 177-178, 1986.

1986-03-JVSTBv4-p515
S. Subbanna, H. Kroemer and J. L. Merz, "Summary Abstract: Growth and selected properties of GaAs Layers and GaAs/(Al,Ga)As superlattices with the (211) orientation," J. Vac. Sci. Technol. B, Vol. 4, pp. 515-516, 1986.

1986-04-JVSTBv4-p641
P. N. Uppal and H. Kroemer, "Summary Abstract: Growth of device-quality GaAs and (Al,Ga)As on (211)-oriented silicon substrates, with thin (0.1 5m) superlattice buffer layers," J. Vac. Sci. Technol. B, Vol. 4, p. 641, 1986.

1986-05-SSv174-p143
A. Forchel, U. Cebulla, G. Trankle, H. Kroemer, S. Subbanna, and G. Griffiths,

"Size-Induced Direct-to-Indirect Gap Transition in GaSb/AlSb Multiple Quantum Well Structures," Surf. Sci., Vol. 174, pp. 143-147, 1986.

1986-06-SSv174-p299
H. Kroemer, "Band Offsets at Heterointerfaces: Theoretical Basis, and Review, of Recent Experimental Work," Surf. Sci., Vol. 174, pp. 299-306, 1986.

1986-07-PRLv5J-p3217
A. Forchel, U. Cebulla, G. Trankle, E. Lach, T. L. Reinecke, H. Kroemer, S. Subbanna, and G. Griffiths, "$2E_g$ - Transitions in GaSb/AlSb Quantum Well Structures," Phys. Rev. Lett., 57, pp. 3217-3220, 1986.

1986-08-JMRv1-p803
D. W. Chung, M. Inada, and H. Kroemer, "Defects in GaP film grown on Si (211) by molecular beam epitaxy," J. Mater. Res., Vol. 1, pp. 803-810, 1986.

1986-09-IOPCSv79-p749
E. Zielinski, H. Schweizer, B. Maile, M. H. Pilkuhn, G. Griffiths, S. Subbanna, and H. Kroemer, "Recombination effects and laser properties of GaSb/AlSb multiple quantum well structures," Inst. Phys. Conf. Ser., Vol.79, pp. 749-750, 1986.

1986-10-PMRSv67-p3
H. Kroemer, "MBE Growth of GaAs on Si: Problems and Progress," Matls. Res. Soc. Symp. Proc., Vol. 67, Heteroepitaxy on Silicon, pp. 3-14, 1986.

1987-01-JVSTBv5-p1150
H. Kroemer, "Sublattice Allocation and Antiphase Domain Suppression in Polar-on-Nonpolar Nucleation," J. Vac. Sci. Technol. B, Vol. 5, pp. 1150-1154, 1987.

1987-02-IOPCSv83-p221
G. Trankle, A. Forchel, E. Lach, H.Leier, M. H. Pilkuhn, G. Weimann, H. Kroemer, and M. Razeghi, "Spectroscopic Investigation of the Properties of III-V Quantum Well Structures at High Densities," Inst. Phys. Conf. Ser., Vol 83, pp. 221-226, 1987.

1987-03-Conf-p573
A. Forchel, U. Cebulla, G. Trankle, T. L. Reinecke, H. Kroemer, S. Subbanna, and G. Griffiths, "Optical Spectroscopy of $2E_g$ - Transitions in GaSb/AlSb Quantum Wells," Proc. 18th Int. Conf. Phys. Semicond. 1986, Stockholm, pp. 573-576; World Scientific Publishing Co. Singapore, 1987.

1987-04-SMv22-p63
H. Tomazawa, D. Braun, S. Phillips, A. J. Heeger, and H. Kroemer, "Metal-Polymer Schottky Barriers on Cast Films of Soluble Poly(3-alkylthiophenes)," Synthetic Metals, Vol. 22, pp. 63-69, 1987.

1987-05-SMv3-p429
U. Cebulla, A. Forchel, G. Trankle, G. Griffiths, S. Subbanna, and H. Kroemer, "Direct-Indirect Band Crossover in Two-Dimensional GaSb-Al-Sb Quantum Well-Structures," Superlattices and Microstructures, Vol. 3, pp. 429-433, 1987.

1987-06-APLv50-p182
A. Forchel, U. Cebulla, G. Trankle, U. Ziem, H. Kroemer, S. Subbanna, and G.

Griffiths, "$E_o + \Delta_o$ transitions in GaSb/AlSb quantum wells," Appl. Phys. Lett., Vol. 50, pp. 182-184, 1987.

1987-07-EDLv8p30
M. A. Rao, E. J. Caine, S. I. Long, and H. Kroemer, "An (Al,Ga)As/GaAs heterostructure bipolar transistor with non-alloyed graded-gap contacts to the base and emitter," IEEE Electron Dev. Lett., Vol. EDL-8, pp. 30-32, 1987.

1987-08-JAPv61-p643
M. A. Rao, E. J. Caine, H. Kroemer, S. I. Long, and D. I. Babic, "Determination of valence and conduction band discontinuities at the (Ga,In)P/GaAs heterojunction by C-V profiling," J. Appl. Phys., Vol. 61, pp. 643-649, 1987.

1987-09-JCGv81-p193
H. Kroemer, "Polar-on-Nonpolar Epitaxy," J. Cryst. Growth, Vol. 81, pp. 193-204, 1987.

1987-10-SMv3-p1
U. Cebulla, U. Ziem, G. Trankle, A. Forchel, G. Griffiths, S. Subbanna, and H. Kroemer, "Optical Spectroscopy on $E_o + \Delta_o$ Transition in GaSb/AlSb Quantum Wells," Superlattices and Microstructures, Vol. 3, pp. 1-4, 1987.

1987-11-QEv23-p977
H. Schweizer, E. Zielinski, S. Hausser, R. Stuber, M. H. Pilkuhn, G. Griffiths, H. Kroemer, and S. Subbanna, "Enhanced T_o Values in GaSb/AlSb Multi Quantum Well Heterostructures," IEEE J. Quant. Electron., Vol. QE-23, pp. 977-982, 1987.

1987-12-PSv35-p517
U. Cebulla, A. Forchel, G. Trankle, S. Subbanna, G. Griffiths, and H. Kroemer, "Verification of Direct-Indirect Cross-Over in GaSb-Al-Sb MQW's by Time Resolved Spectroscopy," Physica Scripta, Vol. 35, pp. 517-519, 1987.

1987-13-PRBv36-p6712
G. Trankle, E. Lach, A. Forchel, F. Scholz, C. Ell, H. Haug, G. Weimann, G. Griffiths, H. Kroemer, and S. Subbanna, "General relation between band-gap renormalization and carrier density in two-dimensional electron-hole plasmas," Phys. Rev. B, Vol. 36, pp. 6712-6713, 1987.

1987-14-JVSTBv5-p1156
J. S. Ahearn, P. Uppal, T.-Y. Liu, and H. Kroemer, "Control of dislocations in GaAs grown on Si(211) by molecular beam epitaxy," J. Vac. Sci. Technol. B, Vol. 5, pp. 1156-1161, 1987.

1987-15-ConfProc-p274
M. A. Rao, S. I. Long, and H. Kroemer, "A Self-Aligned AlGaAs/GaAs Heterostructure Bipolar Transistor With Non Alloyed Graded-Gap Ohmic Contacts to the Base and Emitter," Proc. IEEE/Cornell Conference on Advanced Concepts in High Speed Semiconductor Devices and Circuits, pp. 274-283, 1987.

1987-16-ConfProc-p91
Z. Liliental-Weber, E. Weber, J. Washburn, T.-Y. Liu, and H. Kroemer, "The Structure of GaAs/Si (211) Heteroepitaxial Layers," Matls. Res. Soc. Symp. Proc., Vol. 91, Heteroepitaxy on Silicon II, Edited by J. C. C. Fan and J. M.

Poate, pp. 91-98, 1987.

1987-17-ConfProc-p159
A. Forchel, U. Cebulla, G. Trankle, W. Ossau, G. Griffiths, S. Subbanna, and H. Kroemer, "Temperature and Magnetic Field Induced Band Structure Reversal in GaSb/AlSb Quantum Wells," J. de Phys. C5, Suppl. 11, Vol. 48, pp. C5/159-162, 1987.

1987-18-ConfProc-p385
G. Trankle, E. Lach, A. Forchel, C. Ell, H. Haug, G. Weimann, G. Griffiths, H. Kroemer, and S. Subbanna, "Universal relation between band renormalization and carrier density in two-dimensional electron-hole plasmas," J. de Phys. C5, Suppl. 11, Vol. 48, pp. C5/385-388, 1987.

1988-01-IOPCSv91-p21
H. Kroemer, "Polar-on-Nonpolar Epitaxy: Progress and New Problems," Gallium Arsenide and Related Compounds, Inst. Phys. Conf. Ser., Vol. 91, pp. 21-26, 1988.

1988-02-PRBv37-p6278
U. Cebulla, G. Trankle, U. Ziem, A. Forchel, G. Griffiths, H. Kroemer, and S. Subbanna, "Spectroscopic determination of the band discontinuity in GaSb/AlSb multi-quantum-well structures," Phys. Rev. B, Vol. 37, pp. 6278-6284, 1988.

1988-03-SMv4-p473
E. Zielinski, H. Schweizer, R. Stuber, G. Griffiths, H. Kroemer, and S. Subbanna, "Auger Recombination in GaSb/AlSb Multi Quantum Well Heterostructures," Superlattices and Microstructures, Vol. 4, pp. 473-478, 1988.

1988-04-SMv4-p693
S. Subbanna, A. C. Gossard, and H. Kroemer, "Direct-to-Indirect Band Gap Conversion by Application of Electric Fields in the GaSb/AlSb Quantum Well System," Superlattices and Microstructures, Vol. 4, pp. 693-696, 1988.

1988-05-JVSTvB6-p1378
J. M. Gaines, P. M. Petroff, H. Kroemer, R. J. Simes, R. S. Geels, and J. H. English, "Molecular-beam epitaxy growth of tilted GaAs/(Al,Ga)As superlattices by deposition of fractional monolayers on vicinal (001) substrates," J. Vac. Sci. Technol. B, Vol. 6, pp. 1378-1381, 1988.

1988-06-JAPv64-p2746
M. D. Lind, G. J. Sullivan, T. Y. Liu, and H. Kroemer, "Rotational slip in III-V heterostructures grown by molecular-beam epitaxy," J. Appl. Phys., Vol. 64, pp. 2746-2748, 1988.

1988-07-JAPv64-p6810
T.-Y. Liu, P. M. Petroff, and H. Kroemer, "Luminescence of GaAs/(Al,Ga)As superlattices grown on Si substrates, containing a high density of threading dislocations: Strong effect of the superlattice period," J. Appl. Phys., Vol. 64, pp. 6810-6814, 1988.

1988-08-TNSv35-p1657
G. A. Schrantz, N. W. van Vonno, W. A. Krull, M. A. Rao, S. I. Long, and H. Kroemer, "Neutron Irradiation Effects on AlGaAs/GaAs Heterojunction Bipolar

Transistors," IEEE Trans. Nuclear Sci., Vol. 35, pp. 1657-1661, 1988.

1988-09-JEMv17-p297
S. Subbanna, G. Tuttle, and H. Kroemer, "N-Type Doping of Gallium
Antimonide and Aluminum Antimonide Grown by Molecular Beam Epitaxy
using Lead Telluride as a Tellurium Dopant Source," J. Electron. Matls., Vol. 17,
pp. 297-303, 1988.

1988-10-SPIEv944-p123
J. M. Gaines, P. M. Petroff, H. Kroemer, R. J. Simes, R. S. Geels, and J. H.
English, "Growth of quantum wire superlattices and tilted superlattices by
molecular beam epitaxy," Proc. SPIE, Vol. 944, pp. 123-127, 1988.

1988-11-SSEv31-p507
U. Cebulla, S. Zollner, A. Forchel, S. Subbanna, G. Griffiths, and H. Kroemer,
"Hot Carrier Relaxation and Recombination in GaSb/AlSb Quantum Wells,"
Solid-State Electron., Vol. 31, pp. 507-510, 1988.

1989-01-APLv54-p1893
A. Nakagawa, H. Kroemer, and J. H. English, "Electrical properties and band
offsets of InAs/AlSb n-N isotype heterojunctions grown on GaAs," Appl. Phys.
Lett., Vol. 54, pp. 1893-1895, 1989.

1989-02-PMRSv145-p415
G. Tuttle, H. Kroemer, and J. H. English, "Electron Transport in InAs/AlSb
Quantum Wells: Interface Sequencing Effects," Matls. Res. Soc. Symp. Proc.,
Vol. 145, III-V Heterostructures for Electronic/Photonic Devices, pp. 415-420,
1989.

1989-03-JCGv95-p260
P. M. Petroff, J. Gaines, H. Tsuchiya, R. Simes, L. Coldren, H. Kroemer, J.
English, and A. Gossard, "Band Gap Modulation in Two Dimensions by MBE
Growth of Tilted Superlattices, and Application to Quantum Confinement
Structures," J. Cryst. Growth, Vol. 95, pp. 260-265, 1989.

1989-04-PRBv39-p5857
R. M. Abdelouhab, R. Braunstein, M. A. Rao, and H. Kroemer, "Raman
Scattering in $(GaP)_1(InP)_1$ strained layer superlattices," Phys. Rev. B, Vol. 39,
pp. 5857-5860, 1989.

1989-05-JCGv9S-p96
H. Kroemer, T.-Y. Liu, and P. M. Petroff, "GaAs on Si, and Related Systems:
Problems and Prospects," J. Cryst. Growth, Vol. 95, pp. 96-102, 1989.

1989-06-SMv28-pc687
H. Tomazawa, D. Braun, S. D. Phillips, R. Worland, A. J. Heeger, and H.
Kroemer, "Metal-Polymer Schottky Barriers on Processible Polymers," Synthetic
Metals, Vol. 28, pp. C687-C690, 1989.

1989-07-PMRSv145-p393
T.-Y. Liu, P. Petroff, and H. Kroemer, "Electronic Properties of Dislocations in
Heavily Dislocated Quantum Well Structures: Doping Effects," Matls. Res. Soc.
Symp. Proc., Vol. 145, III-V Heterostructures for Electronic/Photonic Devices,
pp. 393-398, 1989.

1989-08-JVSTvB7-p289
S. Subbanna, J. Gaines, G. Tuttle, H. Kroemer, S. Chalmers, and J. H. English, "Reflection high-energy electron reflection oscillations during molecular-beam epitaxy growth of gallium antimonide, aluminum antimonide, and indium arsenide," J. Vac. Sci. Technol. B, Vol. 7, pp. 289-295, 1989.

1989-09-JVSTvB7-p1357
S. A. Chalmers, A. C. Gossard, P. M. Petroff, J. M. Gaines, and H. Kroemer, "A reflection high-energy electron diffraction study of (100) GaAs vicinal surfaces," J. Vac. Sci. Technol. B, Vol. 7, pp. 1357-1362, 1989.

1989-10-JAPv66-p787
R. M. Abdelouhab, R. Braunstein, K. Barner, M. A. Rao, and H. Kroemer, "Raman scattering in a $Ga_{1-x}In_xP$ strained heterostructure," J. Appl. Phys, Vol. 66, pp. 787-792, 1989.

1989-11-JAPv65-p5239
G. Tuttle, H. Kroemer, and J. H. English, "Electron concentrations and mobilities in AlSb/InAs/AlSb quantum wells," J. Appl. Phys., Vol. 65, pp. 5239-5242, 1989.

1990-01-APLv56-p1905
M. R. Rao, E. J. Tarsa, L. A. Samoska, J. H. English, A. C. Gossard, H. Kroemer, P. Petroff, and E. L. Hu, "Superconducting YBaCuO thin films on GaAs/AlGaAs," Appl. Phys. Lett., Vol. 56, pp. 1905-1907, 1990.

1990-02-APLv56-p490
M. R. Rao, E. J. Tarsa, H. Kroemer, A. C. Gossard, E. L. Hu, and P. M. Petroff, "Molecular beam epitaxial growth of InAs on a TlBaCaCuO superconducting film," Appl. Phys. Lett., Vol. 56, pp. 490-493, 1990.

1990-03-APLv57-p1551
A. Nakagawa, J. Pekarik, H. Kroemer, and J. H. English, "Deep levels in Te-doped AlSb grown by molecular beam epitaxy," Appl. Phys. Lett., Vol. 57, pp. 1551-1553, 1990.

1990-04-APLv57-p1751
S. A. Chalmers, H. Kroemer, and A. C. Gossard, "Step-flow growth on strained surfaces: (Al,Ga)Sb tilted superlattices," Appl. Phys. Lett., Vol. 57, pp. 1751-1753, 1990.

1990-05-APLv57-p87
C. Nguyen, J. Werking, H. Kroemer, and E. L. Hu, "InAs-AlSb quantum well as a superconducting weak link with high critical current density," Appl. Phys. Lett., Vol. 57, pp. 87-89, 1990.

1990-06-APLv57-p905
J. Werking, G. Tuttle, C. Nguyen, E. Hu, and H. Kroemer, "InAs-AlSb heterostructure field-effect transistors fabricated using argon implantation for device isolation," Appl. Phys. Lett., Vol. 57, pp. 905-907, 1990.

1990-07-JAPv67-p3032
G. Tuttle, H. Kroemer, and J. H. English, "Effects of interface layer sequencing on the transport properties of InAs/AlSb quantum wells: Evidence for antisite

donors at the InAs/AlSb interface," J. Appl. Phys., Vol. 67, pp. 3032-3037, 1990.
1990-08-JVSTvB8-p431
S. A. Chalmers, A. C. Gossard, P. M. Petroff, and H. Kroemer, "A reflection
high-energy electron diffraction study of AlAs/GaAs tilted superlattice growth by
migration-enhanced epitaxy," J. Vac. Sci. Technol. B, Vol. 8, pp. 431-435, 1990.
1990-09-SMv9-p35
P. M Petroff, K. Ensslin, M. S. Miller, S. A. Chalmers, H. Weman, J. L. Merz, H.
Kroemer, and A. C. Gossard, "Novel Approaches in 2 and 3 Dimensional
Confinement Structures: Processing and Properties," Superlattices and
Microstructures, Vol. 8, pp. 35-39, 1990.
1990-10-ConfProc-p1717
M. S. Miller, C. E. Pryor, H. Weman, L. A. Samoska, H. Kroemer, and P. M.
Petroff, "Serpentine Superlattice: Concept and First Results," Proc. 20th Int.
Conf. on Physics of Semiconductors, Thessaloniki, Greece, Vol. 2, pp. 1717-1720,
Aug. 1990, World Scientific Publishing Co, Singapore, 1990.
1991-01-APLv58-p1428
P. F. Hopkins, A. J. Rimberg, R. M. Westervelt, G. Tuttle, and H. Kroemer,
"Quantum Hall effect in InAs/AlSb quantum wells," Appl. Phys. Lett., Vol. 58,
pp. 1428-1430, 1991.
1991-02-APLv58-p2003
J. Werking, J. Schramm, C. Nguyen, E. L. Hu, and H. Kroemer,
"Methane/hydrogen-based reactive ion etching of InAs, InP, GaAs, and GaSb,"
Appl. Phys. Lett., Vol. 58, pp. 2003-2005, 1991.
1991-03-APLv58-p684
I. Sela, D. E. Watkins, B. K. Laurich, D. L. Smith, S. Subbanna, and H.
Kroemer, "Excitonic optical nonlinearity induced by internal field screening in
(211) oriented strained-layer superlattices," Appl. Phys. Lett., Vol. 58, pp.
684-686, 1991.
1991-04-APLv59-p1711
I. H. Campbell, D. E. Watkins, D. L. Smith, S. Subbanna, and H. Kroemer,
"Electrooptic modulation in polar growth axis InGaAs/GaAs multiple quantum
wells," Appl. Phys. Lett., Vol. 59, pp. 1711-1713, 1991.
1991-05-APLv59-p846
I. H. Campbell, I. Sela, B. K. Laurich, D. L. Smith, C. R. Bolognesi, L. A.
Somoska, A. C. Gossard, and H. Kroemer, "Far-infrared photoresponse of the
InAs/GaInSb superlattice," Appl. Phys. Lett., Vol. 59, pp. 846-848, 1991.
1991-06-JAPv70-p5108
L. D. Chang, M. Z. Tseng, L. A. Samoska, J. J. OShea, Y. J. Li, E. J. Caine, E.
L. Hu, P. M. Petroff, and H. Kroemer, "In situ $YBa_2Cu_3O_{7-x}$ superconductor
films on GaAs/AlAs superlattices," J. Appl. Phys., Vol. 70, pp. 5108-5110, 1991.
1991-07-JAPv70-p5608
I. Sela, H. Campbell, B. K. Laurich, D. L Smith, L. A. Somoska, C. R. Bolognesi,
A. C. Gossard, and H. Kroemer, "Raman scattering study of InAs/GaInSb
strained layer superlattices," J. Appl. Phys., Vol. 70, pp. 5608-5614, 1991.

1991-08-JCGv111-p323
M. S. Miller, C. E. Pryor, H. Weman, L. A. Samoska, H. Kroemer, and P. M. Petroff, "Serpentine superlattice: concept and first results," J. Cryst. Growth, Vol. 111, pp. 323-327, 1991.

1991-09-JCGv111-p360
P. M. Petroff, M. S. Miller, Y. T. Lu, S. A. Chalmers, H. Metiu, H. Kroemer, and A. C. Gossard, "MBE Growth of tilted superlattices: advances and novel structures," J. Cryst. Growth, Vol. 111, pp. 360-365, 1991.

1991-10-JCGv111-p647
S. A. Chalmers, H. Kroemer, and A. C. Gossard, "The growth of (Al,Ga)Sb tilted superlattices and their heteroepitaxy with InAs to form corrugated-barrier quantum wells," J. Cryst. Growth, Vol. 111, pp. 647-650, 1991.

1991-11-PRBv43-p11884
I. Sela, D. E. Watkins, B. K. Laurich, D. L. Smith, S. Subbanna, and H. Kroemer, "Modulated photoabsorption in strained $Ga_{1-x}In_xAs/GaAs$ multiple quantum wells," Phys. Rev. B, Vol. 43, pp. 11884-11892, 1991.

1991-12-SMv10-p361
G. Fuchs, S. Hausser, A. Hangleiter, G. Griffiths, H. Kroemer, S. Subbanna, "Recombination in GaSb/AlSb multiple QWs under high excitation conditions," Superlattices and Microstructures, Vol. 10, pp. 361-364, 1991.

1991-13-SMv9-p119
K. Ensslin, S. A. Chalmers, P. M. Petroff, A. C. Gossard, and H. Kroemer, "Anisotropic Magnetotransport in an Antiwire Array Inserted in a GaAs Heterostructure," Superlattices and Microstructures Vol. 9, pp. 119-121, 1991.

1991-14-SMv9-p499
B. K. Laurich, D. L. Smith, D. E. Watkins, I. Sela, S. Subbanna, and H. Kroemer, "Nonlinear Optical Absorption in Intrinsic Stark Effect Superlattices," Superlattices and Microstructures, Vol. 9, pp. 499-502, 1991.

1991-15-JEMv20-p945
D. Braun, A. J. Heeger, and H. Kroemer, "Improved Efficiency in Semiconductor Polymer Light-Emitting Diodes," J. Electron. Matls., Vol. 20, pp. 945-948, 1991.

1992-01-APLv60-p1676
S. A. Chalmers, H. Weman, J. C. Yi, H. Kroemer, J. L. Merz, and N. Dagli, "Photoluminescence study of lateral carrier confinement and compositional intermixing in (Al,Ga)Sb lateral superlattices," Appl. Phys. Lett., Vol. 60, pp. 1676-1678, 1992.

1992-02-APLv60-p1854
C. Nguyen, B. Brar, H. Kroemer, and J. H. English, "Surface donor contribution to electron sheet concentrations in not-intentionally doped InAs-AlSb quantum wells," Appl. Phys. Lett., Vol. 60, pp. 1854-1856, 1992.

1992-03-APLv60-p3283
I. Sela, C. R. Bolognesi, L. A. Samoska, and H. Kroemer, "Study of interface composition and quality in AlSb/InAs/AlSb quantum wells by Raman scattering from interface modes," Appl. Phys. Lett., Vol. 60, pp. 3283-3285, 1992.

1992-04-APLv61-p213
C. R. Bolognesi, H. Kroemer, and J. H. English, "Interface roughness scattering in InAs/AlSb quantum wells," Appl. Phys. Lett., Vol. 61, pp. 213-215, 1992.

1992-05-APLv61-p601
M. Z. Tseng, C. Nguyen. E. Tarsa, L. D. Chang, E. L. Hu, and H. Kroemer, "Temperature-dependent mobility of a GaAs/AlGaAs heterostructure after deposition of MgO and superconducting $YBa_2Cu_3O_{7-x}$," Appl. Phys. Lett., Vol. 61, pp. 601-603, 1992.

1992-06-EDLv13-p164
J. D. Werking, C. R. Bolognesi, L.-D. Chang, C. Nguyen, E. L. Hu, and H. Kroemer, "High-Transconductance InAs/AlSb Heterojunction Field Effect Transistors with δ-Doped AlSb Upper Barriers," IEEE Electron Dev. Lett., Vol. 13, pp. 164-166, 1992.

1992-07-JVSTvB10-p1032
J. Pekarik, H. Kroemer, and J. H. English, "An AlSb-InAs-AlSb double-heterojunction P-n-P bipolar transistor," J. Vac. Sci. Technol. B, Vol. 10, pp. 1032-1034, 1992.

1992-08-JVSTvB10-p1769
H. Kroemer, C. Nguyen, and B. Brar, "Are there Tamm-state donors at the InAs-AlSb quantum well interface?," J. Vac. Sci. Technol. B, Vol. 10, pp. 1769-1772, 1992.

1992-09-JVSTvB10-p877
C. R. Bolognesi, H. Kroemer, and J. H. English, "Well width dependence of electron transport in molecular-beam epitaxially grown InAs/AlSb quantum wells," J. Vac. Sci. Technol. B, Vol. 10, pp. 877-879, 1992.

1992-10-JVSTvB10-p898
C. Nguyen, B. Brar, H. Kroemer, and J. H. English, "Effects of barrier thicknesses on the electron concentration in not-intentionally doped InAs-AlSb quantum wells," J. Vac. Sci. Technol. B, Vol. 10, pp. 898-900, 1992.

1992-11-PRBv46-p1480
I. Sela, D. L. Smith, S. Subbanna, and H. Kroemer, "Raman Scattering near the $(E_o + \Delta_o)$ resonance from [211]-oriented $Ga_{1-x}In_xAs/GaAs$ multiple quantum wells," Phys. Rev. B, Vol. 46, pp. 1480-1488, 1992.

1992-12-PRBv46-p16142
I. Sela, C. R. Bolognesi, and H. Kroemer, "Single mode behavior of $AlSb_{1-x}As_x$ alloys," Phys. Rev. B, Vol. 46, pp. 16142-16143, 1992.

1992-13-PRBv46-p7200
I. Sela, L. A. Samoska, C. R. Bolognesi, A. C. Gossard, and H. Kroemer, "Raman Scattering from interface modes in $Ga_{1-x}In_xSb/InAs$ superlattices," Phys. Rev. B, Vol. 46, pp. 7200-7203, 1992.

1992-14-PRLv8-p3464
M. S. Miller, H. Weman, C. E. Pryor, M. Krishnamurthy, P. M. Petroff, H. Kroemer, and J. L. Merz, "Serpentine superlattice quantum wire arrays of (Al,Ga)As grown on vicinal GaAs substrates," Phys. Rev. Lett., Vol. 68, pp.

3464-3467, 1992.

1992-15-PRLv69-p2847

C. Nguyen, H. Kroemer, and E. L. Hu, "Anomalous Andreev Conductance in InAs-AlSb Quantum Well Structures with Nb Electrodes," Phys. Rev. Lett., Vol. 69, pp. 2847-2850, 1992.

1992-16-SSv267-p483

J. Scriba, S. Seitz, A. Wixforth, J. P. Kotthaus, G. Tuttle, J. H. English, and H. Kroemer, "Electronic properties and far infrared spectroscopy of InAs/AlSb quantum wells," Surf. Sci., Vol. 267, pp. 483-487, 1992.

1992-17-SSv267-p549

C. Nguyen, Klaus Ensslin, and H. Kroemer, "Magneto-transport in InAs/AlSb quantum wells with large electron concentration modulation," Surf. Sci., Vol. 267, pp. 549-552, 1992.

1992-18-ConfProc-p105

J. J. Pekarik, H. Kroemer, and J. H. English, "AlSb-InAs-AlSb P-n-P transistors with low turn-on voltage, narrow bases, and low base resistance," Tech. Dig. Device Research Conf., pp. 105-106, 1992.

1992-19-ConfProc-p16

H. Kroemer, "Heterojunction Devices," Tech. Dig. Device Res. Conf., pp. 16-17, 1992.

1993-01-APLv62-p1373

S. J. Koester, C. R. Bolognesi, M. J. Rooks, E. L. Hu, and H. Kroemer, "Quantized conductance of ballistic constrictions in InAs/AlSb quantum wells," Appl. Phys. Lett., Vol. 62, pp. 1373-1375, 1993.

1993-02-APLv62-p2274

J. Spitzer, H. D. Fuchs, P. Etchegoin, M. Ilg, M. Cardona, B. Brar, and H. Kroemer, "Quality of AlAs-like and InSb-like interfaces in InAs/AlSb superlattices: an optical study," Appl. Phys. Lett., Vol. 62, pp. 2274-2276, 1993.

1993-03-APLv62-p2539

L. A. Samoska, B. Brar, H. Kroemer, and J. H. English, "Strong far-infrared inter-subband absorption under normal incidence in heavily n-type doped non-alloy GaSb-AlSb superlattices," Appl. Phys. Lett., Vol. 62, pp. 2539-2541, 1993.

1993-04-APLv62-p3303

B. Brar, H. Kroemer, M. Ibbetsen, and J. H. English, "Photoluminescence from narrow InAs-AlSb quantum wells," Appl. Phys. Lett., Vol. 62, pp. 3303-3305, 1993.

1993-05-APLv63-p1211

K. C. Wong, M. Krishnamurthy, B. Brar, J. C. Yi, H. Kroemer, and J. H. English, "Growth and characterization of serpentine superlattices in the GaSb-AlSb system," Appl. Phys. Lett., Vol. 63, pp. 1211-1213, 1993.

1993-06-APLv63-p2251

C. Nguyen, B. Brar, V. Jayaraman, A. Lorke, and H. Kroemer, "Magneto-transport in lateral periodic potentials formed by surface-layer-induced modulation in InAs-AlSb quantum wells," Appl. Phys. Lett., Vol. 63, pp. 2251-2253, 1993.

1993-07-JCGv127-p752

B. Brar, H. Kroemer, and J. H. English, "Quasi-direct narrow GaSb/AlSb (100) quantum wells," J. Cryst. Growth, Vol. 127, pp. 752-754, 1993.

1993-08-JCGv127-p845

C. Nguyen, H. Kroemer, E. L. Hu, and J. H. English, "Low-temperature (4.2 - 9K) transport along InAs/AlSb quantum wells with δ-doped barriers and superconducting Nb electrodes," J. Cryst. Growth, Vol. 127, pp. 845-848, 1993.

1993-09-JEMv22-p255

C. Nguyen, B. Brar, C. R. Bolognesi, J. J. Pekarik, H. Kroemer, and J. H. English, "Growth of InAs/AlSb Quantum Wells Having Both High Mobilities and High Electron Concentrations," J. Electronic Matls., Vol. 22, pp. 255-258, 1993.

1993-10-JVSTvB11-p1354

H. Kroemer, "Semiconductor Heterojunctions at the Conference on the Physics and Chemistry of Semiconductor Interfaces: A Device Physicist's Perspective," J. Vac. Sci. Technol. B, Vol. 11, pp. 1354-1361, 1993.

1993-11-JVSTvB11-p1706

C. Nguyen, B. Brar, and H. Kroemer, "Surface-layer modulation of electron concentrations in InAs-AlSb quantum wells," J. Vac. Sci. Technol. B, Vol. 11, pp. 1706-1709, 1993.

1993-12-JVSTvB11-p2528

S. J. Koester, C. R. Bolognesi, E. L. Hu, H. Kroemer, M. J. Rooks, and G. Snider, "Design and analysis of InAs/AlSb ballistic constrictions for high-temperature operations and low gate leakage," J. Vac. Sci. Tech. B, Vol. 11, pp. 2528-2531, 1993.

1993-13-JVSTvB11-p868

C. R. Bolognesi, I Sela, J. Ibbetson, B. Brar, H. Kroemer, and J. H. English, "On the interface structure in InAs/AlSb quantum wells grown by molecular-beam epitaxy," J. Vac. Sci. Tech. B, Vol. 11, pp. 868-871, 1993.

1993-14-MSEvB21-p201

A. Simon, J. Scriba, C. Gauer, A. Wixforth, J. P. Kotthaus, C. R. Bolognesi, C. Nguyen, G. Tuttle, and H. Kroemer, "Intersubband transitions in InAs/AlSb quantum wells," Matls. Sci. Eng., Vol. B21, pp. 201-204, 1993.

1993-15-SSCv86-p633

J. Scriba, A. Wixforth, J. P. Kotthaus, C. Bolognesi, C. Nguyen, and H. Kroemer, "Spin- and Landau-splitting of the cyclotron resonance in a nonparabolic two-dimensional electron system," Solid-State Comm., Vol. 86, pp. 633-636, 1993.

1993-16-SSTv8-ps137

Ch. Gauer, J. Scriba, A. Wixforth, J. P. Kotthaus, C. Nguyen, G. Tuttle, J. H. English, and H. Kroemer, "Photoconductivity in AlSb/InAs quantum wells," Semicond. Sci. Technol., Vol. 8, pp. S137-S140, 1993.

1993-17-SSTv8-p133

J. Scriba, A. Wixforth, J. P. Kotthaus, C. R. Bolognesi, C. Nguyen, G. Tuttle, J. H. English, and H. Kroemer, "The effect of Landau quantization on cyclotron resonance in a non-parabolic quantum well.," Semicond. Sci. Technol., Vol. 8, pp. 133-136, 1993.

1993-18-EDLv14-p13
C. R. Bolognesi, J. D. Werking, E. J. Caine, E. L. Hu, and H. Kroemer, "Microwave Performance of a Digital Alloy Barrier Al(Sb,As)/AlSb/InAs Heterostructure Field Effect Transistor," IEEE Electron Dev. Lett., Vol. 14, pp. 13-15, 1993.

1993-19-ConfProc-p149
L. A. Samoska, B. Brar, and H. Kroemer, "Normal incidence quantum well infrared photodetectors utilizing ellipsoidal valley intersubband transitions in n-type GaSb/AlSb multi-quantum wells," Proc. SPIE, Vol. 2021, pp. 149-59, 1993.

1993-20-ConfProc-p48
A. G. Markelz, E. G. Gwinn, M. S. Sherwin, J. Heyman, C. Nguyen, H. Kroemer, P. F. Hopkins, and A. C. Gossard, "Far-infrared harmonic generation from semiconductor heterostructures," Proc. SPIE, Vol. 1854, pp. 48-55, 1993.

1994-01-APLv64-p3392
B. Brar, J. Ibbetson, H. Kroemer, and J. H. English, "Effects of the interface bonding type on the optical and structural properties of InAs-AlSb quantum wells," Appl. Phys. Lett., Vol. 64, pp. 3392-3394, 1994.

1994-02-APLv65-p103
C. Nguyen, H. Kroemer, and E. L. Hu, "Contact resistance of superconductor-semiconductor interfaces: The case of Nb-InAs quantum well structures," Appl. Phys. Lett., Vol. 65, pp. 103-105, 1994.

1994-03-EDLv15-p16
C. R. Bolognesi, E. J. Caine, and H. Kroemer, "Improved Charge Control and Frequency Performance in InAs/AlSb-Based Heterostructure Field-Effect Transistors," IEEE Electron Dev. Lett., Vol. 15, pp. 16-18, 1994.

1994-04-JVSTvB12-p1242
B. Brar, L. Samoska, H. Kroemer, and J. H. English, "Electrical and optical properties of heavily n-doped GaSb-AlSb multiquantum well structures for infrared photodetector applications," J. Vac. Sci. Technol. B, Vol. 12, pp. 1242-1245, 1994.

1994-05-PhyBv23-p298
H. Kroemer, C. Nguyen, E. Hu, E. L. Yuh, M. Thomas, and K. C. Wong, "Quasiparticle transport and induced superconductivity in InAs quantum wells with Nb Electrodes," Physica B, Vol. 23, pp. 298-306, 1994.

1994-06-PRBv49-p8514
S. J. Koester, C. R. Bolognesi, E. L. Hu, H. Kroemer, and M. J. Rooks, "Quantized conductance in an InAs/AlSb split-gate ballistic construction with 1.0 μm channel length," Phys. Rev. B, Vol. 49, pp. 8514-8517, 1994.

1994-07-PRBv50-p5710

S. J. Koester, C. R. Bolognesi, M. Thomas, E. L. Hu, H. Kroemer, and M. J. Rooks, "Determination of one-dimensional subband spacing in InAs/AlSb ballistic constrictions using magnetic-field measurements," Phys. Rev. B, Vol. 50, pp. 5710-5713, 1994.

1994-08-PRBv50-p8746

P. V. Santos, P. Etchegoin, M. Cardona, B. Brar, and H. Kroemer, "Optical anisotropy in InAs/AlSb superlattices," Phys. Rev. B, Vol. 50, pp. 8746-8754, 1994.

1994-09-PRBv50-p8922

Y. Abramovich, J. Poplawski, E. Ehrenfreund, D. Gershoni, B. Brar, and H. Kroemer, "Intersubband L-valley and heavy-hole transitions in undoped GaSb/AlSb superlattices," Phys. Rev. B, Vol. 50, pp. 8922-8925, 1994.

1994-10-PSSDM-p397

H. Kroemer, "Devices for the Future: A Peek into the Next Century," Proc. Int. Conf. on Solid State Devices and Materials, Yokohama, Japan, pp. 397-399, 1994.

1994-11-SMv1S-p489

Y. Abramovich, J. Poplawski, E. Ehrenfreund, D. Gershoni, B. Brar, and H. Kroemer, "Photoinduced Absorption within the Valence Γ- and the Conduction L-subband Manifolds in Undoped GaSb/AlSb Superlattices," Superlattices and Microstructures, Vol. 15, pp. 489-493, 1994.

1994-12-SSEv37-p1021

H. Kroemer, C. Nguyen, and E. L. Hu, "Electronic Interactions at Superconductor-Semiconductor Interfaces," Solid-State Electron., Vol. 37, pp. 1021-1025, 1994.

1994-13-SSEv37-p1243

A. G. Markelz, E. G. Gwinn, M. S. Sherwin, C. Nguyen, and H. Kroemer, "Giant Third-Order Nonlinear Susceptibilities for In-Plane Far-Infrared Excitation of Single InAs Quantum Wells," Solid-State Electron., Vol. 37, pp. 1243-1245, 1994.

1994-14-SSEv37-p1293

M. Sundaram, S. J. Allen, Jr., C. Nguyen, B. Brar, V. Jayaraman, and H. Kroemer, "Infrared Spectroscopy of Lateral-Density-Modulated 2DES in InAs/AlSb Quantum Wells," Solid-State Electron., Vol. 37, pp. 1293-1296, 1994.

1994-15-SSEv37-p575

T. Utzmeier, T. Schlvsser, K. Ensslin, J. P. Kotthaus, C. R. Bolognesi, C. Nguyen, and H. Kroemer, "Lateral Potential Modulation in InAs/AlSb Quantum Wells by Wet Etching," Solid-State Electron., Vol. 37, pp. 575-578, 1994.

1994-16-SSTv9-p1580

C. Gauer, J. Scriba, A. Wixforth, J. P. Kotthaus, C. R. Bolognesi, C. Nguyen, B. Brar, and H. Kroemer, "Energy-dependent cyclotron mass in InAs/AlSb quantum wells," Semicond. Sci. Technol., Vol. 9, pp. 1580-1583, 1994.

1994-17-SSTv9-p634

A. G. Markelz, N. G. Asmar, E. G. Gwinn, M. S. Sherwin, C. Nguyen, and H. Kroemer, "Subcubic power dependence of third-harmonic generation for in-plane far-infrared excitation of InAs quantum wells," Semicond. Sci. Technol., Vol. 9, pp. 634-637, 1994.

1995-01-EDLv16-p548

B. Brar and H. Kroemer, "Influence of Impact Ionization on the Drain Conductance in InAs-AlSb Quantum Well Heterostructure Field-Effect Transistors," IEEE Electron Dev. Lett., Vol. 16, pp. 548-550, 1995.

1995-02-IOPCSv141-p7

H. Kroemer, C. Nguyen, and E. L. Hu, "Ballistic Electron Transport and Superconductivity in Mesoscopic Nb-(InAs/AlSb) Quantum Well Heterostructures," Inst. Phys. Conf. Ser., Vol. 141, pp. 7-12, 1995.

1995-03-JAPv77-p811

J. Spitzer, A. Hvpner, M. Kuball, M. Cardona, B. Jenichen, H. Neuroth, B. Brar, and H. Kroemer, "Influence of the interface composition of InAs/AlSb superlattices on their optical and structural properties," J. Appl. Phys., Vol. 77, pp. 811-820, 1995.

1995-04-JCGv150-p883

T. Makimoto, B. Brar, and H. Kroemer, "Hole accumulation in (In)GaSb/AlSb quantum wells induced by the Fermi-level pinning of an InAs surface," J. Cryst. Growth, Vol 150, pp. 883-886, 1995.

1995-05-MGWLv5-p219

M. Reddy, R. Y. Yu, H. Kroemer, M. J. M. Rodwell, S. C. Martin, R. E. Mueller, and R. P. Smith, "Bias Stabilization for Resonant Tunnel Diode Oscillators," IEEE Microwave and Guided Wave Lett., Vol. 5, pp. 219-221, 1995.

1995-06-PRLv74-p2772+erratum

C. Gauer, A. Wixforth, J. P. Kotthaus, M. Kubisa, W. Zawadski, B. Brar, and H. Kroemer, "Magnetic-Field-Induced Spin-Conserving and Spin-Flip Intersubband Transitions in InAs Quantum Wells," Phys. Rev. Lett., Vol. 74, pp. 2772-2775, 1995.

1995-07-ConfProc-p28

B. Brar and H. Kroemer, Impact ionization in InAs/AlSb field effect transistors Tech. Dig. Device Research Conf., pp. 28-29, 1995.

1995-08-ConfProc-p493

B. Jenichen, H. Neuroth, B. Brar, and H. Kroemer, "Structural Properties of InAs/AlSb Superlattices," Matls. Res. Soc. Symp. Proc., Vol. 379, Strained Layer Epitaxy - Materials, Processing, and Device Applications, E. Fitzgerald, J. Hoyt, K.-Y. Cheng, and J. Bean, Eds., pp. 493-498, 1996.

1996-01-APLv68-p1543

C. K. Sun, G. Wang, J. E. Bowers, B. Brar, H.-R. Blank, H. Kroemer, and M. H. Pilkuhn, "Optical investigations of the dynamic behavior of GaSb/GaAs quantum dots," Appl. Phys. Lett., Vol. 68, pp. 1543-1545, 1996.

1996-02-APLv69-p2080

H.-R. Blank, M. Thomas, K. C. Wong, and H. Kroemer, "Influence of the buffer layer on the morphology and the transport properties in InAs/(Al,Ga)Sb quantum wells grown by molecular beam epitaxy," Appl. Phys. Lett., Vol. 69,

pp. 2080-2082, 1996.

1996-03-JAPv79-p120
B. Jenichen, S. A. Stepanov, B. Brar, and H. Kroemer, "Interface roughness of InAs/AlSb superlattices investigated by x-ray scattering," J. Appl. Phys., Vol. 79, pp. 120-124, 1996.

1996-04-NATO-ASIv323-p1
H. Kroemer, "All that Glitters isn't Silicon - or Steel and Aluminum Re-Visited," NATO Adv. Res. Workshop "Future Trends in Microelectronics: Reflections on the Road to Nanotechnology", Ile de Bendor, France, S. Luryi, J. Xu, and A. Zaslavsky, Eds., NATO ASI Series; Series E: Applied Sciences, Vol. 323, Kluwer Academic Publishers, pp. 1-12, 1995.

1996-05-PRBv53-p13063
S. J. Koester, B. Brar, C. R. Bolognesi, E. J. Caine, A. Patlach, E. L. Hu, H. Kroemer, and M. J. Rooks, "Length dependence of quantized conductance in ballistic constrictions fabricated on InAs/AlSb quantum wells," Phys. Rev. B, Vol. 53, pp. 13063-13073, 1996.

1996-06-PRBv53-p7903
R. J. Warburton, C. Gauer, A. Wixforth, J. P. Kotthaus, B. Brar, and H. Kroemer, "Intersubband resonances in InAs/AlSb quantum wells: Selection rules, matrix elements, and the depolarization field," Phys. Rev. B, Vol. 53, pp. 7903-7910, 1996.

1996-07-PRBv54-pR2311
M. Thomas, H.-R. Blank, K. C. Wong, C. Nguyen, H. Kroemer, and E. L. Hu, "Flux-periodic resistance oscillations in arrays of superconducting weak links based on InAs-AlSb quantum wells with Nb electrodes," Phys. Rev. B, Vol. 54, pp. R2311-R2314, 1996.

1996-08-PSvT68-p10
H. Kroemer, "Band Offsets and Chemical Bonding: The Basis for Heterostructure Applications," Physica Scripta, Vol. T68, pp. 10-16, 1996.

1996-09-SMv19-p241
C. Gauer, A. Wixforth, J. P. Kotthaus, B. Brar, and H. Kroemer, "Spin phenomena in intersubband transitions," Superlattices and Microstructures, Vol. 19, pp. 241-249, 1996.

1996-10-SMv19-p365
R. J. Warburton, C. Gauer, A. Wixforth, J. P. Kotthaus, B. Brar, and H. Kroemer, "Collective effects in the intersubband resonances of InAs/AlSb quantum wells," Superlattices and Microstructures, Vol. 19, pp. 365-374, 1996.

1996-11-SSEv40-p679
R. J. Warburton, B. Brar, C. Gauer, A. Wixforth, J. P. Kotthaus, and H. Kroemer, "Cyclotron Resonance of Electron-Hole Systems in InAs/GaSb/AlSb," Solid-State Electron., Vol. 40, pp. 679-682, 1996.

1996-12-SMv21-p61
H. Kroemer and M. Thomas, "Induced Superconductivity in InAs Quantum Wells with Superconducting Contacts," Superlattices and Microstructures, Vol.

21, pp. 61-67, 1996.

1996-13-SSv352-354-p771
C. Goletti, U. Resch-Esser, J. Foeller, N. Esser, W. Richter, B. Brar, and H. Kroemer, "A reflectance anisotropy spectroscopy study of GaSb(100)c(2×6) surface prepared by Sb decapping," Surf. Sci., Vol. 352-354, pp. 771-775, 1996.

1996-14-SSv361-362-p306
H. Drexler, J. G. E. Harris, E. L. Yuh, K. C. Wong, S. J. Allen, E. G. Gwinn, H. Kroemer, and E. L. Hu, "Superconductivity and the Josephson effect in a periodic array of Nb-InAs-Nb junctions," Surf. Sci., Vol. 361/362, pp. 306-310, 1996.

1996-15-SSv361-362-p315
E. L. Yuh, J. G. E. Harris, T. Eckhouse, K. C. Wong, E. G. Gwinn, H. Kroemer, and S. J. Allen, "Far-infrared studies of induced superconductivity in quantum wells," Surf. Sci., Vol. 361/362, pp. 315-319, 1996.

1996-16-SSv361-362-p472
C. Gauer, M. Hartung, A. Wixforth, J. P. Kotthaus, B. Brar, and H. Kroemer, "Zero-field spin-splitting in InAs/AlSb quantum wells," Surf. Sci., Vol. 361/362, pp. 472-475, 1996.

1996-17-NATO-ASIv323-p237
H. Kroemer, "Superconductor-Semiconductor Devices," NATO Adv. Res. Workshop "Future Trends in Microelectronics: Reflections on the Road to Nanotechnology", Ile de Bendor, France, S. Luryi, J. Xu, and A. Zaslavsky, Eds., NATO ASI Series; Series E: Applied Sciences, Vol. 323, Kluwer Academic Publishers, pp. 237-250, 1996.

1996-18-IOPCSv144-p379
E. L. Yuh, J. G. E. Harris, E. G. Gwinn, K. C. Wong, and H. Kroemer, "Far-infrared studies of InAs quantum wells with Nb electrodes," Narrow-Gap Semiconductors; Inst. Phys. Conf. Ser., Vol. 144, pp 379-383, 1996.

1996-19-ConfProc-p164
G. Wang, C.-K. Sun, H. R. Blank, B. Brar, J. E. Bowers, H. Kroemer, and M. H. Pilkuhn, "Time-resolved optical investigation of spatially indirect excitons in GaAs/GaSb quantum dots," Proc. Quantum Electronics and Laser Science Conference, QELS, pp. 164-165 1996.

1996-20-ConfProc-p1835
M. Hartung, A. Wixforth, J. P. Kotthaus, B. Brar, and H. Kroemer, "Interband and Intersubband Transitions Induced Photoconductivity in InAs/AlSb Quantum Wells," Proc. 23rd Internat. Conf. on the Physics of Semiconductors, Berlin, July 1996, Vol. 3, pp. 1835-1838, World Scientific, Singapore, 1996.

1996-21-ConfProc-p2423
A. Wixforth, C. Gauer, J. P. Kotthaus, M. Kubisa, W. Zawadski, B. Brar, and H. Kroemer, "Spin Phenomena in Quantum Well Inter-Subband Transitions," Proc. 23rd Internat. Conf. on the Physics of Semiconductors, Berlin, July 1996, Vol. 3, pp. 2423-2430, World Scientific, Singapore, 1996.

1997-01-APLv70-p1590

M. E. Rubin, H. R. Blank, M. A. Chin, H. Kroemer, and V. Narayanamurti, "Local conduction band offset of GaSb self-assembled quantum dots on GaAs," Appl. Phys. Lett., Vol. 70, pp. 1590-1592, 1997.

1997-02-APLv70-p759
S. Bhargava, H.-R. Blank, V. Narayanamurti, and H. Kroemer, "Fermi-level pinning position at the Au-InAs interface determined using ballistic electron emission microscopy," Appl. Phys. Lett., Vol. 70, pp. 759-761, 1997.

1997-03-APLv71-p3534
H.-R. Blank, H. Kroemer, S. Mathis, and J. S. Speck, "Structural and electrical properties of low-temperature-grown Al(As,Sb)," Appl. Phys. Lett., Vol. 71, pp. 3534-3536, 1997.

1997-04-EDLv18-p218
M. Reddy, S. C.Martin, A. C. Molnar, R. E. Muller, R. P. Smith, P. H. Siegel, M. J. Mondry, M. J. Rodwell, H. Kroemer, and S. J. Allen, Jr., "Monolithic Schottky-Collector Resonant Tunnel Diode Arrays to 650 GHz," IEEE Elect. Dev. Lett., Vol. 18, pp. 218-221, 1997.

1997-05-JCGv175-176-p894
M. Thomas, H.-R. Blank, K. C. Wong, and H. Kroemer, "Buffer-dependent mobility and morphology of InAs/(Al,Ga)Sb quantum wells," J. Crystal Growth, Vol. 175/176, pp. 894-897, 1997.

1997-06-PRBv55-p15401
U. Resch-Esser, N. Esser. B. Brar, and H. Kroemer, "Microscopic structure of GaSb(001) c(246) surfaces prepared by Sb decapping of MBE-grown samples," Phys. Rev. B, Vol. 55, pp. 15401-15404, 1997.

1997-07-ConfProcv1-p309
H. Kroemer and M. Thomas, "Flux Quantization Effects in Semiconductor-Superconductor Junctions," Proc. 12th Internat. Conf. on High Magnetic Fields in the Physics of Semiconductors, Wurzburg, Vol. 1, pp. 309-318, 1997. World Scientific, Singapore.

1997-08-ConfProcv1-p289
W. Zawadski, M. Kubisa, A. Wixforth, C. Gauer, J. P. Kotthaus, B. Brar, and H. Kroemer, "Spin and Depolarization Effects in InAs/AlSb Quantum Wells," Proc. 8th Internat Conf. on Narrow Gap Semiconductors, Shanghai, China, World Scientific Publishing Co, Singapore, pp. 289-299, 1997.

1997-09-ConfProcv1-p313
A. Wixforth, J. Scriba, A. Simon, C. R. Bolognesi, B. Brar, and H. Kroemer, "Density Dependence of the Spin-Splitting of the Cyclotron Resonance in InAs/AlSb Quantum Wells," Proc. 8th Internat Conf. on Narrow Gap Semiconductors, Shanghai, China, World Scientific Publishing Co, Singapore, pp. 313-315, 1997.

1997-10-ConfProcv1-p387
M. Hartung, A. Wixforth, J. P. Kotthaus, M. Thomas, B. Brar, and H. Kroemer, "Photoconductivity Spectroscopy of InAs/AlSb Quantum Wells," Proc. 8th Internat Conf. on Narrow Gap Semiconductors, Shanghai, China, World

Scientific Publishing Co, Singapore, pp. 387-390, 1997.

1998-01-APLv73-p2684

J. C. Rosa, M. Wendel, H. Lorenz, J. P. Kotthaus, and H. Kroemer, "Direct patterning of surface quantum wells with an atomic force microscope," Appl. Phys. Lett., Vol.73, pp. 2684-2686, 1998.

1998-02-JAPv83-p894

B. Brar and H. Kroemer, "Hole transport across the (Al,Ga)(As,Sb) barrier in InAs-(Al,Ga)(As,Sb) heterostructures," J. Appl. Phys., Vol. 83, pp. 894-899, 1998.

1998-03-JCGv187-p18

H.-R. Blank, S.Mathis, E. Hall, S. Bhargava, A. Behres, M. Heuken, H. Kroemer, and V. Narayanamurti, "Al(As,Sb) Heterobarriers on InAs: Growth, Structural Properties and Electrical Transport," J. Cryst. Growth, Vol. 187, pp. 18-28, 1998.

1998-04-JVSTvB16-p2660

E. Hall, R. Naone, J. E. English, H.-R. Blank, J. Champlain, and H. Kroemer, "Operational experience with a valved antimony cracker source for use in molecular beam epitaxy," J. Vac. Sci. Technol. B, Vol. 16, pp. 2660-2663, 1998.

1998-05-PhysBv256-258-p239

S. Brosig, K. Ensslin, B. Brar, M. Thomas, and H. Kroemer, "Landau and spin levels in InAs quantum wells resolved with in-plane and parallel magnetic fields," Physica B, Vol. 256-258, pp. 239-242, 1998.

1998-06-PhysEv2-p214

S. Brosig, K. Ensslin, B. Brar, M. Thomas, and H. Kroemer, "Scattering mechanism in InAs-AlSb quantum wells," Physica E, Vol. 2, pp. 214-217, 1998.

1998-07-PhysEv2-p682

M. E. Rubin, H.-R. Blank, M. A. Chin, H. Kroemer, and V. Narayanamurti, "Imaging and local transport measurements of GaSb self-assembled quantum dots on GaAs," Physica E, Vol. 2, pp. 682-684, 1998.

1998-08-PhysEv2-p887

H. Kroemer, "Supercurrent flow through a semiconductor: The transport properties of superconductor-semiconductor hybrid structures," Physica E, Vol. 2, pp. 887-893, 1998.

1998-09-PhysEv2-p894

M. Thomas, H. Kroemer, H.-R. Blank, and K. C. Wong, "Induced superconductivity and residual resistance in InAs quantum wells contacted with superconducting Nb electrodes," Physica E, Vol. 2, pp. 894-898, 1998.

1998-10-PRBv58-p11676

M. Thomas, H.-R. Blank, K. C. Wong, H. Kroemer, and E. Hu, "Current-voltage characteristics of semiconductor-coupled superconducting weak links with large electrode separations," Phys. Rev. B, Vol. 58, pp. 11676-11684, 1998.

1998-11-PRLv80-p2185

R. J. Warburton, K. Weilhammer, J. P. Kotthaus, M. Thomas, and H. Kroemer, "Influence of Collective Effects on the Linewidth of Intersubband Resonance,"

Phys. Rev. Lett., Vol. 80, pp. 2185-2188, 1998.

1998-12-ConfProc-p30
E. Hall, J. Kim, R. Naone, H. Kroemer, and L. A. Coldren, "AlAsSb-based distributed Bragg reflectors for 1.55 μm VCSELs using InAlGaAs as high-index layer," Proc. Lasers and Electro-Optics Society Annual Meeting, Vol. 1, pp. 30-31, 1998.

1999-01-APLv74-p1135
S. Bhargava, H.-R Blank, E. Hall, M. A Chin, H. Kroemer, and V. Narayanamurti, "Staggered to straddling band lineup in InAs/Al(As,Sb)," Appl. Phys. Lett., Vol. 74, pp. 1135-1137, 1999.

1999-02-ELv35-p425
E. Hall, H. Kroemer, and L. A. Coldren, "AlAs-Sb-based distributed Bragg reflectors using InAlGaAs as high-index layer," Electronics Lett., Vol. 35, pp. 425-427, 1999.

1999-03-JCGv203-p297
E. Hall and H. Kroemer, "Surface morphology of GaSb grown on (111)B GaAs by molecular beam epitaxy," J. Cryst. Growth, Vol. 203, pp. 297-302, 1999.

1999-04-JCGv203-p447
E. Hall, H. Kroemer, and L. A. Coldren, "Improved composition control of digitally grown AlAsSb lattice-matched to InP," J. Cryst. Growth, Vol. 203, pp. 447-449, 1999.

1999-05-MEv47-p377
K. W. Lehnert, N. Argaman, H.-R. Blank, K. C. Wong, S. J. Allen, E. L Hu, and H. Kroemer, "Nonequilibrium supercurrents in mesoscopic Nb-InAs-Nb junctions," Microelectronic Engineering, Vol. 47, pp. 377-379, 1999.

1999-06-PRBv60-pR13989
S. Brosig, K. Ensslin, R. J. Warburton, C. Nguyen, B. Brar, M. Thomas, and H. Kroemer, "Zero-field spin splitting in InAs-AlSb quantum wells revisited," Phys. Rev. B, Vol. 60, pp. 13989-13992, 1999.

1999-07-PRLv82-p1265
K. W. Lehnert, N. Argaman, H.-R. Blank, K. C. Wong, S. J. Allen, E. Hu, and H. Kroemer, "Nonequilibrium ac Josephson Effect in Mesoscopic Nb-InAs-Nb Junctions," Phys. Rev. Lett., Vol. 82, pp. 1265-1268, 1999.

1999-08-SMv25-p877
H. Kroemer, "Quasiparticle dynamics in ballistic weak links under weak voltage bias: an elementary treatment," Superlattices and Microstructures, Vol. 25, pp. 877-889, 1999.

1999-09-cond-mat-9901016
H. Kroemer, "On the magnetoresistance anisotropy of a 2-dimensional electron gas with large half-integer filling factors," cond-mat 9901016 (not reviewed).

1999-10-IOPCSv166-p1
H. Kroemer, "Heterostructures Tomorrow: From Physics to Moore's Law," Inst. Phys. Conf. Ser., Vol.166, pp. 1-11, 1999.

1999-11-ELv35-p1337

E. Hall, G. Almuneau, J. O. Kim, O. Sjvlund, H. Kroemer, and L. A. Coldren, "Electrically-pumped, single-epitaxial VCSELs at 1.55 μm with Sb-based mirrors," Electron. Lett., Vol. 35, pp. 1337-1338, 1999.

1999-12-ConfProc-p25

E. Hall, G. Almuneau, A. Huntington, R. Naone, L. Chusseau, H. Kroemer, and L. A. Coldren, "Epitaxial long wavelength DBRs on InP-AlAsSb or lateral oxidation," Dig. LEOS Summer Topical Meetings: Nanostructures and Quantum Dots/WDM Components/VCSELs and Microcavaties/RF Photonics for CATV and HFC Systems, pp. 25-26, 1999.

1999-13-ConfProc-p145

E. Hall, G. Almuneau, J. K. Kim, O. Sjolund, H. Kroemer, and L. A. Coldren, "Design considerations in electrically-pumped, single-epitaxial VCSELs at 1.55 μm with Sb-based mirrors," Proc. LEOS, Vol.1, pp. 145-146, 1999.

1999-14-ConfProc-p1

G. Almuneau, E. Hall, J. Kim, O. Sjolund, H. Kroemer, and L. A. Coldren, "1.55 μm room temperature electrically pumped operation of fully lattice-matched Sb-based vertical cavity surface emitting lasers," Proc. CLEO, pp. 1-2, 1999.

2000-01-APLv76-p215

T. A. Eckhause, S. Tsujino, K. W. Lehnert, E. G. Gwinn, S. J. Allen, M. Thomas, and H. Kroemer, "Midinfrared studies of the contact region at superconductor-semiconductor interfaces," Appl. Phys. Lett, Vol. 76, pp. 215-217, 2000.

2000-02-cond-mat-0009311

H. Kroemer, "Large-amplitude oscillation dynamics and domain suppression in a superlattice Bloch oscillator," cond-mat-0009311 (not reviewed)

2000-03-cond-mat-0007482

H. Kroemer, "On the nature of the negative-conductivity resonance in a superlattice Bloch oscillator," cond-mat-0007482 (not reviewed)

2000-04-PhysEv6-p856

T. A. Eckhouse, S. Tsujino, E. G. Gwinn, M. Thomas, and H. Kroemer, "Intersubband absorption in Nb-clad InAs quantum wells," Physica E, Vol. 6, pp. 856-859, 2000.

2000-05-PhysEv1-p191

R. J. Warburton, K. Weilhammer, C. Jabs, J. P. Kotthaus, M. Thomas, and H. Kroemer, "Collective effects in intersubband transitions," Physica E, Vol. 7, pp. 191-199, April 2000.

2000-06-PRBv61-p13045

S. Brosig, K. Ensslin, A. G. Jansen, C. Nguyen, B. Brar, M. Thomas, and H. Kroemer, "InAs-AlSb quantum wells in tilted magnetic fields," Phys. Rev. B, Vol. 61, pp. 13045-13049, 2000.

2000-07-JEMv29-p1100

E. Hall, A. Huntington, R. L. Naone, H. Kroemer, and L. A. Coldren, "Increased lateral Oxidation Rates of AlInAs on InP using Short-Period Superlattices," J. Electron. Matls, Vol. 29, pp. 1100-1104, 2000.

2000-08-ConfProc-p151

S. Nakagawa, E. M. Hall, G. Almuneau, J. K. Kim, H. Kroemer, and L. A. Coldren, "1.55-μm, InP-lattice-matched VCSELs operating at RT under CW," Semiconductor Laser Conf. Dig., pp. 151-152, 2000.

2000-09-ConfProc-p726

S. Nakagawa, E. M. Hall, G. Almuneau, J. K. Kim, D. A. Buell, H. Kroemer, and L. A. Coldren, "1.55 μm, double-intracavity contacted, InP-lattice-matched VCSELs," Proc. LEOS, pp. 726-727, 2000.

2001-01-APLv78-p1337

S. Nakagawa, E. Hall, G. Almuneau, J. K. Kim, D. A. Buell, H. Kroemer, and L. A. Coldren, "88°C, continuous-wave operation of apertured, intracavity contacted, 1.555m vertical-cavity surface-emitting lasers," Appl. Phys. Lett., Vol. 78, pp. 1337-1339, 2001.

2001-02-CPCv2-p490

H. Kroemer, "Quasi-Electric Fields and Band Offsets: Teaching Electrons New Tricks (Nobel Lecture)," ChemPhysChem, Vol. 2, pp. 490-499, 2001.

2001-03-RMPv73-p783

H. Kroemer, "Nobel Lecture: Quasi-electric fields and band offsets: Teaching electrons new tricks," Rev. Mod. Phys., Vol. 73, pp. 783-793, 2001.

2001-04-QEv7-p224

S. Nakagawa, E. M. Hall, G. Almuneau, J. K. Kim, D. A. Buell, H. Kroemer, and L. A. Coldren, "1.55-μm InP-lattice-matched VCSELs with AlGaAsSb-AlAsSb DBRs," IEEE J. Selected Topics in Quantum Electronics, Vol. 7, pp. 224-230, 2001.

2001-05-Nobel-p101

H. Kroemer, "Quasi-Electric Fields and Band Offsets: Teaching Electrons New Tricks," Les Prix Nobel, The Nobel Prizes 2000, The Nobel Foundation, Stockholm, pp. 101-121 (Biography on pp. 95-100), 2001.

2002-01-ConfProc-p7

H. Kroemer, "Speculations about future directions," Proc. International Conference on Molecular Beam Epitaxy, pp. 7-8, 2002.

2002-02-ConfProc-p3

H. Kroemer, "From electron tubes to nanostructures: 60 years of electron device research," Tech. Dig. Dev. Res. Conf., pp. 3-8, 2002.

2002-03-IntJMPBv16-p677

H. Kroemer, "Quasi-Electric Fields and Band Offsets: Teaching Electrons New Tricks," Intl. J. Mod. Phys. B, Vol. 16, pp. 677-697, 2002.

2002-04-UFNv172-p1087

H. Kroemer, Quasi-electric fields and band offsets: Teaching electrons new tricks Uspekhi Fizicheskikh Nauk, Vol. 172, pp. 1087-1102, 2002.

2003-01-JCGv251-p17

H. Kroemer, "Speculations about Future Directions," J. Cryst. Growth, Vol. 251, pp. 17-22, 2003.

2003-02-cond-mat-0310019

H. Kroemer, "Wave Packet Dynamics in a Biased Finite-Length Superlattice," cond-mat-0310019 (not reviewed)

2004-01-AJPv72-p51

H. Kroemer, "The Thomas precession factor in spin-orbit interaction," Am. J. Phys., Vol. 72, pp. 51-52, 2004.

2004-02-PhysEv20-p196

H. Kroemer, "The 6.1 Å Family (InAs, GaSb, AlSb) and its Heterostructures: A Selective Review," Physica E, Vol. 20, pp. 196-203, 2004.

2005-01-JAP98-p043701

R. Magri, A. Zunger, and H. Kroemer, "Evolution of the band gap and band edge energies of the lattice-matched GaInAsSb/GaSb and GaInAsSb/InAs alloys as a function of composition," J. Appl. Phys., Vol. 98, pp. 043701-1, 2005.

2005-02-PSSav202-p957

H. Kroemer, "Nano-whatever: Do we *really* know where we are heading?," Phys. Stat. Sol.(a), Vol. 202, No. 6, pp. 957-964, 2005.

Books and Book Chapters

1959
H. Kroemer, "Negative Effective Masses in Semiconductors," Progress in Semiconductors, Vol. 4, pp. 3-34, Heywood, London, 1959.
1960
H. Kroemer, "Negative Effektive Massen in Halbleitern," Halbleiterprobleme, Vol. 5, pp. 75-86, Vieweg, Braunschweig, 1960.
1961
H. Kroemer, "Microwave Stimulated Resistance Oscillations in Germanium at 4K," Bull. Am. Phys. Soc., Series II, Vol. 6, p. 116, Mar. 1961.
1963
H. Salow, H. Beneking, H. Kroemer and W. v. Munch, "Die physikalischen Grundlagen des Transistors," Der Transistor, Springer, Berlin, 1963.
1972
H. Kroemer, "Gunn Effect - Bulk Instabilities," Topics in Solid State and Quantum Electronics, pp. 20-98, Wiley, New York, 1972.
1980
C. Kittel and H. Kroemer, "Thermal Physics," Freeman, New York, 1980.
1984
H. Kroemer, "Forschungsprobleme an Heterostrukturen zwischen polaren und nichtpolaren Halbleitern," Braunschweigische Wissenschaftliche Gesellschaft, Braun-schweig, Germany, pp. 53-63, Jahrbuch 1984.
1985
H. Kroemer, "Theory of Heterojunctions: A Critical Review," Molecular Beam Epitaxy and Heterostructures, L. L. Chang and K. Ploog, eds., NATO ASI Series E (Applied Sciences), Vol. E 87, pp. 331-379, Martinus Nijhoff Publishers, Dordrecht, 1985.
1985
H. Kroemer, "Heterostructure Device Physics: Band Discontinuities as Device Design Parameters," VLSI Electronics: Microstructure Science, Vol. 10, Surface and Interface Effects in VLSI, Academic Press, pp. 121-165, 1985.
1994
H. Kroemer, "Quantum Mechanics," Prentice Hall, Englewood Cliffs, 1994.
1999
H. Kroemer and E. Hu, "Semiconducting and Superconducting Physics and Devices in the InAs/AlSb Materials System," Nanotechnology, Springer, New York, 1999 (AIP Press), G. Timp. Ed., pp. 629-688, 1999.
2000
H. Kroemer, "Review of: Introduction to Superconducting Circuits by Alan M Kadin," Physics Today, Vol. 53, pp. 58-60, May 2000.
2002
M. R. Beasley, S. Datta, H. Kogelnik, H. Kroemer, and D. Monroe, "Defining scientific misconduct," Science 298 (5598), 1554, 2002.

List of Patents

1 Transferred electron effective mass modulator
   Inventor: SULLIVAN GERARD J (US); PEDROTTI KENNETH D (US);
   KROEMER HERBERT (US)
   Applicant: ROCKWELL INTERNATIONAL CORP (US)
   EC: G02F1/017C IPC: G02B6/10
   Publication info: US5067828 - 1991-11-26

2 Method for growing tilted superlattices
   Inventor: PETROFF PIERRE M (US); KROEMER HERBERT (US)
   Applicant: UNIV CALIFORNIA (US)
   EC: H01L21/20B6; H01L21/203C; (+3) IPC: H01L21/20
   Publication info: US5013683 - 1991-05-07

3 PROCESS FOR THE PRODUCTION OF ANISOTROPIC SOLIDS HAVING
   MONO-CRYSTALLINE SUPRAMOLECULAR STRUCTURES
   Inventor: HOFFMANN MARTIN DR; KAEMPF GUENTHER DR;
   KROEMER HERBERT DR; PAMPUS GOTTFRIED DR
   Applicant: BAYER AG
   EC: C08J3/00B; C08L53/02 IPC: C08F37/00
   Publication info: DE2112142 - 1972-09-21

4 TRANSVERSE NEGATIVE MOBILITY DEVICES
   Inventor: KROEMER HERBERT; PHERSON ALAN H MAC; SHYAM MEGHA
   Applicant: FAIRCHILD CAMERA INSTR CO
   EC: G11C27/00; H01L45/02; (+5) IPC: H03C3/22; H03B7/06
   Publication info: US3571759 - 1971-03-23

5 HOLOGRAPHIC METHOD OF FORMING AND ALIGNING PATTERNS ON A
   PHOTOSENSITIVE WORKPIECE
   Inventor: KROEMER HERBERT
   Applicant: FAIRCHILD CAMERA INSTR CO
   EC: G03H1/00; H01L21/027B6B2 IPC: G03C5/08; G02B27/00
   Publication info: US3526505 - 1970-09-01

6 LIGHT EMITTING HETEROJUNCTION SEMICONDUCTOR DEVICES
   Inventor: LEHRER WILLIAM I; KROEMER HERBERT
   Applicant: WILLIAM I LEHRER; HERBERT KROEMER
   EC: H01L33/00 IPC: H01L3/12
   Publication info: US3488542 - 1970-01-06

7 Halbleiterbauelement mit negativen Leitwerten bei
  Mikrowellenfrequenzen
  Inventor: HERBERT KROEMER
  Applicant: VARIAN ASSOCIATES
  EC: H01L47/00; H01L47/02 IPC:
  Publication info: DE1300178 - 1969-07-31

8 Heterojunction semiconductor devices employing carrier
  multiplication in a high gap ratio emitterbase heterojunction
  Inventor: HERBERT KROEMER; FAIRMAN ROBERT D
  Applicant: VARIAN ASSOCIATES
  EC: H01L21/00; H01L29/737B IPC:
  Publication info: US3413533 - 1968-11-26

9 HETEROJUNCTIONS AND DOMAIN CONTROL IN BULK NEGATIVE CONDUCTIVITY
  SEMICONDUCTORS
  Inventor: KROEMER HERBERT
  Applicant: VARIAN ASSOCIATES
  EC: H01L21/00; H01L47/00; (+1) IPC: H01L3/00; H01L5/00
  Publication info: US3467896 - 1969-09-16

10 Solid state radiation emitters
   Inventor: KROEMER HERBERT Applicant: VARIAN ASSOCIATES
   EC: H01L33/00; H01L33/00D3; (+3) IPC:
   Publication info: DE1278003 - 1968-09-19

11 Drifttransistor
   Inventor: KROEMER HERBERT DR (DE)
   Applicant: SIEMENS AG (DE); ALLG ELEK CITAETS GES AEG TELE (DE)
   EC: H01L21/18B; H01L29/00 IPC: H01L
   Publication info: DE1414089 - 1970-10-22

12 Verfahren zum Herstellen eines Drifttransistors
   Inventor: KROEMER; HERBERT DR
   Applicant: SIEMENS AG; ALLG ELEK ZITAETS GES AEG TELE
   EC: H01L21/22; H01L29/00 IPC:
   Publication info: DE1301862 - 1969-08-28

13 Elektronischer Schalter
   Inventor: DIPL-PHYS ALFONS HAEHNLEIN; NIEDER-RAMSTADT
   (Kr. DARMSTADT); DR. HELMUT SALOW UND DR. HERBERT KROEMER
   Applicant: DEUTSCHE BUNDESPOST
   EC: H03K17/72 IPC:
   Publication info: DE1063204 - 1959-08-13

IEEE History Center - Herbert Kroemer Abstract

Herbert Kroemer Oral History

ABSTRACT

Herbert Kroemer was born 25 August 1928 in Weimar, Germany, and studied at the Universities of Jena and Göttingen. During the Soviet blockade of Berlin in 1948, he worked a summer job at Siemens, and escaped the Soviet Zone to go to the West. After obtaining his PhD, Kroemer worked at house physicist at Central Communications Lab, where he got the idea for heterostructure bipolar transistors. After spending three years at RCA labs, he returned to Germany to head Phillips Semiconductor Group, working on gallium arsenide technology in 1957. Returning to the United States, he worked at Varian Associates, and studied the Gunn effect. Awarded the Nobel Prize in Physics in 2000. In 1968 went to University of Colorado, and in 1975 to University of California Santa Barbara.

TABLE OF CONTENTS

INTERVIEWEE: Dr. Herbert Kroemer, 2000 Nobel Physics Laureate
INTERVIEWER: Dr. John Vardalas
DATE: 12 February 2003
PLACE: Dr. Kroemer's office at the University of California at Santa Barbara

Copyright Statement

Vardalas: It's quite an honor to interview a man of your distinction. I suppose the natural place to start is at the beginning. When were you born and where?

Kroemer: I was born on August 25, 1928 in Weimar, Germany.

Vardalas: Did you have any brothers or sisters?

Kroemer: I had two younger brothers.

Vardalas: Did you spend your formative years in Weimar or in that general area?

Kroemer: I lived in Weimer through high school, and then my first university after high school was the University of Jena just 15 miles from my hometown. Then after one year – I think it was 1948 – I went to Göttingen in West Germany. Weimar was in what was then East Germany.

Vardalas: You grew up during a period of great economic and political turmoil in Germany and Europe. What are your strongest memories during this period?

Kroemer: After 1945 Weimar was occupied by Soviet troops. It had originally been taken over by the Americans. Then after June or July when Germany was divided that area was turned over to Soviet troops. I remember a time when the Americans were leaving and the Soviets coming in as a pretty bad time, because the Soviets tried to enforce a type of government there that was very, very repressive. It was really a Stalinist operation, and Weimar was actually a very liberal town.

Vardalas: Was it?

Kroemer: Yes, and it therefore drew their specific attention. It was also pretty scary at times when I studied physics at the University of Jena for two semesters during 1947/48.

Vardalas: Yes, I wanted to get to that.

Kroemer: The student body at Jena was basically very liberal, as was the faculty, so they really cracked down. I remember one day in a philosophy class the professor made some comments that were clearly in disagreement with the communist party line that got a big applause. When class was over someone grabbed hold of me at the exit and said, "We'll keep track of people like you who applaud."

Vardalas: Really?

Kroemer: It was a really threatening kind of atmosphere.

Vardalas: Did you find yourself being interested in social and political issues at the time?

Kroemer: One couldn't help but be interested in social and political issues, but not on the officially approved side.

Vardalas: Yes, of course.

Kroemer: The summer of 1948 was the Berlin blockade.

Vardalas: Yes.

Kroemer: Four of us went as summer students to the Siemens Electrical Laboratories in West Berlin that year. At that point the blockade was pretty much in force and the airlift was in force, so I then decided I was not going to go back home and left for West Germany.

Vardalas: I would like to go deeper into that. Before I do that I want to find out something about your education. Were you in high school in the midst of the war?

Kroemer: Nowadays it is called a gymnasium. I went through four years of elementary school. Then in the German educational system as it existed at the time students got split up at the 5th grade.

Vardalas: What happens?

Kroemer: About 15 percent go on to "oberschule" or gymnasium. That was intended to ultimately lead to a degree called abitur, which is required for admission to German universities.

Vardalas: I see.

Kroemer: It was really a pre-academic education. The remaining majority of people went on in elementary school for another four years. Then after eight years they went to some kind of a trade school – business school or something like that. But I moved on to the German-style high school or gymnasium.

Vardalas: Was this still during the war years that you were in the gymnasium?

Kroemer: Yes, it was during the war.

Vardalas: How did the war affect the level of education and what was being offered at school?

IEEE History Center - Herbert Kroemer Oral History

Kroemer: Many teachers of course had been drafted, so older teachers who had already officially retired were called back into service. And some subjects were dropped. For example the third foreign language was dropped. We started foreign language in 5th grade, which was the first high school year. We started with English and two years later they added Latin – which I hated. Normally French would have been added two years after that, but that never happened. And a few other things were dropped. However science and the first and second language courses continued. English continued.

Vardalas: English continued?

Kroemer: Yes, at full blast.

Vardalas: How interesting.

Kroemer: We had a very interesting instructor in English. He had spent all of his life as a German language and German literature instructor at one of the two big English universities. I do not remember whether it was Oxford or Cambridge, but it was one of those two. He had retired just before the war and came back to Germany. He was immediately called back into service as a teacher of English. And he had a wonderful knowledge not only of English literature but also of the English parliamentary systems. There was never any commentary with the reading material that we got. He would simply say, "Well, they do things differently." He was really subversive, and he got away with it.

Vardalas: He got away with it.

Kroemer: And after the war, under Soviet Occupation, our Latin teacher was also very interesting, though in a different sense. We were reading Caesar's Gallic War, and we didn't go through the whole darn thing beginning to end. He specifically picked sections from the book that indicated how the Romans as occupation troops were dealing with people who were occupied. And so he said, "You see ladies and gentleman, there is nothing new in this world."

Vardalas: And the students understood that.

Kroemer: We all understood that. And he got away with it actually.

Vardalas: Really? Okay.

Kroemer: This was the high school Latin teacher.

Vardalas: Yes. I gather from reading your autobiographical sketch that you exhibited quite a love of learning from an early age.

Kroemer: Yes. And it was a good environment for that.

Vardalas: The school was a good environment?

Kroemer: Yes. Teachers come in different flavors, but there were some that were very good. Learning was encouraged in the fields that interested me.

Vardalas: I understand that your early inclinations ran towards math, physics and chemistry.

Kroemer: That's right.

Vardalas: At the same time, your parents had little education and knew very little about the world of science as you knew it.

Kroemer: That's right, yes.

Vardalas: How were you first drawn into science? Do you recall? Was it something you read? What first piqued your interest?

Kroemer: I couldn't really answer that. I don't know. There was nothing specific I can point to. It happened fairly naturally.

Vardalas: Could you say that you grew to like science and math because it came easily to you?

Kroemer: It certainly did come easily to me. I was surprised when one day I discovered that math is actually easy. And I couldn't understand why other people found it hard. But I never really seriously thought about making a career out of math. It was a discipline that I enjoyed and was good at, and I was good enough that I actually was bored in the classes and often disrupted the proceedings.

Vardalas: Yes, I want to get to that too.

Kroemer: For my career choice it was a sort of toss-up between chemistry and physics, and eventually physics won out.

Vardalas: At that age in high school or gymnasium – which was before university – how did you express these inclinations? Did you go beyond formal subjects you learned in school? What were your interests in science?

Kroemer: I did a lot of reading on my own. We had a good library in the city and I borrowed books voraciously. Typically I would go twice a week and return with two new books each time.

Vardalas: Were these science books?

Kroemer: Typically they were science, though not science in a hard textbook style. It was science in the Scientific American style but in book form. I remember some of those books quite well. I was genuinely interested. As far as physics was concerned, we got a textbook and I went through that textbook on my own. There was too much memorization. I didn't like that.

Vardalas: I guess that's why you didn't like Latin either then.

Kroemer: Oh boy. Yes.

Vardalas: Did anyone guide you in your readings or were you just on your own?

Kroemer: I had no guidance of any kind – none from my parents and not from anyone else.

Vardalas: Was this a matter of you finding your own way?

Kroemer: Yes.

Vardalas: I see.

Kroemer: Nor did I have any classmates that I was interacting with particularly closely in this field.

Vardalas: Really? There was no sense of sharing your interests?

Kroemer: That's right.

Vardalas: It must have been quite lonely.

Kroemer: It wasn't lonely. It was alone but not lonely.

Vardalas: Okay. I understand.

Kroemer: Those are two different things.

Vardalas: Yes. In reading the Scientific American style kind of material, what did you view or imagine physics to be all about at this point in time? And why was it attracting you? What was being done?

Kroemer: Well, it's hard to pin down.

Vardalas: Did you feel a sense of discovery and adventure in this?

Kroemer: Not really. It was more the realization that from a small set of very fundamental laws one could draw very, very far-reaching conclusions.

Vardalas: And this intrigued you.

Kroemer: I was puzzled about the textbook, why the textbook didn't put the law of conservation of energy on page one and then use this as the hook on which everything gets hung. Today I know about this process, but that gives you an idea of my response. I did have one individual with whom I interacted. It was a cousin of mine. He was a few years older than me, and I spent a few summers with him at his parents' place. And he was a very, very good student at his school, and I asked him all sorts of questions – typically chemistry questions. With infinite patience he gave me all sorts of detailed explanations on how these things were done. I had an encyclopedia and I looked up all sorts of diagrams and equipment. That gave me sort of the illusion that I knew what was going on. My father had bought that encyclopedia, which was something like twenty volumes.

Vardalas: Why had he bought the encyclopedia?

Kroemer: They had bought that specifically for me to read. They didn't say, "Read the encyclopedia." It was simply there, and I started poking around. I enjoyed it and spent hours looking up things in the encyclopedia.

Vardalas: That was very perceptive of them.

Kroemer: Yes. Both of my parents felt very, very strongly that I should get the best education that I could handle. If I had done poorly in school, they would probably have taken me out rather than saying, "You have to do this."

Vardalas: I see. Did you siblings have similar interests or did they go in different directions than you?

Kroemer: The second one had a similar interest. I know very little about the third one. He grew up in the post-war years, born in '41. I do not really know how he responded in school. He never went to university, but that may have been for political reasons.

Vardalas: Did your other brother become a professional?

Kroemer: My other brother studied physics.

Vardalas: Oh, he studied physics too?

Kroemer: Yes.

Vardalas: Did you inspire him?

Kroemer: I don't know whether I inspired him or whether my mother inspired him that the firstborn had obviously done well so the second born should follow in his footsteps. I think he would have made a much better engineer than physicist. He would be a very, very good engineer. He's retired.

Vardalas: Okay. I see. You graduated from gymnasium in 1947 at the age of 19.

Kroemer: Yes.

Vardalas: I want to get to this question of your boredom. You made a point of saying how you found school so boring that you became disruptive.

Kroemer: Yes, particularly in math. I found it boring because I already knew that stuff. I was way ahead of it. I had done a lot of reading. How does a 17- or 18-year-old act when he is bored? He is trying to draw attention to himself so he becomes disruptive and makes smart-alecky remarks.

Vardalas: You said you kept showing off. How did you show off?

Kroemer: By speaking up when others were supposed to answer and that sort of thing. One way I showed off that was in an indirect way was that I taught some classmates mathematical techniques that were not part of the curriculum. They didn't really understand what was going on, but they were sort of cookbook recipes.

Vardalas: Was that seen in a bad light?

Kroemer: The teacher, Willibald Wimmer, was a great man. I later met his two daughters when I visited my hometown. Anyway, he had been an instructor at a junior engineering college so he was used to a somewhat more mature group of students. After the war that college did not reopen, so he became a teacher in math and physics at our school with his principal field being mathematics. He was a good teacher, no doubt about it. It is not uncommon for math and physics to be taught by the same teacher. I'll come back to his physics aspect, but he was very good in math. He treated us as adults, and certainly all of us respected that, so we liked the man very much. He was not flamboyant or anything, but he treated us as if he expected us to be reasonable.

Anyway, he found my behavior a nuisance. He knew that I knew this stuff. He couldn't really tell me not to come to class because I had to be in attendance. Therefore he suggested to me that, "Kroemer, I'll promise you an A in math and you don't have to do the homework — under one condition: you shut up. You have to be here, but you can do anything you want to except you must be quiet and not speak up in class unless asked." I thought it was a good deal. And that's where it stayed. I did my English homework during the math class typically.

Vardalas: Do you ever apply that technique to your students?

Kroemer: I've been tempted at times. But it was very different in his physics class.

Vardalas: In what way?

Kroemer: It was different in the sense that he basically knew what was in the textbook but nothing else. I had done voracious reading on physics. I knew what was publicly available in nuclear physics and all sorts of things.

Vardalas: Really?

IEEE History Center - Herbert Kroemer Oral History

Kroemer: A lesser individual would have resented that. And there was another student named Klaus Meyer who was similarly very much interested and knowledgeable in physics. He [Wimmer] simply enlisted the two of us to help him. One of the forms that this help took was in digging through the collection of physics equipment that had somehow survived the war. During the war our school building had been used as a hospital, so much of the equipment was in bad shape and/or not cataloged. We helped him with that and also with preparing lectures. In one case he said, "Why don't you give the next lecture?"

I remember what it was. There was a thermos bottle with a known amount of water and there was a resistor and we applied a voltage. We knew the voltage and the current, so we knew how many watts had to dissipate, and we were measuring the rise in temperature. Experiments along those lines. I don't remember all the details. He was sitting in the front row and having a ball. I enjoyed that.

Vardalas: You enjoyed that?

Kroemer: I enjoyed that.

Vardalas: Was it the teaching or having attention and being in front?

Kroemer: I enjoyed all of it, the way he was treating me and others, and my own ability to teach – which is a form of showing off. That was an interesting experience.

Vardalas: This experience must have also fueled your motivation to keep reading more and staying on top.

Kroemer: Yes, it certainly was positive feedback By that time I had pretty much decided to go into physics rather than chemistry.

Vardalas: Okay. You write that when you told your father that you wanted to study physics at the university he wasn't sure what that was about.

Kroemer: Yes. He asked, "What is that?" You see, he didn't have a high school education. He didn't know what physics was.

Vardalas: And he was concerned about whether you could earn a living at it.

Kroemer: That's right.

Vardalas: And you replied – and I quote – "I certainly could become a physics teacher at a gymnasium, a fairly respectable profession." Was this an answer you gave your father to ease his mind or was this something you actually contemplated as a possibility?

Kroemer: That was more than just easing his mind. That was certainly the most natural end of studying physics. Remember this was 1947 or so. The idea that there were such things as industrial physicists was not all that obvious. And I certainly had not thought about an academic career at that point in time. But it wasn't really something I was explicitly planning. I simply felt that was an option I would definitely have. I knew that there might be other options but was not terribly concerned about what those were.

Vardalas: As a high school student did you ever have dreams of great discoveries? Did you think, "I'm going to do that some day"?

Kroemer: Not as a high school student, but as a student at the university I certainly had some dreams.

Vardalas: Those dreams had not been formulated during high school?

Kroemer: I do not recall dreams of great discoveries as a high school student. I certainly did not have any crackpot ideas.

Vardalas: What would be an example of a crackpot idea?

Kroemer: High school students often have crackpot ideas.

Vardalas: Such as building a time machine?

Kroemer: Yes, or things that violate known laws – either because they don't know the laws or they feel they can violate them. I wasn't that type.

Vardalas: You stayed at Jena for one year?

Kroemer: Yes.

Vardalas: Did your understanding, perspective or expectations of physics change during that year?

Kroemer: That year had a great deal of influence on me, yes, because I was suddenly confronted with teachers who were really at the top level of their profession – certainly in math and physics. I don't remember any chemistry at Jena though I do remember some chemistry at Göttingen. I found one of the mathematicians at Jena very inspiring. His name was Brödel.

Vardalas: In what way was he inspiring?

Kroemer: He was an absolutely fabulous lecturer. I must admit that I didn't understand everything.

Vardalas: That's the first time you weren't bored.

Kroemer: It was the first time I wasn't bored but challenged. One of the reasons I was challenged was that in the first and second semesters I took courses intended for third and fourth semesters. I simply didn't bother with the first -semester courses.

Vardalas: You had that option? They didn't force you to take the first as prerequisites?

Kroemer: No, we could take anything we wanted. In physics, one of the professors was a gentleman by the name of Friedrich Hund. He was a spectacular lecturer and he had a great influence on my becoming interested in the deep fundamental principals of physics.

Vardalas: Why was this?

Kroemer: He is pretty close to the top of my list of people who should have gotten the Nobel Prize but never did.

Vardalas: What was his area of expertise?

Kroemer: He was one of the leaders —though not one of the founders — of quantum mechanics. There is something called "Hund's rule" which plays an important role in atomic physics even today. He was a wonderful teacher in whatever he touched, including quantum mechanics and thermodynamics. I took a thermodynamics course with him. And he was a wonderful person.

Vardalas: You took this course with him your first year?

Kroemer: That was during that one year when I was at Jena.

Vardalas: Do you recall your curriculum at that time? What did you take?

Kroemer: I don't remember.

Vardalas: You took thermodynamics and quantum mechanics?

Kroemer: I didn't take quantum mechanics in the first semester. That came later. I really didn't have the background for that. In physics I probably took a classical mechanics course the first semester. That wasn't terribly inspiring. And it was probably in the second semester that I took the thermodynamics course from Hund.

Vardalas: In the German system did they have what would be called auxiliary courses here in the United States, such as humanities? Did they make you take any courses besides those that related directly to your major?

Kroemer: We could take anything we wanted.

Vardalas: Anything you wanted?

Kroemer: Yes. To get a degree one had to show certain pieces of paper. I was very much interested in taking courses outside of physics – particularly philosophy.

Vardalas: You did? Interesting.

Kroemer: I took a heavy load of philosophy. That's how I got into that trouble with that communist functionary in that class.

Vardalas: Oh really?

Kroemer: I took a heavy load of philosophy. I don't remember what the different topics were, but more than one course. I remember two of the instructors. One taught metaphysics. He was the one we applauded.

Vardalas: Oh, that was the one you got in trouble for applauding.

Kroemer: Yes, and lots of other things. Another instructor was a philosopher named Max Bense, who taught formal logic. He was fascinating. I have never known anybody who could construct sentences so long that were grammatically correct to the end.

Vardalas: That is also a particular peculiarity of German, isn't it?

Kroemer: Yes.

Vardalas: How long did you have to wait for the verb?

Kroemer: No, no. This is really not true that the verbs are always at the end. That really is bad German. It's allowed, but not a good practice. That is not what I meant.

Vardalas: Did you mean clauses inside clauses and so on?

Kroemer: Yes, yes. And the funny thing was that he said it in a way that it was understandable.

Vardalas: I see.

Kroemer: I found him interesting as a person more than I found the course interesting.

Vardalas: Let me go back for a second. You mention rather brilliant people on the staff and at the University of Jena and you only spent a year there. What do you think the effect of Soviet presence was on that environment?

Kroemer: It didn't do anything to the scientific environment, certainly not in those days, and I do not think it did much in subsequent years. But actually I wasn't there of course. It did nothing whatsoever to the scientific environment. The influence was in the political environment and the everyday environment. I had fallen in with a group of students who were all very liberal and I had the suspicion I was being noticed. One of the depressing things was that over that year every week one or two people would disappear.

Vardalas: Really?

Kroemer: One never knew whether they had left by their own free will or ended up in the gulag.

Vardalas: I see.

Kroemer: The gulag in those days meant the uranium mines at the Czech border.

Vardalas: This must have been an environment of great uncertainty and fear.

Kroemer: Yes, yes. We could not trust anyone we did not know personally.

Vardalas: You said your time at Jena was important in your formation and that it had changed you somewhat.

Kroemer: It really exposed me to physics.

Vardalas: Did it also expose you to other good students?

Kroemer: Yes, absolutely. They played an important role.

Vardalas: Were you considered a bright student at this level yet?

Kroemer: Probably, yes. Nothing indicated this officially, but I think I was one of the better students. There is one interesting difference in that education system compared to what we are doing here at UCSB. For example in the calculus course. Homework is an integral part of the course in the United States, but the way this was done there was that there were problem and homework sessions. They were different courses, and sometimes they didn't have too much in common with one another. I knew calculus pretty well already. We had calculus in high school and I had studied it on my own – but not in the rigorous sense in which it gets done at the university where one worries about all sorts of things that could go wrong. The way I learned it one did not worry about things could go wrong. This makes one better qualified as a physicist.

Vardalas: Yes, I was going to say that.

Kroemer: That instructor was pretty boring, so I went only to the problem course. I needed that certificate that I had attended that course two semesters in a row. And that one was fascinating. There I found a real challenge and I got one of the top grades in the class. They congratulated me. I remember it was a lady who was teaching that one. When she was handing out the certificates at the end she said, "I hope to see you again" in the sense that I was one of the ones they wanted to see. I enjoyed it.

Vardalas: I'm curious. With two courses, one solving problems and one for just theory, how were you tested in both? The test would seem to be the problem solving.

Kroemer: The test is the problem solving, yes. I never went through the mainstream course. First of all I discovered very quickly that I already knew everything that was needed to solve the problems and it required more imagination as to how they could be tackled. That was the good part. Also in the physics courses the problems session typically was separate. In that other course with a mathematician whom I admired there was also much homework.

Vardalas: At this stage, after one year's exposure, did you start to see a sense of what kind of physics you wanted to do?

Kroemer: Not really. It was a wide-open field, but it was very clear that I enjoyed physics.

Vardalas: I want to pursue this thing about you winding up in West Berlin. You said four of you had a summer job at Siemens.

Kroemer: Yes. One of us had arranged that summer job at Siemens.

Vardalas: At this point there was not much impediment to travel from East Berlin to West Berlin and work?

Kroemer: No, it was still completely open although this was during the Berlin blockade. The blockade referred to truck traffic, rail traffic and bringing supplies into Berlin, but there were no limitations on people crossing into Berlin or within Berlin and crossing between East Berlin and West Berlin.

Vardalas: Then you go on to say that at that point you decided you were not going back because of this uncertainty and fear you were experiencing in this environment.

Kroemer: Berlin during the Berlin blockade was an exhilarating experience.

Vardalas: In what way?

Kroemer: First of all the West Berlin government was about as firmly anti-communist as could be, even though the mayor was a former communist. He had turned anti-communist.

Vardalas: What was his name?

Kroemer: Ernst Reuter. He was a fascinating man, and I remember going to one of his big speeches. One of the interesting experiences I had was one of the airports which served the airlift – the airlift was functioning pretty well by late summer – was the Tempelhof Airport, which is basically the airport downtown. There is a railroad track along the edge of the airport and that is somewhat elevated, and in the evenings we often took the train to Tempelhof Station just to watch the aircraft take off. It was a fascinating experience.

Vardalas: What was so great about it?

Kroemer: It was staggering how well it was organized. The Soviets could have stopped this whole thing with one fighter plane, but they didn't dare. There were two runways – one only for takeoffs and one only for landing – and they were bringing in planes roughly at the rate of one every 75 seconds.

Vardalas: Wow.

Kroemer: You could see them along the horizon like pearls on a string. The pilots didn't really know to which gate they had to go, so next to the runway were a number of jeeps and as soon as the plane had slowed down so that a jeep could keep up with it the jeep pulled in front of the plane with a sign, "Follow me."

Vardalas: Really?

Kroemer: That jeep driver knew which gate had just been opened.

Vardalas: Oh, I see.

Kroemer: That made you believe in Germany and America having something in common.

Vardalas: The efficiency?

Kroemer: No. You see, in '47 and '48 it was not clear whether Germany had any future, but it was very clear that if there was a future for Germany it was in alliance with the United States. Amazingly, even though Berlin had been badly bombed out by the Americans and the British there was basically no resentment by the people.

Vardalas: That surprises me. Was that because the communists were there?

Kroemer: Yes, well, the contrast was stark.

Vardalas: Okay. The four of you were watching these planes coming in and leaving and then the thought came to you that you were not going back?

Kroemer: I decided I was not going back.

Vardalas: Were you ever personally frightened? I know the man that stood at the door said, "We're going to watch you." Was it a serious consideration for you that, "Maybe I'll get in trouble if I stay in East Germany"?

Kroemer: Yes, but nothing specific. Yes. It was more than just vigorously disliking the environment. I was also worried that I might be forced to do something.

Vardalas: The physics education was good in itself.

Kroemer: Yes.

Vardalas: It was other considerations.

Kroemer: Göttingen was the top university.

Vardalas: But at that time you didn't know about Göttingen?

Kroemer: I knew about Göttingen.

Vardalas: I mean as an option for you. It wasn't a real option yet when you made the decision to leave?

Kroemer: I had written to several universities in West Germany and applied to each. Göttingen was one of them. Göttingen had actually turned me down, but I never got that mail.

Vardalas: Fortunate for you.

Kroemer: I had a recommendation from one of my Jena professors. I had gone back to Weimar to pick up my baggage before leaving by way of Jena. Professor Buchwald suggested, "Why don't you see Professor König and see what he can do for you?" König was an old friend of his.

Vardalas: I see.

Kroemer: I was staying Kassel, which is where relatives of mine were living. It is just about a half hour from Göttingen by train.

Vardalas: Was it a difficult decision for you to head out and leave your family behind? It must have been. Or was it something that you didn't have to anguish over much because you were too worried about staying?

Kroemer: It wasn't really difficult, but probably one of the reasons it wasn't difficult was because I didn't appreciate how difficult it might turn out to be.

Vardalas: What did you later find?

Kroemer: Well, my parents could no longer support me. The currency differential was such that it was totally impossible. I had to find a job. It was an interesting experience.

Vardalas: After the walls went up and other things that happened you must have been cut off from your family.

Kroemer: I went back for a visit a few times.

Vardalas: Weren't you worried about being caught?

Kroemer: Oh, I was always absolutely terrified.

Vardalas: Was your father supportive of your leaving?

Kroemer: He didn't try to talk me out of it. He probably realized that this was perhaps the best thing for me. He didn't actively encourage me. However neither he nor my mother made any attempt to talk me out of it. They said, "All right. If that is what you want to do, then go and do it."

Vardalas: You also write that you found Göttingen to be a wonderfully stimulating place.

Kroemer: It was absolutely wonderful.

Vardalas: Was it the same kind of stimulation you found in Berlin? What was it intellectual stimulation that you found in Göttingen?

Kroemer: Yes, it was intellectually stimulating. Göttingen was always one of the top universities in physics in Germany, and in fact in science in general. During the twenties and thirties Göttingen, Berlin and München were the leading universities in physics. Göttingen had not been bombed out during the war, and as a result many academic people who were refugees from the East congregated there. There were a few what is now called Max Planck Institutes though they didn't have that name at the time. It was an absolutely fascinating collection, absolutely fabulous.

Vardalas: Tell me about your process of your getting into Göttingen. You said you were initially rejected.

Kroemer: I showed up in the office of Professor König and he said, "Admissions are closed and there is nothing I can do," but somehow he decided to take me on a tour anyway. Maybe he just wanted to be nice or maybe he had some afterthoughts. I don't know. I showed up in the office of Professor Richard Becker and had a nice long conversation with him and he asked me many questions. His assistant, Dr. Günther Leibfried, was also there. They spent a long time with me and then passed me on to Professor Wolfgang Paul [Nobel 1989] and Professor Robert Pohl. Gradually it dawned on me that these were not just social conversation. I was in an examination.

Vardalas: Did you start to worry at that point?

Kroemer: No. I remember one of the exam questions given to me by Professor Paul. I'll never forget it. He said, "Mr. Kroemer, you know that a mirror interchanges left and right." I said, "Yes." He asked me, "Why then doesn't it interchange top and bottom?"

Vardalas: What was your answer to that?

Kroemer: I gave him the answer. The answer is it doesn't invert left and right. That's all he wanted to know. However it must have been very obvious that I had to think about this first. At the end of all of this I was led back to Becker's office and they said that admissions were closed but they had received notification from two people who had been admitted that they were not coming. Therefore they had two openings and that within the next day or two they would decide who would get those two openings. They explained that if I was chosen I would receive notice. That was on a Thursday I think. The next Monday or Tuesday I received a postcard notifying me that I was admitted.

Vardalas: How did you support yourself at that time?

Kroemer: That was difficult. I found a job in a local aluminum cooking ware factory. Lots of students where employed there during the night shift.

Vardalas: What did the students do?

Kroemer: We operated the equipment and did whatever jobs had to be done. It paid well enough that if one was frugal one could survive on it.

Vardalas: That must have been tough working night shift and studying days.

Kroemer: Yes, it was tough. One summer I worked in a coal mine.

Vardalas: I wanted to ask you about that. What were you doing one kilometer down in a coal mine? It wasn't a neutrino experiment, was it?

Kroemer: It was not a neutrino experiment. It was a very interesting experience. I don't remember the proper English terminology for what I was doing. Hard coal was being taken out, and of course on the surface of all of this is the dense population of the cities in the Ruhr Valley. In order to keep the surface from collapsing they tried to refill this after the coal had been taken out. It was being refilled with sand typically, and sorts of other debris. I worked with that crew. And it was strenuous work but it paid well. For he first time in my life, I was embedded in an environment as a lone student amongst coal miners. I suddenly realized that these people didn't have as much.

Vardalas: How did they treat you?

Kroemer: Very, very nicely. They always assured me, "At least you know that you will get out of this one day. We are going to be stuck here the rest of our lives." I suddenly understood why people like this were voting communist. Not that I agreed with them.

Vardalas: Right.

Kroemer: I understood their demoralization and their expectations and their feeling that the political system would not take care of them. That changed later on, but this must have been '48 or '49.

Vardalas: That was a tough time economically. All of Europe was in ruins in a sense.

Kroemer: Yes. Now that was during the Marshall Plan year during which time Germany was recovering rather quickly. The coal industry was one of the driving engines of the recovery. It was a very interesting experience.

Vardalas: Referring back to your autobiographical sketch, you highlighted very briefly the deep influence that Dr. Fritz Sauter had on you. He was your Ph.D. supervisor, wasn't he?

Kroemer: Both my diploma thesis – which is sort of a master's degree – and Ph.D. supervisor, yes.

Vardalas: Would you explain more about his influence on you?

Kroemer: Ours was not a close personal or warm relationship. It was a purely professional relationship.

Vardalas: You wrote, "Under influences such as these I never developed into a hardcore Theorist with a capital T but became basically a conceptualist."

Kroemer: Yes.

Vardalas: What does that mean?

Kroemer: Let me go back a few steps on that one. One course that was very important for me throughout my years in Göttingen was the seminar in theoretical physics. This was not a regular course but a sort of special topics thing where the instructor handed out material and assigned papers. The students were required to read the papers and then report on them. I found this a very stimulating environment. I had participated in these kinds of courses right from the beginning of my Göttingen days. Sauter led one of those courses. I got some assignments, and one of those assignments led to a master's dissertation.

Vardalas: What was that assignment?

Kroemer: I was already going toward solid-state physics. That assignment was studying certain things that happened in periodic potentials when some of the parameters are changed. An interesting aspect of Sauter's style was that he didn't call on students to report once or twice a week or on any regular schedule. He basically was available when he was needed, but he left the students alone. He was basically watching and forming his own opinions as he watched. I remember one time coming to him with an idea I had, a physical concept. He listened and then said with a tinge of sarcasm, "Well, Mr. Kroemer, that's all very nice, but you ought to be able to formulate this mathematically." At another time I would report to him how I would formulate something mathematically. Obviously mathematics was important. Then with a slightly more sarcastic tone he responded, "Mr. Kroemer, this is just a piece of math. What does it mean physically?" Once one goes through this loop a few times one gets the message. One gets the message that one has to be able to move at ease from one to the other in order to live up to Sauter's standards. That was a deep and formative experience for me. I have later on encountered the same thing with Bill [William] Shockley, whom I knew quite well. Shockley had that same style, moving back and forth.

There is something else that happened too. We had already agreed on the topic for my Ph.D. dissertation while I was still working on my diploma thesis. One day Sauter came into my office and told me to stop working on the diploma thesis and simply write up whatever I already had and submit it. I protested, but he said, "Never mind. Let's move on to the real thing." He was also wonderful in the sense that he didn't believe that degrees should be awarded on the basis having served time.

Vardalas: Yes. I like that statement.

Kroemer: First of all he didn't have any money to support his students. Sauter's idea was that he was watching people and to him a degree was to certify that this person could do creative and independent work on a certain level and that as soon as this level had been reached – get out. That was very good for me, because I had to earn a living. I got my Ph.D. a few weeks before my 24th birthday.

Vardalas: That's remarkable. Five years of total experience between entering and leaving. Was that unusual?

Kroemer: That was unusual even for Göttingen, yes. I was one of the youngest Ph.D.s in physics after the war. And that would not have happened under any other professor.

Vardalas: Right.

Kroemer: The way I ended up with Sauter was different. I had originally attached myself to a person who at that time didn't even have the professorial rank. He was a lecturer and instructor: "Privatdozent" Dr. Hellwege. I also had a very close personal relationship with Hellwege and had signed up for a diploma thesis with him. However he had an awfully long waiting list. Then Sauter offered me the opportunity to get my diploma under him. I talked to Hellwege and he said to me, "Kroemer, take it. You will be finished with him before you can start with me." This is how I ended up a theorist. Sauter was a fabulous mathematician, but to him it was a tool. To him ultimately physics mattered.

Vardalas: You made the distinction that you were a theorist but not a hardcore theorist.

Kroemer: I wasn't. I was a conceptualist.

Vardalas: What does that mean?

Kroemer: There are theorists that know only theory and are heavily engaged in mathematical formulas, whereas to me the mathematical formulas were never more than a tool. Mathematics always represented something to express physical ideas and the physical ideas were always related to experimental facts even though I was a theorist. And I never was a good experimentalist.

Vardalas: You were never a good experimentalist?

Kroemer: No. I think I was good at picking good projects.

Vardalas: What does it take to be a good experimentalist?

Kroemer: I don't know. Not having been one I could not say.

Vardalas: What couldn't you do? You must have come to the conclusion you're not good at it or didn't like it.

Kroemer: It is not a matter of not liking it. Some people have the touch. They invent their own equipment and build their own equipment. I was better at thinking about what equipment one should build than at actually building it.

Vardalas: Okay. A conceptualist.

Kroemer: I was a conceptualist.

Vardalas: I see. You made the interesting remark that your role model was more Niels Bohr than any other great physicist.

Kroemer: That's right. Yes.

Vardalas: What was it about Niels Bohr that attracted you to him as a role model?

Kroemer: Conceptual depth combined with very simple mathematics. He was an ideas man.

IEEE History Center - Herbert Kroemer Oral History

Vardalas: I don't want to be too simplistic about it, but what about Einstein who did a lot of mathematics and had conceptual ideas too?

Kroemer: Actually Einstein is really a conceptualist. The mathematics of course came with the general theory of relativity. He got his start from Minkowski on that one. In his later years his work was very, very heavily mathematical. However that didn't really start until 1920s or thereabouts. He was really a conceptualist.

Just for the fun of it, I have been rereading some of Einstein's early papers on statistical thermodynamics. The concept of wave particle duality shows up in Einstein's writing before DeBroglie.

Vardalas: That's amazing.

Kroemer: Not in the same formulas.

Vardalas: Do you see anything else in Einstein's early papers that strike you after all these years? Are there any surprises?

Kroemer: Yes. It surprising in the sense of just how brilliant Einstein was, and what a fabulous instinct he had. He anticipated ideas that were not accepted until much later on and helped others get on their way. For example deBroglie's thesis. I don't remember who his thesis advisor was in France.

Vardalas: I can't remember either.

Kroemer: Thesis advisors played a different role in those days. One turned in one's thesis and the thesis advisor was the person who was supposed to judge it. In a way Sauter was like that with me. He simply watched and when he saw a final draft he criticized it.

Anyway, deBroglie's thesis advisor didn't really know what to do with deBroglie's thesis so he turned it over to Einstein. The remark that Einstein supposedly made in regard to deBroglie's thesis was, "He lifts a great veil from the secrets of physics." I am certain that Einstein had anticipated something like this. The history of physics is fascinating to me.

Vardalas: Yes, it's interesting. And yet Einstein is always attributed to saying that God doesn't play dice. His resistance to quantum mechanics.

Kroemer: His resistance to quantum mechanics was first of all that he definitely did not like the probabilistic interpretation.

Vardalas: Yes.

Kroemer: Probably what disturbed him more than the probabilistic interpretation was what we today call the non-locality of the theory. It contained an aspect of action at a distance. That was probably more disturbing to him. I've done a little bit of reading on what exactly Einstein opposed. There's the famous Einstein-Podolsky-Rosen argument. It was entirely consistent with Einstein's thinking in the early days. Having been a participant in writing a book on thermodynamics, I have seen that what is today attributed to Boltzmann was really first clarified by Einstein.

Vardalas: Really?

Kroemer: Einstein took Boltzmann's ideas and put some rigor into those concepts.

Vardalas: Very interesting.

Kroemer: That is something I discovered only recently. I became interested in Einstein basically while trying to understand more about the history the Nobel Prize. Having received that prize I decided to find out more about what makes that system tick.

Vardalas: Have you found out what makes it tick?

Kroemer: I found out a lot, because I've been reading on that. Of course Einstein's Nobel Prize was an extraordinarily controversial thing.

Vardalas: Was it?

Kroemer: Yes. He had been nominated a number of times for the theory of relativity, and the people who controlled the physics Nobel Prize at the Royal Academy were absolutely opposed to this one. One of the key members of the physics committee is on record as saying, "Einstein will never get the Nobel Prize." Planck had trouble too. Anyway, this is why Einstein ended up getting the Nobel Prize for the photoelectric effect rather than the theory of relativity.

Vardalas: Okay.

Kroemer: The political shenanigans of how this was pulled off have been described rather beautifully in some books. I became interested in exactly what Einstein's role was outside of quantum mechanics and relativity because of his role in thermodynamics.

Vardalas: Einstein had a role in thermodynamics?

Kroemer: Absolutely.

Vardalas: I see. Though I studied Einstein, I was not aware of that. From this kind of exploration you've been doing, do you feel that it is important for physics majors to take a course in the history of physics to understand the development of their profession?

Kroemer: I don't know how important it is. I certainly do not believe that physics should be taught in the historical order at all. The historic development of the field of physics contains a staggering number of blind alleys.

Vardalas: Isn't that important for people to understand?

Kroemer: It is not important for understanding physics. However it is important in a cultural sense. I personally am fascinated by the history of physics – by all the blind alleys and by all the mistakes that were made. And it's a story in human culture. However I am opposed to teaching physics, particularly quantum mechanics, in a historical context.

Vardalas: But as a course in history?

Kroemer: Yes

Vardalas: Would you recommend that physicists also take a course in the history of physics?

IEEE History Center - Herbert Kroemer Oral History

Kroemer: I wouldn't want it to be a required course, but I would highly encourage taking such a course taught by someone who knows what he or she is talking about. Friedrich Hund, who I mentioned earlier, wrote a beautiful book on the history of quantum mechanics. I do not know whether it is available in English translation. I read that one twice, which is something I don't do very often. He actually argued that the natural road to quantum mechanics was through thermodynamics rather than through spectroscopy.

Vardalas: Through thermodynamics?

Kroemer: Yes, and I never accepted this thesis of Hund's – until I read the Einstein papers.

Vardalas: Really? That is the conventional wisdom that the road to quantum mechanics was through spectroscopy.

Kroemer: Yes. Max Planck's blackbody radiation was an exercise in thermodynamics.

Vardalas: I see.

Kroemer: The idea was that electromagnetic energy might be quantized, which Planck himself did not accept initially. That of course blocked his Nobel Prize for many years.

Vardalas: It did?

Kroemer: It was obviously ridiculous. All one had to do was look at the diffraction experiment to realize it was a continuous wave. This is the resolution of this one. Einstein was simply the first one who decided to simply ignore the discrepancy.

Vardalas: You just mentioned earlier about Sauter's views about degrees – not serving time but proving a certain capability to execute independent work. Do you take this approach with your students in the system that you have here in the United States?

Kroemer: To the extent I can, yes. I am of course under some constraints. There are certain rigid requirements. I encourage my students to spend at least part of their time working on projects other than their own research projects in order to get broader experience. The overwhelming majority of the Ph.D. dissertations that I have supervised were experimental work.

Vardalas: Really?

Kroemer: And that doesn't go quite as fast as a theoretical dissertation.

Vardalas: That's an interesting contradiction for someone who said he wasn't going to be a good experimentalist.

Kroemer: I think I was pretty good at identifying projects that are worthwhile, and in a semi-facetious way I would like to add that maybe having been a theorist and thereby not knowing how impossible it was to do those things being proposed has helped in actually getting them done. You see, if I had grown up as an experimentalist I would have had a far greater appreciation of the difficulties of doing the things that I wanted done than I have had as a theorist. This has probably helped me. I do not recommend this as a universal procedure. I remember for example how I got into molecular beam epitaxy.

Vardalas: How was that?

Kroemer: There was a technology that I felt was making things possible that had previously not been possible. Therefore I proposed to put gallium phosphide on silicon. And of course if I had been an experimentalist I would never have proposed to put gallium phosphide on silicon because it was too obvious that this would not work and what the difficulties were. Well, we did it.

Vardalas: I see your point. Sometimes a little ignorance is useful.

Kroemer: Yes, but you shouldn't count on it.

Vardalas: Luck is good too. Along these lines about an educational approach to what becoming a physicist means, the chief editor of the IEEE Spectrum wrote about you in the 2002 issue of that publication. In it he says, "To this day his [Kroemer's] view of education is that accumulating methodology matters more than accumulating subject matter knowledge."

Kroemer: Absolutely.

Vardalas: Would you please explain? In my own mind a methodology is a form of knowledge.

Kroemer: By accumulating I mean accumulating data facts and details. I think the question is of how one goes about solving a problem. How do you estimate what problems you have to solve on the way? How do you estimate the chance of success? This to me is far more important.

Vardalas: Can that be learned or is that something one gains from experience? Is that something that can be formally taught or is it something that can only be gained by experience through trial and error?

Kroemer: I think it can be taught to some extent.

Vardalas: What procedures do you use to teach this to your students?

Kroemer: By always insisting that whatever they are doing is not simply done following the recipes, but that they understand the rationale behind it. In this context I have a story to tell. In January of 2001 I was invited as a keynote speaker to a workshop held by the German Ministry of Research and Education at Stanford University to an audience of something like 150 to 200 German Post Docs.

Vardalas: Were these Post Docs from all fields or just physics?

Kroemer: From all fields, and under the jurisdiction of Madame Edelgard Buhlmann, who was then and is still now the German Minister of Research and Education. She was very, very much interested in university reform. At this workshop somehow it got mentioned that I had received my Ph.D. before my 24th birthday. One of the Post Docs in the audience burst out, "But you didn't know anything yet at that time." Which was referring to the deplorable tendency of having people study a lot of material rather than concentrate.

Vardalas: Just for the sake of it you have to know it all.

Kroemer: Yes. I was tempted to answer, "Well, it didn't stop me from getting the Nobel Prize," but I didn't do that. It would not have been nice. I simply said, "Listen. I had learned how to tackle a problem even if I had no previous background in the details. And I feel that was important."

Vardalas: That's very interesting. That explains the distinction to me.

Kroemer: Knowledge of detailed subject matter becomes obsolete. Why would you want to cram your brain with detailed subject matter long before you actually need it? Learn how to find it.

Vardalas: I see.

Kroemer: Richard P. Feynman once made a similar comment.

Vardalas: He did.

Kroemer: Somewhere in his writing there is a comment that he didn't see much point in reading up on what all the others had done because they had obviously not succeeded. I sympathize. I agree with his point of view. Now if you are interested in the history of the field then of course that becomes a different thing. I do not view the history of science as a tragedy; I view it as a comedy.

Vardalas: But you know tragedy and comedy are very close together.

Kroemer: Yes, I know. Maybe personal tragedy with the blind alleys and the narrow-mindedness with which representatives of the established power structure often suppress things that they don't like.

Vardalas: Do you remember the famous quote from Max Planck, "Theories succeed because their opponents eventually die off"?

Kroemer: The trouble is, it applies to him too. Yes, I remember this from his autobiography. He was advised against studying physics because there were no problems left.

Vardalas: Yes.

Kroemer: Philipp von Jolly was the name of his physics professor at München who advised him that with the discovery of the principal of energy there were no more problems left. All the rest was just working out the details. Fortunately Max Planck went on to pursue physics. He was basically a very conservative man. He struggled for years trying to overcome and undo the revolution he had started.

Vardalas: He did?

Kroemer: Oh yes. For years he tried to invent this back into a classical framework – in total contrast to Einstein. The two respected each other tremendously. Einstein seemed to enjoy overthrowing such concepts.

Vardalas: I see. In '52 you graduated, at 24 you got your Ph.D., and you wrote that there were not many opportunities for advancement in academia at this point even for a bright theoretician like yourself.

Kroemer: Yes. Zilch.

Vardalas: Zilch. Nothing.

Kroemer: First of all there were no new universities, no new departments being formed, and there was a long waiting list of people with first-rate credentials and more seniority waiting for any openings. It was sort of a dream but I ruled it out.

Vardalas: But it was something you would have loved to do if you had the opportunity?

Kroemer: Yes. I would have loved to do it. I simply considered the opportunity to be zero – which was a correct assessment for the day.

Vardalas: It turned out the other option had ramifications for you. You said that at the Central Telecommunications Laboratory you were a "house theorist."

Kroemer: That's a verbatim translation from the German. That's a common concept.

Vardalas: What did this position entail and how much latitude were you given to be house theorist?

Kroemer: The position entailed to be available to try to answer whatever theoretical questions came up. That was the minimum, but more than that, I was really expected to take an active role.

Vardalas: To seek the theoretical issues to what they're doing.

Kroemer: To seek and to try to take an active role in making suggestions. I was encouraged to actively poke my nose into the experimentalists' business.

Vardalas: How did they react to your presence?

Kroemer: I was strictly forbidden to touch any equipment, so I was not a competitor. It was a good relationship.

Vardalas: They valued your input rather than saying, "What do we need this theoretical stuff for?"

Kroemer: Another one of the things that I was expected to do – every week or every other week – was to give a lecture to anybody who wanted to come and listen on any subject of my choosing. Naturally I picked subjects that I felt were relevant to the work that was going on.

Vardalas: Was this well received?

Kroemer: Oh yes. That very often involved me having to learn new things. I had never in my university career seen a metallurgic phase diagram. I discovered very quickly when I tried to understand on what principles these recipes were based that we were using make transistors that I had to learn a little bit of metallurgic phase diagrams. Teaching others is a much better way to learn something, so I thoroughly enjoyed this work and I think I was good at it.

Vardalas: You wrote that this job was an important landmark in your development as a physicist. You commented, "I ceased to be a real theoretical physicist, if I ever was one." I took that in a positive sense in that you found yourself having to bridge between theory and practice a lot.

Kroemer: Yes. First of all I was in an environment where I was basically not interacting with other theorists. I was in an environment where I was interacting with experimentalists. I was the theoretical advisor. And of course I was encouraged to follow-up theoretical developments in the literature. I was never qualified as a "real" theoretical physicist, if I ever was one. Whatever that means. I think when it comes to a Professional Theorist with a capital P and T, I am not considered a theorist. And I do not consider this a negative assessment. The important question is not "Are you a theorist or are you an experimentalist?" The important question is, "Are you doing something that is useful? Are you contributing?"

Vardalas: Let me see if I can provoke you to say something controversial.

Kroemer: In general it's very easy.

Vardalas: Do you feel that in some areas of physics there tends to be an overemphasis on theoretical formalism in pursuing theoretical issues?

IEEE History Center - Herbert Kroemer Oral History

Kroemer: I don't know whether it is characteristic of the area. It's probably characteristic of some individuals in all areas. Some areas of course lend themselves more readily to this than others, but I would not want to disparage any single area of physics in particular. And I shouldn't comment on areas that I don't understand anyway. But it certainly is clear that in all areas you find a broad range of people with different interests, from the pure empiricist experimentalist technologist who has not the foggiest idea of the underlying theoretical principals and not caring. Then there are people at the other extreme who see only mathematical formalism.

Vardalas: They like elegance.

Kroemer: They like elegance. I like elegance too.

Vardalas: But elegance for the sake of elegance as opposed to elegance for the sake of a physical principal.

Kroemer: Yes. It is a broad spectrum.

Vardalas: I was thinking more of the areas of the Grand Unified Theory.

Kroemer: I find this a fascinating human exercise. I can see why it appeals to ambitious young people who are doing theoretical analyses that this is the great problem to be solved. I understand this. The question is what do they do when they realize that there is a fierce competition? Thus far no one is making any progress. I wish they were a little broader.

Vardalas: There are a lot of shipwrecks on that rock.

Kroemer: My recommendation to somebody who really is interested in these deep profound principals is sure, do that, but do not restrict to yourself to just that. View this as a part embedded in something much broader.

Vardalas: I see. Getting back again to the Central Telecommunications Laboratory, in Germany its acronym is FTZ. What does that stand for in German?

Kroemer: Fernmeldetechnisches Zentralamt. Fernmelde is telecommunications, technisches is technical, zentralamt means central office.

Vardalas: Okay. I know you answered this before but I have to ask you this anyway. Did you have close interaction with these experimentalists and technologists?

Kroemer: Yes.

Vardalas: Do you think this sharpened your skills as a physicist in a way you would not have gotten if you had stayed in academia?

Kroemer: Probably yes. I don't know whether it sharpened my skills as a physicist, but it certainly broadened the range of things I was thinking about. It was again methodology, and I was taking an active interest in how one could do this, how I could implement this. I was taking an active interest in trying to understand first of all what we had done and then going on from there and saying, "All right, now that I understand, how could I modify it to do better?" It was like a pendulum swinging back and forth. That has a great influence on me that continued in my subsequent jobs.

Vardalas: Would you say that in a sense this experience brought out your versatile talents as an astute problem-solver -- what you call an opportunist? Was this the first seed of you acting like an opportunist?

Kroemer: Yes.

Vardalas: In a good sense.

Kroemer: Yes. It always shocks people that I call myself an opportunist, because it's always a dirty word. But I deliberately use it – perhaps for the shock value – because an opportunist is somebody who is looking for good opportunities to do something.

Vardalas: The right problem. Yes. I want to get to that, because that's an important theme - choosing the right problem. I want to ask you a broader issue here on the different levels of knowledge and methodology. In your career your research has bridged the realms of theoretical physics with a small "t", applied physics and whatever that means and electrical engineering?

Kroemer: Yes, although I am not really an engineer. My degree in engineering is honorary.

Vardalas: Your understanding of some of these issues has been an important part of your work. Looking back over the growth of science and technology in semiconductors, how would you characterize the body of knowledge in each of these areas and how they interact at the interfaces between theoretical physics, applied physics and electrical engineering? And has that direction changed over the last fifty years?

Kroemer: I don't quite know in which terms to answer that one.

Vardalas: After all, you are in an engineering school now.

Kroemer: I am in an engineering school and I feel very comfortable in an engineering school because I bring to bear my scientific background to solve problems and to contribute to engineering developments. However I am basically still a physicist. Throughout my entire career whenever I was working on something and then discovered, "Hey there's something else you have to learn in order to able to do this," I have always tried to acquire this knowledge. Therefore I have a background – though an extraordinarily sketchy background – in electrical engineering. I know those parts of electrical engineering that I need and I know those parts of theoretical physics that I need and those parts of mathematics that I need. I think the field as a whole has required and has benefited from the broad assortment of people in all of those particular disciplines from theoretical physics to metallurgy to you name it.

Those laboratories where the important contributions were made typically had organizations with a broad range of people from different backgrounds – to the point that an individual's background was often not known or even a concern. I certainly learned that in the three years I spent at RCA Laboratories.

Vardalas: Where the problem is the important thing. How the problem is tackled and one's background doesn't really matter.

Kroemer: That's right.

Vardalas: That's interesting. Is that something that you feel has implications as to how universities should train people in these fields?

Kroemer: To some extent this is a matter of education at the university, but to a large extent it is also a matter of personality. I believe that an in-depth education in some specific discipline is important.

Vardalas: Yes.

Kroemer: In my case it was in theoretical solid-state physics. All the other things got added on later. I do not believe in giving people a superficial education in a lot of different things. An in-depth education in the specific area is important, but it also requires an attitude to be interested in what goes on in adjacent fields, trying to interact with people in other fields, trying to contribute to their experiments or projects and having them contribute to your projects.

Vardalas: I guess the issue then is not the knowledge but the attitude to look to the other people.

Kroemer: Yes. For example I think it is important that physicists not take the attitude that pure physics is something more elevated than applied physics. One sometimes finds that attitude. Then in engineering the opposite attitude can be found which looks down on anyone interested in theory or in something for which an application is not already known.

Vardalas: Yes.

Kroemer: You know how hostile I am to that idea.

Vardalas: Yes. I'll get to that. That's interesting.

Kroemer: I think universities can contribute to this in two ways. They can obviously contribute by providing the specific knowledge that is required. There is always hope that along with that specific knowledge there will be instilled a desire to do more than just a specialty. It is desirable that we combine these two aspects and see this embedded in a broader framework. Of course this will depend on the individual, the faculty and the university environment.

Vardalas: Am I correct in judging from what I read that the embryo of your ideas on heterostructure bipolar transistors first came to you while you were in the Central Telecommunication Laboratory?

Kroemer: That's right.

Vardalas: And then you write "and the rest is history."

Kroemer: Yes.

Vardalas: How did the idea come to you? Was it a natural progression of your dissertation?

Kroemer: It had nothing to do with my dissertation.

Vardalas: Was it the environment you were in that prompted you to think about this?

Kroemer: Yes. We were working on the very early transistors. They were so slow that they were basically useless for the applications on hand at the time. I realized that an earlier incarnation of transistors – the first so-called point contact transistor that didn't have junctions inside – talking about bipolar and not FETs - were significantly faster than the first junction transistors. And there were a variety of reasons for that. One of the thoughts that came to me was, "Well, we do know that in the point contact transistor the collector is very leaky." In other words it draws a rather large current by itself without an injection of holes or electrons or whatever it is from the emitter. Of course this leaky collector film introduces an electric field in the body of semiconductor around the collector in such a way that carriers of the opposite polarity were drawn towards the collector. So that was my point of departure. I'd say, "How could you build an electric field into a junction transistor?"

Vardalas: That was your key idea?

Kroemer: That was the key idea and not by doing s leaky collector. The first idea then was, "All right, we are putting a non-uniform doping into the base," and specifically I looked at an exponential doping profile. You can then show that this leads a built-in field that speeds up the carrier. And that required an understanding of band structure.

Vardalas: This is where theoretical issues came into play.

Kroemer: This is where my understanding of basic semiconductor physics told me that it should work.

Vardalas: I see.

Kroemer: Then at one stage the thought occurred to me that another way of putting in a field is to use a non-uniform energy gap. And this is the theory.

Vardalas: Okay. Did this idea come immediately after the first idea? How did this idea emerge?

Kroemer: I don't know what immediate means here.

Vardalas: Did it take a year?

Kroemer: Less than a year. It was part of the same work.

Vardalas: Okay.

Kroemer: The idea came then that a field could be built in by grading the energy gap.

Vardalas: Did something prompt you to think of it?

Kroemer: It was obvious.

Vardalas: It was obvious. Why hadn't people thought of it earlier if it was obvious?

Kroemer: It was not obvious to others. Well, let me say it was obvious because I had a goal. I wanted to put in a field, I needed a sloping band and I realized that I could create a sloping band. You see I always try to view those things from as fundamental a point of view as possible. The need for a band slope was the key idea. The field needed a slope in the band and I realized fairly quickly that a second way to introduce a slope in the band was to grade the energy gap.

Vardalas: Okay, so that's the succession of ideas. Once you committed to a sloping electric field then that led to the next idea.

IEEE History Center - Herbert Kroemer Oral History

Kroemer: Yes. There is a comment in that first German paper on that one and I estimate what kind of a potential drop could be obtained in the band. It was a bit optimistic. These things are always optimistic. I looked specifically at germanium-silicon, not realizing that there were very, very severe problems there. Anyway, my technological colleague, Mr. Hähnlein, was basically the physicist working on the device technology. He looked at this and said, "There is no way I can do it. The most I can do is perhaps put an emitter with a wider energy gap on the base region, but I cannot put a control field in the base." Of course that would mean a uniform gap in the base region – and that would mean that the field that I was trying to achieve would not exist. So that idea was out.

But then on the way home, out of sheer curiosity, I wondered, "Well, what would be the consequences if that was done?" And I realized that this would have unique benefits of its own. The repulsive barrier at the emitter side would be increased for those carriers flowing from the base back into the emitter. That could now also be traded off with other things. This is how the wide gap emitter and the graded gap arose within days of each other. They are both in that first German paper published in 1954, so the idea probably arose in late '53 — though, in that paper it was unfortunately not accompanied by a band diagram. The idea that one has to be able to draw a band diagram came shortly after that.

Vardalas: It seems like you had considerable freedom at this place.

Kroemer: I could work on anything I pleased.

Vardalas: And you found the work quite stimulating I imagine.

Kroemer: Yes.

Vardalas: And yet you wanted to move on.

Kroemer: Well, I wanted to get involved closer to the action.

Vardalas: I gather you first thought of the opportunity with William Shockley.

Kroemer: Shockley visited our place and I had a long and wonderful long discussion with him. I asked him about the chances of coming to Bell Labs at that time. He was reluctant in his response, because he was an official visitor. Official visitors are not stealing people.

Vardalas: Oh yes, of course.

Kroemer: I think his reluctance regarding such matters later decreased. He said, "You will have to take the first step." And then there was Ed Herold from RCA whom I had met at the Physical Society Meeting in Innsbruck. He didn't even know about the existence of our laboratory but I got to have a long discussion with him. He had presented an invited paper which pretty much confirmed everything I had claimed about how what was going on with the metallurgy that some people didn't believe. It turned out he had a day open on his agenda and so he came also to Darmstadt to the FTZ [Central Telecommunications Laboratory] and I had a long chat with him. When I asked him about the chances he was not under the constraints that Shockley had. A good question is why I ultimately went to RCA rather than Bell Labs.

Vardalas: Yes.

Kroemer: That's a very, very good question. In fact Bell Labs was offering more money when it finally came to the details.

Vardalas: And what was it?

Kroemer: I don't know. Jim Early said to me, "Herb, you may have felt that at Bell Labs you would always be in Shockley's shadow whereas at RCA you would be your own man." That may have played role in my decision. I do not know. That is speculation. This is sort of tabletop psychiatry.

Vardalas: Well, let's go forward. What were your impressions of the people in research when you got to RCA Labs?

Kroemer: Oh, it absolutely fabulous.

Vardalas: Can you recreate what you saw and felt when you first got there?

Kroemer: It was simply a different world. There were lots of people, many of them very good people, working on all sorts of aspects. It was an environment where free discussion across disciplines was very, very much encouraged. A theorist who wanted to do an experiment was not talked out of it because people saw no need for it. I said, "All right. Go ahead." It was a wonderful lab and I was surrounded with wonderful people. The ones who influenced me most were perhaps not the ones who were best known.

Vardalas: Who were they?

Kroemer: One particular guy who influenced me tremendously was an office mate named Lou Pensak. He had been heavily involved in the technology of television tubes. He was a constant discussion partner and also introduced me to the idea that if you want to build something the right tools are needed – and the tools should be built first. He had a great influence on me. It was a fascinating laboratory. Ed Herold was a wonderful boss. It was a great experience.

Now this discussion with Ed Herold in Darmstadt was interesting one. You see I understood the metallurgy of those transistors in those early days. They were pnp transistors that had indium alloy to two sides of a germanium wafer. I was curious about npn transistors. I tried to think about how might one do an npn? I couldn't get anyone interested in actually doing it, but I sort of figured out what I would do if I were asked to build an npn. So I asked Herold whether they had also made npn transistors. He said yes. I asked, "What did you as the alloy metal?" He was a little bit reluctant but answered, "Lead." To me it was clear that it was either lead or tin, and for reasons that I do not remember I thought lead was the likely candidate. I said, "But lead is not a donor, so you must have added something else." Long pause, more reluctance. "Antimony," which I thought all along was they had. Then I asked, "Well, how much antimony and at what temperature did you alloy it?" He clammed up. So I told him "You use 9 percent antimony and alloy it at 600°." And his jaw dropped. It was sufficiently close to what they were actually doing. It was a living exercise in the old rule "Never mind *how* something is done. Knowing *that* it has been done is the biggest secret of all."

Vardalas: Is that what convinced him to hire you?

Kroemer: I don't know whether that convinced him, but certainly that was an interesting exchange we had that I fondly remember.

Vardalas: When you went to RCA did you go there with your own research agenda they had approved?

Kroemer: No. Well, I came with certain ideas.

Vardalas: What did they want you to do? How did this process work?

Kroemer: They gave me an office and told me to work on whatever I wanted.

Vardalas: Oh, that's it?

Kroemer: Yes. Herold explained to me later that when he hired people his interest was in how good they were at whatever they had done previously. What field it was didn't matter to him. His concern was only for the quality of their work. His next question would be, "Is this person interested in working on the problems that I can offer?" If the answer was also positive, he would take the person. He would rather take an individual like that who had no idea about the field than somebody who had previous experience in the field but did second rate work.

Vardalas: That's interesting.

Kroemer: He explained to me, "Listen. In the environment of our lab, the kind of people that I hire like to communicate. If I put them into this lab they are sitting in an intellectual feedback loop, where they cannot help but be influenced in being directed towards the topics in which the lab is interested. They are not going to go off on a wild tangent." I would have been permitted to do so. In fact later on I did go on a fairly wild tangent, but even then it was encouraged on the grounds, "Well, maybe something will come of it."

Vardalas: I see. In a place like this where people were allowed to do what they wanted to do, within I gather rather broad constraints, how does accountability work? How did they decide whether to keep people on and determine whether people are good or not good? By the results?

Kroemer: Basically people were judged by the quality of their work. And it was encouraged to go out of the established conventional wisdom.

Vardalas: Was it a healthy environment where failure was encouraged? If you tried and failed was that not seen as a negative thing? In some places that would be treated as a very negative thing like, "That's terrible. We don't want that."

Kroemer: No. I think failure was encouraged. I remember when I was working on a particular project, I was working very hard and it eventually became clear to me that this was not going to work. It had been clear to others long before that it wasn't going to work, but there was no influence on me. I do remember one day in the morning I drew my conclusion that it was not going to work. I cleaned up my workbench, cleaned up my desk, went to my supervisor and said, "Harwick this is not going to work. I'm going to change subjects." He said, "Yes? Well, that's fine. We knew you would come to that conclusion." That was it. There was no pressure. It was an experience.

Of course I was the kind of person who was willing to drop the subject. There are sometimes people who do not want to let go of it. The dumbest reason for continuing something is because you have made the mistake of starting it. I have always been willing to drop projects. In fact I have never been interested in milking the last bit of juice out of something. If I had succeeded in achieving the key point, suddenly I would discover I had lots of friends who were willing to do the rest. Why not let them do it?

Vardalas: Did you have an opportunity at RCA, the kind of problems, to go back to heterostructure bipolar transistors?

Kroemer: Actually that is something I did try to do. In fact that was the project I decided to drop. The technology was staggering, the equipment primitive. And remember, by that time I knew more metallurgy that most theoretical physicists. So I came up with the idea of putting a silicon/germanium alloy on a germanium base. This was all on germanium bases. How do you do this? I knew my phase diagrams quite well, so I made a silicon-gold eutectic. Silicon-gold has a relatively low melting point 360°C or something like that. I made the silicon-gold eutectic which is an unpleasantly brittle substance, put it on an anvil and smashed it into a powder, and with a pair of tweezers picked up little grains and put them on the germanium wafer and alloyed it at 500 or 600°C, the eutectic would melt, it would eat up germanium and the germanium would then recrystallize. And we had added some dopants.

Vardalas: You did this yourself?

Kroemer: I did the experiments. I didn't have a technician. I did all of this myself. My one complaint at RCA is that they did not give me a technician. I think they had the right instinct. I didn't know how to use a technician.

Vardalas: That must have taught you new things trying to do this stuff yourself.

Kroemer: Yes, yes. This is where Lou Pensak's advice on how one does things came in. We actually made some transistors, but it was clearly a technology that would never amount to anything. So this is that and I dropped it. At that point I decided to get out of transistors and move to something altogether different. That must have been in late '56. I did write two papers, and those papers played an important role.

I then became very much interested in hot electron transport at low temperatures – which is something totally different.

Vardalas: When you look back at all the work you did at RCA, of what are you most proud?

Kroemer: In hindsight I see that those two papers that I wrote clearly had the greatest impact. But I did other things that were at the time certainly important to me.

Vardalas: What gave you the most satisfaction at the time?

Kroemer: I don't know. I don't want to single out any specific things. Of course by hindsight it's very clear the heterostructure work was the important one. And that paper in RCA Review was probably one of the most papers I ever wrote. I made a mistake by publishing it in an obscure journal, with the result that no one read it. However Zhores I. Alferov, the Russian with whom I won the Nobel Prize, read it. He knew the paper.

IEEE History Center - Herbert Kroemer Oral History

Vardalas: I see. I would have imagined that the RCA Lab's journal would have been read by a lot of people just like the Bell Labs' journal.

Kroemer: That was RCA's idea. The RCA Review was sort of an imitation of the Bell System Technical Journal, and it never really acquired the same following.

Vardalas: I see.

Kroemer: It turned out to be a place where a lot of papers were put if you couldn't get them published elsewhere.

Vardalas: Did you think this was an important paper when you were writing it?

Kroemer: This is hard to say. In that paper I clearly spelled out the heterostructure design principal. If I were to rewrite that paragraph that I quoted in both my Nobel lecture and elsewhere I would clean up the English a little bit, but the idea is clearly there. I realized the power of the idea. What was not clear to me at all was how important the idea would become. The real triumph of this idea is the heterostructure laser, which draws exactly on the concepts that were outlined in that RCA paper. Incidentally, this is something very typical that the fruitfulness of a new idea will not come until later.

Vardalas: Right.

Kroemer: There is a wonderful quote by David Mermin in Physics Today a few years ago. He says something like – you have to look up the exact wording – "I'm looking forward to the day that people realize that discovery does not work by deciding what you want and then discovering it."

Vardalas: Isn't that in your Nobel lecture?

Kroemer: I quoted that in my Nobel lecture.

Vardalas: This has implications for those who say funding for science should be based on what the applicant—

Kroemer: That is total nonsense.

Vardalas: Is there anything to be gained by trying to get scientists thinking about problems that have some obvious need and utility in society? In other words even in theoretical issues like attacking an important problem and applying good science to it?

Kroemer: This flows two ways. On the one hand people who have problems that they want solved can and should try to get more fundamental-oriented people interested. I think we all acknowledge that one. However on the other side – and I can only speak for myself – I have always been interested in fundamental principles, but whenever some conceptual advance had been made I have also always asked myself, "What could be done with that?" I did not restrict myself in my explorations to things where I saw applications beforehand, but I periodically asked myself, "What kind of applications might this have?" Sometimes I could outline some and sometimes and I could not. And if I could not, I still went on. It didn't stop me pursuing a certain thing.

Vardalas: You gave applications some thought.

Kroemer: I always gave it some thought. Critical thought. One has be honest about it and not come up with something where a few more minutes of additional thought would show it to be nonsense.

Vardalas: I think you spent three years at RCA Labs.

Kroemer: Yes.

Vardalas: Then you returned to Germany to head up the Phillips Semiconductor Group in Hamburg.

Kroemer: That's right.

Vardalas: After reading some of the things you wrote, I was wondering if being homesick was one of your reasons for returning to Germany.

Kroemer: Yes.

Vardalas: Was there something more to it than that?

Kroemer: No. It was simply homesickness – on the part of both my wife and myself, though more my wife.

Vardalas: Is your wife German?

Kroemer: She is from Berlin.

Vardalas: We never discussed when you met your wife.

Kroemer: I met her when I was a student at Göttingen.

Vardalas: Okay. So she was homesick to get back. She must have left Germany with you almost immediately.

Kroemer: She followed me after a year. She stayed behind because we had a young child. She stayed behind with her parents. Then she followed me, but then she was terribly homesick. And I was offered what looked like a rather attractive job at Philips. So I went back. I must say that I have nothing but good things to say about the Philips Company. They treated me wonderfully.

Vardalas: What was the undertaking?

Kroemer: They had a new research laboratory in Hamburg – a lovely city incidentally – and I was the head of the semiconductor group. I decided to steer this group towards gallium arsenide. That was in 1957. I got the people in Eindhoven who were a bit astonished and saying, 'Yes, why not?" One person was assigned to do this technology. When I decided to go back to the United States this project was shut down.

Vardalas: Why did you leave? Were you unhappy with the way things were going there?

Kroemer: I had no problem with Philips. Germany simply was not the same. Our perception of Germany changed.

Vardalas: What does that mean?

Kroemer: My wife and I both suddenly realized that we wanted to go back to the United States. That was basically it. Philips tried all sorts of things to get me to stay. They were very nice about it. It was a reverse homesickness.

Vardalas: You had to leave the U.S. to realize that you wanted to return to the U.S.

Kroemer: Yes, so I book this under mental health expenses.

Vardalas: Did you think you did anything in terms of your own professional development or ideas in physics there?

Kroemer: While I was in Hamburg? No. Nothing.

Vardalas: That must have been a disappointing time for you professionally.

Kroemer: Well, I did some thinking and I think I wrote a couple of theoretical papers, but it does not show up in my productivity resume.

Vardalas: You came back to the U.S. and Varian Associates. Why Varian?

Kroemer: I knew the head of central research at Varian, Lou Malter. I knew him from RCA. And the people at RCA actually wanted me back.

Vardalas: Oh, they did?

Kroemer: Yes. I don't really quite know why I didn't want to go back to RCA. I asked Lou Malter a little and he said, "Oh yes. Sure. Come." And we never agreed what I would work on.

Vardalas: You never did?

Kroemer: We never talked about it. He said, "Just come. There are plenty of interesting things around." He knew me well enough that he realized it would probably work. And I spent several years then at Varian Associates trying to build up a semiconductor activity there.

Vardalas: You were doing the diode laser, right?

Kroemer: Well, this is how the idea came about. I was actually not allowed to work on it.

Vardalas: Yes, I wanted to get to that. How did you come upon this topic?

Kroemer: I had worked on what we now call heterostructure bipolar transistors.

Vardalas: And you put it aside for a while.

Kroemer: I put it aside, yes. I think it was in 1962 at the Device Research Conference in Durham, New Hampshire, which was the big annual device meeting, that all hell broke loose about semiconductor lasers. The first gallium arsenide laser was reported and that dominated the conference. And I really wasn't interested in those things. I knew that the theoretical principle permitted that and I was still astonished that it actually worked, but I was totally occupied with something else. But a colleague of mine, Sol Miller, was also at that conference and he took a deep interest in it, and so he started working on that one after we returned to Varian. And it must have been in March of '63 that Ed Herold, who had come from RCA, the director of research at Varian.

Vardalas: Ed Herold from RCA?

Kroemer: The same Ed Herold. Lou Malter had hired him too.

Vardalas: He was the director of research?

Kroemer: He [Herold] was not director but vice president of research. Anyway, he insisted that things get done RCA style, and that included weekly colloquia. Therefore all of us had to give talks – which I think is a good idea. Sol Miller was asked to give a talk and he picked that laser topic. He gave a beautiful talk pointing out all that had been done and also that these things didn't work at room temperature and didn't even work continuously. It required very, very short pulses, and certainly a low temperature had to be used.

At the end of the talks Herold said, "That's all very nice, but what are the chances of getting this to operate cw and at room temperature? Because that's where the applications are." And Sol Miller replied suddenly something to the effect that, "No, this has been looked at." Then he quoted some paper. I do not want to repeat the quote, but the quote said that this was fundamentally impossible.

Vardalas: Fundamentally impossible.

Kroemer: Ed Herold was not about to put up with the statement that something was fundamentally impossible without explaining why it was fundamentally impossible. Then Sol basically gave an explanation that boiled down to that, "You first of all need a population inversion, so you need a degenerate doping on both sides, and if you bias it to the point that you actually get stimulated emission the electrons leak out to the p-type side very, very rapidly. Holes leak out to the n-type side."

Vardalas: You can't maintain the population inversion?

Kroemer: You just cannot get a decent population inversion except at low temperatures where the statistics is in your favor or pulse where you have transient effects. And I do not know whether Sol Miller was finished, but I certainly said, "That's a pile of crap."

Vardalas: In those words?

Kroemer: In those exact words. All you have to do is put a wider energy gap on the two sides. It was obvious. The moment I was told that there was a problem the answer was obvious because at that point I had been thinking enough about heterostructures. Anyway, everyone was astonished and we did a number of things. I wrote a paper on this one and submitted it to Applied Physics Letters.

Vardalas: Yes.

Kroemer: They rejected it.

Vardalas: Before you go on, do you have any idea why it was rejected?

Kroemer: All of this correspondence got lost.

Vardalas: Do you recall?

Kroemer: I do not remember exactly why. Ed Herold, who was a big shot in the IEEE, did not like the idea that I had submitted it to Applied Physics Letters in the first place. "Well, send it to *Proceedings of the IEEE*. They will publish anything." That is of course why I hadn't wanted to send it there. The letter section there was not very good in those days.

Vardalas: Do you think that was why that paper was ignored?

Kroemer: I submitted it, it was accepted, and it was published. No one read it. A reviewer had pointed out to Panish and Hayashi, who subsequently did this, that this paper of mine existed. When they published they were gentlemen and acknowledged the idea. And we wrote a patent.

Vardalas: Yes. That patent was assigned to Varian, wasn't it?

Kroemer: Yes. It has safely expired. We wrote a patent. I wasn't allowed to put a band diagram into the patent because the head of the patent department was an electron tube man who did not understand semiconductors. He wouldn't put anything into a patent that he himself did not understand.

Vardalas: Oh my goodness.

Kroemer: He argued however that science really does not matter; all that matters is that the correct prescription of what to do is given.

Vardalas: For a patent, yes.

Kroemer: That is technically correct, but if you look at Bill Shockley's patents there is always a very explicit elaboration on the science.

IEEE History Center - Herbert Kroemer Oral History

Vardalas: Was your work in this paper purely theoretical?

Kroemer: Purely theoretical.

Vardalas: There were no prototypes, no fooling around with experiments, it was all theoretical arguments?

Kroemer: It was all theory. And in fact it wasn't at all clear how one would go about building one. We didn't really have the technology. We got a number of criticisms. There were a few people who said, "Your physics is wacky. It doesn't work." But that wasn't really what bothered me. I had been through this sort of mode of operation before and had learned not to pay attention unless everybody tells me that it does not work. I prefer to rely on my own judgment. I was convinced the physics was right. The other argument of course was, "There is no technology." And that was true. But then came the killer. That was, "There is no point in developing the technology because this device will never have any practical applications." End of statement.

Vardalas: Was that the reason why Varian did not want to pursue this further?

Kroemer: That's why. They didn't say, "It's not in our business field." They said, "It has no applications."

Vardalas: Did you fight this or did you give in to it?

Kroemer: I gave in. I probably would not have given in if the Gunn effect had not come along at the same time.

Vardalas: What is the Gunn effect?

Kroemer: The Gunn effect is the phenomena where if one takes a piece of gallium arsenide, applies a high voltage, then under certain conditions high-frequency oscillations result. That was fascinating physics and I was the first to offer an explanation for the Gunn effect. I worked on that for a number of years. That was my consolation prize for not having been able to work on the laser. Well, I should not say consolation prize. It was sort of an alternative since I could not work on the laser. And we had the technology for the Gunn effect. In a way of course this is regrettable, because as a result of this work I never played any role in the subsequent realization of the DH Laser.

Vardalas: Is that a source of disappointment for you?

Kroemer: A little bit, yes.

Vardalas: You also mentioned that as an alternative to this rejection of the laser rekindled your longstanding interest in high-electron negative resistance effects.

Kroemer: This *is* the Gunn effect.

Vardalas: I see. But you had worked enough for your doctorate dissertation in this broad area.

Kroemer: In the broad area, but the Gunn effect is something quite different.

Vardalas: Okay, but you say "with a longstanding interest in this."

Kroemer: I had a longstanding interest in transport properties under obscure or unusual conditions. In my dissertation I worked on hot electron phenomena. During my last year at RCA I was working on hot electrons and hot holes in germanium. One of the reasons why I became interested in gallium arsenide was because theory predicted that the thing that we were looking for in germanium would be much easier to find in gallium arsenide.

Vardalas: Okay.

Kroemer: That influenced me in starting a gallium arsenide project. Then I went to Phillips, and we wanted to look at high field transport properties as the first application. We did not have transistors or lasers in mind. Typical for my kind of thinking, I was simply convinced that the three-five compounds held a tremendous amount of future promise. So let's be amongst the ones to do it.

Vardalas: Did you go to Phillips and pursue this idea of transport issues?

Kroemer: That's not why I went to Phillips. I wanted to do something that they were not doing in the main lab in Eindhoven. And I wanted to do something that was at the forefront of solid-state technology. I wanted to get into compound semiconductors and I wanted to work with gallium arsenide specifically.

Vardalas: You spent ten years on research and engineering around the Gunn effect. Would that be where your biggest theoretical accomplishments lie?

Kroemer: I certainly have more papers in that area than in anything else. It was a significant accomplishment. I would say I was one of the handful of leaders in this field from day one, and it was only when I came to Santa Barbara that I put an abrupt end to it. I decided to do that.

Vardalas: Was your Nobel Prize award related to the Gunn effect?

Kroemer: No. It was for the development of heterostructures for high-speed- and opto-electronics.

Vardalas: Thinking back, do you think the reason they gave this science such a high significance was because of the practical technological effect it has had?

Kroemer: Absolutely.

Vardalas: And that if that hadn't happened they probably would not have given you that award?

Kroemer: Absolutely. Yes. The 2000 Nobel Prize in Physics really was a break with the tradition. The tradition for ninety years had been to award the prize only for discoveries, even though Nobel's will specified *discoveries or inventions*. There are certain reasons why the restriction to discoveries was made around the time of World War I.

Vardalas: Why was that?

Kroemer: It probably had to do with the Nobel Prize for Gustaf Dalén in 1912 which was purely a technological invention. That caused a great deal of protest. Many felt that the Nobel Prize in Physics should not be for something like this. If one simply looks at the Nobel awards, they are almost all for discoveries and those that were inventions were typically inventions essential for research discoveries.

Vardalas: Right. Okay.

Kroemer: The bubble chamber is a good example.

Vardalas: Instrumentation.

Kroemer: Instrumentation. So when friends and colleagues mumbled about me getting it someday, I looked at the statistics and took the attitude, "Well, they do not know the rules of the game." That was probably good for my mental health not to take those comments too seriously. Then, since people had been talking about it I sort of was thinking about the possibility and thought to myself, "Well, they are not going to give it to the HBT. They are not going to give it to me for the laser. I may have been the first one to spell out the idea, but I did not do the first laser. If however they decide to give it for the heterostructure concept, then I have a chance and I will probably share with Alferov."

Vardalas: You said that to yourself?

Kroemer: Yes, I said that to myself. I never talked about this to anyone. I think you are the first one to whom I have told this. And this is of course what happened. But it was clearly a break with tradition.

Vardalas: But yours is less of a break than it was for them to give it to engineers like Jack Kilby.

Kroemer: Yes, for Kilby it was even more so a break with tradition. However I think it was well deserved. They must have gone back to read Nobel's will.

Vardalas: Okay. I want to cover some more ground before our time runs out. I am very interested in your interpretation of Moore's Law. You gave a plenary talk called "Speculations about Future Directions."

Kroemer: This must have been the MBE paper published last year. All right.

Vardalas: In it you say, "Moore's law reflects the triumph of parallel assembly."

Kroemer: Yes.

Vardalas: Would you please elaborate on that?

Kroemer: "Moore's law is an observation; it's not a law.

Vardalas: Of course. Yes.

Kroemer: It is an observation that spans several decades. The number of devices, per chip and per processing step processed in parallel has increased exponentially or approximately exponentially. In order to be able to do this a reduction in dimensions was necessary. But the reduction of dimensions was an enabler. If the dimensions were reduced and then these devices were done serially, one at a time, then you would not have Moore's Law. This is not going to make a Pentium type chip doing one device at a time. This is my perspective on Moore's Law. In that sense it is a triumph of parallel assembly. I raise this issue periodically whenever people say, "As the dimensions get smaller quantum mechanics will become more important. We will build a quantum device then. Once we have a quantum device Moore's Law continue for a little bit longer." But actually quantum devices are all put together serially.

Vardalas: Aha. That's your point.

Kroemer: This is the wet blanket that I put over the subject. Moore was extraordinarily perceptive, because he didn't formulate it this way but he realized the trend and that we were very, very far away from physics limitations. In my perspective, Moore's Law was based on the development in the infrastructure. If one asks, "Why didn't we do it right away?" that is a very easy question to answer: there was no infrastructure, instrumentation, manufacturing, and there were no crystals of the proper perfection.

Vardalas: I'm interested in this in my own research. Someone told me that Moore's Law is essentially a scaling up issue – not indefinitely, but one can scale up in this way.

Kroemer: A straight line cannot be drawn on log-paper forever.

Vardalas: Yes. But I had the impression that one of the reasons it really took off was the emergence of CMOS; that CMOS technology is geared to this kind of thing whereas bipolar is not.

Kroemer: Yes, that is true. Let's put it this way. Take cause and effect the other way around, it is only with CMOS that we could do it. And of course Moore predicted CMOS.

Vardalas: I'm interested in the relation with bipolar technology.

Kroemer: The principal reason this could not be done with bipolar is that thermal load is unbearable. Bipolar is a hot plate.

Vardalas: Okay. Bipolar cannot scale up the way CMOS can.

Kroemer: No.

Vardalas: I see.

Kroemer: Bipolar is obviously very, very important. Your cell phone is loaded with bipolar. Certainly on the transmitting side it's bipolar. Probably HBT is on the transmitter side, and it may be HBT is on the receiving side.

Vardalas: I was thinking in terms of computer development, because at one point a bipolar was a dominant form of device in processors.

Kroemer: Yes. Big mainframe computers used to run on ECL, Emitter Coupled Logic.

Vardalas: Yes.

Kroemer: And there is not a device that generates more heat per function than Emitter Coupled Logic.

Vardalas: I'm writing a book now, Control Data Corporation, and they were so wed to the bipolar because of the speed advantage they had.

Kroemer: Yes. They had a speed advantage, but that advantage is gone.

Vardalas: Cray didn't stop using bipolar until the late 1990s.

Kroemer: Yes. And of course as the devices got smaller they got faster too.

Vardalas: For someone who is not in the field, does the value of the model of heterostructures have an equal technological importance for the CMOS device?

Kroemer: No. At least I do not see it. I think there is a good reason why CMOS continues to be silicon, though we may see silicon-germanium in silicon technology. I know that many of the silicon houses are working on silicon-germanium technology though I do not know any details. In that sense that is heterostructure.

Vardalas: One of the traditional explanations for the rise of silicon was its importance to the military in terms of its stability to damage.

Kroemer: Yes.

Vardalas: And that is the reason that silicon can dominate more than germanium.

Kroemer: Certainly the military were the ones who initially supported it and helped it get off the ground, but I think what was probably more important was organizations like Bell Labs that realized that germanium was too limited.

Vardalas: Too limited in what respect?

Kroemer: Temperature capability, processing capability. Jack Morton probably did more than anyone else to promote silicon technology early in the game when it was still largely bipolar. Then of course there is the importance of that oxide. Silicon is a rather unusual material. All the attempts to imitate silicon in compound silicons have in my opinion been a waste of time. They fail to realize that silicon is the abnormality.

Vardalas: It's not the rule but the exception?

Kroemer: Not the rule. It's the exception. It has that wonderful stable oxide which serves for protection as a masking aid, it has a good thermal conductivity and it has the right kind of an energy gap.

Vardalas: You can't find that in the other materials?

Kroemer: You do not find this in anything else. It has a lousy mobility. That was the principal reason we thought silicon would never fly – the mobility. I remember back in the days at RCA we didn't believe in silicon. First of all one could not get rid of that stupid oxide. And certainly the mobility is still lousy. This is why we believed in germanium.

Vardalas: Oh, I see.

Kroemer: That was wrong of course.

Vardalas: Do you think there are any limits to the switching speeds of bipolar transistors? Have we reached the limits?

Kroemer: You will have to ask my colleague, Mark Rodwell. He is the leader on this one and he expresses things in terms of f_{max}, which is not the same as switching speed. Of course typically it is not useful to switch applications anyway. And his F-maxes are so high that he does not measure them. These are extrapolated figures.

Vardalas: You left Varian in '66 and joined the University of Colorado in '68.

Kroemer: That's correct.

Vardalas: That leaves a two-year gap. What did you do during those two years? Were you a consultant?

Kroemer: I was at Fairchild.

Vardalas: That's not in your biographical sketch.

Kroemer: Yes. Those were two very unhappy years.

Vardalas: Why was that?

Kroemer: I do not want to talk about it.

Vardalas: All right. You were in a silicon world?

Kroemer: I was a misfit in a silicon world. They were unhappy years, but that did not in any way influence my good relations with Gordon Moore.

Vardalas: Okay. That explains why you went to the University of Colorado. You wanted to get out of Fairchild.

Kroemer: Yes.

Vardalas: This was the first time you worked in a university. After all those years in industry or relating to industrial people, what was it like to come back to academia as a physicist?

Kroemer: I enjoyed it. I was a bit concerned about being on a fixed schedule, at least for teaching, but the intellectual freedom that one has at universities is just wonderful.

Vardalas: Okay. Were there any attitudes in the industrial environment that you would like to have found in the academic environment?

Kroemer: No.

Vardalas: What did you do at the University of Colorado that is significant in your mind?

Kroemer: I went there in '68 and was still very heavily involved in Gunn effect. I was also doing a number of theoretical pieces of work, though I didn't really do anything terribly important at that time. The Gunn effect work was good work, but the best part of the Gunn effect was done already while I was still at Fairchild. That is in fact something that I did at Fairchild. I started working on the Gunn effect at Varian and a couple of my most important papers on that subject were written while I was at Fairchild.

Vardalas: Actual publications?

Kroemer: Yes. They are under the Fairchild byline. And I continued to work on the Gunn effect at the University of Colorado but it was getting into increasingly subtle details. When I came to UCSB I decided, "That's it."

Vardalas: When you left the University of Colorado to go to Santa Barbara you insisted that they choose a specialty in which they could realistically be successful rather than being like all the others. What was that about?

Kroemer: Yes. I had an interview at CU with Ed Stear. We had a long discussion. He was at that time chairman of the department here at Santa Barbara. He had heard that I was interested in returning to the west coast so he visited me in Boulder. I knew a little bit about UCSB. They had a very, very good silicon technology teaching lab. But I wasn't at all impressed by the research they were doing. When Ed Stear visited me in Boulder he said, "Well, you know our solid-state laboratory," and I said, "Yes." He asked me, "What would you do with this laboratory?" I very quickly forgot this was a job interview and said, "Well sure as hell not what you are doing." He acted very upset and asked, "Why? What do you mean?" I criticized their research work. So this led to a horrible, unfriendly discussion for a while. I had decided what the hell. And then at one point he looked at me and said, "Oh, shut up." I thought, "Well, all right. He doesn't want to hear what I have to tell him." But very quietly he said, "Herb, I am looking for someone to rock the boat. You sound like you are my man. Back to my question. What would you do?" So I said, "All right. I know what everybody else is advising. Everybody is advising you to get into the mainstream silicon technology. Don't. It's too late. It's too expensive. And most importantly, the graduate program depends on being able to attract top graduate students. You will not be able to do this one and they will all go to other places."

Vardalas: Right.

Kroemer: I said, "My own interest is in compound semiconductors. I think this is going to be an important field. Different from silicon. It's not something in which everyone has to be involved. This discussion took place in the fall of '75. At this point there were three universities that had critical mass – critical mass meaning more than two professors. Stanford, Illinois and Cornell. There is a place for a fourth. If you are willing to put all your eggs in one basket and you are going to gamble, you have a 50/50 chance of making number four. If you go into silicon technology you have a 100 percent chance of being an also-ran. And that still was going to be a gamble. So that's what we did. We still don't have any silicon technology.

Vardalas: That's what actually brought you here.

Kroemer: I was basically the one who set the strategy for this concept.

Vardalas: I see. Okay. In terms of your work here, what interests and fires you now?

Kroemer: Well, it's a variety of things. My most recent papers have been on superconductor-semiconductor combinations basically taking an indium arsenide/aluminum antimonide quantum well that is heavily modulation-doped, so that there are large electrical concentrations in the quantum well but still very high electron mobility.

Vardalas: Okay.

Kroemer: Then on that indium arsenide is put niobium electrodes. Indium arsenide has the fascinating property of not making Schottky barriers, so these are true ohmic contacts. Down to 9 Kelvin these are simply ordinary resistors. Then as soon as the niobium becomes superconducting the conductivity properties of the indium arsenide itself change. Eventually you have a form of induced superconductivity through the indium arsenide. That's a fascinating phenomenon.

Vardalas: What are the theoretical issues that interest you in that?

Kroemer: The theoretical issues are that we don't understand the details. We do not understand the temperature dependence. I think we understand the basic physics that is behind the electron transport. It's a phenomenon called Andreev reflections that plays an important role. Others have studied Andreev reflections, but in other systems the transport is diffusive. In other words the diffusion process, how the carriers get from one electrode to the other.

Vardalas: Right.

Kroemer: And the indium arsenide is ballistic.

Vardalas: Ballistic?

Kroemer: Yes. Therefore we are building what in the jargon would be called ballistic weak links. That topic is still fascinating. We have done quite a bit of experimental work here, but that was basically finished in around '99. It was finished when the Office of Naval Research pulled the rug.

Vardalas: How much of your funding comes from the defense- and government-related work?

Kroemer: At that point my ONR funding was the basic money. I have done other work under industrial funding.

Vardalas: How does one sell that kind of research to ONR?

Kroemer: You would have to ask ONR why they cancelled it.

Vardalas: Okay.

Kroemer: They initially supported it.

Vardalas: What did they see in it? Obviously they saw something as being useful for them initially.

Kroemer: I do not know whether the original reason for funding it had anything to do with utility. ONR has a strong tradition in supporting things that are quite fundamental. I never made any promise for an application. I simply felt it was a sufficiently crazy phenomenon that it deserved study. That was my last major experimental project here. A couple of Ph.D. dissertations came out of that.

Vardalas: I am fascinated by Kroemer's lemma on new technology. A lot of people quote it in one form or another.

Kroemer: Yes, because I've been citing it often enough now. It was originally formulated at a NATO Advanced Research Workshop in France.

Vardalas: For the purpose of this transcript let me quote it here. Let's see if I got it right. "The principal applications of any sufficiently new and innovative technology always have been and will continue to be applications *created* by that new technology."

Kroemer: And the emphasis is on the word created.

Vardalas: What prompted you to come up with this and how?

Kroemer: What prompted me to come up with this is that, particularly in engineering research projects, too much emphasis is put on the applications that can be envisaged. But in the really big stuff the applications are always created. The computer was created by silicon. The portable computer was created by liquid-crystal development.

Vardalas: Was this a defensive move on your part?

Kroemer: The laser created the compact disc and fiber communications.

Vardalas: Obviously you came up with this to counter some other perception.

Kroemer: Absolutely. Yes.

Vardalas: Which has funding implications.

Kroemer: I grew up in a research environment where you were quite free and it was not difficult to get funding without having to promise immediate applications. To a large extent that has disappeared.

Vardalas: Even in universities?

IEEE History Center - Herbert Kroemer Oral History

Kroemer: In the funding for universities, yes, particularly on the engineering end. I think physics doesn't see that to the same extent. If you realize that any sufficiently new invention or discovery the principal applications always have been created, then you understand the importance of this type of research. In addition, the creation typically happens through somebody other than the original researcher. It takes a different kind of mentality. It takes somebody who says, "Hey, this is an interesting discovery. I know exactly what to do with it." If the researcher himself must tell the potential funding organization the applications, then progress is actually being restricted. It is not being advanced.

Vardalas: That's a good point.

Kroemer: That is what I keep hammering on. The Nobel Prize has helped me in getting a few more people listening to that.

Vardalas: Yes. A certain credibility grows around you now.

Kroemer: My success was basically achieved by following this principle. I never gave a damn what the applications were. I would ask myself what might the applications be, but that did not control what I was doing.

Vardalas: Right. I see. Another lemma of yours is the proof of ignorance.

Kroemer: Kroemer's lemma of proven ignorance. "If, in discussing a semiconductor problem, you cannot draw an Energy Band Diagram, you do not know what you are talking about."

Vardalas: Did you formulate this lemma expressly for your students or for colleagues?

Kroemer: It was not formulated for anyone in particular.

Vardalas: Why was there a need for this? How did you come up with this?

Kroemer: Because I have seen it too often that people cannot draw Energy Band Diagrams, and in reality those same people do not understand how things actually work. I really do feel that knowing how things work is absolutely essential. And without Energy Band Diagrams one does not understand heterostructures. I had this in my Nobel lecture. There is also the corollary that, "If you can draw it but don't, then your audience won't know what you are talking about." The students are always chuckling. If someone gives a talk and doesn't draw an Energy Band Diagram they know Herb Kroemer is going to raise the question, "Would you please draw an Energy Band Diagram?" And my colleagues agree with me. They absolutely agree with me.

Vardalas: This would be something that perhaps would be something to apply to engineers working in the area who might be tempted to discuss things without ever going back to these first principals.

Kroemer: That's right. Engineers are particularly prone to not doing Energy Band Diagrams, but they are not the only ones.

Vardalas: Physicists will do it too?

Kroemer: You will find this among physicists too.

Vardalas: Interesting. I have one last question for you. In generations to come some young physicists will look to the greats for a role model and some will choose you. What traits do you hope they will try to emulate? You chose Niels Bohr for certain reasons. Why would you like them to see in you as a physicist?

Kroemer: That's hard to answer. One of the things that made me tick, certainly skepticism towards what authorities say and a deep interest in really understanding on a fundamental level what I have done. These are probably the two important things.

Vardalas: I want to thank you very much for a most fascinating interview.

Kroemer: You're welcome.

____people

Not Just Blue Sky

From high-speed transistors to solid-state lasers, IEEE
Medal of Honor recipient Herbert Kroemer's theories
have led to a wealth of semiconductor applications

By Tekla S. Perry, Senior Editor

An unusual condition was imposed
on Herbert Kroemer at the start of
his research career 50 years ago.
He was not allowed to touch
anything in his workplace, the
Telecommunications Laboratory of
the German Postal Service. The fear
was that this recent graduate in
theoretical physics would break
something. Far from constraining
him, the restriction expanded his
horizons.

With just pencil and paper, he began
sketching out theories that would
resonate across the entire world of
semiconductor science. And that
work would culminate in a Nobel
Prize in Physics in 2000 and this
year's IEEE Medal of Honor, the
latter for "contributions to high-frequency transistors and hot-electron
devices, especially heterostructure devices from heterostructure bipolar
transistors to lasers, and their molecular beam epitaxy technology."

While his theories led to products that earned their manufacturers
billions of dollars, none of the profits came to Kroemer. "That really
doesn't bug me," he says, sitting in his small and sparsely decorated
office on the Santa Barbara campus of the University of California,
where he is now professor of electrical and computer engineering and
materials.

IEEE Fellow Kroemer never tried to develop applications of his work--or
even predict them. "I like lemmas," he told IEEE Spectrum, "and this
one about applications is perhaps my most important message. It's
called 'The futility of predicting applications,' and states: 'The principal
applications of any sufficiently new and innovative technology always
have been and will continue to be applications created by that new
technology.' " So he doesn't begrudge others the fruits of his ideas.

"I've always called myself an opportunist," he says. "This is supposed
to be a derogatory term, but I'm not one bit ashamed of accepting
opportunities. In the scientific sense, I was an opportunist who was
looking for challenging problems."

Too many lists

In high school in Germany, Kroemer played around with chemistry
experiments but soon turned to physics. "I liked the beautiful logic of a
structure with a relatively small number of fundamental principles from
which you could draw far-reaching conclusions," he says. A university
chemistry course that required rote memorization of lists and lists of
chemical reactions destroyed any remaining interest in that science.

College was a breeze. He entered the University of Jena in East
Germany in 1947, then left for West Germany the next year during the
Berlin airlift and was accepted at the University of Göttingen. Four
years later he received his Ph.D. for a theoretical dissertation on
germanium transistors that discussed electron transport in high
electrical fields. It broke little new ground, and he takes no particular

pride in it. He explained some experiments, he says, but the explanation later proved completely wrong.

As he told Spectrum, his actual knowledge of the subject matter was rather limited. But what his research advisor really cared about was methodology. Does a student know how to tackle a problem with no background in the subject? And does he or she know how to acquire the knowledge needed? And that Kroemer knew how to do. To this day, his view of education is that accumulating methodology matters more than accumulating knowledge of subject matter.

"It was not until a number of years after working with him that I realized how unique this is," says William Frensley, a one-time graduate student of Kroemer's and now professor of electrical engineering at the University of Texas, Dallas. "Other students worked for professors who were specialists and became specialists in the same thing, whereas we said we have a problem, and we are going to master whatever techniques it takes to solve it."

Postal service

In 1952, when Kroemer received his Ph.D., an academic career was out of the question. The lines of succession at existing German universities were long, and no new ones were being established. So he joined the Telecommunications Research Laboratory of the German Postal Service in Darmstadt.

This is less of a stretch than it seems. The postal service ran the telephone system and had a small semiconductor research group--some 10 scientists--in its telecommunications laboratory. That group hired Kroemer to answer any theoretical questions that arose, to give talks on any subject he thought relevant--and to keep his hands off the research equipment.

"I enjoyed this thoroughly," he recalls. For one, he had liked the role of teacher since high school, when his physics teacher asked him to prepare and deliver a lecture to the class. For another, being at the researchers' beck and call presented him with a wide variety of problems in diverse subjects.

In solving one of those problems, he went against the conventional wisdom of the time. Researchers were developing pn junctions of indium and germanium. They did this by depositing a layer of indium on a layer of germanium, then heating the structure to form the pn junction. Kroemer was trying to understand how exactly the junction formed.

Obviously the molten indium dissolved some of the germanium, and the belief was that it diffused into the germanium beyond the layer in which the germanium dissolved. But Kroemer concluded that the process was one of recrystallization--the heated indium dissolves some of the germanium, and then upon cooling the germanium precipitates out and recrystallizes, incorporating some of the indium atoms, which replace some of the germanium atoms in the lattice.

What he didn't know was that researchers in the United States, at General Electric Co. and RCA Corp., had simultaneously reached the same conclusion.

But what he did know was that to be at the research forefront, he needed to leave the German Postal Service and get to the United States. He started looking for a way to get there.

Researchers from other countries occasionally visited the lab in which he worked, curious about this small semiconductor research group. In 1953 one visitor was William Shockley, then at Bell Telephone Laboratories. "I spent about two hours with him," Kroemer said. "We were having a marvelous time. I told him about the work that I'd done for my Ph.D. dissertation, and about some of my ideas of how to make transistors fast by putting an electric field into the base. He seemed intrigued by that."

Kroemer asked him about coming to Bell Labs, but Shockley, as an official visitor, told Kroemer that he would have to go through official channels, starting with informing Postal Service management of his intentions to apply for a job in the United States. The young researcher was so grateful for the job he had at the Postal Service that he was "terribly squeamish about telling my management that I wanted to leave."

Later in 1953, the Darmstadt lab had another U.S. visitor: Ed Herold from RCA. Kroemer asked him whether RCA was working on npn transistors (back then pnp transistors dominated). Herold was guarded in his responses; but Kroemer guessed out loud what the RCA researchers were doing, what alloys they were using (lead-antimony), the percentage of the antimony, and the alloy temperatures. His guesses proved quite close to RCA's experiments, and the impressed Herold didn't hesitate to offer him a job. (All the same, it took a year for Kroemer to obtain a visa, even with RCA's help.)

At RCA in Princeton, N.J., Kroemer did theoretical research on an impurity diffusion process for building transistors. In the diffusion process, the doping of the base region was deliberately graded from a high value at the emitter to a lower value at the collector. Because this gradient introduced a built-in electric drift field into the base, the result was called a drift transistor. The first commercial product to come out of that research--the 2N247--had a high-frequency performance far beyond that of other commercially available transistors of its time. Its power gain cutoff frequency of 132 MHz made it suitable for use in FM radios.

While Kroemer was theorizing about how a drift field could make transistors switch faster, he had an idea about grading the basic semiconductor itself. If an alloy of two semiconductors replaced the single semiconductor, it could be given a continually varying composition to change its band gap, which is a measure of the amount of energy required to move an electron from a semiconductor's valence band to its conduction band. This varying band gap would be another way to introduce a drift field into the base, again in order to improve transistor frequency performance.

He had mentioned varying a material's band gap in a paper while still in Germany, but expanded the idea and in 1957 published two papers about it, one in the RCA Review, another in the Proceedings of the IEEE.

Theory into practice

While Kroemer trusted his theory, he didn't know how to build actual semiconductors using his principles. Building them would require either a base region consisting of a graded mix of different semiconductor materials with varying band gaps or else one material in the base but a different material in the emitter.

He tried to build a transistor with germanium-silicon alloy as the emitter on a germanium base. To this end, a gold-silicon blended mixture was alloyed onto germanium at 600 °C, hot enough for the melted mixture to begin eating up germanium, precipitating the germanium-silicon alloy emitter on cooling. Unfortunately, during the cooling, most of the devices cracked. "It was one of those technological blind alleys where you're not exactly embarrassed that you have tried it, but you're not surprised it didn't work," he says.

At the end of 1957, Kroemer decided to get out of transistor research. He had no interest in traditional transistors, and heterostructure transistors, with existing material technology, could not be built.

"I promised myself," he says, "that if a new technology for building heterostructures arose, I'd get back into it."

Kroemer left RCA in 1957 and returned to Germany; he, and more especially, his wife, was homesick. Becoming head of a semiconductor group at Philips Research Laboratory in Hamburg, he pushed for work on gallium arsenide, looking at what happens when you apply large electric fields to gallium arsenide semiconductors. "I thought GaAs was going to be an important material, so it was worthwhile studying it."

Kroemer feels he did little significant work at Philips and, since his wife quickly concluded she preferred the United States after all, in 1959 he went to Varian Associates (Palo Alto, Calif.), where he did a little research on tunnel diodes before turning to other problems.

Back in the heterostructure game

Then Kroemer's ideas about heterostructure devices, shelved for half a dozen years, came back to his attention with a vengeance.

It was March 1963. The previous summer, Kroemer and a Varian colleague, Sol Miller, had attended the Annual Device Research Conference, at which GaAs lasers had been introduced. Miller was interested and at Varian's weekly colloquium, he gave a talk about the new lasers. Though scientifically fascinating, he said, the devices could only work at very low temperatures and only for very short pulses, and so would never be truly practical. Asked why, Miller explained that the problem was the lack of charge-carrier confinement: at normal temperatures, electrons would diffuse out of one side of the device as quickly as they were supplied from the other side, as would the holes; therefore the electron-hole pair concentration would never become high enough to cause laser action by stimulated emission. Low temperatures suppressed the effect, but only for brief periods of time.

Kroemer disagreed. Based on his work in heterostructures, the solution, to him, seemed obvious--you just vary the device's band gap, putting a narrower gap in the center and a wider gap in the outer regions, so that the electrons and holes would concentrate in the center [see "Heterostructures Explained"]. "My reaction was instantaneous," he told Spectrum. "The moment somebody told me about the problem, it snapped."

He wrote up his idea as a paper and submitted it to Applied Physics

Letters, where it was rejected. Rather than fighting the rejection, he was persuaded to submit it to the Proceedings of the IEEE. There it was accepted, but drew little attention. He also filed for a patent on the technology; issued in 1967, it expired in 1985.

Kroemer wanted to start working on the creation of room- temperature lasers at once, but his superiors at Varian told him that such a device would never have any applications.

"This is the classic mistake--judging something not by what applications it might create, but by how it could fit into applications we've already thought of," Kroemer says. The applications it was useful for turned out to include fiber-optic communications, CD and DVD players, LED traffic lights, and laser pointers--none of which were around at the time.

Though Kroemer wasn't pleased by Varian's decision, the Gunn effect, which had just been discovered, interested him. This is a phenomenon in which microwave oscillations are produced when a certain voltage is applied to opposite faces of a semiconductor. For the next decade and more, Kroemer explored theories of why this occurred, three of those years at Varian, two at Fairchild Semiconductor Corp. (Palo Alto, Calif.), and nearly eight at the University of Colorado in Boulder.

Halls of academia

Kroemer was happy to move from industry to academia. Things at Fairchild had not gone well, because the company was dedicated to silicon technology and Kroemer's interests had long been elsewhere. He looked forward to the research freedom and also to teaching.

But he became dissatisfied. "We had hoped to set up a good solid-state engineering graduate program at Boulder, " he says. During the Vietnam War, many students went on to graduate school to reduce their chances of getting drafted. Stanford University typically recruited the academically top 5 percent of graduate students interested in solid-state research, and Boulder drew on the next 5 percent, who were still extremely good. But when graduate enrollments fell after the war's end, that source dried up. "Our ambitious graduate program would not fly--it was clear to me that I would be professionally dead if I stayed there," Kroemer recalls.

Word went out that he was open to a change, and in the fall of 1975, the University of California at Santa Barbara, in the person of Edward Stear, then head of its electrical engineering and computer sciences department, came calling. Santa Barbara at the time didn't have a very good academic reputation; what it did have was a well-equipped semiconductor device teaching laboratory.

"So, Herb, you know about our laboratory," Stear opened. "What would you do with it?"

Kroemer momentarily forgot that this visit was actually a job interview. "Sure as hell not what you're doing!"

"It was a rather unfriendly and hostile discussion, and Stear eventually snapped, 'Shut up,' " Kroemer recalls. He figured he had blown any chance of being hired. But then Stear told him, "I'm looking for someone to rock the boat; it looks like you're my man."

Kroemer, Stear tells Spectrum, "speaks very directly. He is honest, but can be sharp with people, too. He is intense and demanding. He can be a difficult person at times to work with, but people have ended up loving him." In any case, Stear knew that Kroemer could build the kind of program that Santa Barbara needed, and Kroemer was hired.

By the sea

Kroemer left for Santa Barbara in the summer of 1976. He had persuaded Stear not to compete with Stanford, Berkeley, and other top engineering schools in silicon technology, but instead to focus on compound semiconductors such as GaAs. He gave Santa Barbara even odds for making an impact in that technology.

"You want to be first-rate or absent," Kroemer says.

"I promised myself that if a new technology for building heterostructures arose, I'd get back into it"

Kroemer convinced a few former colleagues that they should join him at Santa Barbara, and he also convinced the U.S. Army it should buy him a molecular beam epitaxy machine. He said at the time that he wanted it for making transistors with a gallium phosphide emitter on a silicon base, a crazy project if there ever was one. It was not enough to put Santa Barbara's engineering school on the map.

In the mid-1980s, however, the chancellor of the university, Bob Huttenback, decided to put all available money into improving the College of Engineering. A new dean was hired, and 15 faculty were added. "We raided Bell Labs," Kroemer recalls. "Today we have one of the best materials departments in the country--and we still don't have any silicon technology."

At 73, Kroemer remains a full-time member of the faculty. One problem he is working on concerns the influence of high electric fields on electron transport in semiconductor superlattices (alternating thin layers of two or more materials with different band gaps but similar crystal structures and lattice constants). More specifically, he is focused on a concept, called a Bloch oscillator, which can in theory generate oscillations up into the terahertz range, potentially opening up that frequency range for numerous applications. So far, it has never been satisfactorily demonstrated as a continuously running device. "I have some ideas, which may or may not be correct, of what to do about it," Kroemer tells Spectrum.

He is also looking at the phenomenon of induced superconductivity in semiconductors, created when superconducting materials are deposited on semiconductors and operated at low temperatures.

Tuesday morning, 3 a.m.

Of all the honors Kroemer has received over the years, the strangest was the naming of an asteroid after him. (Asteroid Kroemer orbits between Mars and Jupiter.) One honor that he thought beyond the grasp of a physicist who dealt in such a down-to-earth area as semiconductors (compared to those who grapple with invisible particles) was the Nobel Prize.

"Oh, my name had been mentioned over the years," Kroemer told Spectrum. But the Nobel Prize is almost invariably awarded for fundamental discoveries, not for applied research, and so I never believed the rumors."

The rumors grew stronger in 1996, when Kroemer was invited to give a talk at a Nobel symposium. "I still didn't catch on," he said. "I looked around at the attendees and saw Horst Stormer, and thought he was the most likely candidate of the group. When he received the prize in 1998, I was enthusiastic and didn't envy him at all--after all, my work was applied." (Stormer and two colleagues received the Nobel Prize for discovering that electrons acting together in strong magnetic fields can form new types of particles with charges that are fractions of the electronic charge.)

Although Kroemer never believed the Nobel would come to him, he did continue to pay attention to it. On 9 October 2000, the Nobel Prize in Physics was to be announced the next day. He went to bed that evening thinking, "Wouldn't it be funny if I would get a 3 a.m. phone call? But then I said to myself, Stop being silly, go to sleep!" (Noon in Stockholm, when Nobel announcements are typically made, is 3 a.m. in California.)

But when the phone did ring shortly before 3 a.m., his first response was panic--were his children all right? Had something happened to his grandson? His wife answered, and passed him the phone, saying "It's Stockholm."

"If my life depended on it, I could not reconstruct the next two or three sentences," Kroemer says. Then the caller put a friend of Kroemer's on the phone, to reassure him that it was not a joke, warning him that he had about 15 minutes before the public announcement was made and the media circus started.

"At that point all hell did break loose and the phone was ringing off the hook, starting with German news agencies, since I'm German and it was already midday there. I literally couldn't put the phone down."

The Nobel Prize in Physics that year was shared by three people--Jack S. Kilby, also an IEEE Fellow, for his part in developing the IC, and Kroemer and IEEE member Zhores I. Alferov, for "developing semiconductor heterostructures used in high-speed- and opto-electronics." Alferov, working in Russia, had made similar discoveries in parallel with, but separately from, Kroemer; the two first met in 1972 and have since become friends, even though they are, in a sense, competitors.

Kroemer's Nobel citation emphasizes the general principle of the heterostructure, not the individual devices. And that suits him just fine, because he has routinely deferred the question of applications. "Certainly, when I thought of the heterostructure laser, I did not intend to invent compact disc players," he says. "I could not have anticipated the tremendous impact of fiber-optic communications. I really didn't give a damn about what the uses were."

"That's not what I do. The person who comes up with applications thinks differently than the scientist who lays the foundation."

And Kroemer laid one fine foundation.

Reprinted with permission from
    H. Kroemer,
    ``Zur Theorie des Germaniumgleichrichters und des Transistors,''
    Zeitschr. f. Phys., Vol. 134, pp. 435-450, 1953.

With kind permission of Springer Science and Business Media.

Zeitschrift für Physik, Bd. 134, S. 435—450 (1953).

Zur Theorie des Germaniumgleichrichters und des Transistors*.

Von

HERBERT KRÖMER.

Mit 10 Figuren im Text.

(Eingegangen am 1. Dezember 1952.)

Bei den in den Randschichten von Halbleiter-Metall-Kontakten herrschenden hohen Feldstärken sind für den Transport der Elektronen und Defektelektronen („Löcher") nicht die normalen Beweglichkeiten maßgebend, die gültig sind, wenn der Potentialabfall längs einer mittleren freien Weglänge klein ist gegen die thermische Energie kT.

In stärkeren Feldern gelangen die Ladungsträger mit merklicher Wahrscheinlichkeit ins Bandinnere, im Grenzfall sehr starker Felder oszilliert ein Teilchen zwischen zwei Stößen mehrere Male im Band hin und her und kommt dadurch langsamer vorwärts („Staueffekt"). Die Beweglichkeit nimmt dann so ab, daß die Teilchendichte proportional zu Feldstärke und Stromdichte ansteigt. Der Wert der Proportionalitätskonstanten („Staukonstante") wird für die Löcher zu $2 \cdot 10^3 \, \text{Watt}^{-1}$ abgeschätzt; für die Elektronen dürfte er erheblich kleiner sein.

Die Dichte der gestauten Teilchen kann von gleicher Größenordnung wie die der Störstellen werden. Das Randschichtpotential weicht dann erheblich von der einfachen SCHOTTKYschen Parabelgestalt ab. Mit dem abgeänderten Potential werden Kennlinien von Ge-Transistoren und -Dioden berechnet. Bei geeigneter Wahl der eingehenden Parameter ergeben sich bei den Transistoren hohe Werte für den Stromverstärkungsfaktor, bei den Dioden das experimentell beobachtete Umbiegen der Sperrkennlinien.

I. Einleitung.

Nach SCHOTTKY [13] bäumen sich an der Grenzfläche zwischen Germanium und einem Metall die Bänder des Halbleiters (künftig abgekürzt HL) ohne angelegte Spannung um das sog. Diffusionspotential V_D, bei angelegter Spannung U um $eU + V_D$ auf. Es entsteht eine die Elektronenbewegung hemmende, von U abhängige Potentialschwelle, deren Höhe über der FERMI-Kante (künftig abgekürzt FK) des Metalls

$$e\Phi_0 = V_D + E_L \tag{1}$$

ist, unabhängig von U (Fig. 1).

Infolge der Bildkraftanziehung des Metalls auf die Elektronen wird die Spitze dieser Schwelle um

$$\Delta V = \sqrt{\frac{e^3 F}{\eta}} \tag{2}$$

____
* Auszug aus einer Göttinger Dissertation.

436   Herbert Krömer:

abgeflacht (Fig. 2), wobei F die Feldstärke am Ende der Randschicht (Randfeldstärke) bedeutet. Anstatt der statischen Dielektrizitätskonstante ε des HL wurde dabei eine „dynamische" DK $\eta < \varepsilon$ eingeführt. Denn da die Elektronen eine Geschwindigkeit von 10^7 bis 10^8 cm/sec haben, und da der Abstand des Maximums der abgerundeten Schwelle von der Grenzfläche

$$x_s = \frac{1}{2}\sqrt{\frac{e}{\eta F}}$$

zwischen einigen 10^{-7} und 10^{-6} cm liegt, kann man bei einer Vorbeiflugzeit von 10^{-13} bis 10^{-15} sec wohl kaum mit dem statischen Wert der DK von $\varepsilon = 16$ [5] rechnen. Auch hinkt die Gitterpolarisation in der Phase nach, was die wirksame DK abermals herabsetzt.

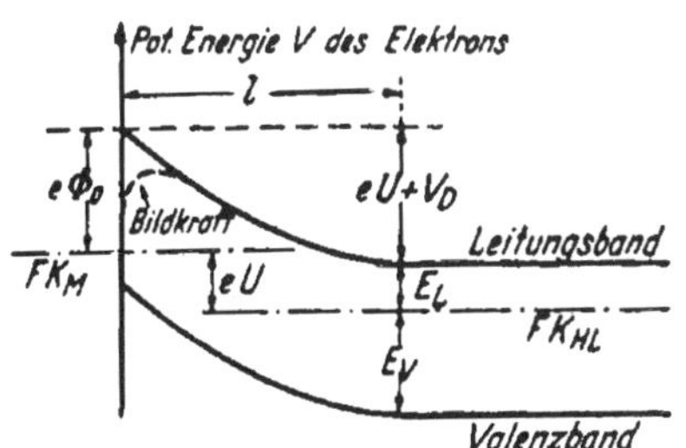

Fig. 1. Randschichtpotential. E_L und E_V sind die Abstände des Leitungs- und Valenzbandes des Halbleiters von der Fermi-Kante.

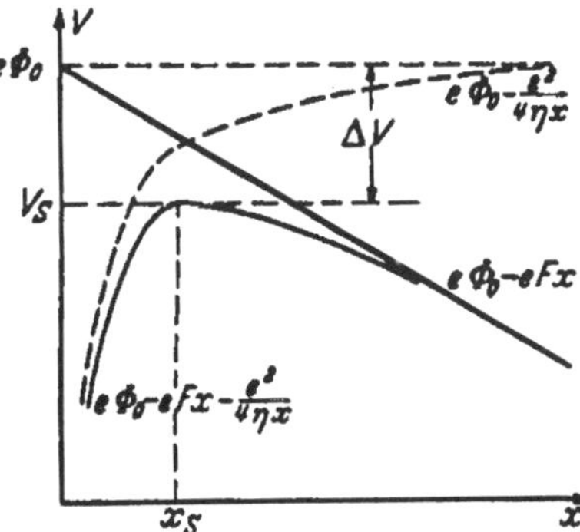

Fig. 2. Abbau der Potentialschwelle durch die Bildkraft.

Unter diesen Umständen muß die Gl. (2) als eine erste Näherung und η als ein Parameter angesehen werden, über den sich zur Zeit theoretisch noch nichts sagen läßt, und der daher vorerst experimentell bestimmt werden muß. Messungen von S. Benzer [3] an Ge-Gleichrichtern lassen sich am besten mit $\eta \approx 1$ beschreiben (s. unten). Ähnliches fand auch W. Oldekop [10] an Selengleichrichtern. Wir rechnen daher im folgenden mit $\eta = 1$.

Mit Berücksichtigung der Bildkraft wird die Gesamthöhe der von den Elektronen zu überwindenden Potentialschwelle

$$V_s = e\Phi_0 - \Delta V = V_D + E_{I.} - \sqrt{\frac{e^3 F}{\eta}}. \tag{3}$$

Nach der Diodentheorie des Kristallgleichrichters [4], [18] beträgt dann die Dichte des in Sperrichtung für $eU \gg kT$ fließenden Elektronenstromes

$$j_n = A T^2 e^{-\frac{V_s}{kT}} \quad \text{mit} \quad A = \frac{4\pi e m k^2}{h^3}. \tag{4}$$

Zur Theorie des Germaniumgleichrichters und des Transistors.   **437**

Im Fall der einfachen Erschöpfungsrandschicht mit der Störstellen-konzentration N_0 gilt nach SCHOTTKY [13]

$$F = \sqrt{\frac{8\pi N_0}{\varepsilon}(eU + V_D)}.\tag{5}$$

Daraus folgt mit (3) und (4) für $eU \gg V_D$

$$j_n = A T^2 \exp \frac{1}{kT}\left[-e\Phi_0 + \sqrt[4]{\frac{8\pi e^7 N_0}{\varepsilon \eta^2}}\sqrt{U}\right].$$

Solche im $(\lg j/j_0, U)$-Diagramm geradlinige Kennlinien wurden z. B. von SEILER [12] an Si-Detektoren gefunden. Aus der Steigung der Geraden läßt sich rückwärts N_0/η^2 bestimmen. SEILER, der mit $\eta = \varepsilon$ rechnete, fand so Werte für N_0, die bis zu einer Zehnerpotenz über anderweitig bestimmten lagen. Wir sehen darin eine Bestätigung unserer Vermutung $\eta < \varepsilon$.

Bei der Herleitung von (5) wird die Annahme gemacht, daß die Dichte der für den Stromtransport verantwortlichen Ladungsträger, die im HL-Innern mit der der ionisierten Störstellen übereinstimmt, gegen die Randschicht hin auf einer kurzen Strecke praktisch völlig abklingt, so daß in der Randschicht nur mit der Ladungsdichte der Stör-stellenionen zu rechnen ist. Wie im folgenden gezeigt wird, ist diese Annahme nur bei nicht zu großen angelegten Spannungen richtig. Wird jedoch die Randfeldstärke hinreichend groß, so wird die effektive Träger-beweglichkeit in der Randschicht so stark herabgesetzt, daß die Träger-dichte stark ansteigt und schließlich von der gleichen Größenordnung wie die der Störstellen werden kann.

Mit der hierdurch bedingten abgeänderten Raumladungsdichte wird die Berechnung der Erschöpfungsrandschicht wiederholt und anschlie-ßend auf die Berechnung von Dioden- und Transistorkennlinien ange-wandt werden.

II. Der Staueffekt.

II, 1. Transportgesetze in schwachen und starken Feldern.

Die Transportgesetze für die Ladungsträger hängen davon ab, wie stark sich die potentielle Energie V der Elektronen längs einer mitt-leren freien Weglänge λ ändert. Letztere ergibt sich aus der Beweglich-keit b gemäß

$$\lambda = \frac{3}{4}\frac{\sqrt{2\pi m kT}}{e}\cdot b \;\dagger.$$

Bei Zimmertemperatur folgt daraus mit

$$b_n = 3600\,\mathrm{cm^2\,V^{-1}\,sec^{-1}}, \qquad b_p = 1700\,\mathrm{cm^2\,V^{-1}\,sec^{-1}}\;[11]:$$

$$\lambda_n = 2{,}5\cdot 10^{-5}\,\mathrm{cm}, \qquad \lambda_p = 1{,}2\cdot 10^{-5}\,\mathrm{cm}.$$

---

† Siehe z. B. W. SHOCKLEY [15], S. 277.

438   HERBERT KRÖMER:

Für $|\operatorname{grad} V| = eF \ll \dfrac{kT}{\lambda}$ gilt:

$$\text{Elektronenstromdichte} \quad \vec{j}_n = b_n\,[n \cdot \operatorname{grad} V + kT \cdot \operatorname{grad} n], \qquad (6a)$$

$$\text{Löcherstromdichte} \quad \vec{j}_p = b_p\,[p \cdot \operatorname{grad} V - kT \cdot \operatorname{grad} p] \qquad (6b)$$

(n bzw. p = Elektronen- bzw. Löcherdichte).

Daraus folgen die Ladungsdichten

$$\varrho_n = -\frac{1}{b_n F}\cdot|\vec{j}_n - b_n\,kT\operatorname{grad} n|; \qquad \varrho_p = +\frac{1}{b_p F}\,|\vec{j}_p + b_p\,kT\operatorname{grad} p|. \qquad (7)$$

Bei festgehaltenem Strom nehmen also die Ladungsdichten mit zunehmendem Feld ab.

Für $eF \gtrsim \dfrac{kT}{\lambda}$ werden die Verhältnisse recht unübersichtlich und ergeben erst im Grenzfall sehr starker Felder wieder einfache Gesetze. Wird

$$|\operatorname{grad} V| \geq eF_s = \frac{2B}{\lambda}, \qquad (8)$$

wo B die Breite des Valenz- bzw. Leitungsbandes ist, so laufen die Ladungsträger im Feld nicht mehr davon, sondern oszillieren vor dem nächsten Stoß zwischen den im Feld gekippten Bandkanten hin und her (Fig. 3), halten sich also länger am gleichen Ort auf, so daß die Teilchendichte wieder ansteigt.

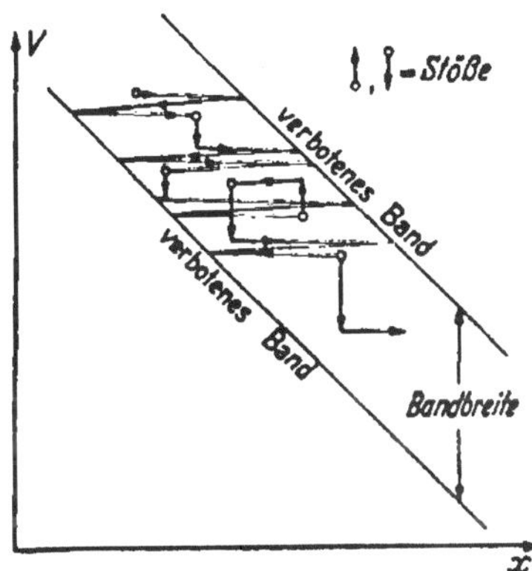

Fig. 3. Elektronenbewegung in sehr starken Feldern.

Über die Breiten B_V und B_L des Valenz- und Leitungsbandes bei Ge sind uns keine experimentellen Angaben bekannt, an theoretischen Untersuchungen nur eine Näherungsrechnung von H. MÜLLER [9] nach einem vereinfachten WIGNER-SEITZ-Verfahren, deren quantitative Brauchbarkeit zweifelhaft ist.

Es scheint jedoch sicher zu sein, daß das Leitungsband erheblich breiter als das verbotene Band ist und mit mehreren höheren Bändern überlappt, während das Valenzband eine Breite von gleicher Größenordnung wie das verbotene Band hat. MÜLLER findet $B_V < E_B$, mit einem allerdings zu großen E_B. Wir möchten annehmen, daß B_V in der Größenordnung von etwa 1 eV liegt, B_L aber $\gg 1$ eV ist. Mit $B_V = 1$ eV und $\lambda_p = 1{,}2 \cdot 10^{-5}$ cm geht (8) über in $F \geq 1{,}7 \cdot 10^5$ cm^{-1}, ein Wert, der bequem erreicht werden kann. Wir untersuchen diesen Fall näher.

Zur Theorie des Germaniumgleichrichters und des Transistors.   439

II, 2. Berechnung der Staukonstanten.

Die Raumladungsdichte der gestauten Teilchen beträgt offenbar

$$\varrho_s = \pm \frac{j \cdot eF}{(\Delta E/\tau)} \qquad \text{für} \begin{Bmatrix} \text{Löcher} \\ \text{Elektronen} \end{Bmatrix}.$$

Dabei ist ΔE der Energieverlust pro Stoß und τ die Zeit zwischen zwei Stößen. Da ΔE und τ statistisch unabhängig sind, gilt

$$\varrho_s = \pm \frac{j \cdot e \cdot F}{\overline{\Delta E} \cdot \overline{(1/\tau)}}. \tag{9}$$

Nach der Theorie der Elektronenstreuung an Gitterwellen lassen sich $\overline{\Delta E}$ und $\overline{(1/\tau)}$ abschätzen.

II, 2a. Danach ändert ein Elektron oder ein Loch bei Streuung an einer Gitterwelle der Frequenz ν seine Energie um $\pm h\nu$. Die Wahrscheinlichkeit für Energieverlust verhält sich zu der für Energiegewinn wie $\dfrac{N+1}{N}$, wo N die mittlere Quantenzahl ist, bis zu der diese Gitterwelle angeregt ist. Auf $2N+1$ Stöße kommt also ein Energieverlust der Größe $h\nu$, also durchschnittlicher Verlust pro ν-Stoß

$$\Delta E_\nu = \frac{h\nu}{2N+1}. \tag{10}$$

Die Wahrscheinlichkeiten für Streuung an zwei Wellen mit verschiedenen Frequenzen ν_1 und ν_2 verhalten sich angenähert wie

$$\frac{W_1}{W_2} = \frac{h\nu_1(2N_1+1)}{h\nu_2(2N_2+1)} \dagger. \tag{11}$$

Führt man noch die Zahl $z(\nu)$ der Gitterschwingungen pro Frequenzintervall 1 ein, so ist wegen (10) und (11)

$$\overline{\Delta E} = \frac{\int \Delta E_\nu \cdot h\nu(2N_\nu+1)\,z(\nu)\,d\nu}{\int h\nu(2N_\nu+1)\,z(\nu)\,d\nu} = \frac{\int (h\nu)^2\,z(\nu)\,d\nu}{\int h\nu \cdot (2N_\nu+1)\,z(\nu)\,d\nu}. \tag{12}$$

Für $z(\nu)$ setzen wir das DEBYEsche Spektrum

$$z(\nu) = \begin{cases} \text{const} \cdot \nu^2 \\ 0 \end{cases} \qquad \text{für} \qquad \nu \lessgtr \nu_g = \frac{k\Theta}{h}$$

ein ($\Theta = $ DEBYE-Temperatur). Beschränkungen auf Temperaturen $T > \Theta$ gibt

$$2N_\nu + 1 = 1 + \frac{2}{e^{\frac{h\nu}{kT}} - 1} \approx 2\frac{kT}{h\nu}. \tag{13}$$

---

† Gl. (10) und (11) folgen aus Gl. (34,33) und (34,36) bei SOMMERFELD und BETHE [*17*] mit $C_1 = C_2$ und $q_1 : q_2 = \nu_1 : \nu_2$.

440   HERBERT KRÖMER:

Damit wird

$$\overline{\Delta E} = \frac{3}{10}\frac{(k\Theta)^2}{kT}.\tag{14}$$

II, 2b. Wenn Gl. (8) erfüllt ist, bewegt sich das Elektron bzw. Loch im Mittel durch das ganze Band hin und her, ehe es das nächste Mal stößt. In den Impulsraum übertragen heißt dies, daß es sich *gleichförmig* quer durch die gesamte, zu diesem Band gehörige BRILLOUIN-Zone bewegt (Fig. 4). Da die Quantenzustände im Impulsraum mit konstanter Dichte verteilt sind, hält es sich in jedem Zustand gleich lange auf, und die Stoßwahrscheinlichkeit $\overline{(1/\tau)}$ ergibt sich einfach als Mittelwert über die ganze BRILLOUIN-Zone. Unter Einführung der freien Weglänge λ und der Gruppengeschwindigkeit $v = \mathrm{grad}_P E(P)$ heißt das:

$$\overline{\left(\frac{1}{\tau}\right)} = \frac{1}{V_P}\iiint \frac{|\mathrm{grad}_P E(P)|}{\lambda(P)}\,dP_x\,dP_y\,dP_z.\tag{15}$$

Für die weitere Auswertung müssen wir die schwerwiegende Annahme machen, daß λ nicht nur am Bandrande, sondern auch im Bandinneren von der Geschwindigkeit und damit von P unabhängig ist. Dann ist

$$\overline{\left(\frac{1}{\tau}\right)} = \frac{1}{\lambda}\overline{|\mathrm{grad}_P E|}.\tag{16}$$

Da uns die genaue Gestalt der Flächen konstanter Energie im P-Raum unbekannt ist, können wir diesen Mittelwert nicht exakt ausrechnen. Wir erhalten aber mindestens die richtige Größenordnung, wenn wir den Gradienten in (16) durch die Größe

$$\frac{\text{Energieunterschied innerhalb der Zelle}}{\text{Halbe Kantenlänge der Zelle}} = \frac{B}{h/2a}$$

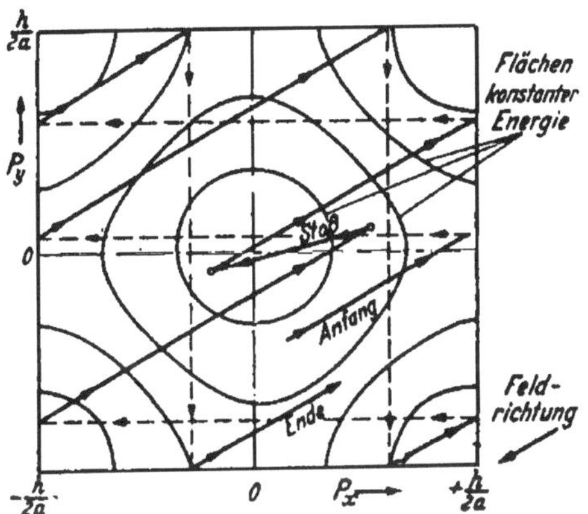

Fig. 4. Schnitt durch den reduzierten Impulsraum eines kubischen Kristalls (a Gitterkonstante). Bewegung eines Elektrons im äußeren Feld.

ersetzen (a = Gitterkonstante). Damit wird

$$\overline{\left(\frac{1}{\tau}\right)} = \frac{a}{h}\cdot\frac{2B}{\lambda}\cdot g,\tag{17}$$

wo g ein von der Zellengeometrie und der Gestalt der Energieflächen abhängiger Zahlenfaktor der Größenordnung eins ist. Wir rechnen mit $g = 1$ weiter; dieser Wert gilt exakt z.B. für eindimensionale Gitter.

II, 2c. Einsetzen von (14) und (17) in (9) gibt

$$\varrho = j\cdot e\cdot F\cdot\frac{10}{3}\cdot\frac{h}{a}\cdot\frac{kT}{(k\Theta)^2}\cdot\frac{\lambda}{2B} = j\cdot eF\cdot s,\tag{18}$$

Zur Theorie des Germaniumgleichrichters und des Transistors.   441

wobei die „Staukonstante" s durch

$$s = \frac{1}{\overline{\Delta E} \cdot \overline{(1/\tau)}} = \frac{10}{3} \cdot \frac{h}{a} \cdot \frac{kT}{(k\Theta)^2} \cdot \frac{1}{eF_s} \tag{19}$$

definiert ist.

Die bei Ge für Defektelektronen einzusetzenden Werte ($B_V = 1$ eV, $\lambda_p = 1{,}2 \cdot 10^{-5}$ cm; $\Theta = 400°$ K[1], $a = 5{,}6 \cdot 10^{-8}$ cm) geben bei Zimmertemperatur

$$F_{s,p} = 1{,}7 \cdot 10^5\,\text{V cm}^{-1}$$

$$s_p = 2 \cdot 10^8 \frac{\text{cm}^{-3}}{\text{V cm}^{-1}\,\text{Amp cm}^{-2}} = 2 \cdot 10^8\,\text{W}^{-1}. \tag{20}$$

Die üblichen Punktkontakte haben Durchmesser von etwa 10^{-3} cm, die im Transistor fließenden Löcherströme sind von der Größenordnung 1 mA. Das gibt Stromdichten von einigen hundert Amp $\cdot$ cm^{-2}. Da die auftretenden Feldstärken bei 10^5 V $\cdot$ cm^{-1} liegen, entstehen somit Staukonzentrationen von derselben Größenordnung wie die Störstellenkonzentrationen.

II, 3. *Diskussion der gemachten Vernachlässigungen.*

a) Gl. (11) gilt nicht streng, da die BLOCHsche Wechselwirkungskonstante C †, die in die Stoßwahrscheinlichkeit eingeht, vom Anfangsund Endzustand nicht ganz unabhängig ist. Da sie jedoch in (12) im Zähler und Nenner auftritt, ist ihr Einfluß gering. Wäre z.B. W proportional zu $(h\nu)^{1+\alpha} \cdot (2N+1)$ mit $\alpha \neq 0$, so träte in (14) nur ein Faktor $\left(1 + \frac{\alpha}{3}\right)\Big/\left(1 + \frac{\alpha}{5}\right)$ vor die rechte Seite, der von der Größenordnung eins ist.

b) Aus demselben Grunde ist die Benutzung des einfachen DEBYEschen Spektrums unerheblich.

c) Die Staukonstante s ist nicht temperaturabhängig! Denn das Integral im Nenner von (12) ist bis auf Konstanten gerade die gesamte Stoßwahrscheinlichkeit, also proportional zu $\overline{(1/\tau)}$, so daß die T-Abhängigkeit herausfällt. Die Näherung (13) ermöglichte nur die einfache Berechnung (λ proportional zu $1/T$ für $T > \Theta$), der Zahlenwert (20) gilt, wenn überhaupt, so für alle Temperaturen.

d) Die Problematik von (18) liegt in der Annahme einer geschwindigkeits- und richtungsunabhängigen mittleren freien Weglänge. Diese Annahme ist am Bandrande richtig[2], im Inneren zweifelhaft. Eine

---

[1] Die angegebenen Werte schwanken stark. Wir schließen uns an HILL und PARKINSON [7] an.

† Vgl. Fußnote S. 439.

[2] Vgl. dazu SHOCKLEY [15], Kap. 17.

442   HERBERT KRÖMER:

genaue Theorie dieses Falles liegt noch nicht vor und kann an dieser
Stelle auch nicht gegeben werden. Hier bleibt daher eine Möglichkeit
für spätere Korrekturen.

II, 4. Mittelstarke Felder.

In mittelstarken Feldern ($kT/\lambda < |\mathrm{grad}\,V| < eF_i$) ist nicht mehr jede
Stelle des Impulsraumes gleichwahrscheinlich. Die Teilchen ziehen sich
mit abnehmendem Feld mehr und mehr auf die Bereiche geringster
Energie, d.h. auf die Mitte des P-Raumes zurück, die bei Löchern dem
oberen, bei Elektronen dem unteren Bandrand entsprechen. In (15)
tritt dann ein gegen die Ränder abnehmender Gewichtsfaktor mit unter
das Integral, dessen genaue Form zur Zeit noch nicht angegeben werden
kann.

Auch die Berechnung von ΔE ist dann nicht mehr so einfach durchführbar, da unmittelbar am Bandrande kein Energieverlust mehr möglich ist.

Für schwache Felder ($eF \ll kT/\lambda$) muß wieder (7) herauskommen. Im
mittleren Feldstärkenbereich sind wir auf eine plausible Interpolation
zwischen (7) und (18) angewiesen. Als einfachste Möglichkeit bietet sich
die Summe beider Ausdrücke, da jeder Teil im „falschen" Bereich gegen
Null geht. Berücksichtigt man beide Trägersorten, so ist dann die gesamte Raumladung

$$\varrho = eN_0 + \left\{ \left[\frac{j_p + b_p \cdot kT \cdot \mathrm{grad}\,p}{b_p} - \frac{j_n - b_n \cdot kT \cdot \mathrm{grad}\,n}{b_n} \right] \cdot \frac{1}{F} + \\ + e\,(s_p j_p - s_n j_n) \cdot F \, . \right\} \quad (21)$$

Wir hatten nur s_p abgeschätzt; s_n dürfte wegen der größeren Band-
breite erheblich größer sein. In späteren Beispielen setzen wir will-
kürlich $s_n = 0.2 \cdot s_p$.

II, 5. Berechnung der Randfeldstärken.

II, 5a. Vernachlässigung des Staugliedes in (21) führt auf die nor-
male Erschöpfungsrandschicht, von der SCHOTTKY [13] gezeigt hat, daß
man den wahren Raumladungsverlauf durch eine konstante Ladungsdichte $\varrho = e \cdot N_0$ von endlicher Dicke l ersetzen kann.

Wir rechnen daher ebenfalls mit einer endlichen Dicke der Randschicht, setzen aber für die Ladungsdichte

$$\varrho = e \cdot N_0 + e\,s_p \cdot j_s \cdot F, \quad (21a)$$

wo

$$j_s = j_p - \frac{s_n}{s_p} \cdot j_n \quad (22)$$

ist.

Zur Theorie des Germaniumgleichrichters und des Transistors. **443**

Bei Flächenkontakten lautet die POISSON-Gleichung dann:

$$\frac{d^2\varphi}{dx^2} = -\frac{4\pi e}{\varepsilon}\left[N_0 + s_p \cdot j_s \cdot \frac{d\varphi}{dx}\right]$$

mit den Randbedingungen

$$e\left[\varphi(l) - \varphi(0)\right] = eU + V_D; \qquad \varphi'(l) = 0.$$

Man findet durch Elimination von l das implizite Gesetz für die Randfeldstärke F_R:

$$eU + V_D = \frac{\varepsilon}{4\pi} \cdot \frac{N_0}{(s_p \cdot j_s)^2}\left[\frac{s_p \cdot j_s \cdot F_R}{N_0} - \ln\left(1 + \frac{s_p \cdot j_s \cdot F_R}{N_0}\right)\right]. \qquad (23)$$

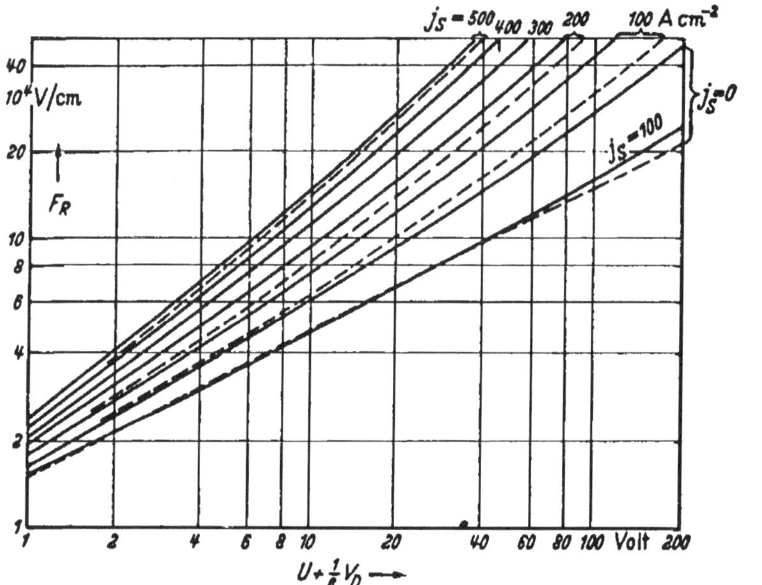

Fig. 5. Randfeldstärke für $N_0 = 10^{11}$ cm⁻³ für Spitzenkontakte mit $r_0 = 10^{-3}$ cm (———) und für Flächenkontakte (— — —).

II, 5b. Die uns mehr interessierenden Kontakte sind Spitzenkontakte von etwa 10^{-3} cm Durchmesser. Wir rechnen daher noch F_R für kugelförmige Kontakte (Radius r_0) mit kugelförmiger Randschicht (Radius R) aus.

Weil $j(r)$ nicht mehr räumlich konstant ist, sondern

$$j(r) = j \cdot \frac{r_0^2}{r^2} \quad \dagger$$

ist, lautet die POISSON-Gleichung jetzt

$$\frac{d^2\varphi}{dr^2} + \frac{2}{r} \cdot \frac{d\varphi}{dr} = -\frac{4\pi e}{\varepsilon}\left[N_0 + s_p \cdot j_s \cdot r_0^2 \cdot \frac{1}{r^2}\frac{d\varphi}{dr}\right]$$

$\dagger$ Wenn j ohne Argument angegeben ist, bedeutet es künftig stets die Stromdichte in der Kontaktfläche.

444   HERBERT KRÖMER:

mit den Randbedingungen

$$e\left[\varphi(R) - \varphi(r_0)\right] = eU + V_D; \qquad \varphi'(R) = 0\,.$$

Dann wird der Zusammenhang zwischen U, F_R und R durch das Glei-
chungspaar

$$F_R = \frac{3\alpha}{r_0^2} \cdot e^{\frac{\beta}{r_0}} \cdot \int_{r_0}^{R} e^{-\frac{\beta}{s}} \cdot s^2\, ds\,, \qquad (24a)$$

$$eU + V_D = \frac{r_0^2}{\beta} \cdot e F_R - \frac{e\alpha}{\beta}\left(R^3 - r_0^3\right) \qquad (24b)$$

gegeben, wobei

$$\alpha = \frac{4\pi e}{\varepsilon} \cdot \frac{N_0}{3} \quad \text{und} \quad \beta = \frac{4\pi e}{\varepsilon} \cdot s_p \cdot j_s \cdot r_0^2$$

ist.

Fig. 5 zeigt $F_R(U, j_s)$, wobei $s_p = 2 \cdot 10^8\,\mathrm{W}^{-1}$ und $N_0 = 10^{16}\,\mathrm{cm}^{-3}$
gesetzt wurde, für einen Spitzenkontakt mit $r_0 = 10^{-3}\,\mathrm{cm}$. Zum Ver-
gleich sind (gestrichelt) einige Kurven für Flächenkontakte ebenfalls
eingetragen.

III. Anwendung auf die Kennlinien.

Durch Kombination von Gl. (3) und (4) und Fig. 5 läßt sich die
Schar der Kennlinien mit j_s als Scharparameter bestimmen.

III, 1. Transistorenkennlinien.

III, 1a. Beim Transistor kommt praktisch der gesamte Löcheranteil
von j aus dem Emittorstrom, und es interessieren die Kennlinien mit
bestimmtem j_p. Diese ergeben sich aus denen mit j_s als Scharparameter,
indem man noch die Schar der waagerechten Geraden

$$j_n = \frac{s_p}{s_n}\left(j_p - j_s\right)$$

in das $(j_n - U)$-Diagramm einzeichnet und die Schnittpunkte zusammen-
gehöriger Kurven verbindet (Fig. 6). Die Kennlinien für den Gesamt-
strom $j_n + j_p$ ergeben sich daraus einfach durch senkrechtes Verschieben
um j_p. Wenn man annimmt, daß der Emittorstrom reiner Löcherstrom
ist, der vollständig zum Kollektor gelangt, sind das die fertigen Transistorkennlinien.

Der Löcherstrom erhöht den Elektronenstrom in mit steigender
Spannung zunehmendem Maße, jedoch höchstens solange, bis die Zu-
nahme der Stauladung des letzteren die des Löcherstromes wieder kom-
pensiert hat. Dann werden die zu verschiedenen j_p gehörigen Kurven
nahezu parallel. Die „innere Stromverstärkung"

$$\alpha_i = \left(\frac{\partial(j_n + j_p)}{\partial j_p}\right)_{U=\mathrm{const}}$$

des Kontakts nimmt also zunächst mit der Spannung zu, bleibt aber schließlich nahezu konstant, und zwar bleibt stets

$$\alpha_i < 1 + \frac{s_p}{s_n}.$$

Die Schnelligkeit des Anstiegs von α_i hängt stark von Φ_0 und T ab, und zwar gilt allgemein: Je höher der Leerstrom des Kontaktes ist, desto rascher steigt α_i an.

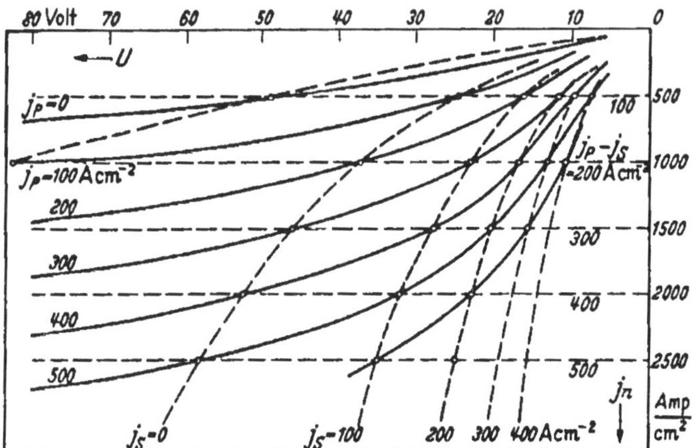

Fig. 6. Konstruktion der Elektronenstromkennlinien für konstanten Löcherstrom. Daten: $r_0 = 10^{-3}$ cm; $e\,\Phi_0 = 0{,}38$ eV; $s_n:s_p = 1:5$; $T = 300°$ K. Sonst wie Fig. 5.

Fig. 7a u. b zeigen zusätzlich zu Fig. 6 zwei Transistorkennlinien mit verschiedenem Φ_0. Die Daten sind: $N_0 = 10^{15}$ cm^{-3}, $r_0 = 10^{-3}$ cm, $s_p : s_n = 5$, $T = 300°$ K und $e\,\Phi_0 = 0{,}38$ und $0{,}50$ eV.

III, 1b. Während im Gültigkeitsbereich der Diffusionstheorie nur α_i-Werte mit $\alpha_i < 1 + \frac{b_n}{b_p} \approx 3$ verständlich sind [*14*], liefert der Staueffekt die Möglichkeit höherer Werte, maximal $1 + \frac{s_p}{s_n}$, vorausgesetzt, daß $\frac{s_p}{s_n} > \frac{b_n}{b_p}$ ist, was wir als sehr wahrscheinlich ansehen möchten. Auch die Allgemeingestalt der Kennlinien wird durch Fig. 7 befriedigend wiedergegeben[1], jedenfalls solange es sich nicht um zu stark durchformierte Kollektoren handelt.

Beim Formieren ändern sich die Kennlinien sehr stark: Die α-Werte steigen und werden dabei gleichzeitig bis zu weit niedrigeren Spannungen als vorher praktisch spannungsunabhängig, so daß die Kennlinienschar oberhalb größenordnungsmäßig 1 V praktisch parallel wird. Was dabei im einzelnen vorgeht, ist noch nicht genau bekannt. Vermutlich tritt sowohl eine Erhöhung von F_R durch eingeschwemmte Donatorionen als

---

[1] Vgl. etwa Bardeen und Brattain [*1*], Fig. 3, oder Shockley [*15*], Fig. 2 und 6.

446   HERBERT KRÖMER:

auch eine Erniedrigung von Φ_0 durch irgendwelche Grenzflächeneffekte
auf [2]. Es ist jedoch fraglich, ob man damit bereits auskommt. Uns
scheint es sehr wahrscheinlich, daß noch irgendwelche Veränderungen
in der Randschicht eintreten, welche die durch die Diodentheorie gegebene
Feld-Stromabhängigkeit [Gl. (3) und (4)] abändern[1].

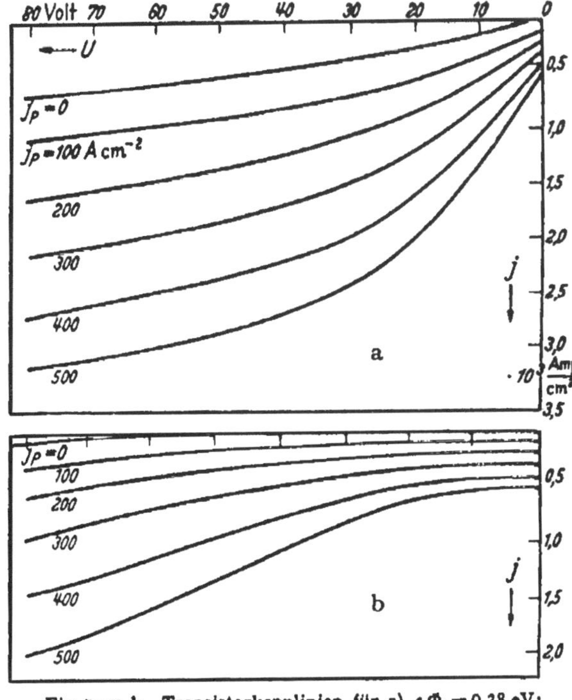

Fig. 7a u. b. Transistorkennlinien für a) $e\,\Phi_0 = 0{,}38$ eV;
b) $e\,\Phi_0 = 0{,}50$ eV. Sonstige Daten wie Fig. 6.

III, 2. Diodenkennlinien.

III, 2a. Bei Dioden ist keine fremde Löcherquelle vorhanden, und der Löcherstrom besteht aus der geringen Anzahl von Löchern die im Halbleiter bereits vorhanden sind, oder die von der Oberfläche oder von der Basiselektrode her einströmen. Diese Löcher gelangen durch Diffusion und durch das Potentialgefälle, das der vom Kollektor ausgehende Elektronenstrom im HL erzeugt, zum Kollektor. Bei hinreichend hohem Elektronenstrom (worauf wir uns hier beschränken) ist die Diffusion vernachlässigbar und der Löcherstrom wird proportional zum Elektronenstrom

$$j_p = G \cdot j_n. \tag{25}$$

Bei homogener Temperaturverteilung und vernachlässigbarer Oberflächenleitung ist

$$G = G_0 = \frac{b_p \cdot p_0}{b_n \cdot n_0} = \frac{b_p}{b_n} \cdot e^{-\frac{E_V - E_L}{kT}}.$$

Im allgemeinen weicht G von diesem Wert aus drei Gründen ab: 1. Die
Temperatur ist infolge der Kontakterwärmung inhomogen. 2. Es existiert eine Oberflächenleitfähigkeit. 3. Bei nicht völlig sperrfreier Basiselektrode kann an der Basis eventuell $j_p/j_n > G_0$ sein; dann wirkt die
Basis wie ein Emittor.

---

[1] Vgl. hierzu das auf S. 450 Gesagte. Eine andere Abweichung von der Diodentheorie stellt z.B. die p-n-Hook-Theorie (SHOCKLEY [14], [15], [16]) dar, die uns jedoch nicht frei von Schwierigkeiten scheint.

Zur Theorie des Germaniumgleichrichters und des Transistors. **447**

Wir verzichten auf eine allgemeine Abschätzung von G und wollen die Kennliniengestalt in Abhängigkeit von G untersuchen.

Die Kennlinien ergeben sich aus der Schar der Kennlinien für festes j_s, indem man diese mit der Schar der Geraden

$$j_n = \frac{j_s}{G - \dfrac{s_n}{s_p}}$$

schneidet und wieder die zusammengehörigen Schnittpunkte verbindet.

III,2b. Bei sehr hochsperrenden Kontakten ist der Strom so niedrig, daß der Staueffekt sich erst

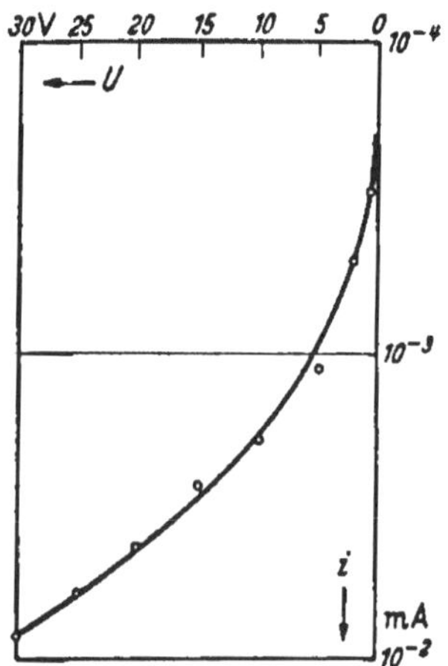

Fig. 8. Vergleich mit einer von BENZER gemessenen Kennlinie.

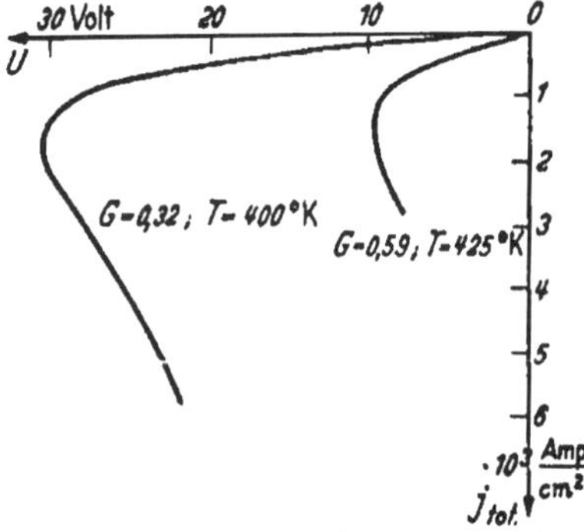

Fig. 9. Umbiegende Kennlinien. Daten außer der Temperatur wie Fig. 7a.

bei hohen Spannungen bemerkbar macht. Man kann dann im Bereich mittlerer Spannungen einfach mit der zu $j_s = 0$ gehörigen Kennlinie rechnen. Das ist in Fig. 8 für $e\,\Phi_0 = 0{,}54\,\mathrm{eV}$, $V_0 = 10^{-3}\,\mathrm{cm}$ und $T = 298°\,\mathrm{K}$ geschehen und mit der zur gleichen Temperatur gehörigen Kurve bei BENZER[1] verglichen. Dabei wurde mit $N_0 = 10^{15}\,\mathrm{cm}^{-3}$ und $\eta = 1$ gerechnet. Die Übereinstimmung ist befriedigend. Für ein höheres η wäre sie wesentlich schlechter; ein höheres N_0, was auch ein höheres η zulassen würde, ist unwahrscheinlich. Sinn dieses Vergleichs ist es, zu zeigen, in welcher Größenordnung Φ_0 bei hochsperrenden Gleichrichtern liegt, und daß man tatsächlich mit $\eta \approx 1$ rechnen muß, um die Experimente erklären zu können.

III,2c. Für $G < s_n/s_p$ ist der Staueffekt der Löcher schwächer als der der Elektronen. Es kommen nur Schnittpunkte mit negativem j_s in Frage; der Strom ist geringer als für $j_s = 0$. Irgendwelche Besonderheiten treten nicht auf.

Es sei jetzt $G > s_n/s_p$. Aus $j_n = j_n(U, j_s)$ folgt wegen (22)

$$dj_n = \left(\frac{\partial j_n}{\partial U}\right)_{j_s} \cdot dU + \left(\frac{\partial j_n}{\partial j_s}\right)_U \cdot \left(dj_p - \frac{s_n}{s_p} \cdot dj_n\right).$$

---

[1] Literatur [*3*], Fig. 10.

448   HERBERT KRÖMER:

Mit $d\dot{j} = d\dot{j}_n + d\dot{j}_p$ und $d\dot{j}_n = \dfrac{d\dot{j}}{1+G}$; $d\dot{j}_p = \dfrac{G \cdot d\dot{j}}{1+G}$ folgt dann

$$\frac{d\dot{j}}{dU} = \left(\frac{\partial \dot{j}_n}{\partial U}\right)_{\dot{j}_s} \cdot \frac{1+G}{1 - \left(G - \dfrac{s_n}{s_p}\right) \cdot \left(\dfrac{\partial \dot{j}_n}{\partial \dot{j}_s}\right)_U} \cdot \qquad (26)$$

Wenn der Nenner verschwindet, erhält die Kennlinie eine senkrechte Tangente und biegt in einen Bereich negativen differentiellen Widerstandes um. Da $\partial \dot{j}_n / \partial \dot{j}_s$ mit zunehmendem U und $\dot{j}_s$ unbegrenzt steigt, ist das für alle Kennlinien mit $G > s_n / s_p$ schließlich der Fall. Fig. 9 zeigt ein Beispiel.

Es erscheint uns nicht ausgeschlossen, daß infolge der Kontakterwärmung und einer Oberflächenleitfähigkeit G hinreichend groß wird, und daß dieser Mechanismus in vielen Fällen für das beobachtete Umbiegen der Kennlinien verantwortlich ist.

III, 3. Thermische Effekte.

III, 3a. Die bisher berechneten Kennlinien waren „isotherme", d.h. zu einer festen Kontakttemperatur gehörige Kennlinien. Tatsächlich erwärmen sich aber die Kontakte ungefähr proportional zur umgesetzten JOULEschen Wärme:

$$T_K - T_0 = \delta \cdot (\dot{j} \cdot U). \qquad (27)$$

Für halbkugelförmige Kontakte in einem nicht zu kleinen HL-Stück gilt

$$\delta = \frac{r_0}{\varkappa} \approx 1{,}7 \cdot 10^{-3}\,°\mathrm{C} \cdot \mathrm{W}^{-1}\,\mathrm{cm}^2$$

bzw.

$$\frac{\delta}{2\pi\,r_0^2} = \frac{1}{2\pi\,\varkappa\,r_0} \approx 270°\,\mathrm{C}\,\mathrm{W}^{-1}.$$

Dabei ist $r_0 = 10^{-3}$ cm und die Wärmeleitfähigkeit $\varkappa$ zu $0{,}14\,\mathrm{cal\,sec}^{-1}\,\mathrm{cm}^{-1}\,°\mathrm{C}^{-1}$ [6] angenommen worden.

Wenn der Strom reiner Elektronenstrom ist, ist bei seiner Berechnung T_K statt T_0 einzusetzen, da aber T_K seinerseits von $\dot{j}$ und U abhängt, werden die Gleichstromkennlinien graphisch ermittelt, indem man die Schar der zu verschiedenen T_K gehörigen isothermen Kennlinien mit der der Hyperbeln

$$\dot{j} = \frac{T_K - T_0}{\delta \cdot U}$$

zum Schnitt bringt und zusammengehörige Schnittpunkte verbindet[1]. Hierbei tritt, auch wenn die Isothermen weiterlaufen, meist ein Umbiegen auf (Fig. 10).

---

[1] HUNTER [8], Die praktische Durchführung geschieht am besten im doppelt logarithmischen Maßstab.

Wenn ein merklicher Löcherstrom auftritt, ist das graphische Verfahren in dieser einfachen Form nicht mehr durchführbar, was wir hier nicht weiter verfolgen wollen.

III, 3b. Die Idee, das beobachtete Umbiegen der Kennlinien in dieser Weise *rein* thermisch zu erklären, findet sich bereits bei HUN-TER [*8*]. Dieser einfachen Deutung steht die Tatsache gegenüber, daß die Hochfrequenzkennlinien selbst bei 30 MHz noch das Umbiegen mit einer Grenzspannung von derselben Größenordnung wie bei Gleichstrom

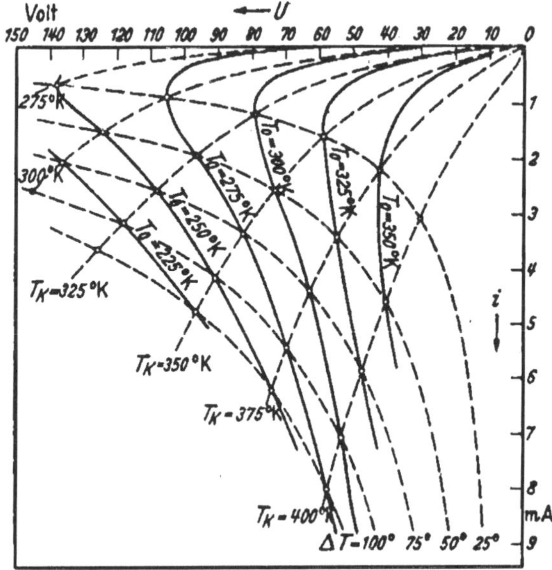

Fig. 10. Konstruktion der Gleichstromkennlinien. Daten: $e\,\Phi_0 = 0,50$ eV; $r_0 = 10^{-2}$ cm; $j_s = 0$; $\delta = 1,7 \cdot 10^{-3}$ °C cm^2 W^{-1}.

zeigen[1], obwohl bei diesen Frequenzen die Temperaturänderung nicht mehr mitkommen kann. Es muß also ein Mechanismus vorhanden sein, durch den bereits die isothermen Kennlinien umbiegen. Einen solchen lieferte oben der Staueffekt.

III, 3c. Die Temperaturabhängigkeit des Umklappgebietes wird durch die hier vorgelegte, eine Erweiterung der Diodentheorie durch den Staueffekt darstellende Theorie noch nicht richtig dargestellt. Während (nach BENZER) das Produkt Strom und Spannung am Umbiegepunkt mit steigender Basistemperatur über einen weiten Bereich linear abfällt und bei Extrapolation des linearen Teils etwa beim Einsetzen der Eigenleitung verschwinden würde, liefert die Theorie in der bisherigen Form nur eine sehr schwache T_0-Abhängigkeit dieses Produktes, es sei denn, man macht die unplausible Annahme, daß G mit steigender Temperatur stark abfällt. Eine nähere Untersuchung zeigt, daß zur Erklärung der

---

[1] S. TORREY-WHITMER [*18*], S. 381.

450 H. Krömer: Zur Theorie des Germaniumgleichrichters und des Transistors.

Experimente im Bereich höherer Spannungen eine weit schwächere T-Abhängigkeit der isothermen Kennlinien gefordert werden muß, als die Diodentheorie (und auch die Diffusionstheorie) sie liefert.

Ob man zur Erfüllung dieser Forderung die Annahme $\frac{d\Phi_0}{dT} > 0$ machen darf, müßte erst untersucht werden. Wir möchten eher annehmen, daß diese Diskrepanz neben dem auf S. 446 Gesagten ein zweiter Hinweis darauf ist, daß das Randschichtmodell der Diodentheorie wesentlich verfeinert werden muß.

Für die Anregung zu dieser Arbeit und für zahlreiche Diskussionen habe ich Herrn Prof. Sauter zu danken.

Literatur.

[1] Bardeen, J., and W. H. Brattain: Phys. Rev. **75**, 1208 (1949). — [2] Bardeen, J., and W. G. Pfann: Phys. Rev. **77**, 401 (1950). — [3] Benzer, S.: J. Appl. Phys. **20**, 804 (1949). — [4] Bethe, H. A.: NDRC-Rep. 14, 23. Nov. 1942. — Bethe, H. A., u. R. G. Sachs: NDRC-Rep. 14, 10. Sept. 1942 u. 15. Juni 1943. — [5] Briggs, H. B.: Phys. Rev. **77**, 284 (1950). — [6] Grieco, A., and H. C. Montgomery: Phys. Rev. **86**, 574 (1952). — [7] Hill, R. W., and D. H. Parkinson: Phil. Mag. **43**, 309 (1952). — [8] Hunter, L. P.: Phys. Rev. **81**, 151 (1951). — [9] Müller, H.: Ann. Phys. (6) **9**, 141 (1951). — [10] Oldekop, W.: Diss. Göttingen 1952. — [11] Pearson, G. L., J. R. Haynes and W. Shockley: Phys. Rev. **78**, 295 (1950). — [12] Seiler, K.: Z. Naturforsch. **5a**, 393 (1950). — [13] Schottky, W.: Z. Physik **118**, 539 (1942). — [14] Shockley, W.: Phys. Rev. **78**, 294 (1950). — [15] Shockley, W.: Electrons and Holes in Semiconduktors. New York 1950. — [16] Shockley, W., M. Sparks and G. K. Teal: Phys. Rev. **83**, 151 (1951). — [17] Sommerfeld, A., u. H. A. Bethe: Handbuch der Physik, Bd. 24/II, Kap. 3. — [18] Torrey, H. C., and C. A. Whitmer: Crystal Rectifiers. New York 1948.

Göttingen, Institut für Theoretischen Physik[1].

---

[1] Jetzt am Fernmeldetechnischen Zentralamt in Darmstadt.

Druck der Universitätsdruckerei H. Stürtz AG., Würzburg

`` I promised myself that if a new technology for building heterostructures arose, I'd get back into it.''

Herbert Kroemer

Theory of a Wide-Gap Emitter for Transistors*

HERBERT KROEMER†, MEMBER, IRE

Summary—In order to obtain a high current amplification factor, it is important in transistors that the ratio of the injected minority carrier current over the total emitter current, γ, be close to unity, or that the quantity $1-\gamma$, called the injection deficit, be as small as possible.

It is shown that the injection deficit of an emitter can be decreased by several orders of magnitude if the emitter has a higher band gap than the base region. This effect can be utilized either in addition to the commonly used high emitter doping in order to eliminate the alpha falloff with current, or to decrease the high emitter doping in order to obtain a lower emitter capacitance.

Decreasing the emitter capacitance in high-frequency transistors may be utilized either to extend their frequency range or to increase their power capabilities by increasing the area.

INTRODUCTION

AN important quantity characterizing the emitter of any transistor is the emitter efficiency, γ, defined as that fraction of the total emitter current that is minority-carrier injection current. Since the current amplification factor, α_{ce}, is proportional to γ, it is desirable that γ be high. Actually, it is very important that γ be close to unity. In the usual grounded-emitter operation the current amplification factor of the transistor is

$$\alpha_{cb} = \frac{\alpha_{ce}}{1 - \alpha_{ce}} . \tag{1}$$

From this it follows that a change of γ such that α_{ce} increases from 0.98 to 0.99 (percentagewise a very small change of about 1 per cent) increases α_{cb} from about 49 to about 99, that is, by a factor of two. It is therefore more appropriate to consider the "injection deficit" $1-\gamma \approx (1-\gamma)/\gamma$ rather than γ itself as a measure of the transistor performance. In a good transistor, the injection deficit ought to be small.

In a p-n-p transistor,

$$\gamma = \frac{j_p}{j_n + j_p} , \tag{2}$$

$$\frac{1 - \gamma}{\gamma} = \frac{j_n}{j_p} , \tag{3}$$

where j_p and j_n are the densities of the hole and electron currents at the emitter junction. In order to obtain a low ratio of electron to hole current, it is necessary in a semiconductor with constant band gap to dope the p side of the junction much more heavily than the n side. There are practical limits, however, as to the magnitude of doping possible, and there are situations where a high

* Original manuscript received by the IRE, April 12, 1957; revised manuscript received, July 29, 1957.
† RCA Labs., Princeton, N. J.

doping of the emitter is undesirable for other reasons. In all cases, an improvement could be obtained if the injection deficit could be decreased by other means instead of, or in addition to, the high doping.

The purpose of this paper is to point out that it is possible to lower the injection deficit by using an emitter material with a wider band gap than the base material (Fig. 1).[1] The reason for this is that in such a case the activation energy (or contact potential) qV_n for electrons flowing from the base into the emitter is higher than the activation energy (or contact potential) qV_p for holes entering the base from the emitter (see Fig. 1). The difference in activation energies is the difference in bandwidth, ΔE. Since the activation energy enters exponentially into the current-flow equations, this means a decrease in the injection deficit by a fraction of $\exp(-\Delta E/kT)$, all other things being equal. This is shown quantitatively in the next section.

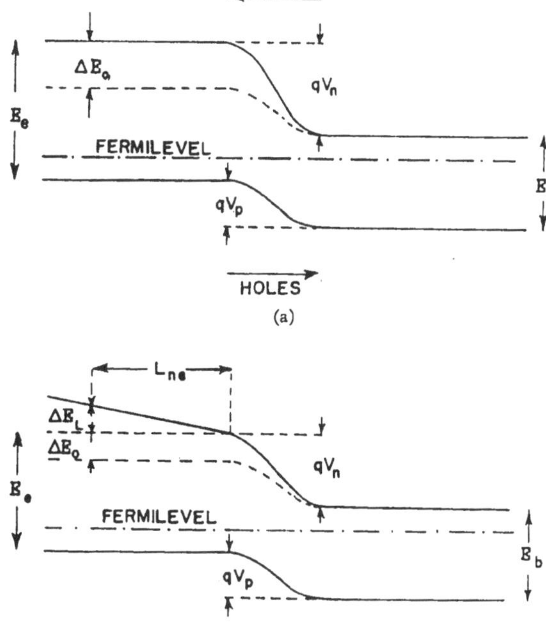

Fig. 1—Band structure of a wide-gap emitter junction. (a) With constant gap outside the depletion region and (b) with linear gap variation inside the emitter.

[1] It has been pointed out to the author that this principle was first suggested by W. Shockley in U. S. Patent No. 2,569,347, issued September 25, 1951.

The Injection Deficit in a Wide-Narrow Junction

We introduce the following notation:

j_n, j_p = electron and hole current densities at the junction.

D_{ne}, D_{pb} = diffusion constants for electrons in the emitter and holes in the base, and

L_{ne}, L_{pb} = corresponding diffusion lengths.

n_{oe}, p_{ob} = equilibrium minority carrier densities in the emitter (electrons) and the base (holes) adjoining to the junction.

V = applied bias.

n_{ie}, n_{ib} = intrinsic carrier densities of the emitter and base semiconductors.

P_e, N_b = net acceptor density in the emitter and net donor density in the base.

$\left.\begin{array}{l} m_{ne}{}^{*}, m_{pe}{}^{*} \\ m_{nb}{}^{*}, m_{pb}{}^{*} \end{array}\right\}$ = effective masses of the electrons and the holes in the emitter region and the base region.

E_e, E_b = emitter band gap and base band gap. $\Delta E_0 = E_e - E_b$.

With this notation for a simple p-n junction,[2]

$$i_n = \frac{qD_{ne}n_{oe}}{L_{ne}}\left(e^{\frac{qV}{kT}} - 1\right) \tag{4a}$$

$$i_p = \frac{qD_{pb}p_{ob}}{L_{pb}}\left(e^{\frac{qV}{kT}} - 1\right) \tag{4b}$$

$$\frac{j_n}{i_p} = \frac{D_{pe}}{D_{pb}}\frac{L_{pb}}{L_{ne}}\frac{n_{oe}}{p_{ob}}. \tag{5}$$

Now,

$$n_{oe} = \frac{n_{ie}^2}{P_e}, \qquad p_{ob} = \frac{n_{ib}^2}{N_b}. \tag{6a, b}$$

Furthermore,

$$\frac{n_{ie}^2}{n_{ib}^2} = \left(\frac{m_{ne}{}^{*}m_{pe}{}^{*}}{m_{nb}{}^{*}m_{pb}{}^{*}}\right)^{3/2} \exp = (\Delta E_0/kT) \tag{7}$$

Therefore,

$$\frac{j_n}{i_p} = \frac{D_{ne}}{D_{pb}}\frac{L_{pb}}{L_{ne}}\frac{N_b}{P_e}\left(\frac{m_{ne}{}^{*}m_{pe}{}^{*}}{m_{nb}{}^{*}m_{pb}{}^{*}}\right)^{3/2}\exp\left(-\Delta E_0/kT\right). \tag{8}$$

This expression differs from that for a junction between semiconductors of equal band gaps by the factor

[2] We treat the case of a simple p-n junction rather than a transistor to obtain more symmetrical notation. In a transistor with a base width $w_b \ll L_{pb}$, one has to replace L_{pb} by w and to omit the "minus one" behind the exponential factor. For the derivation of (4) see W. Shockley, "The theory of p-n junctions in semiconductors and p-n junction transistors," *Bell Sys. Tech. J.*, vol. 28, pp. 435–489; July, 1949.

$$\left(\frac{m_{ne}{}^{*}m_{pe}{}^{*}}{m_{nb}{}^{*}m_{pb}{}^{*}}\right)^{3/2} \exp\left(-\Delta E_0/kT\right) \tag{9}$$

of which the important part is the exponential factor. If, for example, ΔE is 0.2 ev, then at room temperature $kT = 0.025$ ev and $\Delta E/kT = 8$. Assuming the effective masses to be identical, the injection deficit is then decreased by a factor of $e^{-8} = 1:3000$.

If the band gap in the emitter region is not constant but increases linearly with increasing distance from the junctions [Fig. 1(b)], (8) still does not give a full account of the change in injection deficit. If the above ΔE_0 is the gap difference across the depletion layer and if ΔE_L is the gap variation along a diffusion length, L_{ne}, in the emitter, then it can be shown that the factor

$$f(\Delta E_L) = \sqrt{\left(\frac{\Delta E_L}{2kT}\right)^2 + 1} - \frac{\Delta E_L}{2kT} \tag{10a}$$

has to be added to (4a), (8), and (9). For $\Delta E_L \gg 2kT$

$$f(\Delta E_L) \to \frac{kT}{\Delta E_L} \ll 1. \tag{10b}$$

Therefore, a band-gap variation outside the depletion layer also reduces j_n/j_p. However, this is to a smaller degree, namely only linearly rather than exponentially.

Eq. (8) does not hold for arbitrarily large ΔE's, however. This is because (4a) holds only so long as the density of electrons injected into the p-type region remains small compared to the electron density in the source, that is in the n-type region. The analogous statement holds for holes. Mathematically this means

$$V_n - V \gg kT \tag{11a}$$

$$V_p - V \gg kT. \tag{11b}$$

In a p-type wide-gap emitter $V_n > V_p$ and (11a) is fulfilled automatically if (11b) is. Consequently, (8) holds only for voltages that satisfy (11b). But, (11b) also implies that for a workable wide-gap emitter V_p must not

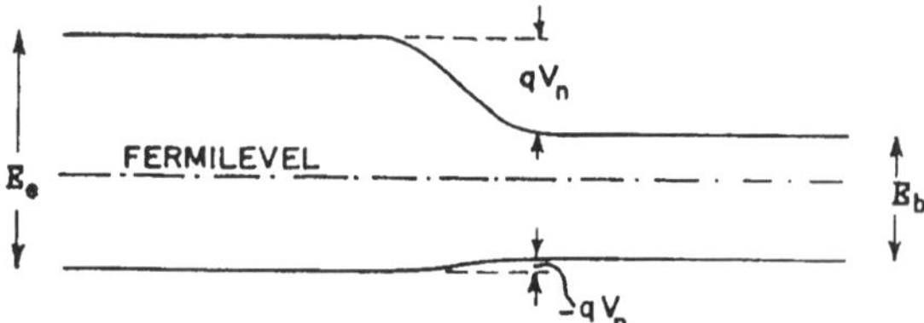

Fig. 2—Wide-narrow junction with negative V_p, due to a low doping ratio. Not suited as wide-gap emitter.

be negative. Therefore, a structure like in Fig. 2 would not be a good wide-gap emitter because the minority carrier density in the base is larger than the majority carrier density in the emitter.

Alpha and Alpha Falloff

In order to estimate the influence of the decreased injection deficit upon α_{cb} in a transistor, one has to take the recombination losses into account. If ρ is the fraction of injected carriers that recombine on the way to the collector, then for $\rho \ll 1$, from (1) and (2)

$$\frac{1}{\alpha_{cb}} \approx \rho + (1 - \gamma) \approx \rho + \frac{j_n}{j_p} . \tag{12}$$

If j_n/j_p is small compared to ρ, α_{cb} is determined solely by the recombination losses. This is the case for many transistors at low injection currents. At high injection currents the injected hole density in the base becomes comparable to the donor density. To maintain electrical neutrality, the electrons in the base increase by the same number. This means that the electron current into the emitter is bigger than the value given by (4a) by the factor by which the electron density has increased. For increasing current, therefore, j_n/j_p is not constant but increases (linearly) with current. Eventually j_n/j_p becomes comparable with and larger than ρ, resulting in the well-known alpha falloff.[3]

If the emitter has a wide band gap, the ratio j_n/j_p increases with current by the same *factor* as for an ordinary emitter. Since on an *absolute* scale, however, j_n/j_p is lower by the factor (9), the alpha-falloff effect sets in at much higher current densities. Since j_n/j_p increases linearly with the total current, alpha falloff sets in at currents which are bigger by the reciprocal of (9), compared to an otherwise identical constant-gap transistor. In our numerical example of $\Delta E = 0.2$ *ev*, the alpha falloff sets in at 3000 times the current. This means the alpha falloff is practically nonexistent.

Capacitance

Another consequence of the exponential factor in (9) is the following:[4] In many transistors for small-signal operation, it is of not primary importance to minimize the falloff effect to the point of vanishing. In these cases the exponential factor may be used to decrease the doping in the emitter by this same factor (9) and still have an unchanged falloff characteristic. It would then be possible to have a usable emitter efficiency with an emitter that has a considerably lower impurity density than the base region. This, however, would imply a reduced emitter transition capacitance.

In the case of audio-frequency large-signal transistors, the emitter transition capacitance is of no great importance while alpha falloff is a very serious effect. In this case one therefore should maintain the high doping

in the emitter. The situation is completely reversed, however, for very-high-frequency transistors, like the drift transistor or the *p-n-i-p* transistor. In these transistors the current amplification factor is, under usual operating conditions, not limited by the injection deficit but rather by transit time effects. The low-frequency alpha falloff therefore is not an important quantity in this case. However, since high-frequency transistors have a rather high impurity density in the base region, the emitter capacitance is rather high. As a result, the emitter capacitance often becomes the limiting factor for the *over-all* frequency behavior of the transistor. In such a case, a wide-gap emitter with a lower doping might improve the over-all frequency limit considerably.

Quantitatively, the capacitance of an abrupt junction is[5] (per unit area)

$$C = \sqrt{\frac{q\epsilon}{8\pi} \frac{NP}{N + P} \frac{1}{V + V_c}} \tag{13a}$$

where V_c is the contact potential. If the two sides have a different dielectric constant,

$$C = \sqrt{\frac{q\epsilon_n N \cdot \epsilon_p P}{8\pi(\epsilon_n N + \epsilon_p P)} \frac{1}{V + V_p}} . \tag{13b}$$

If $P \gg N$ this simplifies to

$$C = \sqrt{\frac{q\epsilon_n N}{8\pi(V + V_c)}} \tag{14a}$$

while for $P \ll N$

$$C = \sqrt{\frac{q\epsilon_p P}{8\pi(V + V_c)}} . \tag{14b}$$

If, in a constant-gap transistor, a doping ratio $P:N = 30$ is assumed as an example, the introduction of a 0.2 *ev* wider emitter band gap allows a reduction of this ratio by $1/3000$, namely to $P:N = 1:100$ without a change in γ. The capacitance, then, would be decreased to one tenth of the original value assuming identical dielectric constants.

A reduction of the emitter capacitance of this order could be utilized either to increase the frequency limit of the transistor or to increase the emitter (and collector) area. In the latter case, one would obtain a higher power capability for the same frequency response.

The Wide-Gap Collector

The use of a wide-gap semiconductor in the collector region would have an advantage only if the collector region at the same time had a lower impurity concentration than the base region.[4] Then one would obtain the lower collector capacitance of a high-resistivity collector region without the increased saturation current that is associated with a higher resistivity collector region in the constant-gap transistor.

[3] W. M. Webster, "On the variation of junction transistor current-amplification with emitter current," Proc. IRE, vol. 42, pp. 914–920; June, 1954.

[4] H. Kroemer, "Zur theorie des diffusions und des driftransistors, part III," *Archiv der Elektrischen Übertragung*, vol. 8, pp. 499–504 November, 1954.

[5] Shockley, *loc. cit.*

`` I was told not to work on light emitting semiconductors, because my ideas were judged on the basis of then existing applications.''

Herbert Kroemer

H. Kroemer, ``A Proposed Class of Heterojunction Injection Lasers,'' Proc. IEEE, Vol. 51(12), pp. 1782-1783, Dec. 1963. [Discussion ibid., Vol. 52(4), pp. 426-427, 1964].

1782 *PROCEEDINGS OF THE IEEE* *December*

Optical Correlation Technique*

This communication describes an optical technique for correlating electronic signals in real time. By this technique 0.5-μsec pulses of 15-Mc carrier were dispersed to 60 μsec and compressed back to 0.5 μsec using linear and pseudo-random frequency modulation. The correlation was accomplished with two identical optical systems, one serving as a transmitter and the other as a receiver. Correlation was also obtained in a single optical system having both transmit and receive capability.

The experimental arrangement is shown schematically in Fig. 1. The operation is as follows: Light from a slit source is collimated by lens L_1 onto an ultrasonic light modulator (ULM). The slit source was produced by focusing the light from a 100-watt mercury arc on a slit. The ULM consists of a small light-transparent tank of water with a quartz transducer mounted at one end. Within the tank and adjacent to the propagation path of the transducer is located a film replica containing a series of lines of varying periodicity similar to a nonuniform diffraction grating.

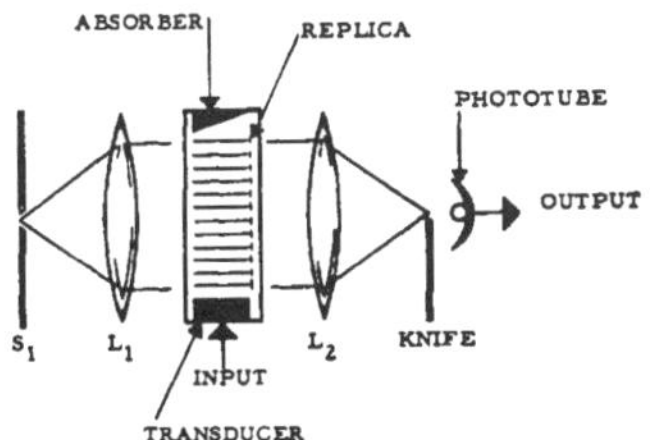

Fig. 1—Optical correlator.

When the transducer is excited by the 0.5-μsec pulse of 15 Mc, it sets up ultrasonic waves in the water which modulate the wavefront of the collimated light and scan the replica. The ultrasonic waves disturb the water in only a narrow region compared to the length of the replica (about 1/120 of effective length). Since the line spacing in the replica corresponds to the ultrasonic wavelengths in the water, the moving ultrasonic "grating" in effect beats with the replica lines and causes fluctuations in the intensity of the diffracted light focused on the photomultiplier tube by lens L_2. The knife in front of the phototube serves to remove the noninformation-carrying components of the light. The variation in the spacing of the replica lines (linear or pseudo-random) corresponds to the signal bandwidth of 2 Mc. Hence, the frequencies contained in the 0.5-μsec pulse are selectively delayed over 60 μsec, the time required for the ultrasonic waves to scan the replica. The output of the phototube is a 60-μsec electronic pulse having a 2-Mc bandwidth about 15 Mc.

For compression, the dispersed pulse, suitably amplified, is applied to the transducer in another identical optical system or,

* Received October 9, 1963.

in the case of single system operation, to a second transducer in the same ULM (used for generating the dispersed pulse). This time, however, the replica is scanned in the opposite direction to reverse the frequency delay contained in the dispersed pulse. Correlation occurs when the ultrasonic waves match the replica over the entire aperture. The phototube output at correlation resembles the initial 0.5-μsec pulse and contains the 15-Mc carrier.

Fig. 2 shows the dispersed pulse and Fig. 3, the compressed pulse obtained with linear frequency modulation. Tapering of the dispersed pulse is attributable to transducer and amplifier band-pass characteristics. The sharp cutoff at edges of the dispersed pulse correspond to the optical system aperture limits.

Fig. 4 shows the dispersed pulse and Fig. 5, the compressed pulse obtained with pseudo-random frequency modulation. No tapering of the dispersed pulse is observed since the frequencies are randomly distributed. Amplitude variation within this pulse is caused mostly by imperfections in the film replica. The replicas used in the experiments

Fig. 2—Dispersed pulse with linear modulation (10 μsec/box).

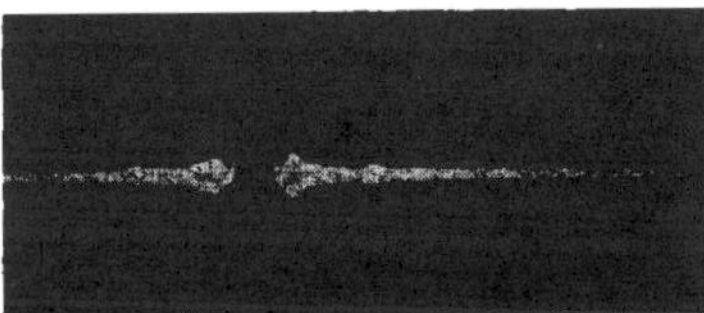

Fig. 3—Compressed pulse with linear modulation (2 μsec/box).

Fig. 4—Dispersed pulse with pseudo-random modulation (10 μsec/box).

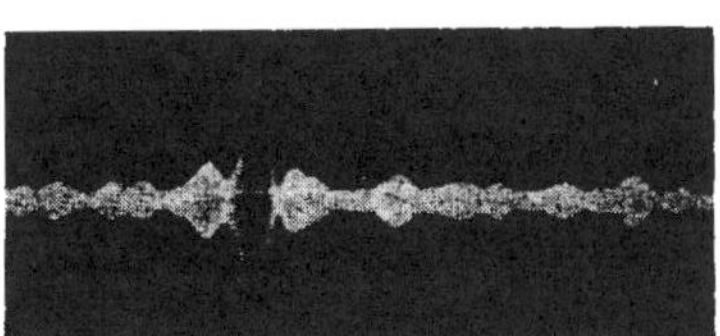

Fig. 5—Compressed pulse with pseudo-random modulation (2 μsec/box).

were produced by a special photographic technique developed by the writer.

Earlier work by Reich and Slobodin[1] discussed real-time optical correlation by a Schlieren technique involving the use of four lenses. The lenses were required to be of very high quality to produce acceptable correlation. In the present system, only the collimating lens need be of high quality; the second lens merely serves to integrate the light onto the phototube. Moreover, positioning the replica adjacent to the acoustic beam path eliminates signal degradation caused by lens field curvature and makes optical alignment much less critical.

Leo Slobodin
Advanced Techniques Dept.
Military Systems Div.
Lockheed Electronics Co.
Plainfield, N. J.

[1] A. Reich and L. Slobodin, "Optical Pulse Expansion/Compression," presented at 1961 Natl. Aerospace Electronics Conf., Dayton, Ohio; May 8, 1961.

A Proposed Class of Heterojunction Injection Lasers*

Laser action in semiconductors has so far been reported only for direct-gap semiconductors like GaAs,[1] GaAs$_x$P$_{1-x}$,[2] InAs,[3] InP,[4] etc., but not yet for indirect gap materials such as Ge, Si and GaP. Even in the direct gap semiconductors it has been necessary, in most cases, to cool the device in order to obtain carrier degeneracy at relizable injection levels.[5] We propose that laser action should be obtainable in many of the indirect gap semiconductors, and improved in the direct gap ones, if it is possible to supply them with a pair of heterojunction injectors. These should consist of heavily doped semiconductor layers with a higher energy gap than the radiating semiconductor and ideally should be of opposite polarity (Fig. 1).

This proposal is based on the assumption that at sufficiently high carrier injection levels laser action could occur, and at higher temperatures than to date, in most semiconductors, including many indirect gap ones. In the latter ones this could take place by spill-over of electrons from their lowest energy valleys into the region of the smallest direct gap. In many semiconductors

* Received October 14, 1963.

[1] R. N. Hall, *et al.*, "Coherent Light Emission from GaAs Junctions," *Phys. Rev. Lett.*, vol. 9, pp. 366–368, Nov., 1962; M. I. Nathan, *et al*, "Emission of Radiation from GaAs pn Junctions," *Appl. Phys. Lett.*, vol. 1, pp. 62–64, Nov., 1962.
[2] N. Holonyak, and S. F. Bevacqua, "Coherent (Visible) Light Emission from Ga(As$_{1-x}$P$_x$) Junctions," *Appl. Phys. Lett.*, vol. 1, pp. 82–83, Dec., 1962.
[3] I. Melngailis, "Maser Action in InAs Diodes," *Appl. Phys. Lett.*, vol. 2, pp. 176–178, May, 1963.
[4] K. Weiser and R. S. Levitt, "Stimulated Emission from Indium Phosphide," *Appl. Phys. Lett.*, vol. 2, pp. 178–179, May, 1963.
[5] An example of room temperature operation at very high current density has been given by G. Burns and M. I. Nathan, "Room-Temperature Stimulated Emission," *IBM J. Res. and Dev.*, vol. 7, pp. 72–73, Jan., 1963.

the direct gap is only a little larger than the indirect gap, for example by 0.14 ev in Ge and by 0.35 ev in GaP, not however in Si (by 1.5 ev). Wide-gap injectors should be capable of providing the necessary injection levels to raise the electron quasi Fermi level above the energy at $k=0$, at least for those semiconductors that have an only slightly raised gap, and certainly for the direct gap ones.

The high injection efficiency of individual heterojunctions as wide-gap emitters in transistors was pointed out earlier.[6] In the device proposed here their effectiveness is greatly multiplied by their use in pairs. We consider the example of Fig. 1. An inner semiconductor base with a thermal energy gap ϵ_B and the thickness w is sandwiched between two outer semiconductor injectors with the gaps $\epsilon_I = \epsilon_B + \delta$. As in a transistor, the base width w shall be large compared to the Debye length, but small compared to the diffusion length in the base. The two injectors shall be heavily doped on both sides, and, in our case, to opposite polarities. For simplicity we also assume: a) The heterojunctions are sufficiently gradual, so that the band picture is as shown in Fig. 1, whith no discontinuities of the band edges, b) the density of states for electrons equals that for the holes and the two are the same in both semiconductors, and c) the Fermi level pentrations Δ_I into the bands are the same in both injectors. These simplifying assumptions could easily be removed.

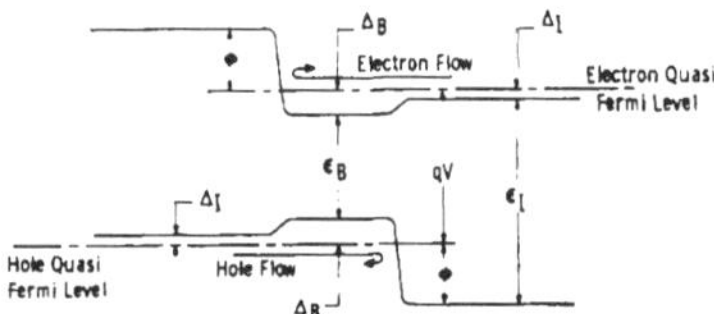

Fig. 1—Heterojunction laser structure with applied bias; $\epsilon_I + \Delta_I > qV > \epsilon_B + 2\Delta_I$.

If a forward bias V is applied to such a structure, both holes and electrons will be injected into the base. Because of the existence of the potential barriers at the heterojunctions these carriers cannot readily flow off, so long as $qV < \epsilon_I + \Delta_I$; they must pile up in the base region. As a result the electrons will be very nearly in equilibrium with the n^+ injector and they will be governed by its Fermi level. Similarly the holes will be governed by the p^+ Fermi level. Electroneutrality requires the two injected densities to be equal. Consequently, the separations of the two band edges from their respective Fermi levels must be equal, too.

For $qV > \epsilon_B$ the two Fermi levels penetrate into the allowed bands by the amount

$$\Delta_B = 1/2(qV - \epsilon_B) \quad (1)$$

and the injected carrier gas becomes degenerate. In the bias range

$$\epsilon_I + \Delta_I > qV > \epsilon_B + 2\Delta_I \quad (2)$$

the Fermi level penetration in the base re-

---

⁶ H. Kroemer, "Theory of a Wide Gap Emitter for Transistors," Proc. IRE, vol. 45, pp. 1535–1537, Nov., 1957.

gion exceeds that in the two injectors. In a homogeneous gap structure the potential barrier opposing the outflow of carriers from the middle region would have vanished at this point. But in the heterojunction structure the barrier still has the height

$$\Phi = \epsilon_I + \Delta_I - qV = \delta + \Delta_I - (qV - \epsilon_B) \quad (3)$$

from the Fermi level in the base. So long as $\Phi > 0$ a degenerately doped injector will be able to maintain the density in the base region against whatever outflow takes place into the opposite injector, which is the reason for the upper voltage limit in (2).

The injected carrier density, then, exceeds the density in the injectors, a situation impossible to achieve with homogeneous-gap junction structures. This is the situation shown in Fig. 1. At the upper end of the voltage range of (2), $\Delta_B = \delta/2$, and this quantity can easily be a sizeable fraction of 1 ev, for example about 0.35 ev for the combination Ge-GaAs. Degenerate spill-over into the Ge central valley would occur already for $\Delta_B \approx 0.14$ ev, which is considerably below $\delta/2$.

Structures with identically doped injector electrodes and an oppositely doped base are also possible and are probably easier to build. In this case the two injectors will have to be at the same potential, forward biased with respect to the base. The heterojunctions will again be efficient injectors of one carrier polarity, preventing the outflow of the neutralizing carriers of opposite polarity. There is no potential barrier for the injected carriers. But their outflow will still be substantially reduced because it would have to take place through the thin base, parallel to the junction. Even for conventional injection lasers the transverse junction dimensions commonly are at least of the order of a fraction of a millimeter. For such or even larger dimensions the resulting concentration gradients would be comparatively shallow, and because of this and because the base is likely to be thin compared to the transverse dimensions the outflow current would be low. Because of inevitable strong surface recombination one could not expect this low outflow if one of the junctions had been replaced by a surface.

The performance of lasers with identically doped injectors would, thus, fall in between that of conventional injection lasers, and that of oppositely doped heterojunction lasers, being much superior to the former but not quite as good as the latter. For such injector materials like ZnSe and ZnTe that appear to be available only in one polarity only the identically doped version is possible.

Because of the wider band gap of the injector the light from a heterojunction laser could be extracted through the injector, transverse to the plane of the junctions. This will not automatically be the case because the base thickness will often be much smaller than the lateral extension of the base and there will therefore be more gain in the parallel escape mode. However, if desired, this difference can obviously be overcome by a suitable device design. If this is done it is advisable to use fairly thin injector lay-

ers, in order to limit losses by free-carrier absorption in the injectors. In this way coherence over a much larger area could be obtained than in an ordinary injection laser.

The predicted injection levels will occur only if radiationless recombination processes do not assume catastrophic magnitudes. The two main processes are volume losses via recombination centers in the base region, and recombination at the heterojunctions proper.

We have considered both processes, and have concluded that, in high-quality Ge at least, recombination currents, say, in excess of 1000 A/cm^{-2} could be caused only by the interface dislocations that arise from the lattice mismatch between base and injectors. For the Ge-GaAs system this mismatch is 0.7×10^{-3}. Of the resulting dislocations all those will contribute to the recombination current that fall onto the lower-gap base side of the interface. The experimental values for the recombination efficiency of dislocation in Ge vary over a 1000:1 range.[7,8] If all the interface dislocations are assumed to contribute and if the most pessimistic values for the recombination efficiency[7] are chosen a rough calculation indicates recombination currents of the order 30,000 to 100,000 A/cm^2, which would be sustainable only pulsed and which would very severely limit the practicality of the proposed device.

Possibly the recombination efficiencies are lower than the most pessimistic values, but in any case these considerations teach the importance of a very close lattice fit and of retaining as many interface dislocations as possible on the high-gap injector side, where they would not contribute to the recombination current. The latter objective can be achieved by epitaxially growing the injector onto a pre-existing base at a sufficiently low temperature, the first by a judicious selection of semiconductor materials and by improving the lattice fit by using alloy mixtures. For example, the already small misfit in the Ge-GaAs system can be made to vanish by alloying either the Ge with about 1.8 per cent Si or the GaAs with about 1.0 per cent GaSb or InAs.

We have investigated the majority of the possible combinations containing Ge, Si, III-V compounds and II-VI compounds. There are at least another two pairs with lattice misfits below 10^{-3} (HgSe-ZnTe and InSb-CdTe) and at least an additional 27 pairs with misfits below 10^{-2}, even without alloying. Besides Ge-GaAs the most interesting combination appears to be GaP-AlP, which might provide an indirect-gap visible laser. Perfect lattice fit could be obtained by alloying 4 per cent GaAs to the GaP. However, at present the Ge-GaAs system appears to be the most immediately realizable one.

Herbert Kroemer
Central Research Lab.
Varian Associates
Palo Alto, Calif.

⁷ G. K. Wertheim and G. L. Pearson, "Recombination in Plastically Deformed Germanium," Phys. Rev., vol. 107, pp. 694–698, Aug., 1957.
⁸ A. D. Kurtz, S. A. Kulin, and B. L. Averback, "Effects of Growth Rate on Crystal Perfection and Lifetime in Germanium," J. Appl. Phys. vol. 27, pp. 1287–1290, Nov., 1956.

PROCEEDINGS OF THE IEEE *April*

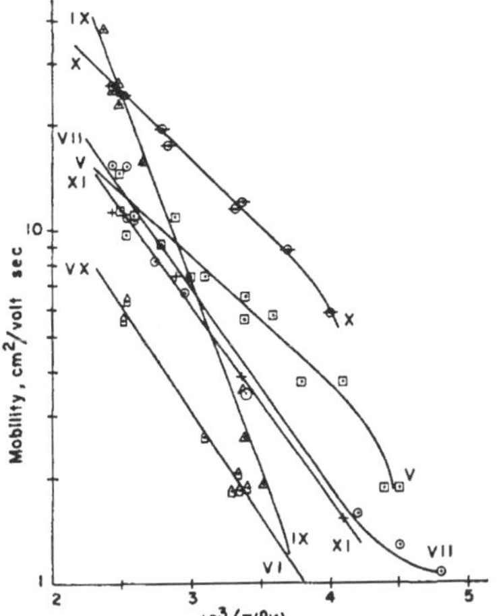

Fig. 1—Temperature dependence of Hall mobility μ_H for CdS films.

TABLE I

Sample	Resistivity (Ω cm) (300 K)	Hall Mobility (cm²/V sec) (300 K)	E_R (eV)	E_μ (eV)	E_ρ (eV)	Source T (°C)	Substrate T (°C)	Processing	Color
5	32	6	0.13	0.07	0.21	—	23	360°C H₂ bake	yellow
6	6	2	0.06	0.12	0.18	—	23	None	black
7	24	4.4	0.07	0.12	0.18	880	100	None	orange
8	10,000	—	—	—	0.32	820	200	None	yel-or.
9	270	3.2	0.05	0.20	0.25	750	200	None	yel-or.
10	1900	12	0.35	0.07	0.42	760	140	None	yel-or.
11	650	4	0.12	0.10	0.22	730	160	None	yel-or.

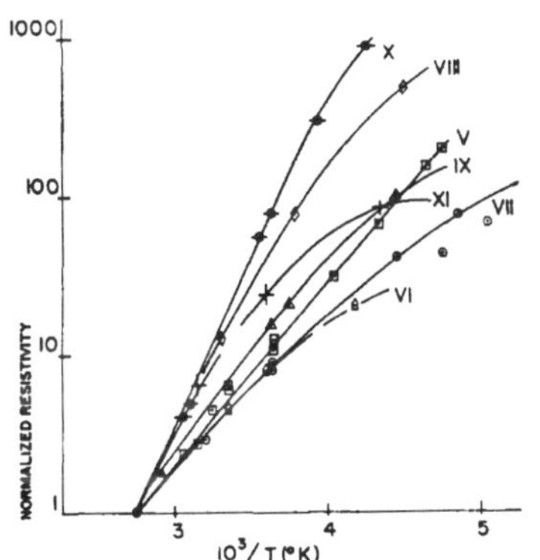

Fig. 2—Temperature dependence of the resistivity ρ of CdS films. The ordinate is normalized at 400°K.

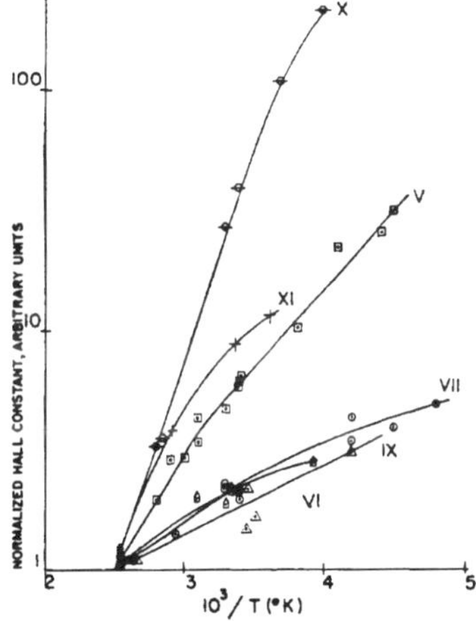

Fig. 3—Temperature dependence of the Hall constant R_H for deposited CdS films. The ordinate is normalized at 400°K.

of the reciprocal of the free-charge density, the sum of the activation energies for the mobility and the Hall coefficient should equal the activation energy of the resistivity, as is observed.

DISCUSSION

The observation of an exponential dependence for Hall mobility on temperature in deposited CdS films was first reported by Berger.[3] Such a dependence has also been found in deposited films of PbS, and, following the analysis of Petritz,[4] it is often ascribed

to scattering at the boundaries between the small crystallites which make up the film. There is reason to doubt this hypothesis chiefly because of the near independence of the observed Hall-mobility value on crystallite size. This view is corroborated by Berger.[3] Work is now going on in this laboratory to ascertain whether or not the observed mobility dependence is not resultant from the large deep-trap densities that are known to characterize these films. This information is being sought through photo-Hall effect measurements, and through thermally-

stimulated trap emptying studies. Present technology in this effort permits the fabrication of films with significantly higher mobilities and resistivities than those discussed in this communication.

R. S. MULLER
B. G. WATKINS
Elec. Engrg. Dept.
University of California
Berkeley, Calif.

Correction to "Relationships between Different Kinds of Network Parameters, Not Assuming Reciprocity or Equality of the Waveguide or Transmission Line Characteristics Impedances"[1]

The following has been called to the attention of the Editor. In the relationship having the S-matrix on the left and expressions involving A, B, C, D, Z_{01} and Z_{02} on the right, a plus sign should appear in the denominator between the terms $(B+CZ_{01}Z_{02})$ and $(AZ_{02}+DZ_{01})$.

R. W. BEATTY
D. M. KEARNS
National Bureau of Standards
Boulder, Colo.

Manuscript received February 11, 1964.
[1] R. W. Beatty and D. M. Kearns, PROC. IEEE (Correspondence), vol. 52, p. 84; January, 1964.

Considerations Regarding the Use of Semiconductor Heterojunctions for Laser Operation

In a recent communication,[1] Kroemer has proposed a new injection scheme using heterojunctions for possible laser action, in which an indirect-gap semiconductor, say Ge, is sandwiched between two direct-gap semiconductors of opposite types, say n- and p-type GaAs. In our laboratory, we also have considered the feasibility of using heterojunctions for laser work based on a different scheme. Kroemer's proposal presupposes that 1) injected electrons and holes would be trapped in the center region by potential barriers at the two heterojunctions and 2) laser action would eventually occur at sufficiently high carrier injection levels. The argument presented in his communication, however, is rather vague and misleading. We would like to discuss theoretical considerations in using heterojunctions for laser operation and to present our scheme in view of these considerations.

Manuscript received January 31, 1964. The research reported herein is made possible through support received from the Departments of Army, Navy and Air Force under grant AF-AFOSR-139-63.
[1] H. Kroemer, "A proposed class of heterojunction injection lasers," PROC. IEEE (Correspondence), vol. 51, pp. 1782–1783; December, 1963.

[3] H. Berger, "Über das Ausheilen von Gitterfehlern frischaufgedampfter CdS-Schichten (I)," *Phys. Status Solidi*, vol. 1, pp. 739–757; July, 1961.
[4] R. L. Petritz, "Theory of photoconductivity in semiconductor films," *Phys. Rev.*, vol. 104, pp. 1508–1516; December, 1956.

The focal point is the lifetime of excess carriers in a radiative recombination process. For degenerate direct-gap semiconductors, the lifetime, given by the reciprocal of Einstein's coefficient of spontaneous emission, is of the order 10^{-11} sec. In indirect-gap semiconductors, however, the radiative recombination process must be accompanied by phonon or impurity scattering to conserve momentum and, consequently, the lifetime of such a process is much much longer. This is manifested by the fact that the quantum efficiency of recombination radiation in GaAs diodes is close to unity while that in Ge diodes is less than 10^{-4}. It is to be recognized that energy pumped into a diode is ultimately converted into lattice heat if not into coherent radiation. Therefore, the second assumption made by Kroemer not only needs theoretical scrutiny but may become academic in practical reality.

Two principal schemes [1], [2] have been proposed to achieve laser action in indirect-gap semiconductors; first, to tunnel electrons into the (000) valley and second, to get an admixture of the various conduction band minima and the (000) valley states through proper impurity states. The scheme to be presented here falls into the first category. Consider an *n-n* heterojunction of Ge-GaAs with its energy band diagram shown in Fig. 1(a).

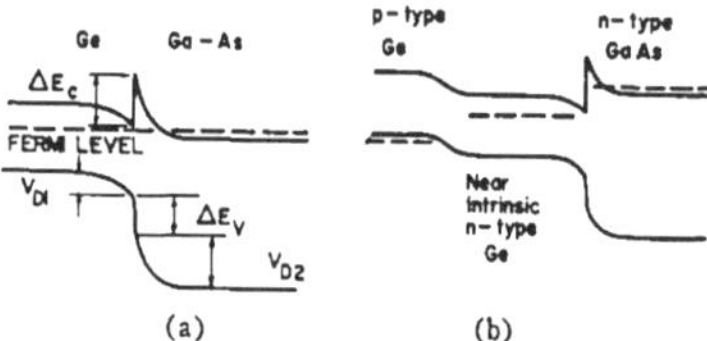

Fig. 1—(a) Energy band structure of a Ge-GaAs heterojunction where V_{D1} and V_{D2} are the built-in electrostatic potentials, and ΔE_c and ΔE_v are the conduction and valence band edges as a result of difference in electron affinity. (b) The proposed laser structure under a forward bias condition.

The nature of current flow across the junction has been analyzed in detail by Anderson [3]. For electrons going from Ge to GaAs or vice versa (a combination of electron emission and tunneling), both the momentum and energy of the electron are conserved. That means, if GaAs is negatively biased with respect to Ge, electrons from GaAs will go to the (000) valley of the conduction band of Ge as in direct tunneling. To prevent the (000) valley electrons from being scattered into the conduction band minima, the device must be operated at low temperatures and the doping concentration of Ge must be low. No mention is made in Kroemer's paper about scattering processes; apparently he is concerned with the indirect transition while we are interested in the direct transition in Ge. The present heterojunction scheme of getting (000) valley electrons may be superior to the tunnel scheme proposed earlier [1] because high doping concentration in the indirect-gap semiconductor is not required here. The proposed laser structure is shown schematically in Fig. 1(b), which consists of, from left to right, regions of degenerate *p*-type Ge, near intrinsic, *n*-type Ge and degenerate *n*-type GaAs. The Ge *p-n* junction and the Ge-GaAs *n-n* junction are to supply excess holes and electrons with $k(000)$ for radiative recombination in the center region. Degeneracy in outer regions is necessary for heavy injection to shorten the lifetime of excess carrier so that the radiative recombination process may compete favorably with scattering processes. A detailed discussion may be found in [1].

It should be pointed out that even in the scheme proposed by Kroemer, excess carriers can not pile up indefinitely in the center region. After a quasi-equilibrium state is reached, the flow of electrons and holes from one outer region to the other can be described directly by quasi-Fermi levels without reference to the center region if recombination in the center region is neglected. That means, the flow of current is governed by the law of diffusion of excess minority carriers in GaAs and the situation is no better or no worse than that in a homogeneous diode. The statement made by Kroemer is misleading because it implies that great benefit can be derived from potential barriers at heterojunctions.

In summary, we believe that the function of heterojunctions in laser work is to supply the right kind of electrons with $k(000)$ to an indirect semiconductor. In view of Anderson's work, the *n-n* heterojunction seems suitable for such a purpose. The question still remains, however, as to how long electrons will stay in the (000) valley. Therefore, phonon and impurity scattering should be minimized to enhance the probability of radiative recombination.

S. WANG
C. C. TSENG
Dept. of Elec. Engrg.
University of California
Berkeley, Calif.

REFERENCES

[1] S. Wang, "Proposal for a two stage semiconductor laser through tunneling and injection," *J. Appl. Phys.*, vol. 34, pp. 3443–3450; December, 1963.
[2] P. Aigrain, private communcation.
[3] R. L. Anderson, "Experiments on Ge-GaAs heterojunctions," *Solid-State Electronics*, vol. 5, pp. 341–351; Pergamon Press, New York, N. Y.; 1962.
[4] ——, "Germanium-Gallium Arsenide Contacts," Ph.D. dissertation, Department of Electrical Engineering, Syracuse University, N. Y.; 1959.

Author's Reply[2]

I wish to comment only on those parts of the above communication that seem to contain a criticism of my proposal, not upon Wang and Tseng's proposal that is intimately intertwined with this criticism.

1) I stated in my paper that in indirect gap semiconductors laser action "could take place by spill-over of electrons from their lowest energy valleys into the region of the smallest direct gap." Obviously, then, I was referring to direct transition, under conditions where the electrons in the central valley are essentially in equilibrium with those in the outlying valleys. In such a case it is irrelevant how short the interband relaxation time in the central valley is, and into which valley the electrons are originally injected. In fact, if the original injection should take place into the lower outlying valleys, a short interband relaxation time might be desirable. The total injection level necessary to reach degeneracy in the central

[2] Manuscript received February 14, 1964.

valley in germanium is just over 10^{20} cm^{-3}. This admittedly is a high number but an extension of the Hall-Shockley-Read recombination theory into the degenerate range indiates that the nonradiative recombination at this level would not be prohibitive, provided this injection level can be obtained in the first place. The latter is the function of the heterojunction injectors.

2) I have never claimed that charge carriers can pile up indefinitely in the base region, but merely that degenerate injection levels can be obtained that are much higher than with a homojunction structure, and high enough to cause degenerate spill-over into the central valley of Ge. Wang and Tseng state correctly that "the flow of electrons . . . if recombination in the center region is neglected . . . is governed by the law of diffusion of excess minority carriers in GaAs" but they do not show why this fact should contradict rather than support my statement and why the situation should be "no better or no worse than that in a homogeneous diode." In the absence of a justification for their claim, I must maintain that indeed "great benefit can be drived from potential barriers at heterojunction" and must refer for the proof to my previous correspondence and to my 1957 paper quoted therein.

HERBERT KROEMER
Central Research Lab.
Varian Associates
Palo Alto, Calif.

WWV and WWVH Standard Frequency and Time Transmissions

The frequencies of the National Bureau of Standards radio stations WWV and WWVH are kept in agreement with respect to each other and have been maintained as constant as possible since December 1, 1957 with respect to an improved United States Frequency Standard (USFS).[1] The corrections reported here were arrived at by means of improved measurement methods based on transmissions from the NBS stations WWVB (60 kc) and WWVL (20 kc). The values given in the table are 5-day running averages of the daily 24-hour values for the period ending at 1800 UT of each day listed.

The time signals of WWV and WWVH are also kept in agreement with each other. Since these signals are locked to the frequency of the transmissions, a continuous departure from UT2 may occur. Corrections are determined and published by the U. S. Naval Observatory. The time signals are maintained in close agreement with UT2 by properly offsetting the broadcast frequency from the USFS at the beginning of each year when necessary. This new system was commenced on January 1, 1960.

Manuscript received February 24, 1964.
[1] See "National Standards of time and frequency in the United States," PROC. IRE (*Correspondence*), vol. 48, pp. 105–106; January, 1960.

``If, in discussing a problem in semiconductor device physics, you cannot draw an energy band diagram, then you don't know what you're talking about.''

Herbert Kroemer

Reprinted from

H. Kroemer, ``Heterostructures for Everything: Device Principle of the 1980's?'', Japan. J. Appl. Phys., Vol. 20 (Suppl. 1), pp. 9-13, 1981.

Proceedings of the 12th Conference on Solid State Devices, Tokyo, 1980;
Japanese Journal of Applied Physics, Volume 20 (1981) Supplement 20-1, pp. 9–13

(Invited) Heterostructures for Everything:
Device Principle of the 1980's?

Herbert KROEMER

*Department of Electrical and Computer Engineering,
University of California,
Santa Barbara, California 93106, USA*

One of the dominant themes of semiconductor device R & D during the 1980's
will be the incorporation of heterostructures into most existing kinds of devices,
and the emergence of new kinds of devices made possible by heterostructures. In
this paper the power of heterostructures as a design tool is illustrated by discussing
several ways in which the incorporation of heterostructures can improve the
bipolar transistor. The dominant idea is that energy gap variations are a powerful
way to control carrier flow; in bipolar structures they permit the control of
electrons and holes independently. Several applications of this principle are
discussed, going beyond the familiar wide-gap emitter concept, and including
several concepts not previously discussed in the literature. The paper closes with
a brief discussion of non-bipolar applications and speculative future applications.

§1. Introduction

It has now been ten years since the experimental realization of the double heterostructure laser. Such lasers are today used in actual communications systems, and the heterostructure technology is rapidly spreading to other devices. I believe that one of the dominant themes of semiconductor device R & D during the 1980's will be the incorporation of heterostructures (HS's) into every kind of semiconductor device whose performance can be improved by such an incorporation, and for which the improvement is sufficiently desirable to justify the technology. Such improvements can be made in almost all classes of devices. Finally, new kinds of devices made possible by HS's are rapidly emerging, and will assume an increasing role toward the end of this decade.

Rather than attempting to cover every conceivable application of HS's, I shall try to illustrate the power of HS's as a design principle by concentrating on the variety of ways how their incorporation can drastically improve the familiar bipolar transistor. The underlying idea is that energy gap variations are a powerful way to control carrier flow, in addition to the control exerted by the electrostatic potentials generated by doping and bias. In bipolar structures, energy gap variations can be used to control electron and hole flows *separately*. Various

examples of non-bipolar applications will be given in the last part of the paper.

§2. Heterojunction Bipolar Transistors

2.1 *Wide-gap emitters*

The idea that the performance of a bipolar transistor could be improved by increasing the energy gap in the emitter relative to the base is as old as the bipolar transistor itself.[1] Consider an npn transistor with an energy band diagram as shown in Fig. 1. The operating principle of the device is the injection of electrons from the emitter into the base, and their subsequent collection by the collector. Asso-

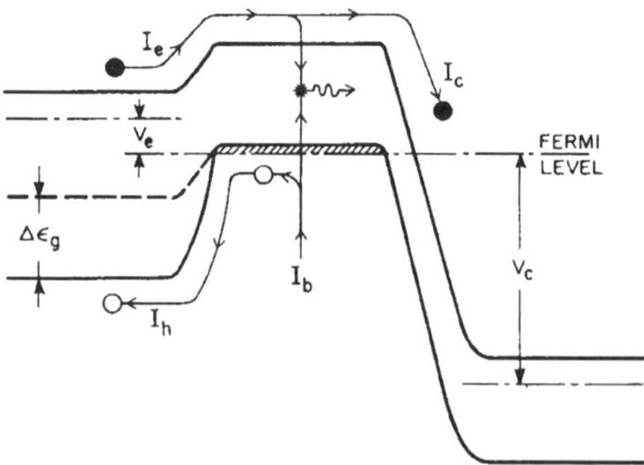

Fig. 1. Energy band diagram and carrier flow in an
npn transistor with a wide-gap emitter. The heavy
broken line in the emitter shows the valence band
edge as it would exist in a homojunction transistor
with the same emitter doping.

10   Herbert KROEMER

ciated with the desired emitter-to-base electron injection is an undesirable base-to-emitter hole injection. This hole injection current is part of the base current; at high current levels it is often the dominant part. In order to have a desirable current gain $\beta = I_c/I_b$ of about 100, the hole injection current from base to emitter must be kept below 1 % of the emitter-to-base electron injection current. In conventional transistors this is achieved by a high emitter-to-base doping ratio of typically about 100 : 1. This is the dominant design constraint in conventional bipolar transistors.

Suppose now that the emitter band gap is increased beyond that of the base. If the emitter doping is (initially) kept unchanged, all the increase in energy gap goes into depressing the valence band edge, introducing an additional energy barrier into the path of the hole flow, but not into the path of the electron flow. The result is a reduction of the base-to-emitter hole injection current by a factor $\exp\left(-\Delta\varepsilon_g/kT\right)$. This is incredibly effective: Energy gap difference of several tenths of 1 eV are readily available, and their effect is so large that the base-to-emitter hole injection current becomes negligible, regardless of the emitter-to-base doping ratio. The current gain β will be limited only by the recombination currents. It was recognized by Kroemer[2] that this could be utilized to improve the transistor by using a much higher base doping and a much lower emitter doping. One result would be a greatly reduced β-falloff with current than in conventional transistors. If the emitter doping were reduced below the value that the base doping has in a conventional transistor ("super-inverted" doping), the emitter capacitance would be reduced, with benefits for the high-frequency performance. It was subsequently recognized[3-5] that an even greater improvement in frequency response would result from the reduction in base resistance that could be obtained by a drastic increase in base doping. Maximum oscillation frequencies f_{max} of 100 GHz and more have been predicted.[3,4,6]

As a result of rapid progress in the heteroepitaxial growth of III/V compound semiconductors, especially GaAs/(Al, Ga)As, it appears that these predicted improvements are about to become realized in practice. Several authors[7-10] have reported β-values over 1000, much larger than the β-values of even the best conventional transistors.* Some of these results were on phototransistors, which are easier to construct than true three-terminal devices.

The high-frequency performance still lags appreciably behind that of Si transistors, and even more behind the theoretical possibilities, largely due to non-optimized technologies. For true three-terminal devices, only three papers[10-12] have so far reported performance above 100 MHz, up to $f_t \cong 1$ GHz. The fastest HS transistors so far are phototransistors.[8,13,14] In one case,[13] response times as short as 1 nsec have been reported. Considering the high gain of these devices ($\gg 100$), this would correspond to sinusoidal operating frequencies up to many gigahertz. Perhaps more significant: This particular device is the first high-performance bipolar transistor reported in the literature that was prepared by MOCVD rather than LPE. With the rapid progress in MOCVD and MBE, one may expect the future progress in transistor frequency performance to be rapid.

The technology of HS transistors is likely to be dominated by III/V compounds, because of the comparative ease with which the new epitaxial technologies permit the preparation of defect-free HS's in lattice-matched III/V compound pairs. Promising lattice-matched III/V systems in addition to (Al, Ga)As-on-GaAs are (Ga, In)P-on-GaAs and InP-on-(Ga, In)As.[14]

Because of the extremely high state of development of Si-IC technology, there is a strong incentive to develop an HS-IC technology for Si, even if its ultimate performance is less than for an all-III/V compound technology. Very promising results in this direction have been obtained by Matsushita et al.[15] who used an emitter made from amorphous SiO_x, which has a wider energy gap than Si. It remains to be seen what the high-frequency potential of this combination is; the low-frequency current gain ($\beta \gtrsim 500$) is excellent.

A potentially very promising system is GaP-on-Si. The two semiconductors are fairly well lattice-matched (within 0.4 %); however, good

---

*This field lacks a good recent review. References 7–14 give only recent results for which $\beta > 1000$ or $f > 100$ MHz. For references to most of the earlier pioneering papers see ref. 5.

epitaxy of GaP on Si appears to be hard to achieve. The first GaP-on-Si transistor, prepared by VPE, has been reported by Katoda and Kishi,[16] with (so far) very low β-values. In our own laboratory, we are attempting to grow GaP emitters on Si by MBE. A detailed theoretical estimate[6] suggests that npn Gap-on-Si transistors with a maximum oscillation frequency f_{max} up to about 100 GHz should be achievable. It remains to be seen what will come of these GaP-on-Si efforts.

2.2 Wide-gap collectors

In most bipolar logic families (ECL is the dominant exception) the collector is forward-biased during part of the logic cycle. If the base is more heavily doped than the collector, this causes major base-to-collector injection of holes, which increases dissipation and slows down the switching speed. Using a wider energy gap on the collector as well as on the emitter side, suppresses this highly undersirable effect.[6] In I^2L, this is not a problem, but I^2L has problems of its own, which also suggest HS's as a solution; see below.

2.3 Utilizing turn-on voltage differences
2.3.1 Double base layers

We may view the role of HS's as providing barriers to control the flow of electrons and holes independently of each other. The idea of the wide-gap emitter and collector was to block the flow of holes (in an npn transistor, electrons in a pnp). The idea can be extended to control the flow of the carriers with opposite polarity as well. Ladd and Feucht[3] pointed out that it is desirable, in an HS transistor just as in a conventional transistor, to have a thick base region outside the emitter, and that the realization of the full promise of the HS transistor might well hinge on achieving a suitable device geometry. It was pointed out by this writer[6] that a natural solution to this problem lies in a thick planar design in which the outer base consists of two layers, the upper one of which has the same wide energy gap as the emitter and is very heavily doped (Fig. 2). In a homostructure transistor, such a design would lead to a disastrous loss of β, emitter-base tunneling effects and a high emitter capacitance. In an HS transistor, because the vertical part of the emitter junction has a wide energy gap on both

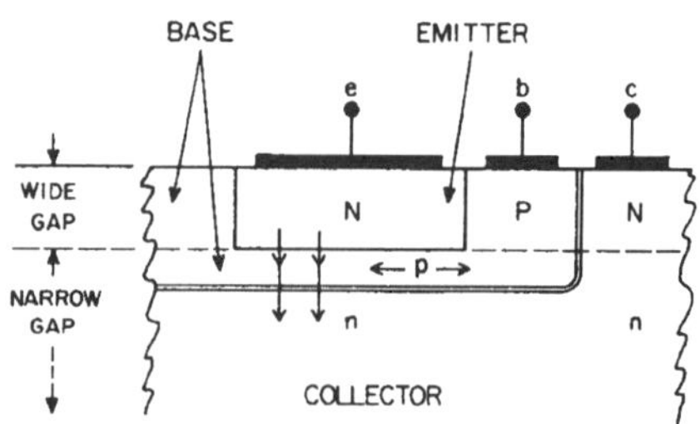

Fig. 2. Two-base layer design in a heterojunction transistor. The external base layer has, on top of a thin narrow-gap layer, a thick very heavily doped second base layer extending right to the emitter junction, minimizing the external base resistance. Because of the wider energy gap, negligible currents will flow through this part of the emitter junction.

sides, this part will have its current density reduced by the same Boltzmann factor by which the base-to-emitter hole injection is depressed. Using a different terminology: the vertical part of the junction is biased below its turn-on voltage, which is higher than that of the horizontal part by approximately $\Delta \varepsilon_g / q$. Because the emitter would be weakly doped, the effect of a deep sidewall on the emitter capacitance is small, and emitter-base tunneling cannot occur at all. Yet the high conductivity of the upper base layer in effect brings the base contact as close to the emitter-base junction as is physically possible. In fact, this design has been used in the HS transistor reported in ref. 11.

2.3.2 Wide-gap sidewalls and injection barriers

The idea to use energy gap variations to suppress carrier injection into portions of the base region where no injection is desired, is an important new concept, the power of which does not appear to have been widely recognized. The lateral pnp transistor in I^2L is an example of a device that could be greatly improved by incorporating this idea. In homostructure designs, this transistor is always a poor transistor. It could be greatly improved by an HS geometry as, for example, in Fig. 3. The actual transistor is the all-narrow-gap p^+np structure shown embedded between the two wide-gap layers. The two wide-gap p^+np transistors above and below it are biased below turn-on; their n-type base regions simply act as walls to confine the injected holes in the true n-base. Because of the small hole diffusion lengths in III/V compounds, the implementation of such a structure will require submicron technologies.

12   Herbert KROEMER

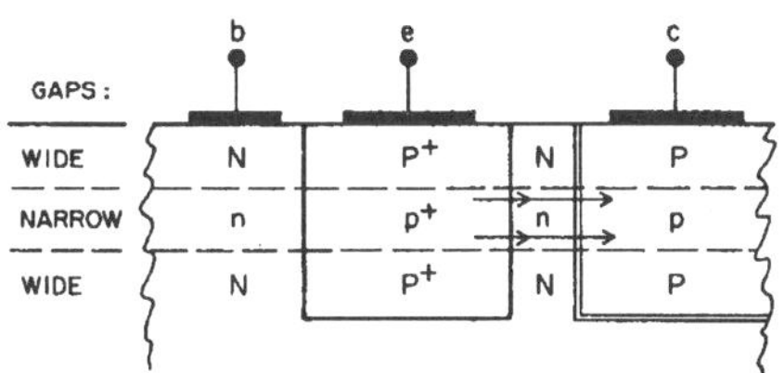

Fig. 3. Injection current in a lateral pnp transistor, as in I²L. Only the narrow-gap portion (n) of the base carries current.

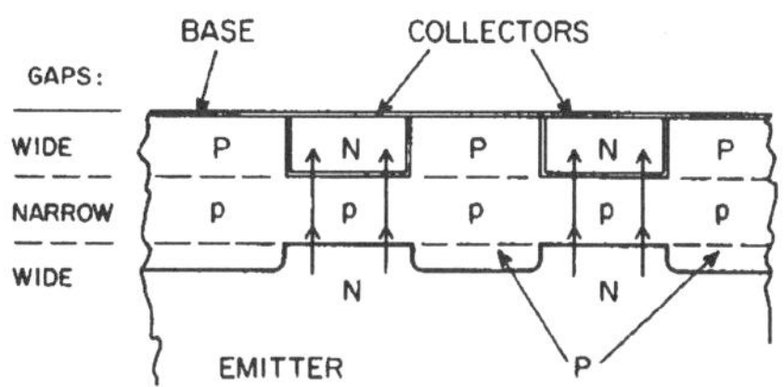

Fig. 4. Supression of electron injection into selected portions of the base region in I²L, by means of wide-gap injection barriers, achieved by pulling the emitter-base junction into the wide-gap region of the structure.

Another example of suppression of undesired injection is the following. In I²L, a significant fraction of the base area is in contact with the emitter, but not with one of the collectors. Electron injection into those portions of the base creates stored charge that wastes power and slows down the switching speed. In an HS-I²L design, this could be avoided by simply pulling the emitter-base pn junction below the hetero-interface, into the wide-gap region, as shown in Fig. 4.

§3. Beyond Bipolar Transistors

The detailed discussion of the bipolar transistor was intended as an example. There probably does not exist a kind of device that cannot be similarly improved by the incorporation of HS's. Space does not permit me to say more than a few words about these other possibilities. Perhaps the most powerful general design principle applicable to many non-bipolar devices involves the use of thin ($\ll 1000$ Å) alternating layers of different semiconductors. At sufficiently abrupt HS's, there usually occurs a sharp discontinuity in the conduction band edge, similar to the discontinuity at Schottky barriers, giving the alternating layers an energy band structure as in Fig. 5. Because

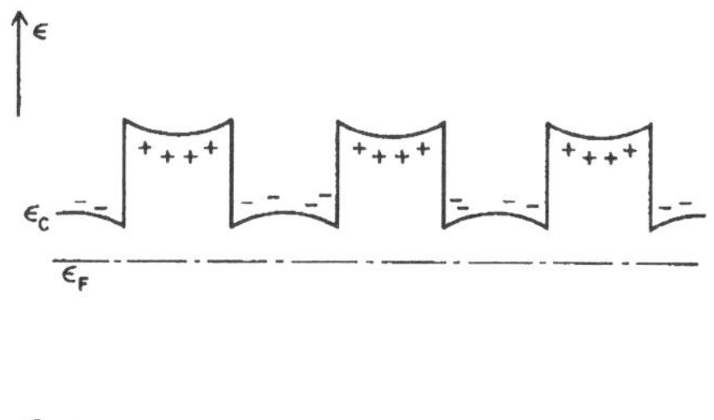

Fig. 5. Energy bands in a modulation-doped multilayer structure. All donors are contained in the wide-gap layers, all electrons in the narrow-gap layers.

of the band edge steps, electrons from the high-ε_c layers will drain into the potential wells inside the low-ε_c layers, provided the layers are thin enough to minimize the built-in voltages created by the resulting space charges. If the low-ε_c layers are left undoped, with all donors placed into the high-ε_c layers, one obtains a high net electron concentration, without any impurity scattering. Such structures have an enhanced electron mobility compared to a bulk semiconductor of the same net electron concentration, especially at low temperatures.

This is the concept of **modulation doping**[17] (MD), one of the most important new ideas to have emerged from the new hetero-epitaxial technologies. Its first utilization in an FET has been reported,[18,19] leading to a major improvement in low-T performance. The use of MD is a very general concept applicable to many devices besides FET's, and it may even give rise to new kinds of devices.

One of the most promising aspects of MD is that it should permit the utilization, in the low-ε_c layers, of narrow-gap semiconductors such as InAs. These have much higher mobilities to begin with. Because of the absence of impurity scattering, MD structures would retain these mobilities; they would also minimize several other drawbacks of low-gap semiconductors that have prevented their device use in bulk structures. I believe that their use at low temperatures will lead to a re-appraisal of the use of semiconductors compared to Josephson devices at low temperatures.

Of particular interest in MD structures will be the non-linear high-field transport properties along the potential wells, including especially negative-differential-conductance(NDC) effects. They should be more pronounced than in bulk semiconductors. Scattering of hot electrons out of the potential wells should lead to a new and strong NDC mechanism,[20] similar to that of the Gunn effect, and of great promist for millimeter wave applications.

Finally, there will be **quantum wells**, structures as in Fig. 5, but with dimensions to small that discrete quantum levels are formed within the wells. Many future semiconductor lasers will go beyond present HS designs, towards quantum well designs.[21] Even more exotic possibilities exist for the use of large numbers of quantum wells in periodic arrays, as artificial **superlattices.**[22]

The list could easily be extended, especially if one were to include light-sensing devices other than phototransistors, and various ultra-specialized applications. In fact, many future applications of HS's have probably not even been perceived yet. But considering the major examples I have given, and doing so against the background of the proven role of HS's in lasers, I have probably made my point: Heterostructures for Everything? Well . . .!?

References

1) W. Shockley: U.S. Patent 2569347, filed 26 June 1948, issued 26 Sept. 1951.

2) H. Kroemer: Proc. IRE **45** (1957) 1535.

3) G. O. Ladd and D. L. Feucht: IEEE Trans. Electron Devices ED-17 (1970) 413.

4) W. P. Dumke, J. M. Woodall and V. L. Rideout: Solid-State Electronics **15** (1972) 1339.

5) For an excellent review of the various potential benefits of a heterojunction transistor (to the extent they had been recognized at the time) see Milnes and D. L. Feucht: *Heterojunctions and Metal-Semiconductor Junctions* (Academic Press, New York, 1972) Chap. 3.

6) H. Kroemer: Device Research Conference, 1978, Santa Barbara; see IEEE Trans. Electron Devicse **ED-25** (1978) 1339; Bull. Am. Phys. Soc. **24** (1979) 230.

7) B. W. Clark, H. G. B. Hicks, I. G. A. Davis and J. S. Heeks: *Gallium Arsenide and Related Compounds, 1974* (Inst. Phys. Conf. Series, Bristol, 1975) Vol. 24, pp. 373 ff.

8) H. Beneking, P. Mischel and G. Schul: Electron. Lett. **12** (1976) 375.

9) M. Konagai, K. Katsukawa and K. Takahashi: J. Appl. Phys. **48** (1977) 4389.

10) P. W. Ross, H. G. B. Hicks, J. Froom, J. G. Davies, F. J. Probert and J. E. Carroll: Electron. Engin. **49** (1977) 35.

11) D. Ankri and A. Scavennec: Electron. Lett. **16** (1980) 41.

12) J. P. Bailbe, A. Marty, P. H. Hiep and G. E. Rey: IEEE Trans. Electron. Devices **ED-27** (1980) 1160.

13) R. A. Milano, T. H. Windhorn, E. R. Anderson, G. E. Stillman, R. D. Dupuis and P. D. Dapkus: Appl. Phys. Lett. **34** (1979) 562.

14) M. Tobe, Y. Amemiya, S. Sakai and M. Umemo: Appl. Phys. Lett. **37** (1980) 73.

15) T. Matsushita, N. Oh-uchi, H. Hayashi and H. Yamato: Appl. Phys. Lett. **35** (1979) 549.

16) T. Katoda and M. Kishi: J. Electron. Matls. **9** (1980) 783.

17) R. Dingle, H. L. Störmer, A. C. Gossard and W. Wiegmann: Appl. Phys. Lett. **33** (1978) 665.

18) T. Mimura, S. Hiyamizu, T. Fugi and K. Nanbu: Jpn. J. Appl. Phys. **19** (1980) L225.

19) For a review of other HS benefits in FET's, see D. Boccon-Gibod, J.-P. André, P. Baudet and J. P. Hallais: IEEE Trans. Electron. Devices **ED-27** (1980) 1141.

20) K. Hess, H. Morkoc, H. Schichijo and B. G. Streetman: Appl. Phys. Lett. **35** (1979) 469.

21) J. P. van der Ziel, R. Dingle, R. C. Miller, W. Wiegmann and W. A. Nordland, Jr.: Appl. Phys. Lett. **26** (1975) 463. For a recent review of the developments since this first paper, see N. Holonyak, R. M. Kolbas, R. D. Dupuis and D. D. Dapkus: IEEE J. Quantum Electron. **16** (1980) 170.

22) For a recent compact review of the very extensive literature see L. L. Chang and L. Esaki: Prog. Crystal Growth Charact. **2** (1979) 3.

``If you can draw an energy band diagram and don't, no one else will know what your are talking about.''

Herbert Kroemer

PROCEEDINGS OF THE IEEE, VOL. 70, NO. 1, JANUARY 1982

Heterostructure Bipolar Transistors and Integrated Circuits

HERBERT KROEMER, FELLOW, IEEE

Invited Paper

Abstract—Two new epitaxial technologies have emerged in recent years (molecular beam epitaxy (MBE) and metal-organic chemical vapor deposition (MOCVD)), which offer the promise of making highly advanced heterostructures routinely available. While many kinds of devices will benefit, the principal and first beneficiary will be bipolar transistors. The underlying central principle is the use of energy gap variations beside electric fields to control the forces acting on electrons and holes, separately and independently of each other. The resulting greater design freedom permits a re-optimization of doping levels and geometries, leading to higher speed devices. Microwave transistors with maximum oscillation frequencies above 100 GHz and digital switching transistors with switching times below 10 ps should become available. An inverted transistor structure with a smaller collector on top and a larger emitter on the bottom becomes possible, with speed advantages over the common "emitter-up" design. Double-heterostructure (DH) transistors with both wide-gap emitters and collectors offer additional advantages. They exhibit better performance under saturated operation. Their emitters and collectors may be interchanged by simply changing biasing conditions, greatly simplifying the architecture of bipolar IC's. Examples of heterostructure implementations of I^2L and ECL are discussed. The present overwhelming dominance of the compound semiconductor device field by FET's is likely to come to an end, with bipolar devices assuming an at least equal role, and very likely a leading one.

"What is claimed is:

1) . . .

2) A device as set forth in claim 1 in which one of the separated zones is of a semiconductive material having a wider energy gap than that of the material in the other zones."

Claim 2 of U.S. Patent 2 569 347 to W. Shockley,
Filed 26 June 1948,
Issued 25 September 1951,
Expired 24 September 1968.

I. INTRODUCTION

THIS IS A PAPER about an idea whose time has come: A bipolar transistor with a wide-gap emitter. As the introductory quote shows, the idea is as old as the transistor itself. The great potential advantages of such a design over the conventional homostructure design have long been recognized [1]–[3], but until the early 70's, no technology existed to build practically useful transistors of this kind, even though numerous attempts had been made [3], [4]. The situation started to change with the emergence of liquid-phase epitaxy (LPE) as a technology for III/V-compound semiconductor heterostructures, and in recent years reports on increasingly impressive true three-terminal heterostructure bipolar transis-

tors (HBT's) have appeared at an increasing rate [5]–[14]. In addition, there is also a rapidly growing literature on *two*-terminal *photo*transistors with wide-gap emitters [15]. Many of the phototransistors employ InP emitters with a lattice-matched (Ga, In) (P,As) base.

Since the mid-70's, two additional very promising heterostructure technologies have appeared: molecular beam epitaxy (MBE) [16] and metal-organic chemical vapor deposition (MOCVD) [17]. Impressive results on MOCVD-grown (Al,Ga)As-GaAs phototransistors have already been published [18]; HBT's grown by MBE have also been achieved [19].

Because of the pre-eminence of silicon in current IC technology, there exists a strong incentive to incorporate wide-gap emitters into Si transistors, in a way compatible with existing Si technology. A possible approach—and the most successful one so far—has been the use of heavily doped "semi-insulating polycrystalline" silicon (SIPOS) as emitter [20], utilizing the wider energy gap of "polycrystalline" (really: amorphous) Si compared to crystalline Si. An alternate approach has been the use of gallium phosphide, which has a room-temperature lattice constant within 0.3 percent of that of Si, grown on Si either by CVD [21] or by MBE [22]. But the results reported for the GaP-Si combination have so far been disappointing.

Finally, the first reports have recently appeared, in which HBT's have been integrated on the same chip with other devices, such as double-heterostructure (DH) lasers [23] or LED's [24].

In view of these recent developments it appears that Shockley's vision is about to become a reality. In fact, one of the purposes of this paper is to show that the possibilities for HBT's go far beyond simply replacing a homojunction emitter by a heterojunction emitter.

To appreciate these possibilities, it is useful first to view the wide-gap emitter as a simple example of a more general *central design principle* of heterostructure devices; it is discussed in Section II of this paper. Discussions of future device possibilities must be based on technological premises; they are discussed in Section III. In Section IV and V the concept and the high-speed benefits of the wide-gap emitter are reviewed, including some recent conceptual developments that do not appear to have been widely appreciated. Section VI discusses the promising concept of an inverted transistor design, in which the collector is made smaller than the emitter and placed on the surface of the structure, similar to I^2L, but using a heterostructure design applicable to all transistors. In Section VII the idea of a single-heterostructure transistor with a wide-gap emitter is generalized to DH transistors with both wide-gap emitters and wide-gap collectors. Such a design appears to

Manuscript received June 30, 1981; revised August 31, 1981. This work was supported in part by the Army Research Office and by the Office of Naval Research.

The author is with the Department of Electrical and Computer Engineering, University of California, Santa Barbara, CA 93106.

PROCEEDINGS OF THE IEEE, VOL. 70, NO. 1, JANUARY 1982

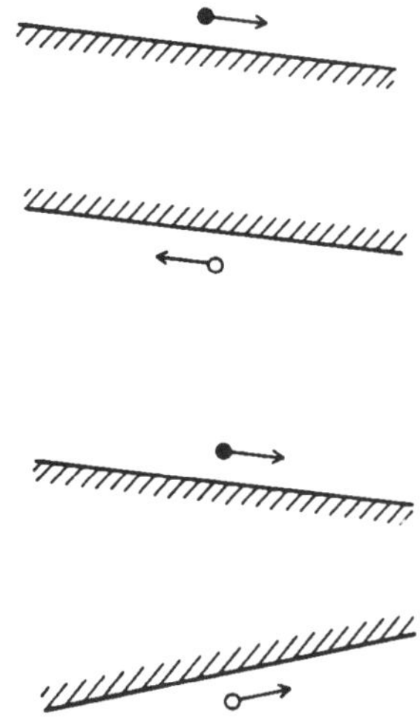

Fig. 1. Forces on electrons and holes. In a uniform-gap semiconductor (top) the two forces are equal and opposite to each other, and equal to the electrostatic force $\pm q\vec{E}$. In a graded-gap structure, the forces in electrons and holes may be in the same direction.

offer surprisingly large advantages for both microwave and digital devices, and especially for digital IC's. As examples of potential IC advantages, heterostructure modifications of both I^2L and ECL architecture are discussed. Finally, Section VIII offers some speculations on the question of FET's-versus-bipolars, and related questions.

In line with the character of this Special Issue (integrated) digital HBT's are emphasized over (discrete) microwave devices, but not to the point of exclusion of the latter. It would be artificial to attempt a complete separation: Not only was much of the past development of HBT's oriented towards discrete microwave devices, but several of the newer concepts originating in a digital context would improve microwave transistors as well.

II. THE CENTRAL DESIGN PRINCIPLE OF HETEROSTRUCTURE DEVICES

If one looks for a general principle underlying most heterostructure devices, one is led to the following considerations. If one ignores magnetic effects, the forces acting on the electrons and holes in a semiconductor are equal (except for a sign in the case of electrons) to the slopes of the edge of the band in which the carriers reside (Fig. 1). In ideal homostructures the energy gap is constant; hence the slopes of the two band edges are equal, and the forces acting on electrons and holes are necessarily equal in magnitude and opposite in sign. In fact they are equal to the ordinary electrostatic force $\pm q\vec{E}$ on a charge of magnitude $\pm q$ in an electric field $\vec{E}$. In a heterostructure, the energy gap may vary; hence the two band edge slopes and with it the magnitudes of the two forces need not be the same, nor need they be in any simple way related to the electrostatic force exerted by a field $\vec{E}$. In fact, the two slopes may have opposite signs (Fig. 1), implying forces on electrons and holes that act in the same direction, despite their opposite charges.

In effect, heterostructures utilize energy gap variations in addition to electric fields as forces acting on electrons and holes, to control their distribution and flow. This is what I would

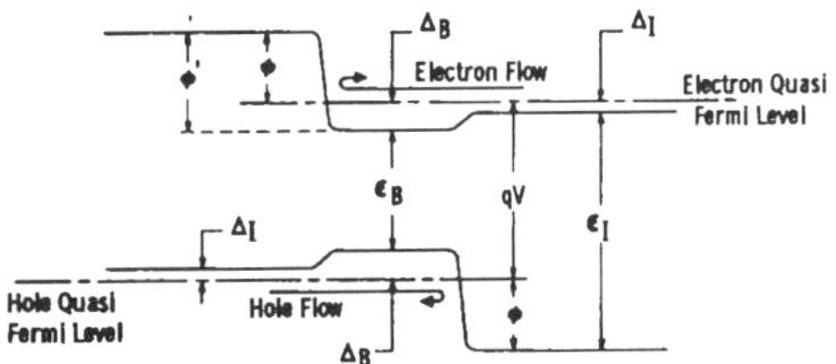

Fig. 2. Energy band diagram of a DH laser, showing the confinement forces driving both electrons and holes towards the active layer, on both sides of the latter. (From [25].)

like to call the *Central Design Principle* of heterostructure devices. It is a very powerful principle, and one of the purposes of this paper is to give examples that show just how powerful it is.

Although by no means restricted to bipolar devices, the principle is especially powerful when, as in a bipolar transistor, the distribution and flow of *both* electrons and holes must be controlled. By a judicious combination of energy gap variations and electric fields it then becomes possible, within wide limits, to control the forces acting on electrons and holes, *separately and independently of each other*, a design freedom not achievable in homostructures.

The central design principle plays a role in almost all heterostructure devices, and it serves both to unify the ideas underlying different such devices, and as guidance in the development of new device concepts. No device demonstrates the central design principle better than the oldest and so far most important heterostructure device, the DH laser. This point is illustrated in Fig. 2, which shows the energy band structure of the device under lasing conditions, as anticipated (with only slight exaggeration) in the paper in which this device was first proposed [25], and from which Fig. 2 is taken. The drawing shows band edge slopes corresponding to forces that drive *both* electrons and holes towards the inside of the active layer, at *both* edges of the latter. This is the principal reason why the DH laser works, although it is not the only reason. The difference in refractive indices between the inner and outer semiconductors also plays an important role. Such a participation of additional concepts is not uncommon in other heterostructure devices either.

III. THE TECHNOLOGICAL PREMISE

Throughout its history, heterostructure device design has chronically suffered from a technology bottleneck. Even LPE, whatever its merits as a superb laboratory technology, has outside the laboratory been largely limited to devices, such as injection lasers for fiberoptics use, which could simply not be built without heterostructures, but which were needed sufficiently urgently to put up with the limitations of LPE technology. Already for the "ordinary" three-terminal transistor (i.e., excepting phototransistors), the necessary high-performance combination of LPE and lithography was never developed to the point that the resulting heterostructures would reach the speed capability of state-of-the-art Si bipolars, much less reach their own theoretical potential exceeding that of Si.

As a result of the emergence of two new epitaxial technologies in the last few years, the heterostructure technology bottleneck is rapidly disappearing, to the point that the incorporation of heterostructures into most compound semiconductor devices will probably be one of the dominant themes of compound semiconductor technology during the remainder of the present decade.

The two new technologies are MBE [16] and MOCVD [17]. Although differing in many ways, for the purposes of this paper the commonalities of the two technologies are more important than their differences, and there is no need to enter here into the debate as to which of the two technologies will eventually be best for doing what.

Both technologies are capable of growing epitaxial layers with high crystalline perfection and purity, comparable to state-of-the-art results with LPE and halide-CVD. Highly controlled doping levels up to 10^{19} impurities per cm^3 and more can be achieved, and highly controlled changes in doping level are possible during growth without interrupting the latter, and with at most a minor adjustment in growth parameters. The doping may be changed either gradually or abruptly. Because of the comparatively low growth temperatures (especially for MBE), diffusion effects during growth are weak, and with certain dopants much more abrupt doping steps can be achieved than with any other technique, not only when doping is "turned on," but also when it is "turned off."

Most important in our context of heterostructures, it is possible in both technologies to change from one III/V semiconductor to a different (lattice-matched) III/V semiconductor with greater ease than in any other technique. In both techniques, a change in semiconductor and hence in energy gap is not significantly harder to achieve than a change in doping level! In particular, the change can again be accomplished during growth without interruption, either gradually or abruptly and, if abruptly, over extremely short distances.

Finally, in both techniques the growth rates and hence the layer thicknesses can be very precisely controlled. Because the growth rates themselves are low (or can be made low), extremely thin layers can be achieved, to the point that effects due to the finite quantum-mechanical wavelengths of the electrons can be readily generated. It is in the context of the study of such quantum effects that both techniques have demonstrated their so far highest capability level. With both MOCVD and MBE, GaAs-(Al,Ga)As structures with over 100 epitaxial layers have been built [26], [27], and essentially arbitrary numbers appear possible. With MOCVD, layer thicknesses below 50 Å have been achieved, with MBE, below 10 Å. In either case, the capability far exceeds anything needed in the foreseeable future for transistor-like devices.

So far, these are laboratory results, mostly on GaAs-(Al,Ga)As structures. But it is the consensus of those working on the two technologies that much of this performance can be carried over into a production environment, with high yields and at an acceptable cost. Acceptable here means a cost low enough that it will not deter the use of the new technologies in most of those high-performance applications that need the performance potential of heterostructure devices.

An extension of both technologies to lattice-matched III/V-compound heterosystems beyond GaAs-(Al,Ga)As is an all but foregone conclusion, including GaAs-(Ga,In)P, InP-(Ga,In)(P,As), and InAs-(Al,Ga)Sb.

In view of these developments, the following scenario for the III/V-compound heterostructure technology of the 1990's is likely. Epitaxial technologies will be routinely available in which both the doping and the energy gap can be varied almost at will, over distances significantly below 100 Å, and covering

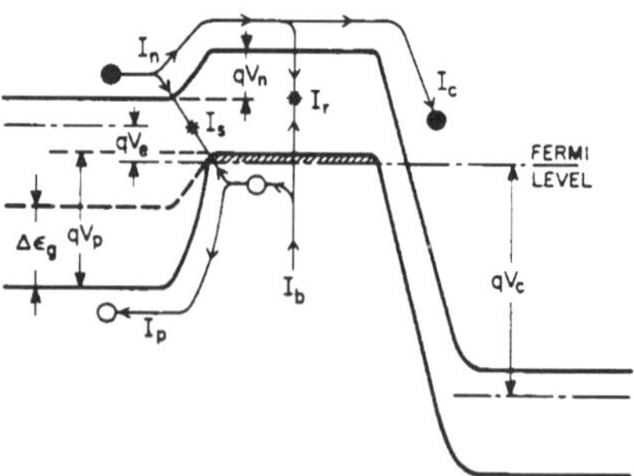

Fig. 3. Band diagram of an n-p-n transistor with a wide-gap emitter, showing the various current components, and the hole-repelling effect of the additional energy gap in the emitter.

a large fraction of their physically possible ranges, by what is essentially a software-controlled operation within a given growth run. The cost of the technology will be sufficiently low to encourage the development of high-performance devices that utilize this capability. The cost will be essentially a fixed cost per growth run, depending on the overall tolerance level but hardly at all on the number of layers and what they contain, similar to the cost of optical lithography, which has largely a fixed cost per masking step, almost independent of what is on the mask (at a given tolerance level). In particular, there will be only a negligible cost increment associated with using a heterojunction over using a homojunction (or no junction at all), and hence there will be only a negligible economic incentive *not* to use a heterojunction.

What *will* be expensive, just as with masking, are multiple growth runs, in which the growth is interrupted and the wafer removed from the growth system for intermediate processing, with the growth to be resumed afterwards. Hence there will be a strong incentive to accomplish the desired device structure with the minimum number of growth runs, no matter how complicated the individual run might become.

The above scenario is the technological premise of the remainder of this paper. Although presented here in the context of bipolar transistors and IC's, this scenario, as well as the central design principle of Section II, obviously go far beyond these specific devices. Together, the two concepts might form the starting point for a fascinating speculation about the future of semiconductor devices beyond simple bipolar structures. However, such a discussion would go beyond the scope of this paper as well as of this Special Issue.

IV. THE WIDE-GAP EMITTER

A. Basic Theory

The basic theory behind a wide-gap emitter is simple [1]. Consider the energy band structure of an n-p-n transistor, as in Fig. 3. In drawing the band edges as smooth monotonic curves we are implicitly assuming that the emitter junction has been graded sufficiently to obliterate any band edge discontinuities or even any nonmonotonic variations of the conduction band edge. We will return to this point later. There are the following injection-related dc currents flowing in such a transistor:

a) A current I_n of electrons injected from the emitter into the base;

b) A current I_p of holes injected from the base into the emitter;

c) A current I_s due to electron–hole recombination within

16   PROCEEDINGS OF THE IEEE, VOL. 70, NO. 1, JANUARY 1982

the forward biased emitter-base space charge layer.

d) A small part of I_r of the electron injection current I_n is lost due to bulk recombination.

The current contribution I_n is the principal current on which the device operation depends; the contributions I_p, I_s, and I_r are strictly nuisance currents, as are the capacitive currents (not shown in Fig. 3) that accompany any voltage changes. We have neglected any currents created by electron–hole pair generation in the collector depletion layer or the collector body.

Expressed in terms of these physical current contributions, the net currents at the three terminals are:

Emitter current: $I_e = I_n + I_p + I_s$ (1a)

Collector current: $I_c = I_n - I_r$ (1b)

Base current: $I_b = I_p + I_r + I_s$.   (1c)

A figure of merit for such a transistor is the ratio

$$\beta = \frac{I_c}{I_b} = \frac{I_n - I_r}{I_p + I_r + I_s} < \frac{I_n}{I_p} \equiv \beta_{max}. \quad (2)$$

Here, β_{max} is the highest possible value of β, in the limit of negligible recombination currents. It is the improvement of β_{max} to which the wide-gap emitter idea addresses itself.

To estimate β_{max} we assume that emitter and base are uniformly doped with the doping levels N_e and P_b. We denote with qV_n and qV_p the (not necessarily equal) heights of the potential energy barriers for electrons and holes, between emitter and base. We may then write the electron and hole injection current densities in the form

$$J_n = N_e v_{nb} \exp\left(-qV_n/kT\right) \quad (3a)$$

$$J_p = P_b v_{pe} \exp\left(-qV_p/kT\right). \quad (3b)$$

Here v_{nb} and v_{pe} are the mean speeds, due to the combined effects of drift and diffusion, of electrons at the emitter-end of the base, and of holes at the base-end of the emitter. In writing (3a, b) with simple Boltzmann factors, we have implicitly assumed that both emitter and base are nondegenerate. In a homojunction transistor the emitter might be degenerate; in a heterojunction transistor the base might be degenerate, as is in fact assumed in Fig. 3. This requires small corrections either in (3a) for the homojunction case, or (3b) for the heterojunction case, which we neglect here for simplicity. We have also neglected correction factors allowing for the differences in the effective densities of states of the semiconductors.

We are interested here only in the ratio of the two currents. If the energy gap of the emitter is larger than that of the base by $\Delta\epsilon_g$, we have

$$q(V_p - V_n) = \Delta\epsilon_g \quad (4)$$

and we obtain

$$\frac{I_n}{I_p} = \beta_{max} = \frac{N_e}{P_b}\frac{v_{nb}}{v_{pe}} \exp\left(\Delta\epsilon_g/kT\right). \quad (5)$$

For a good transistor, a value $\beta_{max} \gtrsim 100$ is desirable.

Of the three factors in (5), the ratio v_{nb}/v_{pe} is least subject to manipulation. As a rule

$$5 < v_{nb}/v_{pe} < 50. \quad (6)$$

To obtain $\beta_{max} \gtrsim 100$ it is therefore necessary that either

$$N_e \gg P_b \quad (7)$$

or that $\Delta\epsilon_g$ is at least a few-times kT.

Energy gap differences that are many-times kT are readily obtainable. As a result, very high values of I_n/I_p can be achieved *almost regardless of the doping ratio*. This does not mean that arbitrarily high β's can be obtained. It simply means that the hole injection current I_p becomes a negligible part of the base current compared to the two recombination currents: $I_b \cong I_s + I_r$. To have a useful transistor, we must still have $I_r \ll I_n$. If we approximate I_e by I_n, we obtain

$$\beta = \frac{I_n}{I_r + I_s}. \quad (8)$$

Based on the evidence from high-β HBT's that have been reported ($\beta \geqslant 10^3$),[1] the emitter–base hetero-interface can be made sufficiently defect-free to keep the interface recombination current I_s below $10^{-3}I_n$, at least at sufficiently high current levels I_n. At the same time, the base doping in a properly designed heterostructure transistor will be very high, and hence the minority carrier lifetime correspondingly low, to the point that the bulk recombination current I_r, rather than the interface recombination current I_s will dominate, in contrast to the situation in many homojunction transistors. We therefore neglect I_s beside I_r.

The bulk recombination current density may be written

$$J_r = \gamma n_e(0) w_b/\tau. \quad (9)$$

Here $n_e(0)$ is the injected electron concentration at the emitter end of the base, w_b is the base width, and τ the average electron lifetime in the base. The factor γ is a factor between 0.5 and 1.0, indicating by how much the average electron concentration differs from the electron concentration at the emitter end. If we insert (3a) and (9) into (8), and neglect I_s, we obtain

$$\beta \cong \frac{1}{\gamma}\frac{v_{nb}\tau}{w_b}. \quad (10)$$

This depends on the base doping only through the effect of the base doping on the lifetime. For heavy base doping levels the lifetimes may be short indeed.[2] Nevertheless, even for very short lifetimes, high β's should be achievable in transistors with a sufficiently thin base region, which is the case of dominant interest in any event. As an example, assume $w_b \cong 1000$ Å $= 10^{-5}$ cm. In such a transistor the electron velocity is likely to approach values close to bulk limited drift velocities $v_{nb} \cong 10^7$ cm $\cdot$ s^{-1}. Even for a lifetime as short as 10^{-9} s, this would lead to $\beta \cong 10^3$, a value that should satisfy even the most stringent demands. Evidently, no serious problems from reduced minority carrier lifetimes arise unless the latter drop to the vicinity of 10^{-10} s or lower, at least not for plausible base widths not exceeding 1000 Å.

Much of the remainder of this paper will deal with the tradeoffs made possible when high β-values can be obtained without a high emitter-to-base doping ratio. Before turning to these tradeoffs, it is instructive to return to (5) and to apply it to

<hr>

[1] See, e.g., [7], [8], [9], [14], [18]. Even higher values have been found in some phototransistors. See [15] for further references.

[2] For GaAs, injection laser experience suggests lifetimes between 10^{-10} and 10^{-9} s for degenerate doping levels, slightly longer for nondegenerate doping.

energy gap variations in the conventional silicon transistor. The energy gap of Si, like that of the other semiconductors, is not strictly constant, but decreases slightly at the high doping levels that are desirable in the emitters of a homojunction transistor. As a result, a Si transistor is not strictly a uniform-gap transistor; it is itself a heterojunction transistor, but with a small yet highly undesirable *negative* value of $\Delta\epsilon_g$. The best available data, taken on actual transistor structures [28], indicate a gap shrinkage beginning at a doping level $N_d \sim 10^{17}$ cm^{-3}, and reducing the gap approximately logarithmically with doping level, reaching a gap shrinkage between 75 and 80 meV ($>3kT$) at $N_d \sim 10^{19}$ cm^{-3}. According to (5), an emitter gap shrinkage of $3kT$ reduces the ratio I_n/I_p by a factor $e^{-3} \sim 1/20$. The overall effect at this doping level is the same as if the emitter were doped only to 5×10^{17} cm^{-3}, without gap shrinkage. To obtain β-values larger than the ratio v_{nb}/v_{pe} (<50), the base region must be even less heavily doped than this value, which is far below what is metallurgically possible, and far below what would be desirable in the interest of almost all other performance characteristics, especially base resistance. Increasing the emitter doping beyond 10^{19} cm^{-3} improves β only very slowly, roughly proportionally to $N_e^{0.33}$. By pushing everything to the limit, state-of-the-art microwave transistors with P_b-values (averaged over the base region) of about 1×10^{18} cm^{-3} have been achieved [29]. But this is still far below what would be desirable.

Evidently, the conventional Si bipolar transistor behaves far less well than the naive uniform-gap textbook model would predict. In fact, the energy gap shrinkage and its consequences represent one of the dominant performance limitations of the device.

B. Graded Versus Abrupt Emitter Junctions

In Fig. 3, and in the discussion accompanying it, we had assumed that the emitter/base junction is compositionally graded, so as to yield smoothly and monotonically varying band edges. Such graded transistors are easily achieved, but unless the appropriate measures are taken to do so, the modern epitaxial technologies tend to produce abrupt transistors in which band edge discontinuities are present. As a rule, the conduction band on the wider gap side lies energetically above that on the narrower gap side. Applied to the wide-gap emitter in a transistor, this leads to the "spike-and-notch" energy band diagram shown in Fig. 4(a). Because the emitter-to-base doping ratio in an HBT tends to be low, most of the electrostatic potential drop will occur on the less heavily doped emitter side, and the potential spike will project above the conduction band in the neutral portion of the base, leading to a potential barrier of net height $\Delta\epsilon_B$. Such a barrier has both advantages and disadvantages, and a brief discussion is in order.

Consider first the potential notch accompanying the barrier on the base side. Such a notch will collect injected electrons, and therefore enhance recombination losses, a highly undesirable effect. Because of the low emitter-to-base doping ratio expected in an HBT, the notch will be quite shallow, with a depth given approximately by

$$\Delta\epsilon_N = (P_e/N_b)\, q V_n \qquad (11)$$

which will typically be of the order 5 meV $\ll kT$. Nevertheless, because of the danger of interface recombination defects, it would be desirable to eliminate the notch altogether, and perhaps even replace it by a slightly repulsive potential, as

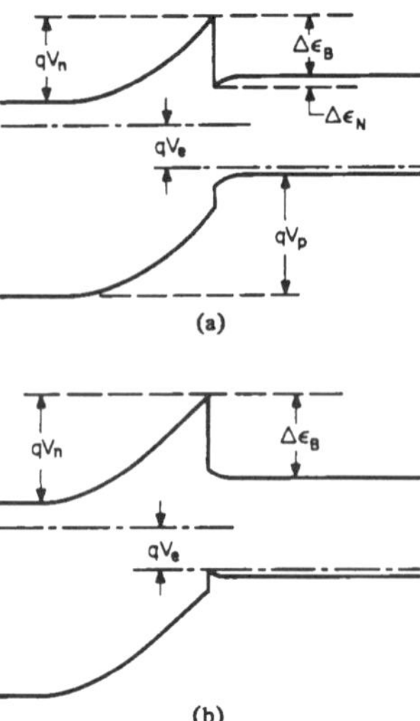

Fig. 4. Band structure of an abrupt wide-gap emitter, showing the spike barrier and the accompanying electron trapping notch (a) in the conduction band structure. The notch can be removed (b) by the incorporation of a planar acceptor doping sheet into the heterojunction.

shown in Fig. 4(b). This is easily accomplished by incorporating a very thin sheet with a very high acceptor concentration right at the interface. Typical required sheet doping concentrations will be of the order 10^{11} acceptors per *square* centimeter. The feasibility of such "planar doping" sheets has been demonstrated [30], at least with MBE, and there is little doubt that it can be accomplished by MOCVD as well.

As to the barrier itself, one minor drawback of its existence is the accompanying increase of the order $\Delta\epsilon_B/q$, in required emitter voltage to yield a given current density. More severe is the (related) drawback that the potential barrier $\Delta\epsilon_B$ drastically reduces the ratio J_n/J_p, from the value in (5), by a factor exp $(-\Delta\epsilon_B/kT)$. Instead of (4a), we now have

$$q(V_p - V_n) = \Delta\epsilon_g - \Delta\epsilon_B \cong \Delta\epsilon_V. \qquad (4b)$$

The last equality results if the notch depth is small compared to kT, in which case $\Delta\epsilon_B = \Delta\epsilon_C$. Here $\Delta\epsilon_C$ and $\Delta\epsilon_V$ represent the conduction and valence band discontinuities. Instead of (5a), we obtain

$$\frac{I_n}{I_p} = \beta_{max} = \frac{N_e}{P_b}\frac{v_{nb}}{v_{pe}} \exp\left(\Delta\epsilon_V/kT\right). \qquad (5b)$$

If the valence band discontinuity is sufficiently large, a major improvement remains. Unfortunately, in the system of largest current interest, the (Al,Ga)As/GaAs system, the valence band discontinuity is quite small, $\Delta\epsilon_V = 0.15\,\Delta\epsilon_g$ [31], and the reduction of the spike by grading is probably essential. A detailed discussion of the detrimental effects of the spike is found in a paper by Marty *et al.* [10].

The above drawbacks of the extra potential barrier accompanying an abrupt emitter/base junction are partially compensated by the fact that such a barrier would inject the electrons into the base region with a substantial kinetic energy, and hence with a very high velocity ($\sim 10^8$ cm/s). Because of the directional dependence of the polar optical phonon scattering that is the dominant scattering process in III/V-compounds,

PROCEEDINGS OF THE IEEE, VOL. 70, NO. 1, JANUARY 1982

several collisions are required before the electrons have lost their high forward velocity. The result should be a highly efficient and very fast near-ballistic electron transport through the base. Such ballistic transport effects have been of great interest recently, and although their discussion has been largely in an FET context [32], [33], much of this discussion applies as well (or even more) to bipolar transistors with an emitter junction barrier that represents, in effect, a ballistic launching ramp.

Exactly what the balance between drawbacks and benefits will be for the abrupt emitter/base junction versus the graded one, remains to be seen. But it appears likely that ballistic effects will find their way into future transistors specifically designed around them.

An extreme case of high-energy electron injection into the base was discussed some time ago by Kroemer [34], in the form of a so-called *Auger transistor*. If the conduction band discontinuity $\Delta\epsilon_C$ becomes larger than the energy gap in the base, the electron injection may lead to Auger multiplication of electrons, and hence to a transistor with true current amplification in a grounded-base configuration $\alpha > 1$. Such a transistor might be of interest for power switching applications at very high microwave speeds. It remains to be seen what will come of this idea.

V. Speed Tradeoffs

A. The Emitter Capacitance Tradeoff

High beta-values above, say, 100 are of limited interest by themselves, except perhaps in phototransistors. The principal benefit of a wide-gap emitter is therefore not the ability to achieve high β-values, but the freedom to change doping levels in emitter and base without significant constraints by injection efficiency consideration, and thereby to re-optimize the transistor at a higher performance level.

We start our discussion with the choice of emitter doping. A wide-gap emitter permits a drop in emitter doping by several orders of magnitude without a deterioration of β, a prediction [1] that has been confirmed experimentally in almost all HBT's built. Now it is well known that the junction capacitance of a highly unsymmetrically doped p-n junction depends only on the doping level of the less heavily doped side. Suppose the base doping is initially kept fixed. If the emitter doping is now dropped below the base doping, the emitter capacitance of the transistor then depends principally on the emitter doping and drops with a decrease of the latter, roughly as

$$C_e \propto N_e^{1/2}. \tag{12}$$

Evidently, by dropping the emitter doping sufficiently far below the (initial) base doping a large reduction in emitter capacitance can be obtained [1], and this reduction remains if the base doping is subsequently increased. The result is an improvement in speed, but this effect is usually small, because the emitter capacitance is only one of several capacitances. The true significance of the reduction of the capacitance per unit area lies in two different facts. First, it permits an increase in the capacitive emitter area in the inverted transistor design discussed later, without increase in total emitter capacitance. Second, in HBT's for small-signal microwave amplification, a reduction in emitter capacitance will reduce the noise significantly [35].

Obviously, the doping in the emitter cannot be lowered arbitrarily far. Even if achievable crystal purities permitted it, the emitter series resistance would eventually become excessive, at least for a thick emitter body. However, under the technological scenario envisaged earlier, the weakly doped part of the emitter can always be kept very thin (say, a few-times 10^{-5} cm) to permit a drop in emitter capacitance per unit area by at least a factor 10 before emitter series resistance effects become serious.

A minor advantage of reduced emitter doping, mentioned by Milnes and Feucht [3], might be that the resulting emitters would have a significant reverse breakdown voltage. It is not clear how much of an advantage this would be.

B. The Base Resistance Tradeoff: Microwave Transistors

The most important single change made possible by a wide-gap emitter is a drastic increase in base doping, limited only by technological constraints and by the need to keep the minority lifetime in the base significantly above 10^{-10} s. The principal benefit is a major reduction in base resistance, which, in turn, increases the speed significantly [2]. A second benefit is a major improvement in overall transistor performance at high current densities [1], [3]–[5], including specifically an improvement in the speed-versus-power tradeoffs of microwave transistors.

Because we are principally interested in low-power speed aspects, we concentrate here on the effect of base resistance reduction. This effect is somewhat different in microwave transistors and in switching transistors.

For microwave transistors, Ladd and Feucht [2] have given a very detailed analysis, using the maximum oscillation frequency $f_{\max}$ as the figure of merit. It may be written in the form

$$f_{\max} = \tfrac{1}{2}(f_t f_c)^{1/2} \tag{13}$$

where f_t has its familiar meaning as the frequency at which the current gain is reduced to unity, and f_c is the frequency equivalent of the RC time constant of the combination base resistance–collector capacitance,

$$f_c = 1/(2\pi R_b C_c). \tag{14}$$

Evidently, a reduction in R_b causes an increase in f_c and with it a smaller increase in $f_{\max}$.

Ladd and Feucht's work was done in the late 60's and they give numerical values only for the "best" system known at the time, a GaAs emitter on a Ge base, of a construction previously demonstrated by Jadus and Feucht [36]. Because of severe limitations inherent in the then-available technology, the *external* base resistance (between the emitter edge and the base contact) could not be significantly decreased, and as a result, Ladd and Feucht concluded that only a negligible improvement in frequency could be achieved with the then-existing technology. If, however, the external base resistance problem could be solved, maximum oscillation frequencies $f_{\max}$ around 100 GHz would be achievable. Similarly high values can be predicted for other heterosystems such as (Al,Ga)As-on-GaAs or GaP-on-Si [37], [38]. There is little point in quoting more exact values, becaue the predictions depend noticeably on both technological and operating parameters whose choice would be applications-dependent. To pursue these matters in detail would lead us too far away from our principal interest in digital switching transistors.

C. The Base Resistance Tradeoff: Digital Switching Transistors

The quantity of interest in digital switching transistors is not the maximum frequency of oscillation but the (somewhat vaguely defined) switching time. Although one would expect that any structural measures that improve the maximum oscillation frequency will also improve the switching speed, there is no simple one-to-one relationship between the two. The modes of operation are just too different. For example, in microwave transistors a high output power is usually of interest, while in highly integrated digital switching transistors the opposite is the case.

A comparison is further complicated by the fact that switching time depends on the circuit, and no standard measure for switching time, comparable to the frequencies f_t and $f_{\max}$ for oscillatory operation, has been agreed upon. Probably the best measure of switching time applicable to HBT's is the estimate by Dumke, Woodall, and Rideout (DWR) [5], who estimate the switching time as

$$\tau_s = \frac{5}{2} R_b C_c + \frac{R_b}{R_L} \tau_b + (3C_c + C_L) R_L. \qquad (15)$$

Here R_b is the base resistance, C_c the collector capacitance, and τ_b the base transit time, while R_L and C_L are load resistance and capacitance of the circuit. The result (15) is based on Ashar's analysis [39] of a two-transistor circuit, modified by Dumke. Dumke's modification simply consists of the following [40]. The load resistance must be large enough to develop a potential change equal to the necessary emitter swing ΔV on the next stage. Therefore, $R_L = \Delta V/I = R_E$, where I is the current that is switched to. Making the appropriate substitutions in Ashar's expression yields (15). Dumke *et al.* apply (15) to estimate the switching time of a hypothetical (Al, Ga) As-on-GaAs transistor with the following parameters. Base width: 1200 Å; base doping: 3×10^{18} cm^{-3}; base and emitter stripe widths: 2.5 μm, separated by 0.5-μm gaps; collector doping: 3×10^{16} cm^{-3}; load resistance: 50 Ω; load capacitance: negligible compared to collector capacitance. These values lead to the following values for the three terms in (15): 8.3 ps, 1.4 ps, and 8.3 ps, combining into an overall switching time of ~18 ps. The authors state that this is "roughly a factor of 5 or 8 faster than that which might be realized from the current post alloy diffused Ge or double diffused Si technologies respectively." Today, nearly 10 years later, post-alloy diffused Ge technology is all but forgotten (it never made it into IC's), and much of the then-predicted advantage over Si remains.

Just as in the case of Ladd and Feucht's estimate of $f_{\max}$, much of the improvement is due to the reduction in base resistance that is associated with the high base doping possible in an HBT. In fact, two of the three terms in (15) depend linearly on R_b rather than with the square root as does $f_{\max}$. This means that as long as those terms dominate τ_s, a reduction of R_b is even more effective in a digital switching transistor than in a microwave transistor. Only after the base resistance reduction has been carried so far that the $R_L C_L$ term dominates, does a further reduction in R_b lead to no further benefit. The hypothetical device analyzed by DWR lies at the borderline between the two regimes.

The specific numerical values quoted above should be viewed as approximations. To obtain an expression as simple as (15), Ashar and Dumke had to make numerous simplifications, just as the expression (13) for $f_{\max}$ is based on gross simplifications. The importance of the Ashar–Dumke result (15) is that it indicates the relative significance of the most important transistor parameters. A more detailed analysis is certainly needed, in particular, one that investigates the extent to which the various approximations made in deriving (15) remain applicable in HBT's that have been drastically modified from conventional design.

The assumption of different structural transistor parameters would, of course, have led to different values of τ_b. But the values assumed by DWR were quite reasonable in 1972; they are easily within the range of today's technology, and hence conservative. Further reductions in τ_b to below 10 ps appear readily achievable.

One possibility for improvement is to strive for a lower load resistance than the ad hoc value of 50 Ω assumed by DWR. One sees readily from (15) that the switching time goes through a minimum for

$$R_L = [R_b \tau_b/(3C_c + C_L)]^{1/2} \qquad (16)$$

for which (15) reduces the

$$\tau_s = \frac{5}{2} R_b C_c + 2[(3C_c + C_L) R_b \tau_b]^{1/2}. \qquad (17)$$

For the structural values assumed in DWR one would need $R_L \cong 21$ Ω, which would yield $\tau_s \cong 15$ ps. The improvement is not large, and the low load resistance might not be easy to achieve [40]. A much larger improvement would result from a reduction of the collector capacitance, obtained by inverting the transistor. This possibility will be discussed later.

D. The External Base Resistance Problem

In their detailed analysis of the (microwave) performance potential of HBT's, Ladd and Feucht go to great lengths to discuss the special problem posed by the highly detrimental external portion of the base resistance. Because their considerations also apply to digital switching transistors, and because they appear not to have been fully appreciated by subsequent workers on heterostructure bipolar transistors [41], it appears proper to re-emphasize the problem raised by Ladd and Feucht here, and to offer a remedy.

In all real transistors only part of the base resistance lies underneath the emitter, part lies between the edge of the emitter and the base contact. Usually, the outer region of the base is appreciably thicker than the inner region, and the near-surface portion of the outer base is more heavily doped than the remainder (Fig. 5(a)). This design minimizes the outer base resistance. If one wishes to obtain the postulated advantages of a wide-gap emitter, it is essential that the outer base resistance is not permitted to dominate the overall base resistance. This is easier said than done. For example, suppose that technological changes associated with the change from a (diffused or implanted) homojunction emitter to a heterojunction emitter, forced a change in geometry from that in Fig. 5(a) to that in Fig. 5(b) with a thin outer base. This is in fact the geometry used in the HBT's reported in the literature, except for the transistors reported by Ankri *et al.* [11], [14] and by Katz *et al.* [23]. Even though the doping level in the

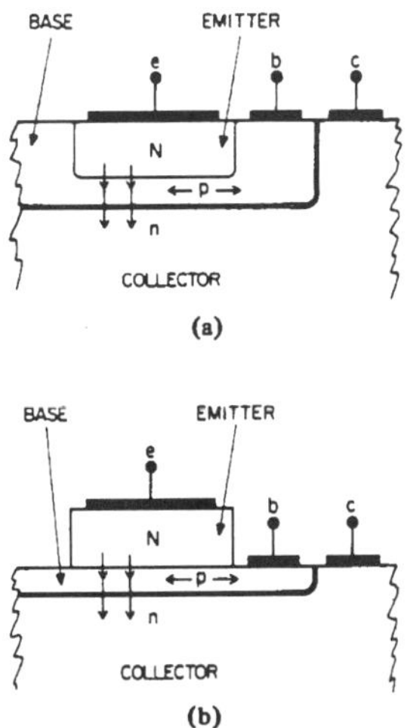

(a)

(b)

Fig. 5. In homojunction transistors of current technology (a), the base region is usually much thicker and more heavily doped outside the emitter than between the emitter and collector, reducing the external base resistance. This desirable feature would be lost in heterostructure transistors with the emitter island design shown in (b). To appreciate this point fully, recall that in actual structures the horizontal dimensions greatly exceed the vertical ones. In this drawing (and in Fig. 6) the vertical dimensions have been greatly exaggerated relative to the horizontal ones.

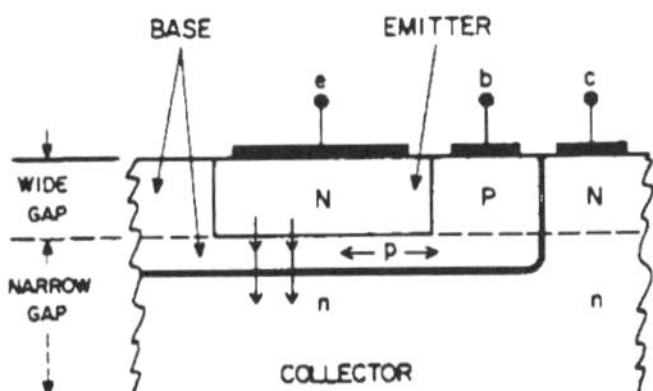

Fig. 6. Desirable emitter structure in which the p-n junction does not follow the planar hetero-interface, but is pulled up towards the surface.

outer base may have been increased, the beneficial effect of this change would be at least partially compensated by the reduction in thickness of the outer base. In unfavorable cases the outer base resistance might even have increased. Ladd and Feucht fully recognized the importance of this problem. They wrote "... *it is clear that the advantages of the low base resistance of the heterojunction devices will only be exploited if suitable geometries can be developed*."

It is now important to realize that the wide-gap emitter configuration contains a built-in design possibility to keep the outer base resistance low [37], [38], [41], [42]. The design is shown in Fig. 6. Rather than constructing the wide-gap emitter as an island riding by itself on the top of a uniformly thin narrow-gap base layer, the wide-gap semiconductor may be extended beyond the emitter edge, forming part of the outer base region, with the emitter-base p-n junction pulled away from the heteroboundary and towards the surface. Such a configuration should be easily achievable by first growing the top wide-gap layer with the same relatively low n-type doping as the emitter, and then converting the region outside the emitter to heavy p-type doping by diffusion or ion implantation.

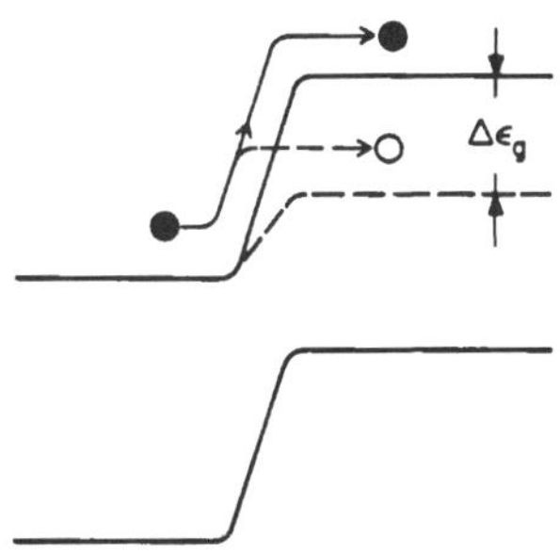

Fig. 7. Blocking of injection of electrons into the wide-gap portion of the base region in Fig. 6, due to the extra repulsive force generated by the wider energy gap.

In such a design the portion of the emitter that lies within the wide-gap region carries only a negligible current, compared to the wide–narrow portion. The reason for this is illustrated in Fig. 7. For injection into the wide-gap p-region, the electrons would have to climb a barrier that is higher by the energy gap difference $\Delta \epsilon_g$. But this reduces the injection current density by the same factor $\exp(-\Delta \epsilon_g/kT)$ that also reduces the hole injection into the wide-gap emitter.

This possibility does not appear to have been as widely recognized as it deserves; it *has* been used in the devices reported by Ankri *et al.* [11], [14], and by Katz *et al.* [23]. In both cases diffusion was used to convert the wide-gap portion of the base region to p-type.

VI. The "Inverted" Transistor

Since the first days of the alloy transistor, bipolar transistors have been built with a larger collector than emitter area, in the interest of efficient charge collection. In planar technology, the two junctions are necessarily of different area. The need for efficient charge collection then enforces the familiar configuration with the collector at the bottom and the emitter at the top. The exception to this rule is, of course, integrated injection logic (I^2L), where other considerations override this rule—at a price. I will say more about I^2L below. But apart from the I^2L exception, the "emitter-up rule" is so pervasive that it has become hard to imagine that a useful transistor could be built with the inverse order.

Now we have just seen that with a wide-gap emitter the emitter junction can be designed in such a way that part of the emitter-base junction does not inject carriers. Evidently, with such a design the need for efficient carrier collection can be met even with an emitter larger than the collector, IF those portions of the emitter-base junction that are not immediately opposite to a part of the collector-base junction are inactivated by pulling them onto the high-gap side of the hetero-interface. Once this is done, the transistor might just as well be "flipped," with the emitter on the substrate side and the collector on top, as shown in Fig. 8. The inverted configuration has several advantages, to the point that it might very well turn out the "canonical" configuration of future heterostructure bipolar transistor design [43].

The principal (but not the only) advantage of the inverted transistor is that it permits the use of a significantly smaller collector area, with an appropriately smaller collector capacitance. The consequences for the high-speed performance are obvious. Modern high speed transistors, both digital and (inter-

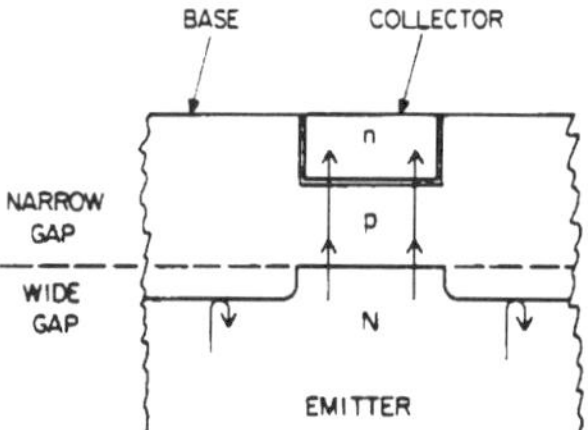

Fig. 8. Inverted "collector-up" transistor structure in which the emitter has a larger area than the collector, but the external portions of the emitter do not contribute to the injection, because there the p-n junction has been pulled into the wide-gap portion of the structure.

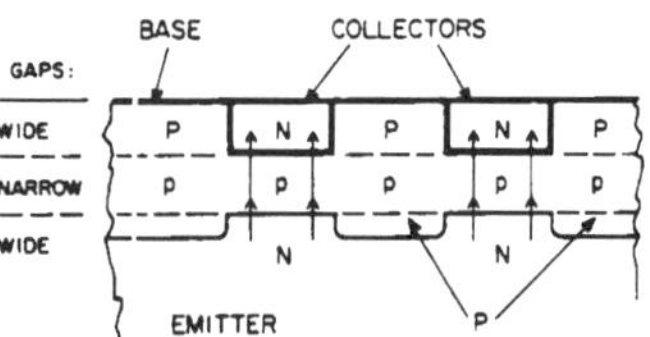

Fig. 9. A DH implementation of I^2L, combining wide-gap collectors with noninjecting emitter regions between the collectors.

digitated) microwave transistors, typically have a collector area close to three-times the active (emitter) area. Inverting the structure thus permits a reduction of the collector capacitance by close to a factor of 3. For example, in the hypothetical switching transistor analyzed by Dumke *et al.* [5], the emitter area was 3.4-times the emitter area. If, in that device, one reduces the collector area by a factor $\frac{1}{3}$ and leaves all other quantities unchanged, the two dominant terms in (15) are reduced by the same factor, and the switching time is reduced from ~18 ps to ~7 ps. Similar improvements would occur in microwave power transistors.

However, some care is in order: Because now the total emitter area is larger than the active area, the emitter junction capacitance will increase, at least compared to a heterostructure transistor of conventional emitter-up configuration. But, as we saw earlier, the emitter junction capacitance per unit area of a heterojunction transistor can in any event be made significantly less than for a homojunction transistor. Hence, compared to the latter, a net reduction in emitter capacitance may result even in the face of a larger (inactive) emitter area.

A second advantage of the inverted configuration is the possibility of a major reduction of the large lead inductance in series with the emitter that is present in the conventional emitter-up configuration. Again an improvement in high-frequency properties will result.

A third advantage of an inverted transistor configuration, for digital switching transistors, will emerge later.

Technologically, the inverted structure should be achievable in essentially the same way as the pulled-up emitter junction: By first growing the top layer lightly n-type doped throughout, and then converting the region outside the collector to heavy p-doping by diffusion or ion implantation. Obviously, the collector layer must be chosen thick enough to support the intended collector bias voltage. Converting part of the surface inside the collector region to n$^+$ might be desirable.

VII. DH TRANSISTORS

A. Introduction: The Wide-Gap Collector

A reading of Shockley's patent quoted at the beginning of this paper leaves no doubt that the *"one... zone ... having a wider energy gap than...the other zones"* is the emitter of the transistor. The question was soon raised whether there might also be advantages to a wide-gap collector [1]; but only the trivial and insignificant advantage of a reduction in the reverse-biased collector saturation was recognized.

This assessment must be revised in the light of the anticipated technological scenario discussed in Section III of this paper, and particularly in the light of the increased interest in highly integrated digital switching transistors. It appears that there are in fact several excellent reasons urging a wide-gap collector design, to the point that DH transistors with a wide-gap collector might very well be the rule rather than the exception for future bipolar transistor designs.

I give in this Section three examples that illustrate advantages to be gained by such a design. They fall into three groups:

a) Suppression of hole injection from base into collector in digital switching transistors under conditions of saturation;

b) Emitter/collector interchangeability in IC's;

c) Separate optimization of base and collector, especially in microwave power transistors.

The presentation does not attempt to give a complete and systematic critical evaluation of all aspects of DH transistor design. Its purpose is to initiate a discussion, not to end it.

B. Suppression of Hole Injection into the Collector under Saturated Conditions

In many digital logic families the collectors of the transistors are forward-biased during part of the logic cycle. If the base region is more heavily doped than the collector, as would normally be desirable, a copious injection of holes from the base into the collector takes place, which increases dissipation and slows down the switching speed. In a heterostructure technology, this highly deleterious phenomenon is easily suppressed the same way hole injection into the emitter is suppressed: By making the collector a wide-gap collector [40]. Such a design is an attractive alternative to the Schottky clamp in Schottky-TTL. Just as the wide-gap emitter, the wide-gap collector should be fairly lightly doped, in the interest of a low collector capacitance, and the base should remain heavily doped, in the interest of low base resistance. This choice of relative doping levels remains both possible and desirable in the inverted I^2L configuration, rather than calling for a weakly doped base to suppress collector injection, with its high base resistance penalty. In fact, in a recent paper [42], Kroemer has proposed a DH implementation of I^2L, which combines this idea with the idea of a selectively injecting emitter, discussed earlier. The structure is shown in Fig. 9. It avoids both the electron injection into those portions of the base where such injection is undesirable because of the absence of a collector opposite to the emitter, and the injection of holes into either collector or emitter. Even electrons spilling over at the edge of the active portion of the base region would not be able to penetrate into the upper part of the inactive portion of the base, because they would be repelled by the heterobarrier in the conduction band at the p-P interface. Because of the essentially complete suppression of parasitic charge storage, combined with greatly reduced RC-time constant effects due to the reduced base resistance, such an implementation of I^2L can be expected to have a much higher speed than the notoriously

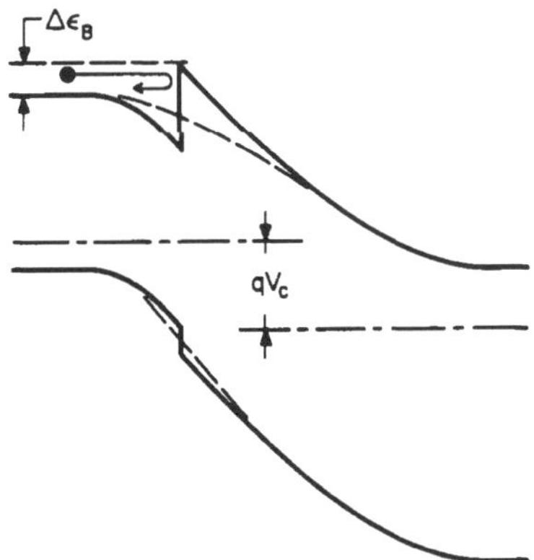

Fig. 10. Electron blocking action for low reverse bias at an abrupt p-n heterojunction collector. The blocking action can be prevented by grading the heterojunction, as indicated by the broken line.

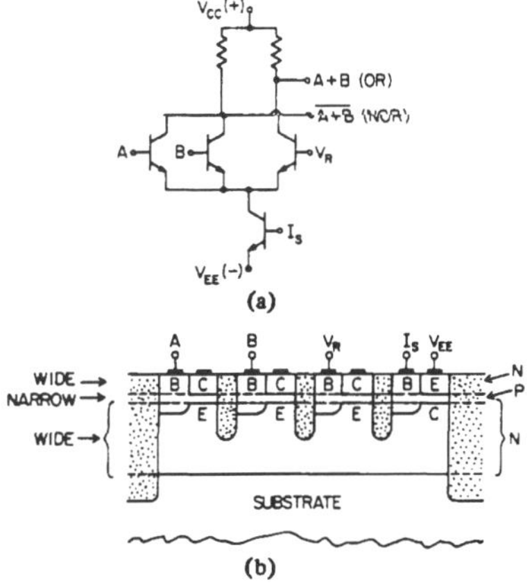

Fig. 11. Input stage of a DH implementation of ECL. The four transistors shown are implemented by three inverted and one noninverted transistor of identical structure, differing only in biasing. The dotted regions are isolation regions, prepared by proton bombardment or equivalent techniques.

slow homostructure implementations of I^2L, without increasing the highly desirable low dissipation levels of I^2L. Unfortunately, no quantitative estimates of the expected performance improvement have so far been published, but the possible improvements appear to be large.[3]

The referenced paper [43] also shows that the pnp horizontal transistor that serves as a current source in I^2L is easily incorporated into a DH design. It emerges as a rather peculiar structure that is basically a homostructure transistor with heterostructure sidewalls, which confine the current and improve the performance of the device.

There is one important restriction in the use of wide-gap collectors, which must not be overlooked. It is important that the free collection of electrons by the reverse-biased collector not be impeded by any heterobarrier due to a conduction band discontinuity (Fig. 10). Such barriers are easily eliminated by grading the heterostructure [44], [45].

C. Emitter/Collector Interchangeability

The advantages of a DH design for bipolar transistors are not restricted to the suppression of hole injection into the collector in saturating logic. A different advantage lies in the possibility of designing transistors in which the role of emitter and collector can be interchanged by simply changing the biasing conditions, while retaining the advantages of a wide-gap emitter regardless of which of the two terminal n-regions is used as the emitter. To achieve this freedom, the transistor need not be geometrically symmetrical: In the inverted structure shown earlier in Fig. 8, in which the active portion of the lower p-n junction covered the same area as the upper p-n junction; either the upper junction or the lower junction could be used as the emitter. While this might be no more than a mildly esoteric advantage in a discrete transistor, it offers a major new option in the architecture of digital IC's, be they of the saturating or nonsaturating variety: The DH design makes it possible, within a common three-layer n-p-n epitaxial layer structure, to integrate high-performance wide-gap emitter transistors having the conventional emitter-up configuration, with similar transistors

---

[3] I have been informed by an anonymous reviewer that K. T. Alavi, in an unpublished M.S. thesis (M.I.T., 1980) has estimated that "over a 10-fold improvement in speed–power product can be anticipated." I did not have access to this work.

having the I^2L-like inverted emitter-down configuration discussed previously.

The full power of this new option can probably not be appreciated without an example. The input stage of emitter-coupled logic (ECL), a nonsaturating logic family, serves admirably. Fig. 11(a) gives the basic circuit diagram of the parts of interest here. The top three transistors serve as a differential switch that compares the voltage levels of two logic signals A and B with a reference voltage V_R. The bottom transistor serves essentially as a constant-current source. (In some simpler versions of ECL it is replaced by a resistor.)

Evidently, the configuration calls for tying together the emitters of three transistors with the collector of a fourth. In a DH design, this integration is achieved easily, without sacrificing a high transistor performance, by implementing the top three transistors as inverted transistors, and the current supply transistor as a conventional emitter-up transistor, as shown in Fig. 11(b). The emitters of the three top transistors and the collector of the bottom transistor come together in a buried n-layer on top of the substrate. All four transistors are structurally identical; they differ merely in their biasing. Those readers who are familiar with ECL and its notorious integration difficulties will undoubtedly recognize the great integration advantages offered by what I would like to call HECL, for Heterostructure ECL.

A complete discussion of various other heterostructure modifications of ECL is intended for another place; the purpose of the present discussion was merely to demonstrate the central idea of the interchangeability of emitter and collector in a DH IC design.

D. Separate Optimization of Base and Collector

Except for the interrelated needs of a high mobility and a high saturated drift velocity for the electrons, the semiconductor properties desired for the base of a transistor are quite different from those for the collector and for the base/collector depletion layer. This is especially true in microwave power transistors. Evidently the different needs of base and collector regions can, at least in principle, be optimized best by selecting different materials in the two regions, that is, by a heterostructure collector. In practice, this tends to mean a semiconductor with a wider energy gap in the collector and in the base/collector layer, compared to the base region.

Again, an example is called for to illustrate this idea. Consider the question as to the semiconductor combination offering the highest speed in a room-temperature microwave power transistor. One can argue that the fastest possible such transistor would be a GaAs-Ge-GaAs transistor [46] –IF such a transistor could in fact be built, which is by no means certain.

The reason for the choice of Ge as the ideal semiconductor for the base region is its high hole mobility, unexcelled by any other group-IV or III/V-compound semiconductor. Also, Ge is easily doped very heavily p-type. Taken together, the two properties assure a much lower base resistance than any other known useable semiconductor.

Admittedly, Ge has a lower electron mobility than several III/V compounds one might consider. But in a microwave power transistor with its necessarily fairly thick collector depletion layer (in the interest of a high breakdown voltage and a low collector capacitance) the transit time through the base is only a minor speed limitation compared to that through the collector depletion layer. Hence the beneficial effects of the high hole mobility in a Ge base layer are much larger than the detrimental effects of the lower electron mobility compared to, say, GaAs. On the other hand, Ge is hardly a desirable semiconductor for the collector and the base/collector depletion layer: Apart from a somewhat low saturated electron drift velocity ($v_s \cong 5 \times 10^6$ cm/s) and a high dielectric constant ($\epsilon \cong 16$), its low energy gap would lead to a low breakdown field and high thermally generated currents. Here a wider gap semiconductor is needed. Lattice-matching considerations suggest GaAs, which would be near-ideal in any event. One might be inclined to argue that the narrow gap of Ge also rules Ge out as a base region material of acceptably low thermal current generation rate. However, this is not the case: In a practical GaAs-Ge-GaAs transistor the Ge base region would be so thin and so heavily p-type doped that the thermal generation of electrons in the base would not contribute an unacceptably high collector saturating current.

Unfortunately, it is not at all clear whether or not GaAs-Ge-GaAs transistors with an acceptably low density of interface defects can be grown. Our own work at UCSB with the MBE growth of GaAs on Ge, and GaP on Si, has shown that the defect-free growth of a polar semiconductor such as GaAs on a nonpolar substrate such as Ge faces a number of quite fundamental difficulties, which have so far not been surmounted, and which may, in fact, be insurmountable [47].

However, none of the experimental uncertainties affect the principal point of our discussion here: The desirability of different semiconductors for base and collector, implying a heterostructure collector, is likely to be the rule rather than the exception in the technology of the future.

VIII. Some Speculations About the Future of Compound Semiconductor Devices

A. Bipolar Transistors versus FET's

If one ignores injection lasers and other optoelectronic devices, today's compound semiconductor device world is a pure FET world with essentially no bipolar inhabitants. A paper that predicts what amounts to a bipolar revolution in this FET world cannot simply ignore FET's. This is true even more once one realizes that the same technologies that promise to revolutionize bipolar transistors will also improve FET's

[48]. In fact, very active and successful research into heterostructure FET's is already under way. However, on balance, heterostructures can be expected to benefit bipolar devices much more than they benefit FET's, and if so, this will naturally tend to shift the balance between the devices much more towards bipolars than past developments might suggest. There are several reasons for these expectations:

a) As was pointed out already in Section II, the Central Design Principle permits one to control the flow of electrons and holes separately and independently of each other. This makes heterostructures a very major advantage in bipolar devices (including lasers) in which there are in fact both kinds of carriers present. It does little for an FET, although a related benefit is obtained in FET's through the concept of modulation doping [49].

b) Every device has a dimension in the direction of current flow that controls the speed of the device. In FET's (other than VMOS) the current flow is parallel to the surface, and the critical control dimension is established by fine-line lithography. In a bipolar transistor, the speed-determining part of the current path is perpendicular to the surface (and to the epilayers), and to the first order, speed is governed by the layer thicknesses. Because vertical layer thicknesses can be easily made much smaller than horizontal lithography dimensions, there is, for given horizontal dimensions, an inherently higher speed potential in bipolar structures than in FET's. The two qualifiers "to the first order" and "for given horizontal dimensions" are important, though: Small horizontal dimensions are still needed to minimize speed-limiting *second-order* effects caused by horizontal resistive voltage drops in the thin base layers. These second-order effects are actually reduced in HBT's, due to the much higher base doping levels, and they are not as severe as the first-order limiting effects of the horizontal dimensions in FET's. But in any event, there is nothing in bipolar technologies that would require or even suggest the use of larger horizontal dimensions than in FET's. The same fine-line lithography technologies that are used for FET's, can and will be used for bipolar devices. The capability offered by the new epitaxial technologies is an *additional* capability, not an alternate.

c) Once sufficiently small dimensions have been achieved, "ballistic" effects become important [32], [33], and they are in fact extensively studied, so far predominantly in an FET context. On the whole, ballistic effects improve device performance by minimizing electron scattering. To obtain this benefit, two conditions must be satisfied. First, the electrons must be accelerated very quickly [32]. The most effective way to do this is by launching the electrons with a high kinetic energy from the conduction band discontinuity in a heterostructure, as discussed earlier. This is much more effective than acceleration by an ordinary nonuniform electric field, the rate of nonuniformity of which is limited by Debye-length considerations. Second, the path along which ballistic effects are to be utilized, must be short, at most a few thousand Angstrom units long. Evidently, both the abrupt launching and the short current paths call for a current flow perpendicular to the epitaxial layers rather than parallel to them, once again favoring the geometry of bipolar designs.

d) All digital switching transistors have a critical bias voltage (often called turn-on voltage), in the vicinity of which the switching action takes place. For high-performance digital IC's, especially VLSI circuits, it is important that this critical voltage be as reproducible as possible, not only across the chip in a single VLSI circuit, but also from wafer to wafer. This repro-

PROCEEDINGS OF THE IEEE, VOL. 70, NO. 1, JANUARY 1982

24

ducibility is easier to achieve in bipolar transistors than in FET's. In bipolar transistors the turn-on voltage is almost fixed for a fixed energy gap of the semiconductor in the base region. It depends logarithmically on the base doping and, apart from temperature, on hardly anything else. Hence it is easy to keep stable. One might say with little exaggeration that it is close to being a natural constant. The turn-on voltage in an FET is, by contrast, purely "man-made," depending at least linearly on both the electron concentration in the channel and the channel thicknesses. To achieve reproducible turn-on voltages, at least two separate quantities must be controlled tightly. Considering that processing differences tend to be very important in IC technology, this particular difference between bipolars and FET's might well turn out to be as important as the more fundamental differences, strongly favoring bipolars [50].

The above arguments suggest strongly that bipolar devices will play a much larger role in the future that they have in the past, eventually assuming a leading role ahead of FET's. Exactly where the border between the two technologies will be, is something too hazardous to predict.

B. A Change in Technological Philosophy?

We have witnessed, since about 1964, a steady growth in III/V-compound semiconductor devices, principally GaAs devices. The driving force behind this development has been the high performance of such devices, not attainable with mainstream Si devices. If we ignore once again lasers and other optoelectronic devices, and restrict ourselves to purely electronic amplifying and switching devices, high performance has been largely synonymous with high speed, made possible by the high electron mobility of GaAs, and by the availability of semi-insulating GaAs as a substrate. However, not even the most ardent advocate of GaAs ever claimed that GaAs was used because it had an attractive technology. We used GaAs despite its technology, not because of it, and the threat was never far away that Si devices, with their much simpler and more highly developed technology, would catch up with GaAs performance, the fundamental advantages of GaAs notwithstanding.

It is exactly this imbalance between fundamental promise and technological weakness that is being removed by the new epitaxial technologies. If the technological scenario postulated in Section III of this paper is even remotely correct, it means nothing less but that the great future strength of III/V-compounds lies precisely in their new technology, which permits an unprecedented complexity and diversity in epitaxial structures, going far beyond anything available in Si technology! This new technological strength is thus emerging as more important than the older fundamental strengths of high mobilities and semi-insulating substrates. It is a remarkable reversal of priorities indeed.

None of this means even remotely that III/V compounds will replace Si. They will not do so any more than aluminum, magnesium, and titanium replaced steel. The analogy of Si to steel is due to M. Lepselter, who called Si technology "the new steel" [51], to bring out the similarity in the role of Si in the new industrial revolution of our own days, to the role of steel in the industrial revolution of the early-19th century. I would like to carry this excellent analogy a bit further. Just as the *structural* metallurgy of the 19th century found it necessary eventually to go beyond steel, to aluminum, magnesium, titanium, and others taking their place beside steel, so the

electronic metallurgy of our own age is going beyond Si, to the III/V-compounds and probably further, to take their own place beside Si.

We continue to build locomotives, ships, and automobiles from steel, but if it is airplanes and spacecraft we want, we need the other metals besides. And, of course, it took us a while to go from locomotives to spacecraft. The analogy to semiconductors is too obvious to require elaboration; only the time scale will be compressed.

All along the way from steel to titanium there were those who argued that the next step, while perhaps possible, was one for which no foreseeable need existed: All foreseeable needs of man could presumably be met by improvements of the technologies already in hand. Well, this too has not changed.

Acknowledgment

The work of this paper has greatly benefitted from uncounted discussions, over several years, with numerous individuals. Foremost amongst them were R. C. Eden, D. G. Chen, and S. I. Long, all (at the time) at the Rockwell Electronics Research Center. Others at Rockwell to whom I am indebted for discussions are J. S. Harris, R. Zucca, D. L. Miller, and P. Asbeck. I am grateful to W. P. Dumke (IBM) for a copy of his unpublished work on the switching time of bipolar transistors, which clarified many questions I had about that difficult problem. The final version of the paper benefitted from intense discussions with Prof. H. Beneking (Aachen) and from comments made by two anonymous reviewers.

References

[1] H. Kroemer, "Theory of a wide-gap emitter for transistors," *Proc. IRE*, vol. 45, no. 11, pp. 1535–1537, Nov. 1957.

[2] G. O. Ladd and D. L. Feucht, "Performance potential of high-frequency heterojunction transistors," *IEEE Trans. Electron Devices*, vol. 17, pp. 413–420, May 1970.

[3] For a review see A. G. Milnes and D. L. Feucht, *Heterojunctions and Metal-Semiconductor Junctions*. New York: Academic, 1972. (See especially ch. 3.)

[4] For additional references see B. L. Sharma and R. K. Purohit, *Semiconductor Heterojunctions*. Elmsford, NY: Pergamon, 1974. (See especially sect. 7.6.)

[5] W. P. Dumke, J. M. Woodall, and V. L. Rideout, "GaAs-GaAlAs heterojunction transistor for high frequency operation," *Solid-State Electron.*, vol. 15, no. 12, pp. 1339–1334, Dec. 1972.

[6] a) M. Konagai and K. Takahashi, "Formation of GaAs-(GaAl)As heterojunction transistors by liquid phase epitaxy," *Elect. Eng. Japan*, vol. 94, no. 4, 1974;
b) ——, "(GaAl)As-GaAs heterojunction transistors with high injection efficiency," *J. Appl. Phys.*, vol. 46, no. 5, pp. 2120–2124, May 1975.

[7] B. W. Clark, H. G. B. Hicks, I. G. A. Davies, and J. S. Heeks, "A (GaAl)As-GaAs heterojunction structure for studying the role of cathode contacts on transferred electron devices," *Gallium Arsenide and Related Compounds 1974* (Deauville), Inst. Phys. Conf. Ser., vol. 24, 1975, pp. 373–375.

[8] M. Konagai, K. Katsukawa, and K. Takahashi, "(GaAl)As/GaAs heterojunction phototransistors with high current gain," *J. Appl. Phys.*, vol. 48, no. 10, pp. 4389–4394, Oct. 1977.

[9] P. W. Ross, H. G. B. Hicks, J. Froom, L. G. Davies, F. J. Probert, and J. E. Carroll, "Heterojunction transistors with enhanced gain," *Electron Eng.*, vol. 49, no. 589, pp. 36–38, Mar. 1977.

[10] A. Marty, G. Rey, and J. P. Bailbe, "Electrical behavior of an n-p-n GaAlAs/GaAs heterojunction transistor," *Solid-State Electron.*, vol. 22, no. 6, pp. 549–557, June 1979.

[11] D. Ankri and A. Scavennec, "Design and evaluation of a planar GaAlAs-GaAs bipolar transistor," *Electron. Lett.*, vol. 16, no. 1, pp. 41–47, Jan. 1980.

[12] J-P. Bailbe, A. Marty, P. H. Hiep, and G. E. Rey, "Design and fabrication of high-speed GaAlAs/GaAs heterojunction transistors," *IEEE Trans. Electron Devices*, vol. ED-27, pp. 1160–1164, June 1980.

[13] H. Beneking and L. M. Su, "GaAlAs/GaAs heterojunction microwave bipolar transistor," *Electron. Lett.*, vol. 17, no. 8, pp. 301–302, Apr. 1981.

[14] D. Ankri, A. Scavennec, C. Besombes, C. Courbet, F. Heliot, and J. Riou, "High frequency low current GaAlAs-GaAs bipolar transistor," presented at Dev. Res. Conf., Santa Barbara, June 1981, unpublished.

[15] For an up-to-date account, containing essentially complete earlier references, see three of the most recent papers on the subject:
a) M. Tobe, Y. Amemiya, S. Sakai, and M. Umeno, "High-sensitivity InGaAsP/InP phototransistors," *Appl. Phys. Lett.*, vol. 37, no. 1, pp. 73–75, July 1980;
b) M. N. Svilans, N. Grote, and H. Benking, "Sensitive GaAlAs/GaAs wide-gap emitter phototransistors for high current applications," *IEEE Electron Devices Lett.*, vol. ED-11, pp. 247–249, Dec. 1980;
c) J. C. Campbell, A. G. Dentai, C. A. Burrus, Jr., and J. F. Ferguson, "InP/InGaAs heterojunction phototransistors," *IEEE J. Quant. Electron.*, vol. QE-17, pp. 264–269, Feb. 1981.

[16] For two reviews see:
a) A. Y. Cho and J. R. Arthur, "Molecular beam epitaxy," *Prog. Solid State Chem.*, vol. 10, pt. 3, pp. 157–191, 1975;
b) K. Ploog, "Molecular beam epitaxy of III-V compounds," in *Crystals: Growth, Properties, and Applications*, H. C. Freyhardt, Ed. New York: Springer-Verlag, 1980, vol. 3, pp. 73–162.

[17] For a review with complete references to earlier work, see: R.D. Dupuis, L. A. Moudy, and P. D. Dapkus "Preparation and properties of Ga$_{1-x}$Al$_x$As-GaAs heterojunctions grown by metalorganic chemical vapor deposition," *Gallium Arsenide and Related Compounds 1978* (St. Louis), Inst. Phys. Conf. Ser., vol. 45, pp. 1–9, 1979.

[18] R. A. Milano, T. H. Windhorn, E. R. Anderson, G. E. Stillman, R. D. Dupuis, and P. D. Dapkus, "Al$_{0.5}$Ga$_{0.5}$As-GaAs heterojunction phototransistors grown by metalorganic chemical vapor deposition," *Appl. Phys. Lett.*, vol. 39, no. 9, pp. 562–564, May 1979.

[19] D. L. Miller, personal communication.

[20] a) T. Matsushita, N. Oh-uchi, H. Hayashi, and H. Yamoto, "A silicon heterojunction transistor," *Appl. Phys. Lett.*, vol. 35, no. 7, pp. 549–550, Oct. 1979;
b) N. Oh-uchi, H. Hayashi, H. Yamoto, and T. Matsushita, "A new silicon heterojunction transistor using the doped SIPOS," *IEDM Dig.*, pp. 522–524, Dec. 1979;
c) T. Matsushita, H. Hayashi, N. Oh-uchi, and H. Yamamoto, "A SIPOS-Si heterojunction transistor," *Japan. J. Appl. Phys.*, vol. 20, suppl. 20-1, pp. 75–81, Jan. 1981 (Proc. 12th Conf. Solid-State Devices, Tokyo, Aug. 1980).

[21] T. Katoda and M. Kishi, "Heteroepitaxial growth of gallium phosphide on silicon," *J. Electron Mat.*, vol. 9, no. 4, pp. 783–796, Apr. 1980.

[22] S. L. Wright and H. Kroemer, to be published.

[23] J. Katz, N. Bar-Chaim, P. C. Chen, S. Margalit, I. Ury, D. Wilt, M. Yust, and A. Yariv, "A monolithic integration of GaAs/GaAlAs bipolar transistor and heterostructure laser," *Appl. Phys. Lett.*, vol. 37, no. 2, pp. 211–213, July 1980.

[24] H. Beneking, N. Grote, and M. N. Svilans, "Monolithic GaAlAs/GaAs infrared-to-visible wavelength converter with optical power amplification," *IEEE Trans. Electron Devices*, vol. ED-28, pp. 404–407, Apr. 1981.

[25] H. Kroemer, "A proposed class of heterojunction injection lasers," *Proc. IEEE*, vol. 51, pp. 1782–1783, Dec. 1963.

[26] See, e.g., J. J. Coleman, P. D. Dapkus, N. Holonyak, Jr., and W. D. Laidig, "Device-quality epitaxial AlAs by metalorganic-chemical vapor deposition," *Appl. Phys. Lett.*, vol. 38, no. 11, pp. 894–896, June 1981. This paper quotes only structures containing about 80 layers; much larger numbers have been achieved in unpublished work (personal communication).

[27] For two recent reviews see:
a) L. L. Chang and L. Esaki, "Semiconductor superlattices by MBE and their characterization," *Prog. Cryst. Growth Charact.*, vol. 2, no. 1, pp. 3–12, 1979;
b) A. C. Gossard, "Molecular beam epitaxy of superlattices in thin films," in *Thin Films: Preparation and Properties*, K. N. Tu and R. Rosenberg, Eds. New York: Academic, to be published.

[28] J. S. Slotboom and H. C. de Graaf, "Measurement of bandgap narrowing in Si bipolar transistors," *Solid-State Electron.*, vol. 19, no. 10, pp. 857–862, Oct. 1976.

[29] See, e.g.,
a) J. A. Archer, "Design and performance of small-signal microwave transistors," *Solid-State Electron.*, vol. 15, no. 3, pp. 249–258, Mar. 1972.
b) J. M. Gladstone, P. T. Chen, P. Wang, and S. Kakihana, "Computer aided design and fabrication of an X-band oscillator transistor," Int. Electron Devices Meeting (IEDM) 1973, *IEDM Dig.*, pp. 384–386, Dec. 1973.
c) T. W. Sigmon, "Characteristics of high performance microwave transistors fabricated by ion implantation," Int. Electron Devices Meeting (IEDM) 1973, *IEDM Dig.*, pp. 387–389, Dec. 1973.

[30] C.E.C. Wood, G. Metze, J. Berry, and L. F. Eastman, "Complex free-carrier profile synthesis by atomic-plane doping of MBE GaAs," *J. Appl. Phys.*, vol. 51, no. 1, pp. 383–387, Jan. 1980.

[31] R. Dingle, "Confined carrier quantum states in ultrathin semiconductor heterostructures," *Festkörperprobleme/Advances in Solid State Physics*, vol. 15, pp. 21–48, 1975.

[32] For a review, see H. Kroemer, "Hot electron relaxation effects in devices," *Solid-State Electron.*, vol. 21, no. 1, pp. 61–67, Jan. 1978.

[33] M. S. Shur and L. F. Eastman, "Ballistic and near ballistic transport in GaAs," *IEEE Electron Devices Lett.*, vol. EDL-1, pp. 147–148, Aug. 1980.

[34] H. Kroemer, "Heterojunction device concepts," U.S. Air Force Tech. Rep. AFAL-TR-65-243, Oct. 1965, unpublished. A published description is found in Milnes and Feucht [3], pp. 28–29.

[35] R. E. Yeats, personal communications.

[36] D. K. Jadus and D. L. Feucht, "The realization of a GaAs-Ge wide band gap emitter transistor," *IEEE Trans. Electron Devices*, vol. 16, pp. 102–107, Jan. 1969.

[37] H. Kroemer, Dev. Res. Conf. 1978, Santa Barbara; see *IEEE Trans. Electron Devices*, vol. ED-25, p. 1339, Nov. 1978.

[38] ——, *Bull. Amer. Phys. Soc.*, vol. 24, p. 230, Mar. 1979.

[39] K. G. Ashar, "The method of estimating delay in switching circuits and the figure of merit of a switching transistor," *IEEE Trans. Electron Devices*, vol. ED-11, pp. 497–506, Nov. 1964.

[40] W. P. Dumke, personal communication, unpublished.

[41] The concept at issue here *has* been widely discussed in the DH laser literature. See, e.g., W. Susaki, H. Namizaki, H. Kan, and A. Ito, "A new geometry double-heterostructure injection laser for room-temperature continuous operation: Junction-stripe-geometry DH lasers," *J. Appl. Phys.*, vol. 44, no. 6, pp. 2893–2894, June 1973.

[42] H. Kroemer, "Heterostructures for everything—device principle of the 1980's?," *Japan. J. Appl. Phys.*, vol. 20, suppl. 20-1, pp. 9–13, Jan. 1981 (Proc. 12th Conf. Solid-State Devices, Tokyo, Aug. 1980).

[43] Inverted transistors (with a Schottky collector) with a wide-gap emitter, but without the idea of inactivating the "uncovered" part of the emitter area, have already been reported: H. Beneking, N. Grote, W. Roth, L. M. Su, and M. N. Svilans, "Realization of a bipolar GaAs/GaAlAs Schottky-collector transistor," *Gallium Arsenide and Related Compounds 1980* (Vienna), Inst. Phys. Conf. Ser., vol. 56, pp. 385–392, 1981.

[44] W. G. Oldham and A. G. Milnes, "n-n semiconductor heterojunctions," *Solid-State Electron.*, vol. 6, no. 2, pp. 121–132, Mar./Apr. 1963.

[45] D. T. Cheung, S. Y. Chiang, and G. L. Pearson, "A simplified model for graded-gap heterojunctions," *Solid-State Electron.*, vol. 18, no. 3, pp. 263–266, Mar. 1975.

[46] Some of the ideas on this subject were independently developed by Dr. Daniel G. Chen, to whom I owe several detailed discussions on this subject.

[47] For a discussion of those problems, see H. Kroemer, K. J. Polasko, and S. C. Wright, "On the (110) orientation as the preferred orientation for the molecular beam epitaxial growth of GaAs on Ge, GaP on Si, and similar zincblende-on-diamond systems," *Appl. Phys. Lett.*, vol. 36, no. 9, pp. 763–765, May 1980.— Unfortunately, even the switch to the (110) orientation has not solved the problems satisfactorily.

[48] See, e.g., D. Boccon-Gibod, J-P. André, P. Baudet, and J-P. Hallais, "The use of GaAs-(Ga, Al)As heterostructures for FET devices," *IEEE Trans. Electron Devices*, vol. ED-27, pp. 1141–1147, June 1980.

[49] See, e.g., S. Judaprawira, W. I. Wang, P. C. Chao, C.E.C. Wood, D. W. Woodard, and L. F. Eastman, "Modulation-doped MBE GaAs/n-Al$_x$Ga$_{1-x}$As MESFETs," *IEEE Electron Devices Lett.*, vol. EDL-2, pp. 14–15, Jan. 1981.

[50] The importance of this advantage of bipolar devices was first pointed out to me by Dr. R. C. Eden.

[51] M. Lepselter, "Integrated circuits—the new steel," Int. Electron Dev. Meeting (IEDM) 1974, *IEDM Dig.*, Dec. 1974.

``It is the beginning of the end of the light bulb.''

Herbert Kroemer

20

IEEE ELECTRON DEVICE LETTERS, VOL. EDL-4, NO. 1, JANUARY 1983

Staggered-Lineup Heterojunctions as Sources of Tunable Below-Gap Radiation: Operating Principle and Semiconductor Selection

H. KROEMER, FELLOW, IEEE, AND G. GRIFFITHS, MEMBER, IEEE

Abstract—Heterojunctions in which both the conduction and the valence band edges of one semiconductor are shifted upward relative to those of the other, can exhibit adjacent dual quantum wells for electrons and holes on the two sides of the interface. Tunneling-assisted radiative recombination between the wells should be an efficient, bias-tunable source of radiation at below-gap quantum energies. Several semiconductor combinations that exhibit the proper lineup are available.

THE band lineups at abrupt semiconductor heterojunctions vary over a wide range [1]. We are interested here in staggered lineups, in which both the conduction and the valence band edges of one semiconductor are shifted upward relative to those of the other, but with a residual gap E_R at the interface (Fig. 1(a)). This case has drawn surprisingly little attention in the past. One of the exceptions is the recent work by Osbourn, Gourley, and Biefeld [2, 3] on dislocation-free superlattices made from elastically-strained lattice-mismatched semiconductors. The system chosen by these authors for investigation, GaP/Ga(P, As), is one for which they claim (and present experimental evidence) that the semiconductors exhibit a staggered lineup, a circumstance that played a significant role in their investigation.

This paper, although applicable to superlattices, is directed more towards single lattice-matched staggered-lineup heterojunctions. The purpose of the paper is twofold: (a) To point out that externally-biased staggered-lineup heterojunctions offer the ability to emit bias-tunable interface recombination radiation with a quantum energy significantly below the energy gap of both participating semiconductors. (b) To show that there are several III/V compound combinations for which a staggered lineup can be predicted with a high degree of confidence, even if the constraint of lattice matching is imposed.

Some of the ideas under point (a) are related to ideas developed by Döhler, Ploog et al. [4, 5] in the context of doping ("nipi") superlattices, not based on heterostructures, but also exhibiting bias-tunable below-gap recombination radiation.

We consider a staggered-lineup heterojunction in the presence of band bending (Fig. 1(b)), exhibiting potential wells for both electrons and holes on the two sides of the interface. We assume (temporarily) that the semiconductor on the downside of the band edge step is n-type doped, that on

Manuscript received Oct. 21, 1982.

The authors are with the Department of Electrical and Computer Engineering, University of California, Santa Barbara, CA 93106.

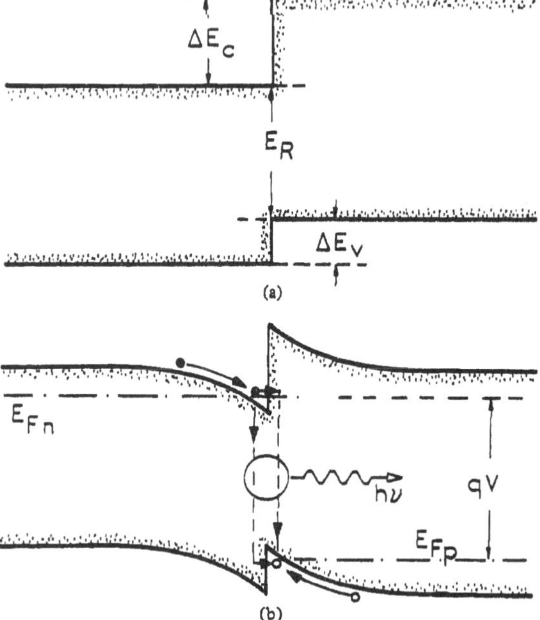

Fig. 1. (a) Flatband energy band diagrams of staggered-lineup heterojunctions, showing the conduction band offsets ΔE_c and ΔE_v, and the residual energy gap E_R. (b) n(down)/p(up) junction under forward bias, leading to formation of adjacent dual quantum wells for electrons and holes, with a possibility of tunneling-assisted radiative recombination.

the upside, p-type. We refer to this configuration as n(down)/p(up). A structure with this doping will develop dual wells if sufficient forward bias is applied. Associated with the well formation is the accumulation of free electrons and holes in the two wells; their space charge is in fact the charge necessary to support the band bending forming the wells.

If the interface is sufficiently abrupt, the wave functions of the electrons and holes will overlap strongly, due to tunneling across the interface. This makes possible the efficient radiative recombination of the high concentration of accumulated electron-hole pairs. Even if the downside semiconductor is an indirect-gap semiconductor with conduction band minima at a set of nonzero wave vectors $\{\vec{k}\}$, radiative recombination becomes possible if one pair of $\vec{k}$-vectors is perpendicular to the interface.

Evidently, staggered lineup heterojunctions offer promise as radiation sources (coherent or incoherent) whose quantum energy is substantially smaller than the energy gaps of both participating semiconductors. Most important, the exact value of the quantum energy should be bias-tunable over a finite range, for the following reasons. In general, the two potential wells will be sufficiently narrow that the electron (and hole) motion perpendicular to the plane of the well is quantized. To the first order, the wells may be approximated as triangular quantum wells with infinitely high walls on one side and linear potential ramps on the other. The energy level calculations for such wells are found in many quantum mechanics texts [6]. The expression for the ground state energy of such a well can be written

$$E_O = 2.388 \left[\hbar^2 q^4 \sigma^2 / 2m^* \epsilon^2 \right]^{1/3}. \tag{1}$$

Here σ is the net electric charge density inside the well, expressed as the number of elementary charges per unit area, and ϵ is the permittivity of the semiconductor. The remaining symbols have their familiar meaning. The effective energy gap of such a structure would be

$$E_g \text{ (eff)} = E_R + E_{On} + E_{Op}. \tag{2}$$

Through its dependence on σ, this gap is bias-dependent, hence the radiation would be tunable over a substantial range. To obtain an estimate of the tunability range assume, as a representative order-of-magnitude example, $\sigma = 10^{12}$ cm^{-2}, $\epsilon = 10\epsilon_O$ and $m^* = 0.1\,m_e$. This yields $E_O = 117$ meV. In reality, the ground state energy would be somewhat less because the vertical wall is not infinitely high and the sloping side is sublinear, but our example should give the overall magnitude of the tuneability to be expected. Evidently, a tuning range of at least 0.2 eV appears readily possible. This compares with the tunability range achieved in doping superlattices [5].

We discussed the n(down)/p(up) structure, because for it the achievement of a high electron-hole pair concentration is especially easy, by electrical biasing, which accumulates (majority) carriers. The principle is readily generalized to other doping combinations such as p(down)/n(up), or to isotype junctions, with at least one of the two carrier species being a minority carrier. For example, in the p/p structure (Fig. 2), minority-carrier electrons placed into the conduction band of the downside semiconductor would collect in the electron well where they could recombine with majority-carrier holes on the other side. The electrons may be generated either by electrical injection (Fig. 2) or by illumination. By varying both the excitation intensity and the p/p bias, luminescence intensity and wavelength can be varied independently.

There is no shortage of semiconductor pairs that exhibit a staggered lineup, even if one imposes the additional constraint of lattice matching. Ignoring lattice matching for the moment, it is clear that any two semiconductors with the same energy gap could be non-staggered only by accident. If the two semiconductors have different anions, staggering is assured, because of the well-established anion rule [7] of band lineups, which states that the valence band energies depend principally on the

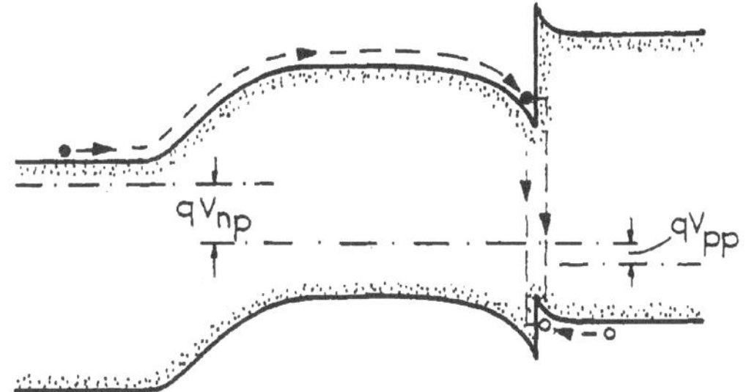

Fig. 2. A postulated n/p/p structure in which an n/p junction injects electrons towards a staggered-lineup p/p junction at which the minority-carrier electrons recombine radiatively with majority carrier levels. The n/p junction may itself be a heterojunction.

electronegativity of the anion, in such a way that, for III/V compounds,

$$E_v(\text{P}) < E_v(\text{As}) < E_v(\text{Sb}). \tag{3}$$

Hence, if in a phosphide/arsenide pair the energy gap of the arsenide is equal to that of the phosphide, or somewhat larger, staggering is assured. The extension to arsenide/antimonide and phosphide/antimonide pairs is obvious. Mixed-anion staggered pairs are easily found, even if one imposes the additional constraint that each pair is lattice matched. Table I lists three pairs for which we have estimated the actual band lineups. The valence band offsets were obtained from Harrison's theory [8], the conduction band offsets by adding the experimental 300 K energy gaps [1]. Because Harrison's lineup theory is only an approximation, the quantitative validity of these predictions is uncertain, but whatever errors are present are not believed to affect the prediction that these three systems have staggered lineups.

We referred earlier to the relation of our concept to the earlier work of Osbourn et al. [2, 3], and of Döhler et al. [4, 5] and a comparison is in order. (a) Addressing themselves not to staggered-offset superlattices but to heterostructure superlattices in general, Döhler et al. have argued that technologically it is a great advantage of the doping superlattice that it employs only one semiconductor. If technology were the only consideration, this would undoubtedly be true. But such a structure deliberately deprives itself of the most powerful force to confine electrons and holes in alternating layers, namely band offsets. Confinement by "ordinary" electrostatic fields, achieved by heavy doping, is much less effective. As a result, the electrons and holes are always separated by an appreciable distance, which greatly weakens their tunneling interaction and hence their radiative recombination probability. While this might not matter for other applications of doping superlattices, it is a disadvantage if efficient below-gap light emission is desired. (b) Osbourn has pointed out that lattice matching, a requirement for low-defect single heterostructures, is far less critical in superlattice structures, hence allowing a wider range of semiconductors to be used. However, at least for the purposes of tunable sub-gap light emitters, the residual advantage of the wider choice of materials in a superlattice is likely to be outweighed by alternate advantages of

IEEE ELECTRON DEVICE LETTERS, VOL. EDL-4, NO. 1, JANUARY 1983

TABLE I

Semiconductors		Band Offsets		Residual Gap
downside	upside	E_v [eV]	E_c [eV]	E_R [eV]
InP	$Al_{.48}In_{.52}As$	0.26	0.38	1.1
InP	$AlAs_{.56}Sb_{.44}$	0.44	0.65 (?)	0.91
$Ga_{.50}In_{.50}P$	AlAs	0.26	0.52	1.64

Estimated band lineup energies for three lattice-matched heterojunction pairs for which a staggered lineup is predicted. The valence band offsets were obtained from Table 10-1 on p. 253 of ref. [8], interpolating linearly for the ternary alloys. Similar lineups are predicted by the Frensley-Kroemer theory [9]. The residual gaps and the conduction band offsets were obtained by adding the experimental 300 K energy gaps to the valence band data. The energy gap of $AlAs_{.56}Sb_{.44}$ is uncertain; the quoted value is our own interpolation estimate between AlAs and AlSb. This alloy may be unstable (see [1]) and hence hard to prepare.

single-interface structures. One such advantage is the direct electrical access to both sides of the heterointerface, greatly facilitating the application of an electrical bias. In a superlattice the forward bias current must be applied from the edges of the layers and would have to flow through the high series resistance along the necessarily very thin layer, causing current crowding effects qualitatively similar to those in thin-base bipolar transistors at high current densities, but quantitatively much more severe. Another expected advantage of the single-interface design is the following. Any radiative recombination process must compete with non-radiative recombination via interface defects. For a given level of excitation, it appears preferable to force this excitation at a high current density through a single interface, swamping the non-radiative processes, rather than to dilute it over many interfaces.

ACKNOWLEDGMENT

It is a pleasure to acknowledge several useful discussion with Dr. Khalid Mohammed, Dr. Steven L. Wright, and Mr. G. Sullivan. This work was supported by the Office of Naval Research.

REFERENCES

[1] For a review see Chapters 4 and 5 of H. C. Casey and M. B. Panish, *Heterostructure Lasers*, New York: Academic Press, 1972.

[2] G. C. Osbourn, "Strained-layer superlattices from lattice mismatched materials," *J. Appl. Phys.*, vol. 53, no. 3, pp. 1586–1589, March 1982; "Electronic structure of $GaAs_xP_{1-x}/GaP$ strained-layer superlattices with $x < 0.5$," *J. Vac. Sci. Technol.*, vol. 21, no. 2, pp. 469–472, July/Aug. 1982.

[3] P. L. Gourley and R. M. Biefeld, "Growth and photoluminescence characterization of a $GaAs_xP_{1-x}/GaP$ strained-layer superlattice," *J. Vac. Sci. Technol.*, vol. 21, no. 2, pp. 473–475, July/Aug. 1982; G. C. Osbourn, R. M. Biefeld, and P. L. Gourley, "A $GaAs_xP_{1-x}/GaP$ strained-layer superlattice," *Appl. Phys. Lett.*, vol. 41, no. 2, pp. 172–174, July 1982.

[4] Döhler, "Electron states in crystals with 'nipi' superstructure," *Phys. Stat. Solidi*, vol. B52, no. 1, pp. 79–92, July 1972; "Electrical and optical properties of crystals with 'nipi' superstructure," *Phys. Stat. Solidi*, vol. B52, no. 2, pp. 533–545, Aug. 1972.

[5] For a complete up-to-date review see K. Ploog and H. Künzel, "Growth and properties of new artificial doping superlattices in GaAs," *Microelectron J.*, vol. 13, no. 3, pp. 5–22, July/Aug. 1982.

[6] See, for exampe, S. Flügge, *Practical Quantum Mechanics*, pp. 101–105, Berlin: Springer-Verlag, 1974.

[7] J. O. McCaldin, T. C. McGill, and C. A. Mead, "Correlation for III–V and II–VI semiconductors of the Au Schottky barrier energy with anion electronegativity," *Phys. Rev. Lett.*, vol. 36, no. 1, pp. 56–58, Jan. 1976. These authors expressed the correlation between valence band lineup and anion electronegativity for Schottky barriers; the approximate applicability of their result to heterojunctions is discussed by Frensley and Kroemer, ref. [9] below.

[8] W. A. Harrison, *Electronic Structure and the Properties of Solids: The Physics of the Chemical Bond*, San Francisco: Freeman, 1980. See especially Sec. 10F. See also W. A. Harrison, "Elementary theory of heterojunctions," *J. Vac. Sci. Technol.*, vol. 14, no. 4, pp. 1016–1021, July/Aug. 1977.

[9] W. R. Frensley and H. Kroemer, "Prediction of semiconductor heterojunction discontinuities from bulk band structures," *J. Vac. Sci. Technol.*, vol. 13, pp. 810–815, July/Aug. 1976; "Theory of the energy-band lineup at an abrupt semiconductor heterojunction," *Phys. Rev. B*, vol. 16, no. 6, pp. 2642–2652, Sept. 1977.

IEEE ELECTRON DEVICE LETTERS, VOL. EDL-4, NO. 10, OCTOBER 1983

Rebuttal to "Response to 'Critique of Two Recent Theories of Heterojunction Lineups'"

H. KROEMER

Abstract—Recent arguments by Nussbaum against an earlier critique of the Adams–Nussbaum (AN) heterojunction lineup theory are refuted.

IN 1979, Adams and Nussbaum [1] published a theory of heterojunction band lineups, which I attacked in a critique in the January 1983 issue of these letters [2]. In the August 1983 issue, Nussbaum published a response [3] in which he attempts to refute my critique. The present work is a rebuttal to Nussbaum's response. I maintain that Nussbaum assumes what he claims to derive, and I shall attempt to show the fallacy contained in his "derivation."

Nussbaum points out (correctly) that it is not only the dielectric displacement D that must be continuous across the interface, but the electrostatic potential V as well. However, he claims, quite incorrectly, that others have neglected this requirement of a continuous electrostatic potential, and that the Adams-Nussbaum (AN) lineup rules follow directly from an integration of the Poisson-Boltzmann (PB) equation, together with the continuity requirement $\Delta V = 0$. I assert that it is the AN theory that violates this (indeed essential) requirement, and that the AN lineup rules do *not* follow by such an integration: In their 1979 paper [1], AN never carry the second of the required *two* integrations across the interface itself, but simply make an implicit *ad-hoc* assumption about what the results of that second integration would be *if* it had been performed.

Inasmuch as this mistake in AN is not obvious, an elaboration is called for. Electrostatic potentials are potentials relative to some agreed-upon zero. In a homostructure it is always possible to choose this zero such that the intrinsic Fermi level throughout an unbiased structure occurs at an energy $E_i = -eV$, as shown in Fig. 1 of the 1979 Adams-Nussbaum paper [1]. In a heterostructure the same choice may be made on *one* of the two sides of the interface, but the potential on the *other* side must then be obtained by integrating the PB equation twice *across* the interface! Adams and Nussbaum omit the second integration and instead simply take it for granted that $E_i = -eV$ on both sides. There is no basis for such an assumption. Unless this assumption happens to be satisfied

accidentally, the quantity u in the Adams-Nussbaum paper [1] does not vary like eV/kT throughout *both* semiconductors, but contains an additional nonzero step at the interface. Requesting that u be continuous in effect violates the requirement that V be continuous. The AN lineup rules simply reflect the assumption that the step in u be zero. Determining the height of this step is the *real* task of any theory of heterojunction lineups; it requires external physical considerations, like those provided by the theories quoted in [2], rather than simply setting the step to zero.

Nussbaum's claim [3] that in the case of CdS/InP [4] and InAs/GaSb [5] "a continuous intrinsic level explain(s) the results equally well" flies in the face of the data contained in [4] and [5]. The InAs/GaSb discrepancy is especially blatant: The experimental separation between the two intrinsic Fermi levels is about 0.7 eV (give or take 0.1 eV), which is twice as large as the entire energy gap of InAs, and almost as large as the gap of GaSb.

The rest of the discussion in Nussbaum's note, pertaining to a criticism of Anderson's Electron Affinity Rule (EAR), is orthogonal to the issue of the validity of Nussbaum's own theory. One need not be an adherent of the EAR to reject the AN theory: Harrison's theory [6] fits the experimental data vastly better than the AN theory, a fact Nussbaum continues to ignore even though it has been pointed out to him [7]. Nussbaum's statement "Kroemer regards the voltage-current measurements as experimental support for the Anderson model" is a misrepresentation of my statements on this matter.

Manuscript received July 20, 1983; revised August 9, 1983. This work was supported by the U.S. Army Research Office and by the Office of Naval Research.

The author is with the Department of Electrical and Computer Engineering, University of California, Santa Barbara, CA 93106.

REFERENCES

[1] M. J. Adams and A. Nussbaum, "A proposal for a new approach to heterojunction theory," *Solid-State Electron.*, vol. 22, pp. 783–791, 1979.

[2] H. Kroemer, "Critique of two recent theories of heterojunction lineups," *IEEE Electron Device Lett.*, vol. EDL-4, pp. 25–26, 1983.

[3] A. Nussbaum, "Response to 'critique of two recent theories of heterojunction lineups,'" *IEEE Electron Device Lett.*, vol. EDL-4, pp. 267–268, 1983.

[4] J. L. Shay, S. Wagner, and J. C. Phillips, "Heterojunction band discontinuities," *Appl. Phys. Lett.*, vol. 28, pp. 31–33, 1976.

[5] H. Sakaki, L. L. Chang, R. Ludeke, C.-A. Chang, G. A. Sai-Halasz, and L. Esaki, "In$_{1-x}$Ga$_x$As-GaSb$_{1-y}$As$_y$ heterojunctions by molecular beam epitaxy," *Appl. Phys. Lett.*, vol. 31, pp. 211–213, 1977.

[6] W. A. Harrison, "Elementary theory of heterojunctions," *J. Vac. Sci. Technol.*, vol. 14, pp. 1016–1021, 1977.

[7] See ref. [12] of Nussbaum [3].

Reprinted with permission from

H. Kroemer, ``Heterostructure Bipolar Transistors: What Should We build?''
J. Vac. Sci. Technol. B, Vol. 1(2), pp. 126-130, 1983.

Heterostructure bipolar transistors: What should we build?

Herbert Kroemer

Department of Electrical and Computer Engineering, University of California, Santa Barbara, California 93106

(Received 21 October 1982; accepted 11 January 1983)

The paper discusses likely future developments in heterostructure bipolar technology, especially by MBE. This written version concentrates on two new conceptual developments extending earlier concepts. One of these pertains to the problem of emitter/base junction grading. A grading scheme is proposed that extends the grading through the base region and creates a graded-gap base. The other proposes an extension of permeable base transistor technology to bipolar transistors in what is called a gridded-base bipolar transistor. Both promise a further increase in device speed, largely by addressing themselves to the persistent problem of base resistance reduction. Several other topics, already contained in the author's January 1982 review paper and presented orally at the Workshop, have been omitted from this printed version.

PACS numbers: 85.30.Pq, 73.40.Lq

I. INTRODUCTION

This is a sequel to a recent paper[1]—hereafter called HK82—which discussed the mid-1981 state of the development of the heterostructure bipolar transistor (= HBT). That paper stressed the post-1978 development of ideas rather than specific technological achievements implementing those ideas. It preceded the first published reports of HBT's prepared by MBE, but was written in anticipation of extensive MBE developments in this area. The first published reports on MBE-prepared HBT's have by now appeared.[2-4] However, the intent of the present paper is not to review those early MBE achievements: The situation is developing so rapidly that such a review would be outdated by the time it appears in print. The intent is rather to contribute to further advances in the conceptual development of the HBT. Much of this is technology independent, yet is a highly appropriate subject matter for an MBE Workshop: It is hoped that the paper will help the MBE technologist in deciding what to attempt building. As shall be seen, some of this will be quite specific, and perhaps unexpected.

Because of space limitations, the paper will discuss only two selected areas, in which significant conceptual progress has been made since HK82: (a) the question of abrupt versus graded emitter/base heterojunctions, culminating in a proposal for grading not only the emitter/base junction, but to resurrect the old idea[5,6] of grading the entire base region as well (Sec. II), (b) the concept of a gridded-base bipolar transistor, an application of permeable base transistor (PBT) technology[7] to HBT's (Sec. III).

II. THE EMITTER/BASE GRADING PROBLEM
A. The problem

Although the first proposal for a wide-gap emitter, by Shockley,[8] assumed an abrupt heterojunction, until recently most subsequent work, following Kroemer,[9] assumed a graded junction in which both band edges vary monotonically across the emitter/base interface. If all parameters other than the emitter energy gap were kept constant, the effect of the wide-gap emitter would then simply be to increase the

injection ratio J_n/J_p by the inverse Boltzmann factor involving the energy gap difference

$$(J_n/J_p)_{\text{hetero}} = (J_n/J_p)_{\text{homo}} \times \exp(\Delta\epsilon_g/kT). \qquad (1)$$

Here, J_p is the hole current density injected from the base into the electrically neutral part of the emitter body, excluding that portion of the total hole outflow from the base that recombines with electrons in the emitter/base space charge region. Similarly, J_n is the electron current density injected from the emitter into the base. For large values of $\Delta\epsilon_g/kT$, which are readily obtainable, the increase in injection ratio can be many orders of magnitude. As was discussed extensively in HK82, the central idea of HBT design is to trade off this increase in injection ratio for various other design changes that lead to several major improvements in device performance. The first and most important design change made possible is a large increase in base doping, which lowers the base resistance and thereby drastically improves both the high current and the high frequency properties of the device. This base resistance problem continues to play an important role in the present paper.

Heterojunctions grown by MBE tend to be quite abrupt, unless specific measures are taken to grade the transition. As was discussed in HK82, an abrupt emitter/base junction introduces a potential barrier $\Delta\epsilon_B = \Delta\epsilon_c - \Delta\epsilon_N$ into the path of the electron flow (Fig. 1), the height of which tends to be very close to the conduction band offset $\Delta\epsilon_c$ that is characteristic of the semiconductor pair employed. As a result, the improvement in injection ratio is much less than for a graded junction. To the first order, for a sufficiently thin and sufficiently heavily doped base region, the energy gap difference $\Delta\epsilon_g$ in Eq. (1) is simply replaced by $\Delta\epsilon_g - \Delta\epsilon_B = \Delta\epsilon_g - \Delta\epsilon_c + \Delta\epsilon_N \approx \Delta\epsilon_g - \Delta\epsilon_c$, that is, by the valence band offset $\Delta\epsilon_v$, which tends to be much smaller than $\Delta\epsilon_g$, leading to a large reduction in the magnitude of the exponential factor in Eq. (1), and hence in the injection ratio. The simple first-order replacement $\Delta\epsilon_g \rightarrow \Delta\epsilon_v$ is *quantitatively* valid only if the base is sufficiently thin that the speed of diffusive electron flow across the base is supply limited, as is the flow across the spike barrier. This is a reasonable approximation for transistors of greatest interest, with base thicknesses of the order 10^{-5} cm.

The reduction of injection ratio by the conduction band spike barrier is particularly severe for $Al_xGa_{1-x}As/GaAs$ heterojunctions, because of their small valence band discontinuity, only about $\Delta\epsilon_v \cong 0.15\Delta\epsilon_g$ (this equals $\cong 0.176\Delta\epsilon_c$, or about 1.87 meV per percent of Al in the Al:Ga ratio). Now, in order to avoid injection of electrons into the low mobility upper valleys of the GaAs band structure, it is necessary to keep the height of this spike below 0.3 eV. In an *abrupt* $Al_xGa_{1-x}As/GaAs$ emitter with $\Delta\epsilon_c = 0.3$ eV, corresponding to $x \cong 0.28$, this implies a valence band offset $\Delta\epsilon_v$ of only $\cong 0.053$ eV ($\cong 2kT$), which is too small to permit the full range of device design adjustments (especially the high base doping) that are the essence of HBT design.

One possible way out of this dilemma has been to deliberately grade the emitter/base junction, which reduces the height of the electron barrier. In fact, the first successful MBE-grown HBT's, reported by Asbeck *et al.*,[2] incorporate just such grading. We will return to this point shortly.

On the other hand, it has been argued[1,10] that the injection of electrons into the base from a "ballistic launching ramp" would actually be desirable, so long as this does not lead to transfer into higher valleys. The high speed of such ballistic electrons would lead to higher speed transport through the base. Such ballistic effects would make up for some (but not all) of the disadvantages of the greatly reduced injection ratio. However, there is actually no need to sacrifice any of these desirable properties to obtain the others: They can be obtained simultaneously, and it is evidently desirable to do so.

To this end it is necessary either to choose a different semiconductor pair with a higher $\Delta\epsilon_v{:}\Delta\epsilon_c$ ratio, or to modify the energy gap grading procedure.

B. (Ga, In)P/GaAs as an alternate

The undesirably small valence band offset in the (Al,Ga)As/GaAs system is not an accident: It is a direct consequence of the two semiconductors having the same anion, namely arsenic. Replacing the emitter by a semiconductor that contains a different anion with a higher electronegativity, namely phosphorus, would automatically increase the valence band offset.[11] The ternary alloy $Ga_{0.5}In_{0.5}P$ is lattice

matched to GaAs. Using the Harrison theory[12] of heterojunction band lineups, we estimate a valence band offset of 0.29 eV ($\cong 11kT$), which should yield an injection ratio equal to that of a graded $Al_xGa_{1-x}As$ emitter with $x \cong 0.27$. But in contrast to the latter, a (Ga, In)P abrupt emitter would still retain a conduction band offset estimated at $\cong 0.16$ eV, available as a significant ballistic launching ramp.

The first few published reports[13-16] of MBE growth of $Ga_{0.5}In_{0.5}P$ on GaAs have appeared. Although the papers reveal the usual startup problems one expects for a new materials system, these difficulties appear, if anything, less severe than those exhibited by (Al, Ga)As at the same stage of its development, and they contain nothing suggesting any really serious problems. I propose that we take (Ga, In)P *very* seriously for HBT's.

C. Graded emitter/base junctions

If the conduction band spike barrier is lowered, through compositional grading, by an amount $\Delta\epsilon_l$, the effect on the injection ratio is the same as if $\Delta\epsilon_l$ were added to the valence band offset, leading to the energy $\Delta\epsilon_v + \Delta\epsilon_l$ to appear instead of either $\Delta\epsilon_g$ or $\Delta\epsilon_v$ alone in the exponential in Eq. (1). The HBT's reported by the Rockwell group[2] have employed this principle with results that appear promising. However, such barrier lowering also reduces the benefits of ballistic carrier injection. One may obtain both a large injection ratio *and* strong ballistic effects by combining the grading with an increase of the Al fraction, to somewhere near 40%. In the absence of grading, this would yield conduction and valence band offsets of about 0.42 and 0.07 eV. Reducing the conduction band barrier by grading, to between 0.20 and 0.25 eV, would increase the effective energy in the injection ratio enhancement factor to between 0.24 and 0.29 eV, a more than adequate value.

D. Graded-base long-gradient HBT

Given the desirability of grading the energy gap, what should be the ϵ_g-vs-position profile? We argue here in favor of a design that grades not only the emitter/base junction proper, but that extends grading through the base, to the edge of the base/collector depletion layer, as shown in Fig. 2. The idea of base grading is to introduce a strong *quasielectric field* [6] into the base to aid the minority carrier transport. Such a design was first proposed by this writer in 1954[5] and 1957.[6] It was recently taken up again by Levine *et al.*,[17] and by Asbeck *et al.*[18] For a given base thickness, such a quasielectric field can greatly reduce the base transit time τ_b. Now, in a well-designed bipolar transistor, the base transit time is only a relatively small fraction of the total signal propagation delay. Hence, the improvement obtainable by incorporating a drift field into a base with *fixed thickness* is quite limited. But the high electron drift velocity can be traded off to retain a *fixed transit time* for a much thicker base region, which would have a much lower base resistance, which in turn increases the speed of the transistor.

If one keeps the total potential energy drop $\Delta\epsilon_b$ within the base below the energy at which electrons can transfer into higher low mobility valleys, the electric field in the base may be allowed to exceed the threshold field above which such

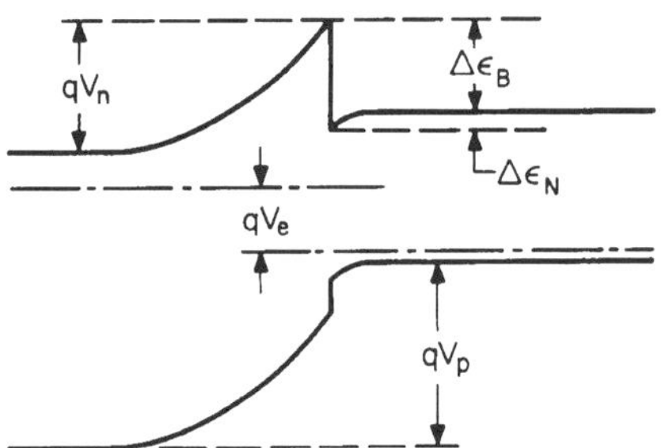

FIG. 1. Band structure of an abrupt wide-gap emitter, showing the electron spike barrier. No interface charge is assumed. (From Ref. 1.)

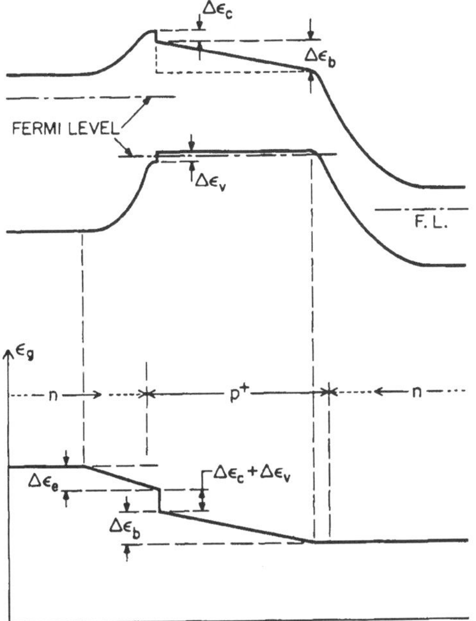

FIG. 2. Proposed heterostructure bipolar transistor structure with graded-gap base region. (a) Band diagram. (b) ϵ_g-vs-position profile.

transfer would take place in a long sample. It is then justified to use the below-threshold mobility μ_b to estimate the drift velocity. For given values of τ_b, $\Delta\epsilon_b$, and μ_b one easily finds a base width

$$w_b = (\mu_b \Delta\epsilon_b \tau_b / q)^{1/2}. \qquad (2)$$

It is not obvious which mobility value to use. Because of the high base doping, it would be unrealistic to assume low field mobilities as high as those in high purity GaAs. At the same time, because the electrons in such a structure tend to be hot electrons less subject to impurity scattering then lattice-temperature electrons, it would also be incorrect to assume the low field mobilities of heavily doped GaAs. Assuming the *adhoc* value $\mu_b = 2500$ cm^{2}V^{-1}s^{-1}, together with $\Delta\epsilon_b$ $= 0.25$ eV and $\tau_b = 1$ ps, one finds $w_b = 2.5 \times 10^{-5}$ cm, far above current design practices for high speed bipolar transistors. The associated quasielectric field is $E_b = \Delta\epsilon_b/$ $qw_b = 10^4$V/cm, far above the intervalley transfer field in a long GaAs specimen. The drift velocity would be $v_b = \mu_b E_b$ $\cong 2.5 \times 10^7$ cm/s.

Because of the high drift field inside the base, any ballistic launching ramp at the entrance into the base should be kept low. In effect the electrons are traveling under near-ballistic conditions in the high field anyway, and a high launching ramp would only introduce the danger of transfering electrons to the higher low mobility valleys. What *is* probably desirable is a shallow ($\cong 1kT$) ballistic "kicker," to bring the electrons up to the desired high base drift velocity instantaneously. If the energy gap gradient in the base is achieved by controlled temperature ramping of the MBE beam sources during growth, such a shallow kicker may be easily obtained

by simply shuttering the column-III sources for an appropriate time while the temperature ramping proceeds.

The remainder of the energy gap variation shown in Fig. 2 takes place inside the emitter/base space charge layer, possibly extending into the neutral emitter body itself. It is only this energy gap variation to the left of the kicker step that contributes to the desirable high injection ratio enhancement factor in Eq. (1), and it should therefore be kept large compared to kT, preferably no less that $9kT$ ($e^9 \sim 8000$). In Fig. 2 we have shown linear gap variations with position (with different slopes on the two sides of the ballistic kicker step). The width of the transition on the emitter side was chosen to coincide with the width of the emitter/base space charge layer; the optimal grading width is expected to be close to this value.

The structure discussed here differs from the ballistic HBT recently discussed by Ankri and Eastman.[10] In their structure the initial electron launching energy is just below the intervalley transfer energy, but the base region is of uniform gap. The initial injection velocity in the Ankri/Eastman device is much higher, but electrons of such a high energy can lose energy very rapidly. Following the earlier proposal in HK82, Ankri and Eastman argue that, because of the forward-directed nature of polar scattering, the electron retains a high forward velocity for a large distance. This would presumably make a fairly thick base possible. But it is not clear to what extent this remains valid in a base region as heavily doped as would be desirable in an HBT, in which other scattering mechanisms may be very important. In our present long-gradient design, a natural "sloping floor" limiting the energy loss is placed underneath the electron. It is believed that in this way a larger distance can be traversed in a given transit time, and at higher doping level, two factors that should combine to yield a most desirable lower base resistance.

III. THE GRIDDED-BASE BIPOLAR TRANSISTOR
A. Bipolars vs FET's

No discussion on the merits of HBT's can ignore their principal competitors, field effect transistors (FET's). Hence, HK82 contained an extensive discussion on this comparison. Two new developments have entered and changed the picture since then: The sudden emergence of the new High Electron Mobility Transistor[19] (HEMT), and the striking technological progress that has been made in implementing the permeable base transistor (PBT) idea.[7]

It is assumed that the reader is sufficiently familiar with the HEMT to be aware that it is basically a FET with higher electron mobilities and hence higher speed than conventional FET's, and that the mobilities and the device speeds increase drastically with decreasing temperature. Therefore, if highest raw speed at any cost is desired, and if "at any cost" includes a willingness to go to cryogenic operation, the HEMT is unquestionably preferable to any form of bipolar transistor, and probably even superior to Josephson devices. However, for operation under more common conditions, at or above room temperature, the comparisons between HBT's and FET's made in HK82 largely carry over to the

HEMT version of FET's, with only some quantitative shifts in favor of the HEMT in borderline cases.

B. The PBT: The best of both worlds?

From a bipolar perspective, the progress in PBT technology[7] is perhaps more important. The PBT is basically a vertical FET, in which the current flows vertically through a thin epitaxial layer, and in which the controlling gate electrodes have been embedded into the semiconductor in the form of narrow and very closely spaced metal figures. The vertical flow geometry permits a much closer source-to-drain spacing and hence a higher speed than in horizontal FET's, including HEMT's. In some ways the PBT resembles a bipolar device, but without the base resistance that is the nemesis of the latter. However, this speed improvement comes at a very high technological cost: The finger spacing must not only be very close, requiring x-ray or electron beam lithography but—worse—it must be very uniform within a device, and in IC's also from device to device. Even small fractional variations in this spacing cause large variations in turn-on voltage. Variations within a single device smear out the turn-on characteristics and reduce the transconductance of the device. Variations from device to device reduce their integratability. Quite possibly, there does not, at this time, exist a device that is more demanding on horizontal lithography.

C. Applying PBT technology to bipolars

If one contemplates the reaons for this difficulty, one realizes that it is not so much related to the PBT technology itself as to the fact that the PBT is a vertical FET rather than some other kind of device less sensitive to lithography tolerances. But this connection need not exist! Divorced from its connection to a specific kind of device, PBT technology may be viewed as being simply a technology to embed a conductive metal grid into a single-crystalline semiconductor body without compromising the device quality of the semiconductor. The remarkable fact that the PBT works as well as it does means nothing less than that such embedded-metal structures must be considered as very serious contenders for all kinds of future device structures. This includes specifically their use to create improved internal electrical access to the base region of a bipolar transistor, significantly improv-

ing what has always been the main bottleneck of bipolar transistor design, the base resistance. Figure 3 shows the envisaged configuration, which I would like to call a *gridded bipolar transistor*. It differs from the PBT principally by the insertion of a very heavily *p*-type doped base region in low resistance electrical contact with the metal grid. And of course the emitter region has been changed to a wide-gap emitter, making the structure a heterostructure bipolar device. It may (or may not) be desirable to surround the grid metal by a very thin heavily *p*-type doped wide-gap "sheath," as shown. This would suppress parasitic current flow across the forward-biased grid-to-emitter Schottky barrier, as well as loss of injected electrons from the base into the grid. It should be readily possible to create such a sheath—if in fact necessary—by outdiffusion of a suitable dopant (Be?) from the metal during subsequent semiconductor growth.

Except for possible unforseen difficulties with this *p*-type sheath—which might not be needed anyway—the structure should not be much more difficult to build than a conventional PBT, and it might have significantly better properties: (a) Being a true bipolar, it should have the higher transconductance of a bipolar compared to an FET. (b) The turn-on voltage should no longer be highly sensitive to the exact value of the grid finger spacing, but should depend principally on the well-defined energy gap of the base region semiconductor.

In contrast to halide–VPE, MBE has so far exhibited difficulties in growing high quality crystals on top of a metal. Progress in this direction has recently been made,[20] but it is not all clear whether, say, semi-insulating overgrowth between the emitter contact and the base grid fingers would even be a drawback.

ACKNOWLEDGMENTS

I have greatly benefitted from numerous discussions with more individuals than can be listed here, but none more than Dr. Peter Asbeck and Dr. David Miller at Rockwell. This work was supported by the U.S. Army Research Office and by the Office of Naval Research.

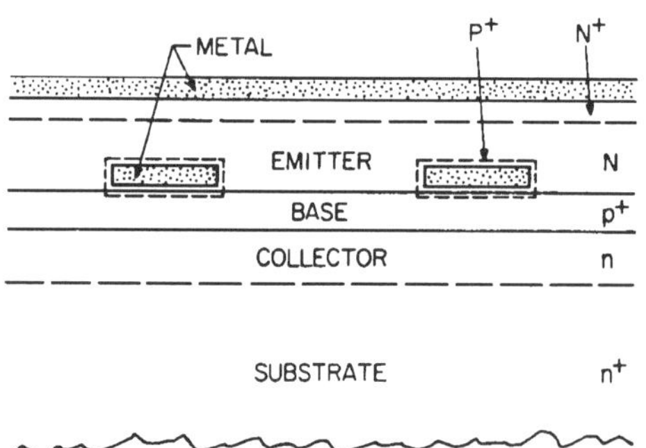

FIG. 3. Proposed gridded-base bipolar transistor.

[1] H. Kroemer, Proc. IEEE **70**, 13 (1982). For a shorter version, containing, however, some additional material, see also: H. Kroemer, Jpn. J. Appl. Phys. Suppl. **20-1**, 9 (1981).

[2] P. M. Asbeck, D. L. Miller, R. A. Milano, J. S. Harris, Jr., G. R. Kaelin, and R. Zucca, *IEEE Internat. Electron Devices Meeting*, (IEEE, New York, 1981), p. 629; J. S. Harris, Jr., D. L. Miller, and P. M. Asbeck, *14th Conf. on Solid State Devices, Tokyo 1982*, (The Japan Society of Applied Physics, Tokyo, 1982) Digest of Papers, p. 199 (1982). D. L. Miller, P. M. Asbeck, and W. Petersen, *2nd Internat. Symposium on Molecular Beam Epitaxy and Related Clean Surface Techniques, Tokyo 1982*, (The Japan Society of Applied Physics, Tokyo, *1982*) Collected Papers, p. 121 (1982).

[3] W. V. McLevige, H. T. Yuan, W. M. Duncan, W. R. Frensley, F. H. Doerbeck, H. Morkoc, and T. J. Drummond, IEEE Electron Device Lett. **EDL-3**, 43 (1982).

[4] D. Ankri, W. Schaff, C. E. C. Wood, and L. F. Eastman, *GaAs and Related Compounds* (to be published).

[5] H. Kroemer, Arch. Elect. Übertraguug **8**, 499 (1954).

[6] H. Kroemer, RCA Rev. **18**, 332 (1957).

[7] For a review, see C. O. Bozler and G. D. Alley, Proc. IEEE **70**, 46 (1982).

[8] W. Shockley, U.S. Patent No. 2 569 347 (25 Sept. 1951).

[9] H. Kroemer, Proc. IRE **45**, 1535 (1957).

[10]D. Ankri and L. F. Eastman, Electron. Lett. **18**, 751 (1982). See also Ref. 4.

[11]J. O. McCaldin, T. C. McGill, and C. A. Mead, Phys. Rev. Lett. **36**, 56 (1976).

[12]W. A. Harrison, J. Vac. Sci. Technol. **14**, 1016 (1977).

[13]G. B. Scott and J. S. Roberts, Inst. Phys. Conf. Ser. **45**, 181 (1979); G. B. Scott, J. S. Roberts, and R. F. Lee, Appl. Phys. Lett. **37**, 30 (1980); J. S. Roberts, G. B. Scott, and J. P. Gowers, J. Appl. Phys. **52**, 4018 (1981); G. B. Scott, G. Duggan, and J. S. Roberts, J. Appl. Phys. **52**, 6312 (1981).

[14]Y. Kawamura, H. Asahi, and H. Nagari, Jpn. J. Appl. Phys. **20**, L807 (1981).

[15]P. Blood, J. S. Roberts, and J. P. Stagg, J. Appl. Phys. **53**, 3142 (1982).

[16]H. Asahi, Y. Kawamura, and H. Nagai, J. Appl. Phys. **53**, 4928 (1982).

[17]B. F. Levine, W. T. Tsang, C. G. Bethea, and F. Capasso, Appl. Phys. Lett. **41**, 470 (1982). This paper gives apparently complete references to earlier work on this topic.

[18]P. M. Asbeck, D. L. Miller, R. Asatourian, and C. G. Kirkpatrick, IEEE Electron. Device Lett. **EDL-3**, 403 (1982).

[19]See, for example: S. Hiyamizu and T. Mimura, J. Cryst. Growth **56**, 2 (1982); T. Mimura, Surf. Sci. **113**, 454 (1982); D. Delagebeaudeuf and N. T. Linh, IEEE Trans. Electron. Devices **ED-29**, 955 (1982).

[20]A. R. Calawa, B. A. Vojak, G. M. Metze, and M. J. Manfra (these proceedings).

Reprinted from

H. Kroemer, ``Heterostructure Devices: A Device Physicist Looks at Interfaces,'' Surf. Sci., Vol. 132, pp. 543-576, 1983.

Surface Science 132 (1983) 543–576   543
North-Holland Publishing Company

HETEROSTRUCTURE DEVICES: A DEVICE PHYSICIST LOOKS AT INTERFACES

Herbert KROEMER

Department of Electrical and Computer Engineering, University of California, Santa Barbara, California 93106, USA

Received 18 October 1982; accepted for publication 30 December 1982

The band offsets occurring at abrupt hetero-interfaces in heterostructure devices serve as potential steps acting on the mobile carriers, in addition to the macroscopic electrostatic forces already present in homostructure devices. Incorporation of hetero-interfaces therefore offers a powerful device design parameter to control the distribution and flow of mobile carriers, greatly improving existing kinds of devices and making new kinds of devices possible. Unusual device requirements can often be met by band lineups occurring in suitable semiconductor combinations. Excellent theoretical rules exist for the semi-quantitative ($< \pm 0.2$ eV) prediction of band offsets, even unusual ones, but no quantitatively accurate ($< \pm 1\,kT$) purely theoretical predictive rules are currently available. Poorly-understood second-order nuisance effects, such as small interface charges and small technology-dependent offset variations, act as major limitations in device design. Suitable measurements on device-type structures can provide accurate values for interface physics parameters, but the most widely used measurements are of limited reliability, with pure $I-V$ measurement being of least use. Many of the problems at interfaces between two III/V semiconductors are hugely magnified at interfaces between a compound semiconductor and an elemental one. Large interface charges, and a strong technology dependence of band offsets are to be expected, but can be reduced by deliberate use of certain unconventional crystallographic orientations. An understanding of such polar/nonpolar interfaces is emerging; it is expected to lead to a better understanding and control of III/V-only device interfaces as well.

1. Introduction

This paper takes a look at interfaces in submicron structures, from the point of view of a device physicist who is interested in incorporating semiconductor hetero-interfaces into future *high-performance* semiconductor devices.

A significant fraction of such devices will be compound semiconductor rather than silicon devices. Before long, most compound semiconductor devices will involve heterostructures [1,2]. Homostructure devices made from a single compound semiconductor will probably be relegated to the low-performance/low-cost end of compound semiconductor technology, although silicon device technology will very likely continue to be dominated by homostructure devices. Furthermore, high performance in devices usually means minimizing the non-active part of the device volume, to the point that the device turns from a collection of semiconductor regions separated by interfaces, to a collection of interfaces with a minimum of semiconductor between them.

As this development progresses, it calls for a constant interchange of ideas between the device physicist and the more fundamentally-oriented "basic" surface/interface physicist. This interchange goes both ways: On the one hand, the device physicist (even if inclined to do so) can less and less rely on "cookbook empiricism"; instead he must closely follow the basic physicist in assimilating and utilizing the new fundamental knowledge that the latter has acquired. On the other hand, device physics constantly poses new problems to the basic physicist; and experiments on device-type structures (sometimes deliberately "misdesigned" as devices) offer themselves as powerful tools for basic research. One of the purposes of this paper is to contribute to this necessary interchange of ideas between the device physicist and the basic physicist.

Throughout the paper, the term *heterostructure device* is to be understood in the sense that the hetero-interface plays an essential role in the operation of the device, rather than just serving as a passive interface between what is basically a homostructure device and a chemically different substrate as in silicon-on-sapphire structures. In many cases, the interface *is* the actual device. The emphasis must therefore be on "good" interfaces made by "good" technology. Various kinds of interface defects, although never totally absent, can then at least be assumed to be present in only such small densities that their effect can be treated as a perturbation of a defect-free interface model, rather than as dominating the physics. These assumptions are by no means unrealistic "academic" ones, made to simplify the problem in neglect of practical realities: They spell out the conditions that a heterointerface must satisfy to be of interest for incorporation into the active portion of a high-performance device. This poses stringent demands on the concentrations of these defects, to the point that they can rarely be neglected altogether.

The main device physics problems of hetero-interfaces can be roughly divided into problems of the static energy band structure, and problems of the electron transport *within that structure*. I shall concentrate here on the band structure aspects, and ignore the transport aspects. This is not because I consider transport problems less interesting or important (heaven forbid!), but simply because the transport aspects of the device physics are well covered by others at this Symposium. Instead, I will address myself at the end to an area of electronic structure that is not yet in the mainstream of heterostructure device development: The problems of achieving device-quality polar/nonpolar interfaces, involving such pairs as GaAs-on-Ge or GaP-on-Si. This is already an area of active interest to the basic physicist, but so far only from the structural point-of-view, largely neglecting the electrical properties that are the essence of device. Currently, the device physicist is disenchanted about the consistently miserable electrical properties that have resulted whenever device-type structures of this kind have been attempted. I believe that device-quality interfaces in such systems *can* be achieved, but only if both structural and electrical considerations are pursued jointly. This raises some new kinds of problems that simply do not exist in III/V-only systems, but the understanding of which is likely to have benefits far beyond these esoteric mixed systems themselves, feeding back even on such much simpler systems as the familiar GaAs/(Al, Ga)As systems.

2. Energy band diagrams of hetero-interfaces

2.1. Band offsets: the Shockley–Anderson model

From a device physics point-of-view the most important aspect of a semiconductor hetero-interface, and the point of departure for all subsequent considerations is the energy band diagram of the interface. We assume that the transition from one semiconductor takes place over at most a few lattice constants. For such abrupt interfaces the "canonical" energy band model is the *Shockley–Anderson model* [3–6], (Fig. 1). Its characteristic feature is an abrupt change in energy gap at the interface, leading to discontinuities or *offsets* in the conduction and valence band edges. The magnitudes of these offsets are assumed to be characteristic properties of the semiconductor pair involved, essentially independent of doping levels and hence of Fermi level considerations, but possibly dependent on the crystallographic orientation and on other factors influencing the exact arrangement of the atoms near the interface. Far away from the interface, the band energies are governed by the requirement that a bulk semiconductor must be electrically neutral, which fixes the band energies relative to the Fermi level. Except for certain fortuitous doping levels, the combination of specified band offsets with specified band energies at infinity calls for band bending, accommodated by space charge layers near the interface, similar to the space charge layers at p–n homojunctions. The calculation of the exact shape of this band bending is an exercise in electrostatics and Fermi statistics, not of interest here [5].

The band diagram shown in fig. 1 is for an n–n structure (often written n–N structure, to indicate the change in energy gap). As the figure shows, the conduction band offset then leads to a shallow potential notch and a Schottky-barrier-like potential spike barrier, both of which play large roles in the electrical properties of such junctions. Fig. 2 shows two other possibilities, an N–p junction and an n–P junction.

From the device physics point-of-view, the band offsets are the dominant aspect of heterostructure interfaces, and their existence is in fact the principal

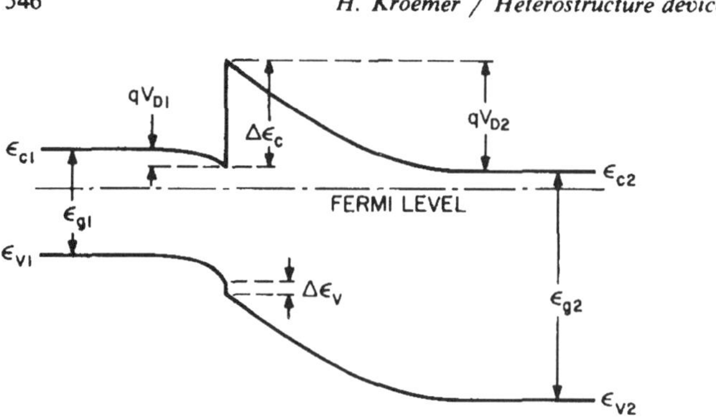

Fig. 1. Band diagram of the Shockley–Anderson model for an abrupt unbiased n–N heterojunction, showing the band edge discontinuities (or offsets) that are the characteristic feature of the model. The specific lineup shown is the "normal" lineup, for whch the narrower forbidden gap falls within the wider gap at the interface.

reason why heterostructures are incorporated into semiconductor, devices: The band offsets act as potential barriers, exerting very strong forces on electrons and holes. These quantum-mechanical "quasi-electric" forces exist in addition to those purely classical electrostatic forces that are due to space charges and applied voltages, which govern carrier flow and distribution in homostructures made from a single semiconductor. The band offset forces may be made either to assist or to counteract the classical electrostatic forces. This gives the device

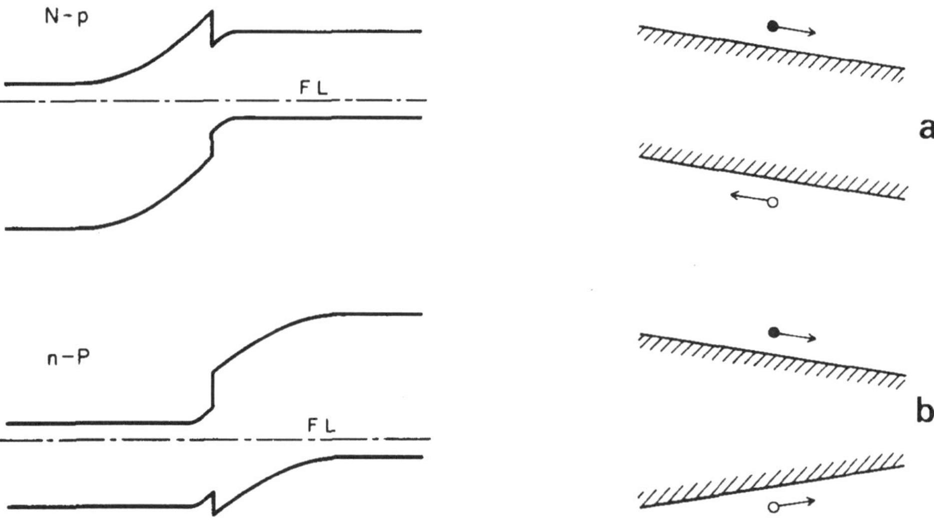

Fig. 2. Band diagrams for N–p (top) and n–P (bottom) heterojunctions.

Fig. 3. Forces on electrons and holes. In a uniform-gap semiconductor (top) the two forces are of equal magnitude but opposite direction, equal to the electrostatic forces $\pm qE$. In a graded-gap structure (bottom) the forces on electrons and holes may be in the same direction. From ref. [2].

physicist an extraordinary new degree of design freedom in controlling the distribution and flow of carriers, to improve the performance of existing devices, and to make possible new kinds of devices.

Basically, it is not the electrostatic force $\pm qE$ that acts as force on the carrier, but the slope of the band edge of the band containing the carrier, multiplied by the sign of the charge of the carrier. In a homostructure, the slopes are necessarily equal to each other and to qE (fig. 3a). But in a heterostructure, energy gap variations cause the slopes of the conduction and valence bands to differ from each other and from the electrostatic force. The case of abrupt band offsets is simply a limiting case; the underlying physics is perhaps clearer by considering the more general case of a graded energy gap, as in fig. 3b, in which only band edge slopes are visible, with no hint as to the magnitude or even the direction of the electric field.

This *general heterostructure design principle* [1,2] may be used in many different ways. A judicious combination of classical electrostatic forces and band gap variations (fig. 3b) makes it possible in a bipolar structure on control the flow of electrons and holes separately and independently. This principle is the basis of operation of the double-heterostructure laser [7,6] that serves as the heart of emerging light-wave communications technology. It also forms the basis of new kinds of improved bipolar transistors [2], and probably of other future devices.

In unipolar devices only one kind of carriers, usually electrons, are present. Here the band offset force has been used with great success in at least two different ways: (a) to confine electrons in quantum wells [8] that are much narrower and have much steeper walls than would be achievable by classical electrostatic forces ($=$ doping) alone; (b) to spatially separate electrons from the donors, against their mutual Coulomb attraction [9]. The latter possibility forms the basis of a rapidly developing new class of field effect transistors [10]. Quantum well structures form the basis of new classes of lasers [11], and they will probably also be responsible for fundamentally new kinds of future device that would not exist at all without quantum wells.

In the energy band diagrams shown in figs. 1 and 2 the signs and magnitudes of the two band offsets were such that at the interface the narrower of the two gaps fell energetically within the wider gap. This "straddling" lineup is the most common case. The most extensively studied of all hetero-interfaces, $GaAs/Al_xGa_{1-x}As$, is of this kind, and its lineup is known to a higher accuracy than that of any other system: For $x < 0.45$, the range in which (Al, Ga)As is a direct-gap semiconductor, the conduction band offset is $85\% \pm 3\%$ of the total energy gap discontinuity ("Dingle's rule" [8]), which translates into a conduction band offset of 10.6 meV per percent of Al. For higher Al concentrations see Casey and Panish [6].

Although the "straddling" lineup, with varying ratios of $\Delta\epsilon_c : \Delta\epsilon_v$, appears to be the most common case, "staggered" lineups, as in fig. 4a, can also occur.

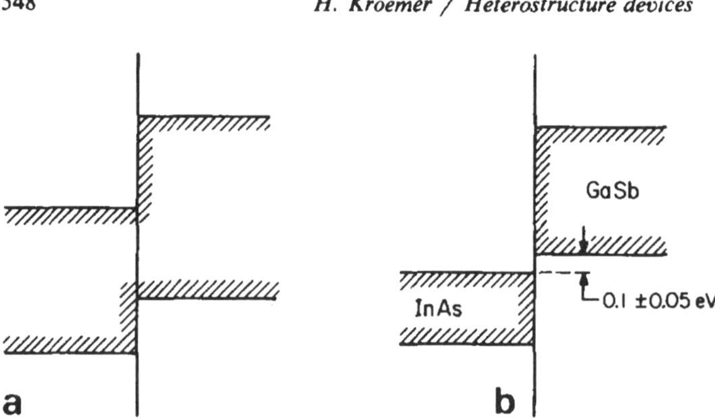

Fig. 4. (a) "Staggered" lineups are expected to occur in many semiconductor pairs. (b) The InAs/GaSb lineup has a broken gap, as shown.

One of the most extreme (and most interesting) lineups is the "broken-gap" lineup at the InAs/GaSb interfaces (fig. 4b): The conduction band edge of InAs falls below the valence band edge of GaSb, by an amount somewhere between 60 and 150 meV [12].

Such different kinds of lineups give the device physicist a powerful device design tool. One of the purposes of this paper is to give a few examples illustrating this point, another is to give some guidance about what governs the lineups in several basic heterosystems. But first we must turn to some of the nuisance effects that complicate considerably the simple Shockley–Anderson model.

2.2. Interface charges

The Shockley–Anderson model in its simplest form described above, is an oversimplification in that it neglects the possibility that there might be interface charges associated with the hetero-interface. Any such interface charge would deform the energy band diagram from that in figs. 1 and 2. Fig. 5 shows the results for an n–N heterostructure, for both signs of the charge. A negative interface charge raises the height of the spike barrier, a positive charge lowers it, and if the positive charge is large enough, the barrier is obliterated altogether, creating instead a potential well. Evidently, interface charges – if strong enough – can have a significant effect on the overall barrier heights seen by the carriers, and hence on the properties of any heterostructure device employing the offset barriers.

Interface charges may arise either from the accumulation of chemical impurities at the interface during growth, or from various kinds of structural defects at the interface. An additional mechanism discussed in detail in section 5 occurs at hetero-interfaces that combine two semiconductors from different columns of the periodic table (example: GaAs/Ge), in which case there will often exist a large net interface charge due to non-cancellation of the ion core charges at the interface.

Major modifications of the band diagram occur already for interface charge densities that are still small compared to monolayer densities. Hence, interface charges can play a non-negligible role even at hetero-interfaces which by any other criterion might be considered interfaces with a high degree of perfection.

Consider GaAs, with a lattice constant $a = 5.653$ Å and a dielectric constant $\epsilon_r = 13$. The density of atoms in a monolayer is $2/a^2 = 6.23 \times 10^{14}$ atoms per cm^2. Suppose the GaAs is doped to a level of 10^{17} cm^{-3}, and a region of $d = 10^{-5}$ cm thickness is depleted at a heterojunction, corresponding to $\sigma = 10^{12}$ charges per cm^2, a number certainly very small compared to a monolayer. The electric field supported by such a charge is $E = q\sigma/E\epsilon_0 \cong 1.4 \times 10^5$ V/cm. The accompanying band bending is $\Delta\epsilon_c = \frac{1}{2}qEd = 0.7$ eV, about twice the band bending occurring at a typical GaAs/(Al, Ga)As n–N heterojunction. Evidently, an interface charge density due to defects of, say, 10^{12} charges per cm^2, equivalent to 1.6×10^{-3} monolayer charges, will change the energy band diagram of such a heterojunction completely, and with it the electrical properties of any device containing this heterojunction. Even much smaller interface charge densities, of the order 10^{-4} monolayers, will still have a significant effect. Unfortunately, effects apparently attributable to interface charges of such small but non-negligible magnitude appear to occur frequently

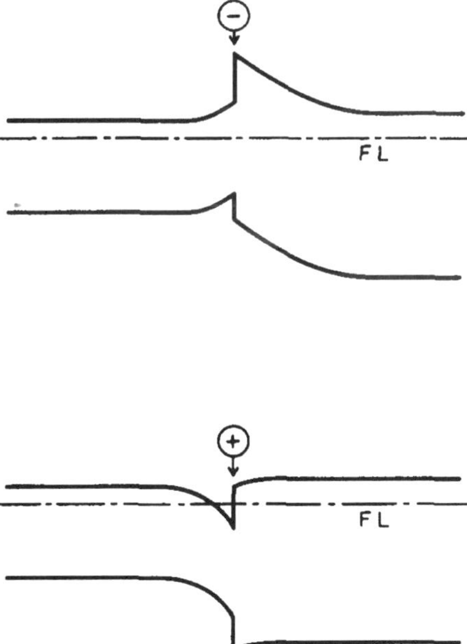

Fig. 5. Band deformation due to a negative (top) or positive (bottom) interface charge.

[13–15]. Evidently, the interface charge is an example of the high degree of sensitivity of the performance of heterojunction devices on the exact atomic structure at or near the hetero-interface, and hence an example of the interrelation between "nanostructure" and device performance.

To a basic physicist, an interface charge of, say, 10^{-3} monolayers may be all but indistinguishable from a "perfect" interface with zero interface charge. But to a device physicist such a small change is a major effect, whose neglect would be unrealistic, and which must be considered along with the band offsets. Still, the roles of the two effects are different: Whereas the band offsets are fundamental and are usually the reason for using heterostructures in devices, the interface charges are almost always a nuisance. Hence we will continue to stress the effects of offsets, raising the issue of interface charges only where necessary.

Unfortunately, interface charges are not the only nuisance: The band offsets themselves appear to be at least somewhat sensitive to exactly how the heterostructure is grown [16], on a level that is not negligible for the device properties, even though it may again be of minor concern to the basic physicist. This introduces another element of uncertainty into the device design, about which we will have to say more later.

3. Band offsets as central device design parameters

3.1. General comments

The extent to which band offsets influence device performance varies tremendously from device to device. At one extreme, the abrupt band offsets may be a nuisance. The heterojunctions in double heterostructure lasers are a good example: Although a varying energy gap is an essential ingredient of the device, a gradual variation would, for various reasons, be greatly preferable over an abrupt step [6,7]. Similar considerations apply to the p–n heterojunctions in heterostructure bipolar transistors [2]. If the semiconductors involved exhibit a continuous mutual solid solubility, the abrupt offsets are easily eliminated by gradient the transition, and this is frequently done.

Of greater interest in the context of this Symposium are devices that call for the retention of the sharp band edge discontinuities, usually with a highly specific kind of mutual band lineup. Many of the more recent heterostructure device concepts are of this kind. Such devices call for a good understanding and knowledge of the band offsets, but exactly what is needed in the way of understanding and knowledge varies greatly from case to case. It depends strongly on the nature of the device; for a given device it changes with the state of development of that device; and more often than not, the needs of the device physicist are again quite different (usually much more severe) than those of the basic physicist. Roughly, the device physicist needs three different levels of knowledge about band offsets:

(a) Semi-quantitative theoretical predictions of the band offsets for as wide a range of semiconductor pairs as possible, to assist in the selection of promising semiconductor pairs to implement new device concepts.

(b) Quantitative data about band offsets, much more accurate than ± 0.1 eV, for those semiconductor pairs that are of clear interest for practical devices, to assist in the detailed development of such devices. Ideally, this should not be restricted to accurate empirical data, but would include a theoretical understanding on a level permitting theoretical predictions with this accuracy.

(c) Data about, and a theoretical understanding of, such nuisance effects as offset variations and interface charges.

In the following three sub-sections of this paper (3.2 through 3.5), these three items are taken up, one by one. Only with respect to item (a) does a satisfactory solution exist, and only with respect to this item have the needs of the device physicist been fully met by the interests of the basic physicist. One of the hopes of this writer is that this paper might stimulate the basic physicist to take up a similar interest in the other two problem areas, to contribute to a satisfactory resolution to those problems as well.

3.2. Rough device design: semi-quantitative theoretical offset rules

New heterostructure device concepts, especially the truly novel ones, usually start out as a hypothetical energy band diagram which, if it could be realized in an actual semiconductor structure, would presumably lead to the desired device properties. The solid state photomultiplier proposed by Williams, Capasso and Tsang ($=$ WCT) [17], and discussed by Capasso earlier at this Symposium, is an excellent example. It requires a highly unsymmetric band lineup, with a conduction band offset that is larger than the gap of the narrower-gap semiconductor, and a valence band offset as small as possible. In such cases, in which the choice of semiconductors is not obvious, the first task is to determine whether the needed energy band diagram is in fact achievable by a real semiconductor combination, and whether or not any such combination is compatible with whatever other constraints may be present (lattice matching, mobilities, overall energy gap constraints, etc.). To this end, semi-quantitative predictive lineup rules are required.

The oldest and still widely used such rule is Anderson's *Electron Affinity Rule* [4–6], according to which the conduction band offset should equal the difference in electron affinities between the two semiconductors. Although the rule has been repeatedly criticized on various grounds [18–20], it is better than nothing at all. In fact, it has found vocal defenders [21,22], and it continues to be widely used despite all criticism, largely because its principal competitors, the Frensley–Kroemer theory [23] and the Harrison theory [19,24] are not so overwhelmingly superior to have caused its abandonment.

Although none of these three rules or theories are accurate enough to base a quantitative device design on their predictions, all of them are very useful as semi-quantitative guides. In fact, in simple cases, such as the WCT device [17], even rougher guides may be useful, such as the *Equal Anion Rule* [25]. It states that, for heterojunctions in which the anion atom (the column V or VI element) is the same on both sides, most of the energy gap discontinuity occurs in the conduction band, and the valence band offset is small compared to the conduction band offset. The GaAs/(Al, Ga)As pair has a common anion, and the comparatively small valence band discontinuity in that system $\Delta\epsilon_v \sim 0.15$ $\Delta\epsilon_g$ (for an Al fraction less than 0.45) demonstrates both the rule itself and its approximate nature. The rule has a theoretical foundation: For the III/V and II/VI semiconductors, the valence band wave functions are heavily concentrated around the anion atoms, with only a small part of the wave function being near the cation atom. Equal anion atoms thus naturally mean similar valence band energies [26].

Inasmuch as the WCT solid state photomultiplier calls for as small a valence band offset as possible, it naturally calls for a semiconductor pair that shares the anion species, such as a pair of phosphides, arsenides, or antimonides. Lattice matching is an additional important consideration, and because all Al and Ga compounds with the same anion tend to have very similar lattice constants [6], we can restrict the consideration further to the pairs AlP/GaP, AlAs/GaAs, and AlSb/GaSb, or related alloys. A look at the energy gaps eliminates all but the last pair, which remains as the natural candidate. With energy gaps of 1.60 eV (AlSb) and 0.72 eV (GaSb) [27], the equal anion rule predicts a conduction band offset of 0.88 eV, more than enough to exceed the gap of GaSb, and making some allowance for the approximate nature of that rule. In fact, the Harrison theory [19,24,28] predicts a valence band offset of only 0.02 eV, with the GaSb valence band edge actually the lower of the two semiconductors, that is, a very slightly staggered arrangement. Such a 20 meV prediction should not be taken seriously – the whole theory is probably not better than ± 0.2 eV – but it certainly suggests that the predictions of the equal-anion rule cannot be far off, and it makes AlSb/GaSb a natural candidate for the WCT device. This is in fact one of the two systems discussed by WCT [17] for their device; the foregoing discussion was intended to illustrate by what simple considerations one arrives at this kind of selection. Because AlSb and GaSb do not lattice-match perfectly (2.66 versus 2.65 Å), the addition of a few percent of As to the AlSb is desirable and probably necessary, but this is a refinement going beyond the semi-quantitative considerations discussed here [17].

The Frensley–Kroemer theory [23] (without the doubtful dipole corrections of that theory) predicts an only slightly different band lineup: $\Delta\epsilon_v \cong 0.05$ eV, with AlSb having the lower valence band. Evidently, this changes little. The widely-uded electron affinity rule [4–6] cannot be applied to this system, because the electron affinity of AlSb is unknown, and we do not consider the use of Van Vechten's theoretical values [29] – suggested by Shay et al. [21] and by Philips [22] – as a reliable substitute: The Harrison theory tends to give more accurate values.

The equal-anion rule can be extended into a prediction of how valence band edges vary as the anion is changed: With increasing electronegativity of the anion, the valence bands tend to move to lower energy [25], essentially because the increase in electronegativity reflects a lowering of the valence electron states within the anion atomic potential. In the case of Au Schottky barriers, a quantitative correlation was found [25] between valence band energies relative to the Fermi level, and the anion electronegativity. In the case of semiconductor heterojunctions, no *quantitative* correlation exists, but the anion electronegativity rule remains a useful *qualitative* predictor – see the broken-gap lineup in InAs/GaSb [12,30] – especially if one compares semiconductors whose energy gaps are not too dissimilar. In such cases the valence bands of the phosphides should be lower than those of the arsenides, which in turn should be lower than those of the antimonides.

This kind of prediction can be of great help if – for whatever reasons – a staggered band lineup is desired. As a good example, consider a superlattice with staggered band lineup as shown in fig 6. There has recently been a strong interest in such superlattices [31], for the following reasons. In a staggered structure, any electrons would accumulate in the low-ϵ_c layers, any holes in the high-ϵ_v layers. If both kinds of layers are thin enough (≤ 100 Å), there would be significant tunneling of both electrons and holes, and the entire superlattice would behave essentially as a homogeneous substance with an overall energy gap smaller than that of either constituent compound, slightly larger than the separation between the highest valence band and the lowest conduction band. Suppose next that the low-ϵ_c layer is n-type doped, and the high-ϵ_v layer p-tape. If selective contacts are made to the n-type and p-type layers, and a bias voltage applied, the effective energy gap is varied. But a voltage-adjustable

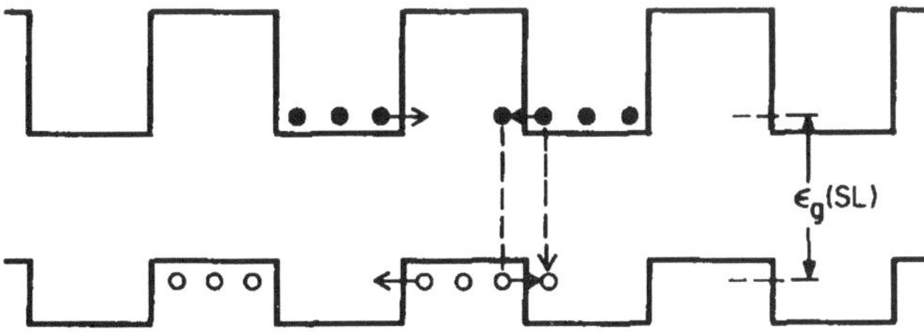

Fig. 6. Staggered-offset superlattice, in which electrons and holes (if present) accumulate in alternating layers. Because of electron tunneling, such structures can have an effective gap narrower than the gaps of both bulk semiconductors.

energy gap would of course be an extremely valueable new phenomenon.

The whole concept is simply an elaboration of the n–i–p–i superlattice concept of Döhler and Ploog [32], except that the spatial separation of the high concentrations of electron and hole from each other is now achieved very easily by the band offset forces, rather than purely electrostatically, by heavy doping.

The occurrence of a broken gap in the InAs/GaSb system suggests that less extreme cases of staggering are indeed achievable, but are they achievable in semiconductors with much larger energy gaps? The anion electronegativity rule [25] suggests that combinations of a phosphide with an antimonide form a promising point of departure. Because phosphides tend to have smaller lattice constants (and larger energy gaps) than antimonides, it is advisable to start with the phosphide that has the largest lattice constant (and the smallest gap), InP, and the combine it with the largest-gap antimonide, AlSb. For this system the Harrison theory [28] predicts indeed staggered band offsets, with a conduction band well depth $\Delta\epsilon_c = 1.20$ eV, a valence band well depth $\Delta\epsilon_v = 0.97$ eV, and a net band separation

$$\epsilon_g(\text{SL}) > \epsilon_c(\text{InP}) - \epsilon_v(\text{AlSb}) \cong 0.4 \text{ eV}.$$

The actual superlattice gap should be somewhat larger, increasing with decreasing superlattice period.

Although the estimate was rough, the message is clear: Staggered superlattices with usefully large gaps should be achievable! Whether or not the simple InP/AlSb pair is indeed a promising pair, remains to be seen, but it is certainly a useful point of departure. If anything, the staggering is larger than needed and the effective gap (≥ 0.4 eV) too small to be useful. Evidently, the conditions to obtain staggering may be relaxed somewhat. Now, one of the drawbacks of the InP/AlSb pair is a large lattice mismatch ($\cong 4.6\%$). Such a lattice mismatch, would almost certainly be fatal to device performance in a single-interface heterostructure device due to inevitable misfit dislocations. But it might be quite acceptable in a short-period superlattice, where the lattice misfit can be taken up by elastic strain, a point recently elaborated upon by Osbourn [31] in the context of strained-layer staggered superlattices based on the GaP/Ga(P, As) system. If necessary, the lattice misfit could be reduced by replacing AlSb with an Al(Sb, As) alloy. This would make the valence band well shallower and increase the net gap, but the Harrison theory predicts that even for perfect lattice match to InP, that is, for $AlAs_{0.56}Sb_{0.44}$ [33], a valence band well of 0.46 eV and a net gap of 0.91 eV should remain. Further fine-tuning could be achieved by replacing some of the Al by Ga [6].

Two other lattice-matched pairs for which staggered lineups can be safely predicted are $InP/Al_{0.50}In_{0.50}As$ ($\epsilon_g \geq 1.1$ eV) and $Ga_{0.52}In_{0.48}P/AlAs$ ($\epsilon_g \geq 1.6$ eV).

There is some evidence [34] that the GaP_xAs_{1-x} system for $x > 0.5$ leads to staggered lineups with large net gaps, but for this system the theoretical predictions are not as clear-cut as for the above combination. We will return to this point later.

3.3. Quantitative device design: the absence of theoretical guidance

Although semi-quantitative lineup prediction rules are very useful in identifying promising hetero-pairs for hypothetical device applications, a detailed device design requires far more accurate values. In any device in which current flows *across* a heterostructure barrier, the current depends on the barrier height $\Delta\epsilon$ at least like a Boltzmann factor $Exp(-\Delta\epsilon/kT)$, implying a factor e for every change in barrier height by $1\ kT$ ($\cong 26$ meV at 300 K). If the current is tunneling rather than thermionic current, the dependence tends to be even steeper. There is no need to discuss here whether a prediction to some fraction of kT is necessary or whether $\pm 1\ kT$ or even $\pm 2\ kT$ would be sufficient: None of the predictive theories comes anywhere near even the less demanding limit. Those physicists (not involved in actual device design) who have expressed their satisfaction with either the electron affinity rule [21,22] or the Harrison theory [20], quote examples of "excellent agreement" between theory and experiment, in which the predicted offsets vary by 0.2 eV ($\cong 8\ kT$) or more from reliable experimental data. Presumably, then, this is roughly the level of reliability of existing predictive rules or theories. This degree of agreement may indeed be very satisfactory to the fundamental physicist, who wants a general understanding of band offsets; it is totally unsatisfactory as a quantitative basis for device design.

Nor is the need for an accurate prediction significantly less demanding in those devices in which current does not flow *across* a hetero-barrier, but *along* it, as in the new high electron mobility transistor (HEMT) [10] which represents one of the most active areas of heterostructure device research and development, also discussed (from a physics- rather than device-oriented point-of-view) by Störmer at this Symposium. One of the most important design parameters in these devices is their threshold voltage, that is, the gate voltage at which the conductance along the 2D conducting channel is effectively turned on or off (it may be either a positive or a negative voltage, depending on the desired design). To be useful in future high-performance IC's (their dominant area of interest), the threshold voltages of these devices must be predictable much more accurately than ± 0.1 V, preferably to ± 0.1 V, which calls for a knowledge of the band offsets to within a similar accuracy.

As the HEMT case shows, the absence of any purely theoretical predictive tools with the desired accuracy is not preventing the design of this particular device to go forward. The band offsets at the (Al, Ga)As-on-GaAs (100) interface *are* known to the required degree of accuracy [6]. But this accurate knowledge is the result of accurate *experimental* measurements [8], not of an accurate predictive theory. Once the evolution of a new heterostructure device has progressed beyond the initial speculative stage, to the point of practical device development, it is necessary that the band offsets be accurately *known*, but the knowledge need not come from a predictive theory; knowledge from accurate experimental data may actually be preferable to a theoretical prediction. This de-facto status of the band offsets is similar to that of energy gaps: Whenever available, we use accurate experimental values of energy gaps, rather than theoretical values. Only when experimental data are missing, will we use theoretical ones.

Does any of this mean, however, that the attempts to predict band offsets theoretically have no value beyond the crude semi-quantitative value discussed earlier? Far from it! First of all, the purpose of theories of band offsets (e.g. electron affinity rule, Harrison's theory, etc.) is only partially to provide the device physicist with quantitative design data. A more important role is to test the assumptions that go into each theory, and thereby to test our fundamental understanding of what determines the band offsets. This is similar to the way band structure calculations test our understanding of band structures more than providing accurate theoretical gap values when accurate experimental values are already available. All these are *retrodictive* theories more than *predictive* ones! By that standard, neither the electron affinity rule not the Harrison theory, with their ± 0.2–0.3 eV accuracy, are doing badly (nor does the Frensley–Kroemer theory, which is of similar accuracy). Inasmuch as the present paper is to describe a device physicist's view of hetero-interfaces, it does not provide a suitable forum to discuss exactly how well these theories meet the needs of the basic physicist, and much less to discuss critically the enthusiastic support that Shay et al. [21] and Philips [22] have expressed for the electron affinity rule, and Margaritondo and his co-workers [20] for the Harrison theory.

A second reason why more accurate theoretical predictions could be useful as quantitative rather than merely semi-quantitative predictive tools occurs whenever the accuracy of the existing theories is insufficient to yield a clear-cut yes–no decision about a speculative device, but in which experimental data would require the prior development of an elaborate technology. A theoretical guidance on whether or not the development of this technology is worthwhile would be highly useful in such cases [18].

A good example is once again at hand. There has been considerable speculation [31] that a $GaP/GaP_{0.6}As_{0.4}$ superlattice would be of the interesting staggered variety shown in fig. 6. This speculation is partially based on the electron affinity rule, using the electron affinity value of 4.3 eV quoted by Milnes and Feucht [5] without giving any source. Partially it is based on a highly indirect claim by Davis et al. [35] (contradicting other data) that the conduction band offset in the GaP/GaAs system should be near zero. A very careful measurement of the electron affinity has recently been performed by Guichar et al. [36], yielding 3.70 ± 0.05 eV. Using this value, and the known electron affinity for GaAs, 4.07 eV, and making due allowance for the change from direct gap to indirect gap in going from GaAs to GaP, one predicts a conduction band offset for the superlattice of only 0.02 eV, just very slightly staggered. The Harrison theory predicts the same value [28]. With the reliability of both the electron affinity and the Harrison theory rule being no better than ± 0.2 eV, this prediction is simply a draw. Inasmuch as the development of an entire superlattice technology hinges on this prediction, it is an excellent example of why more accurate predictions would indeed be desirable.

Recent experiments suggest [34] that the superlattice is indeed staggered, by about 0.2 eV. If future measurements confirm this result, this would show that both theoretical predictions are indeed incorrect by about 0.2 eV.

3.4. The nuisance effects: offset variations and interface charges

In the preceding discussion we pointed out the device physicist's need for knowing band offsets to an accuracy much better than ± 0.1 eV. But this request implicitly assumed that the band offsets are in fact constants that characterize a given semiconductor pair, rather than being variables themselves. As was pointed out by Bauer [37] and by Margaritondo [20] at this Symposium, evidence is accumulating [16,38] that the offsets are process-dependent, changeable over a finite range outside of the tolerance limits of the device designer. A dependence on crystallographic orientation is almost to be expected, and while it might be a nuisance, it does not introduce any problems into device design. Nor do we need to be surprised about large offset variations in systems in which a compound semiconductor (GaAs, GaP) is grown on an elemental semiconductor (Ge, Si), or vice versa, the cases of particular interest to Bauer [37] and Margaritondo [20]. We shall argue in section 5 that in such systems technology-dependent offset variations and interface charges *are to be expected*. What *is* disturbing are offset variations and interface charges in such supposedly well-behaved lattice-matched systems as GaAs/(Al, Ga)As. It was found by Waldrop et al. [16] that for {110}-oriented MBE growth at a substrate temperature of 580°C the band lineup depends noticeably on whether AlAs is grown on GaAs ($\Delta\epsilon_v \cong 0.15$ eV), or GaAs on AlAs ($\Delta\epsilon_v \cong 0.40$ eV). By comparison, the {100}-lineup data of Dingle [8] for GaAs/(Al, Ga)As, extrapolated to Gas/AlAs, corresponds to an in-between value of $\Delta\epsilon_v \cong 0.20$ eV.

Although differences between {100} and {110} might have been expected, the strong growth sequence dependence for the {110} orientation comes as a rude shock. For a given orientation, band offsets can depend on growth sequence only through differences in the exact atomic arrangement near the interface. Evidently the atomic arrangements for {110} interfaces depend strongly on growth sequence. Put bluntly: At least for this orientation the offsets depend quite strongly on technology [39] rather than being a fundamental materials parameter! The question naturally arises whether or not this might quite generally be the case. Might there be a similar growth sequence dependence for {100} growth? I find it hard to believe that any significant growth sequence asymmetry of {100} band offsets would leave intact the superb fit of Dingle's superlattice data (which automatically involve both growth sequences) to a single-offset model, especially considering Dingle's wide range of layer thicknesses. Yet there exists strong evidence that, if not the band offsets, at least the transport properties in the 2D electron gas along GaAs/(Al, Ga)As {100} heterojunctions, depend quite strongly on the growth sequence [38], with higher mobilities occurring for (Al, Ga)As-on-GaAs than for GaAs-on-(Al, Ga)As. In fact, it appears that in structures containing multiple interfaces, the properties of the interfaces grown first differ from those grown later [40]!

One frequently hears the argument that effects such as these are somehow artifacts of the growth process, reflecting "bad" interfaces. While in a practical sense this might be true, it avoids the fundamental issue: Even a "bad" interface must have some atomic configuration that causes these effects, and which configuration constitutes "badness"? And can this "badness" in fact be avoided under the numerous constraints imposed upon the growth of an actual device?

We clearly need an understanding of these effects, and this may indeed by one of the most urgent research topics in which the device physicist would like to see the basic physicist take an active interest. To the basic interface physicist, offset variations of $\equiv 50$ meV might be a minor nuisance, negligible to the basic understanding of the interface physics. But the degree to which these offset variations can be controlled, may be decisive for the role heterostructure FET's will play in future high-speed VLSI technology.

A return to the earlier example of HEMT threshold voltages will illustrate the urgency. As we stated, these threshold voltages depend on several structural parameters, one of which is the conduction band offset. Now the most important envisaged applications of this transistor is in future very fast large-scale digital integrated circuits which may contain anywhere from 10^3 to 10^6 identical FET's per chip. For a variety of reasons, it is necessary that the threshold voltages of all transistors on the same chip have essentially the same value, *and* that this design value can be technologically maintained from chip to chip and even from wafer to wafer. Threshold voltage variations far below ± 0.1 V on a single chip are essential, or else the IC will simply not work, and variations below 10 mV are desirable. Worse, the variations from chip to chip should not be much larger. Evidently this calls for tight tolerances on the band offsets and on residual interface charges.

To achieve these tolerances requires an understanding of what causes offset variations and interface charges, not just purely empirical tight process control. In fact, it is probably more important to develop a physical understanding of offset *variations* on the ± 5 meV level than to be able to *predict* the exact magnitude of these offsets to better than ± 0.1 eV.

4. On measuring band offsets experimentally

4.1. Introductory comments

There does not exist any experimental technique to determine band offsets that is simultaneously simple, reliable, and universally applicable.

The most careful and presumably most accurate determination of any band lineup is Dingle's well-known work [8] on the infrared absorption spectra of superlattices of weakly-coupled multiple GaAs/(Al, Ga)As quantum wells. Dingle was able to fit large numbers of data, for wells of various widths, to a single model in which the conduction band offset is $85\% \pm 3\%$ of the energy gap difference.

For sufficiently narrow wells, the method is fairly insensitive to errors by small interface charges. Major distortions in the well shape would quickly destroy the excellent fit of the experimental data to the simple square-well model. Dingle's data prominently include transitions involving the higher energy levels in the wells, which would be especially sensitive to any distortions of the well shape. It is hard to believe that the large number of observed transitions, over a wide range of well widths, could be fitted just as well to a significantly different well shape. This same quality-of-fit argument also speaks against various kinds of modifications in the band offsets, such as growth sequence asymmetries, etc. Certainly, the burden of the proof for any such modifications lies with those who would propose such modifications. Note, however, that Dingle's data, being strictly {001} data, in no way rule out any offset dependence on crystallographic orientation.

A second widely used technique to determine band offsets is based on photoelectron spectroscopy [20,41], executed with various levels of sophistication. It is even less sensitive to interface charges, and is in principle capable of giving quite accurate offsets, perhaps more directly than Dingle's technique. Especially the Rockwell group of Kraut, Grant, Waldrop and Kowalczyk [41] has cultivated this technique to a high level of perfection, to the point that in favorable cases offsets with (believable) uncertainties of ± 0.03 eV were obtained. Inasmuch as Margaritondo, another practitioner of this technique, has discussed it at this Symposium, we refer to his paper [20] for more information and references.

Both the superlattice absorption technique and the photoelectron spectroscopy technique are "physicist's techniques", rather than device-type techniques. Now we argued earlier in this paper that the properties of heterostructure devices depend sensitively on band offsets. It should therefore be possible to extract accurate band offsets from measurements on devices. Because of the simplicity of purely electrical measurements, such attempts have indeed often been made [5], and many band offsets found in the literature were in fact obtained from purely electrical measurements, usually on simple p–n or n–n heterojunctions. Unfortunately, such measurements are sensitive not only to band offsets; they are just as sensitive to other phenomena that deform the band diagram, especially interface charges. Most of the electrical measurements have difficulty separating these effects. More often than not the data are merely fitted to the simple Shockley–Anderson model ignoring such complications, which can lead to grossly inaccurate band offsets.

Inasmuch as this paper represents a review of hetero-interfaces from the device physicist's point-of-view, a critical review of the main techniques is in order.

4.2. Capacitance–voltage profiling

Probably the best of the purely electrical measurement techniques is based on a powerful adaptation of conventional C–V impurity profiling, recently developed by Kroemer et al. [14,42]. It can, under favorable circumstances, give reliable *separate* values for both the band offsets and any interface charges. The method requires an n–n heterojunction whose doping profile is known, a condition often satisfied for junctions grown by highly developed technologies such as MBE. A Schottky barrier is placed on the outer surface of the heterostructure, parallel to the hetero-interface, and the C–V relation of the Schottky barrier rather than of the heterojunctions itself is measured. The method works best with heterojunctions exhibiting poor rectification, which are particularly hard to evaluate by other means. An *apparent electron concentration* $\hat{n}$ is determined by the conventional interpretation of C–V profiling theory [42],

$$\frac{\mathrm{d}}{\mathrm{d}V}\frac{1}{C^2} = \frac{2}{q\epsilon}\frac{1}{\hat{n}(x)},\tag{1}$$

where C is the capacitance per unit area, and $x = \epsilon/C$. The $\hat{n}(x)$ profile will differ both from the doping profile $n_d(x)$ and from the true electron concentration $n(x)$. But if the doping distribution $n_d(x)$ is known, the interface charge is easily obtained by integrating the apparent difference distribution $\hat{n}(x) - n_d(x)$, and the conduction band offset is obtained from the first moment of this difference distribution. The true electron distribution is not needed! The method is simple and powerful, and readily applicable to any technology that permits the growth of heterostructures in which the doping level can be kept accurately constant on both sides of the interface, with an abrupt switch at the interface. The two constant doping levels need not even be predetermined; they may be extracted from the C–V profile itself.

The method may be made self-checking, by using the two doping values, the interface charge, and the band offset, to simulate on a computer the C–V profile that *should* have been seen experimentally, and by comparing this reconstructed profile with the profile actually observed.

H. Kroemer / Heterostructure devices 561

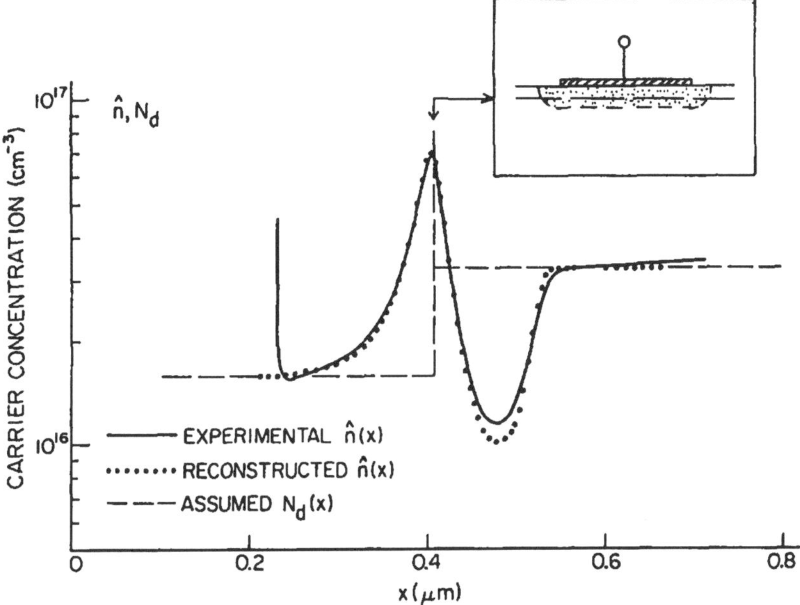

Fig. 7. $C-V$ profiling through an LPE-grown GaAs/(Al, Ga)As n–N junction, after ref. [14]. From the measured apparent electron concentration $\hat{n}(x)$ (solid curve) and the assumed donor distribution $n_D(x)$ (broken curve) one can calculate a conduction band offset $\Delta\epsilon_c = 0.248$ eV and an interface charge density $\sigma_1 = +2.7 \times 10^{10}$ cm^{-3}. The inset shows the basic test arrangement.

Fig. 7 shows an example, from ref. [14], for an LPE-grown n–N heterostructure, not ideally suited for the purpose, but so far the only published result in which the method has been used for a quantitative determination of both a band offset and an interface charge, including the self-consistency check. The technique should be even better suited to MBE- or MOCVD-grown interfaces, in which an abrupt transition with flat adjacent doping levels is more easily achieved, and this writer does in fact expect that it will be widely used in the future.

4.3. The C–V intercept method

When the doping level n_d and hence the electron concentration n in an n-type semiconductor is position-independent, the $C-V$ profiling theorem (1) yields a linear C^{-2}-versus-V plot. This remains true for the capacitance of a p–n junction, including a p–n heterojunction, if the carrier concentrations on both sides are constant. This has led to the *C–V intercept method*, which claims that the intercept voltage V_{int} in such a linear C^{-2}-versus-V plot is exactly equal to the total built-in voltage of the heterojunction (fig. 8), sometimes called the *diffusion voltage*,

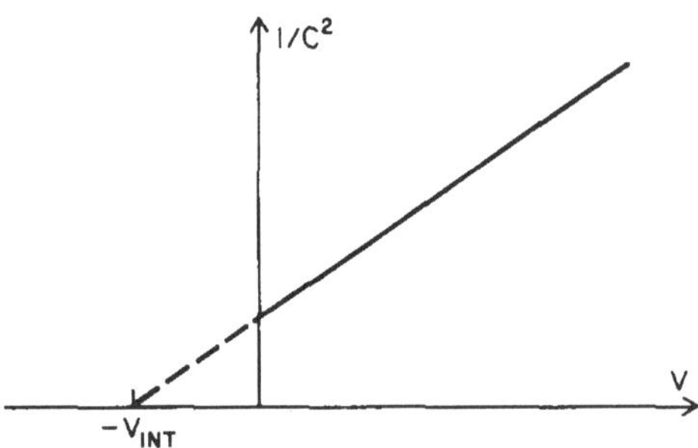

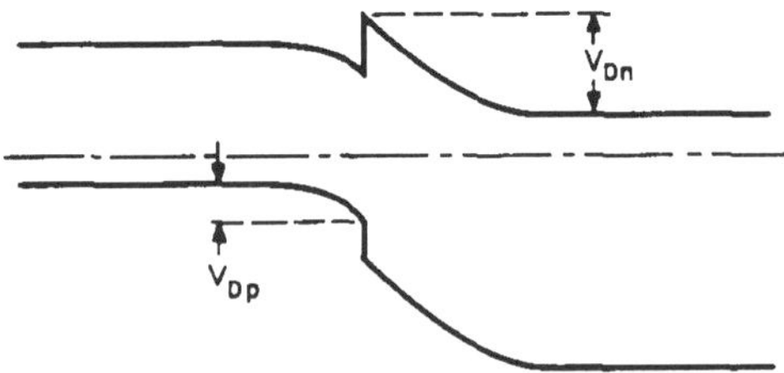

Fig. 8. The C–V intercept method of determining the band offsets at p–N heterojunctions. *If* the heterojunction is abrupt, with constant doping levels right to the interface (no grading), and without any interface charges, then the intercept voltage V_{int} in a C^{-2}-versus-V plot is related to the two diffusion voltages V_{Dn} and V_{Dp} via eq. (3). If both doping levels (and hence both Fermi energies) are known, this permits a determination of the band offsets. The method is sensitive to errors caused by grading or interface charge effects.

$$V_{int} \overset{?}{=} V_{Dn} - V_{Dp}. \tag{2}$$

For known doping levels, the energy separations between the bulk band edges and the Fermi level are known, and hence the band offsets are known if $V_{Dn} + V_{Dp}$ is known. Unfortunately, the accuracy of eq. (2) is largely a (persistent) muth. First of all, (2) neglects the so-called Gummel–Scharfetter correction [43]; it should really read

$$V_{int} = V_{Dn} + V_{Dp} + 2kT/q, \tag{3}$$

a small correction, but not a negligible one. More important: Even in the form (3), the intercept rule is strictly valid only if both doping levels are constant right to the hetero-interface, forming an abrupt transition there, and if no interface charges are present [42,44]. Interface charges tend to lower the intercept voltage, whereas impurity grading effects raise it. A small region right at the interface always remains inaccessible, even if C–V profiling is extended to forward bias values. Any space charge re-adjustments entirely inside this region will not affect the linearity of the C^{-2}-versus-V plot unless the charge inside the depletion region somehow depends on the applied voltage (which may be the case for deep levels, but not otherwise). Although these facts have been established for some 25 years [44], they remain strangely ignored except by a small fraternity of semiconductor device physicists intimately familiar with C–V profiling theory. Even as astute a researcher as Phillips [22] writes in a recent paper: "The great merit of this technique is that it is self-checking, i.e., when chargeable traps are present at the interface, C^{-2} is not a linear function of V_a. The deviations from linearity automatically provide estimates of the accuracy of the determination of V_D and from it the accuracy of ΔE_C and ΔE_V." Well, they don't. To get experimental access to the charges located right near the interface, one must profile *through* the heterojunction from the outside, as described earlier, not from the interface outward.

Considering this inherent weakness of the intercept method, it is not surprising that the offset values determined by it have fluctuated widely whenever data from more than one investigator have been available, and often even for the data from the same group. Two examples are provided by the chaos in the offset data reported for GaP/GaAs and Ge/GaAs. In most of these measurements, C–V intercept data were not used alone, but in conjunction with current–voltage (I–V) data. However, this hardly excuses the failure of the intercept method to "catch" the ever greater inadequacies of the I–V techniques.

In the case of GaP/GaAs, the reported conduction band offsets vary by at least 0.65 eV: Weinstein et al. [45] claim $\Delta\epsilon_c \cong 0.22$ eV, Alferov et al. [46], $\Delta\epsilon_c \cong 0.65$ eV, and Davis et al. [35], $\Delta\epsilon_c \cong 0$. It is anybody's guess which of these values is least far away from the truth – if there is in fact a single "true" value.

The situation for Ge/GaAs is, if anything, even worse. Conduction band offsets varying from 0.09 to 0.54 eV can be found in the literature, a range corresponding to 68% of the energy gap of the narrower-gap semiconductor, Ge. The reason is probably only partially due to erratic measurements. As we shall see later, for polar/nonpolar systems such as GaAs/Ge, an erratic technology-dependence of the offsets should be expected.

Despite this history of unreliable results, the intercept method should be capable of yielding accurate offsets *if* the uncertainties inherent in it are treated with due respect, and are eliminated by suitable complementary data, especially for interfaces grown by one of the better and more tractable technologies, such as MBE or MOCVD. There is something inherently satisfactory about C–V profiling measurements: They are essentially purely electrostatic measurements of equilibrium charge distributions versus position, almost completely unencumbered by transport effects.

4.4. Current–voltage measurements

Whatever criticisms one might have of band offsets based primarily on $C-V$ intercepts, most of those based on current–voltage ($I-V$) data on p–n or n–n heterojunctions are even less well-founded. Exceptions tend to occur for systems with unusual band lineups, in which the $I-V$ data on heterojunctions differ already qualitatively in drastic ways from those of ordinary p–n homojunctions. The outstanding (but not the only) example is the striking broken-gap lineup at the InAs/GaSb interface (fig. 4b), for which the first experimental evidence was obtained [12] from systematic rectification experiments with lattice-matched Ga(As, Sb)/(Ga, In)As p–n heterojunctions of varying (lattice-matched) alloy compositions. As the GaSb/InAs end was approached, all rectification effects suddenly disappeared, due to the "uncrossing" of the forbidden gaps.

But $I-V$ data on p–n heterojunctions without special lineup feature tend not to contain enough qualitatively different detail to be useful for quantitative offset determination, although they may be useful to supplement other data.

Worst, $I-V$ data on n–N rather than p–n heterojunctions, although they could in principle be quite informative, have in the past been largely worthless. For example, the claim that the conduction band offset of GaP–Si interfaces is essentially zero, is based on nothing more than the failure to observe any rectification effects in Si-on-GaP n–n junctions even at liquid nitrogen temperature [47]. More recent data on this system show [48,49] this claim to be quite false. How erroneous such absence-of-rectification data can be, is illustrated by what is now the best understood heterostructure of all, the GaAs/(Al, Ga)As structure: Most early data on this system showed a more or less complete absence of rectification in n–N junctions [50]. The explanation in terms of zero conduction band offset flatly contradicted Dingle's lineup data. The problem seems to have gone away with subsequent improvements in technology; it was almost certainly due to donor-like defects at the interface, as first proposed by Kroemer et al. [13]. Similar donor-like defects were probably responsible for the lack of rectification in Si/GaP heterojunctions [47].

5. Polar / nonpolar heterostructures

5.1. Motivation

Almost all heterostructure *device* structures currently under active investigation employ heterostructures between III/V compounds only. There are strong incentives to extend heterostructure device technology to other systems, especially to combinations of a III/V semiconductor with one of the elemental semiconductors, Ge or Si. Natural pairs, because of their close lattice match, would be GaAs/Ge and GaP/Si. The latter is particularly interesting. If device-quality interfaces between GaP and Si could be achieved, this would be a major advance towards bridging the wide gap between highly-developed Si technology and the rapidly developing technology of III/V compounds, with potentially far-reaching device applications.

A number of attempts to grow such polar/nonpolar heterostructures have led to disappointing results: These systems are clearly far more difficult than III/V-only heterosystems. However, a physical understanding of these systems is beginning to emerge that explains why many of the earlier purely empirical "cookbook" approaches *should* have failed, and which suggests that a better understanding of both the growth mechanism and the electronic structure of these interfaces might make possible substantial progress towards the elusive goal of device-quality polar/nonpolar heterostructures.

In fact, the incentives to achieve such a better understanding go far beyond the device utilization of polar/nonpolar interfaces themselves: It would also advance the understanding of more "ordinary" III/V-only interfaces. Many of the problems that occur at polar/nonpolar interfaces are simply hugely magnified versions of problems that occur already at the GaAs/(Al, Ga)As interface. Examples: Residual interface charges, offset variations, crystallographic orientation dependence, and technology dependence. The difference is purely quantitative: In the III/V-only cases these problems are second-order nuisances, in the polar/nonpolar cases they dominate. I believe this dominance is the reason why polar/nonpolar interfaces have so far proven so intractable. It is reasonable to expect that a better understanding of these effects, leading to control in the polar/nonpolar case, will also greatly benefit the III/V-only case.

5.2. Interface neutrality and crystallographic orientation

In 1978, Harrison, Kraut, Waldrop and Grant (HKWG) published a classical paper [51] that forms the point of departure for any rational understanding of the problems of polar/nonpolar interfaces. The authors studied the electrostatics of the simplest possible atomic configurations for the three lowest-index orientations of an ideal GaAs/Ge hetero-interface. They showed that for both the {100} and {111} orientations these atomic configurations correspond to a huge net electrostatic interface charge, of the order of one-half of a monolayer charge. The argument is brought out in fig. 9 for the (001) interface, viewed in the [$\bar{1}$10] direction. The black circles represent Ga atoms, the white circles As atoms, and the shaded ones, Ge. An alternate possibility has Ga and As interchanged. An important point in the HKWG argument is a point emphasized earlier by Harrison [52]: The tetrahedral bond configuration *guarantees* that each of the bonds connecting each atom to its four nearest neighbors contains exactly two electrons, just as in Ge, and regardless of whether the bonds are Ge–Ge, Ga–As, or mixed Ga–Ge or Ge–As bonds. Only the electron distribution along each bond depends on these details, not the overall bond charge. This means that the net electrical charge associated with the overall interface region can be determined by simply counting each column-V atom as having one extra proton charge relative to a neutral column-IV atom, and each column-III atom as missing one such charge. The overall interface charge is easily obtained by a fictitious process, whimsically called "theoretical alchemy", in which one pretends that the GaAs portion of the heterostructure has been obtained from a Ge single crystal by moving a proton lattice from one-half of the Ge atoms to the other half of the Ge atoms, creating Ga and As in the process. Depending on whether the fictitious proton motion is away from the interface or towards it, a negative or positive charge imbalance is thereby created at the interface. The bottom half of fig. 9 shows the electrostatic potential resulting from a proton transfer away from the interface, with the electron distribution along the bonds initially kept fixed. The potential staircase on the GaAs side is evident. The average slope of this staircase represents a net electric field, which is easily shown to be that of a charge of $-q/2$ per interface atom. With an interface atom density of $2/a^2$, this is a charge density $-q/a^2$. The important point is now that the bond charge relaxation following the proton transfer does not change the net

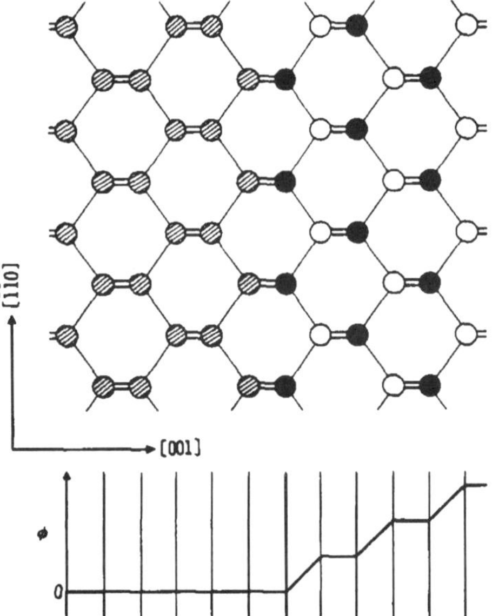

Fig. 9. Atomic arrangement and electrostatic potential at an idealized unreconstructued Ge/GaAs(001) interface, from ref. [51]. The idealized atomic arrangement exhibits a large charge imbalance at the interface, leading to a staircase potential with a large net electric field on the GaAs side. The full circles represent Ga atoms, the open circles As atoms.

interface charge, even though it is strong enough to actually reverse the sign of the net charge on the Ga and As atoms inside the GaAs side. But the total charge per bond always remains at exactly two electrons; no net charge crosses the Ga and As atomic planes inside the GaAs side, implying conservation of net interface charge during the relaxation. In terms of the potential diagram in fig. 9, the shape of the individual steps in the staircase changes, but the net *average* slope remains unchanged.

As HKWG point out, the field supported by the net interface charge is huge ($E = q/a^2\epsilon \cong 4 \times 10^7$ V/cm, assuming the dielectric constant of GaAs), sufficient to guarantee an atomic re-arrangement during the crystal growth itself, to minimize those interface charges. The authors give two specific atomic configurations which lead to zero interface charge, shown in figs. 10 and 11. The first of these contains one mixed-composition layer, but it retains a finite interface dipole. In the second configuration, containing two mixed-composition layers, the interface dipole has also been obliterated. The authors speculate that the second configuration might actually arise during epitaxial growth.

It is at this point that we must differ from HKWG. Although there can be no doubt that a drastic atomic re-arrangement will take place, and almost certainly in the general direction postulated by HKWG, it appears inconceiva-

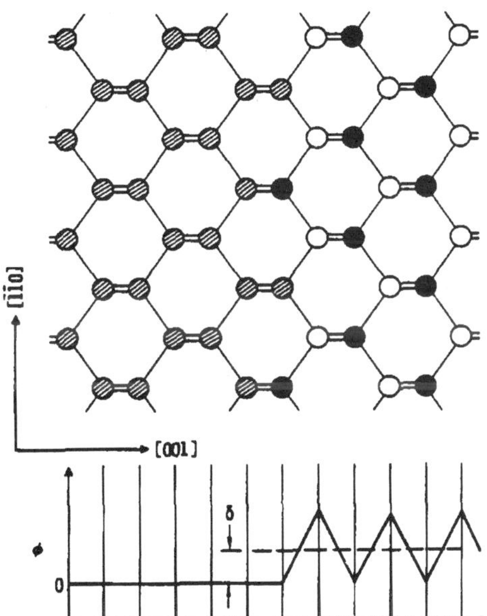

Fig. 10. Modified atomic arrangement and electrostatic potential at a Ge/GaAs(001) interface containing one atomic plane of mixed composition, with zero net interface charge, but retaining finite interface dipole. From ref. [51].

                    H. Kroemer / Heterostructure devices

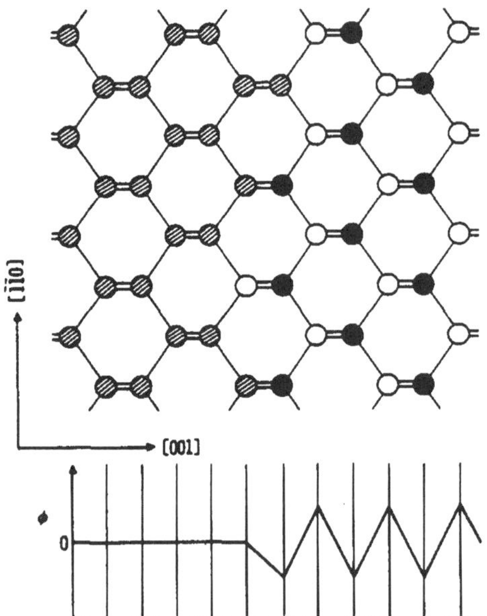

Fig. 11. Further modification of the atomic arrangement at a Ge/GaAs (001) interface, containing two atomic planes of mixed composition yielding both zero interface charge and a zero interface dipole. From ref. [51].

ble that any such re-arrangement goes sufficiently far towards completion that the remaining interface charge becomes negligible for device purposes. We recall that even a charge of only 10^{-3} monolayers is still a large interface charge for device purposes; even if the interface atomic re-arrangement goes 99% towards completion, this would still leave an intolerably large charge five times as large.

We therefore conclude that, at least for the {100} orientation, large residual interface charges must be expected at GaAs/Ge and similar polar/nonpolar interfaces. Worse, the exact amount of interface charge left must be expected to depend on the growth process. Hence the interface charges will not only be large, but technology-dependent. Finally, because even for zero interface charge the residual interface dipoles still depend on exactly which atomic re-arrangement was created, the band offsets must also be expected to be technology-dependent and hence poorly reproducible.

There are mitigating circumstances present if the growth sequence is non-polar-on-polar. Harrison has pointed out [53] that the electrostatic arguments of HKWG also apply, with some modification, to the free surface of a compound semiconductor. A GaAs {001} surface terminating in complete Ga or As planes is electrostatically just as unfavorable as an ideal GaAs/Ge interface. The actual atomic configuration present at a free GaAs {100} surface will already be such that the net surface charge is minimized. If all dangling surface bonds dimerize, apparently a good first-order approximation, an atomic arrangement leading to a neutral surface will also lead to a neutral Ge/GaAs interface, if the vacuum is subsequently replaced by Ge.

But this argument does not apply if GaAs is grown on Ge. Thus we are led to a second prediction: Polar/nonpolar interfaces must be expected to exhibit drastic growth sequence dependences, much stronger than those observed in the GaAs/(Al, Ga)As system. Unfortunately, the more difficult polar-on-nonpolar growth sequence is demanded in the majority of device applications. In my opinion, attempts to grow GaAs/Ge or similar polar-on-nonpolar {100} heterojunctions or – worse – polar/nonpolar superlattices with this orientation, in the hope that device-quality interfaces will somehow result, are likely to be little more than a waste of time. The fact that this orientation is so successful for III/V-only growth is quite irrelevant. The likely answer – if any – to the quest for successful polar-on-nonpolar growth lies in the use of one of the nonpolar orientations to be discussed presently.

The HKWG argument is by no means restricted to the {100} orientation. Qualitatively similar arguments with only minor quantitative modifications can be made for {111}-oriented interfaces, and in fact for all interface orientations except those in which the interface is parallel to one of the ⟨111⟩ bond direction.

The condition for this can be expressed as a mathematical condition on the Miller indices (*hkl*) of the interface [54]. Let [*hkl*] be the direction perpendicular to the interface plane. The plane is parallel to one of the ⟨111⟩ bond

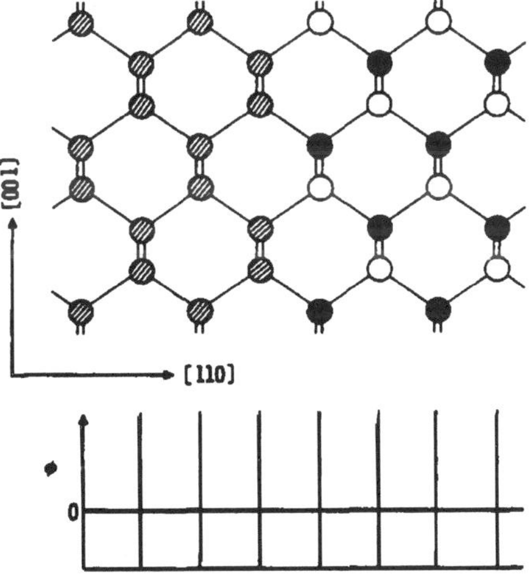

Fig. 12. Atomic arrangement and electrostatic potential at an ideal Ga/GaAs(110) interface. Each GaAs plane parallel to the interface contains an equal number of Ga and As atoms and is hence electrically neutral. From ref. [51].

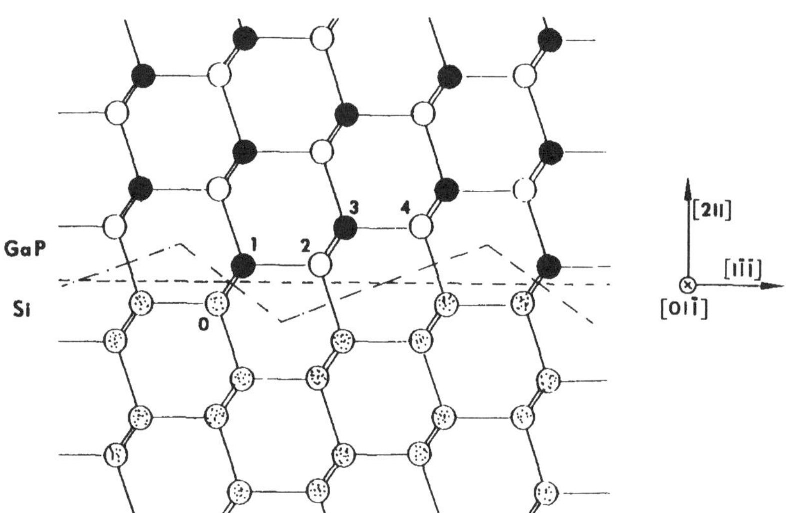

Fig. 13. Atomic arrangement at idealized GaP/Si(211) interface, from ref. [54]. As in the {110} case, each GaP plane parallel to the interface contains an equal number of Ga and P atoms and is hence electrically neutral. But in addition, the bonding of the "black" sublattice sites across the interface is much stronger (two bonds) than that of the "white" sublattice sites (one bond). When GaP is grown on Si, this bonding difference can be utilized to achieve growth free of antiphase disorder, with the "black" sublattice occupied by P atoms, the white by Ga atoms.

directions if $[hkl]$ is perpendicular to that direction. This implies

$$[hkl] \cdot \langle 111 \rangle = \pm h \pm k \pm l = 0,$$

for at least two of the eight possible independent sign combinations. The simplest such orientation is the {110} orientation, already recognized as such and intensively discussed by HKWG. The next-simplest orientation is {112}, followed by {123}, etc. Figs. 12 and 13 show the atomic arrangements at a (110) and at a (112)-oriented polar/nonpolar interface, both viewed again in the $[\bar{1}10]$ direction.

In the absence of specific reasons to do otherwise, it is probably advisable to use the lowest-index orientation for the epitaxial growth. If only the nonpolar-on-polar growth sequence is needed for a particular device, the {110} orientation may indeed be the preferred orientation. Inasmuch as the {110} planes are the natural cleavage planes of III/V compounds, this happily coincides with the natural interest of the surface physicist in this orientation: Most of the non-device studies of the initial growth of Ge on GaAs have indeed used these planes. However, if the polar-on-nonpolar growth sequence is demanded (which automatically induces polar/nonpolar superlattices), altogether new considerations intervene.

5.3. Polar-on-nonpolar growth: the site allocation problem

When, in a polar/nonpolar heterosystem, the polar (compound) semi-conductor is to be grown on the nonpolar (elemental) one, a new problem arises [54,55]: Avoiding antiphase disorder in the growing compound semi-conductor. This problem does not exist at all in element-on-compound growth, and it is at most a minor problem in compound-on-compound growth. But for compound-on-element growth it is as severe and fundamental as the interface neutrality problem at {001} polar/nonpolar interfaces, and it totally dominates the problem of polar-on-nonpolar growth for nonpolar orientations, such as {110} and {112}.

When a binary compound with two different atoms per primitive cell (e.g. GaAs, GaP) is grown on an elementary substrate (e.g. Ge, Si) in which the two atoms are identical, there exists an inherent ambiguity in the nucleation of the compound, with two different possible atomic arrangements, distinguished by an interchange of the two sublattices of the compound. If different portions of the growth exhibit different sublattice ordering, antiphase domains result, separated by antiphase domain boundaries, a defect similar to grain and twin boundaries. For high-performance devices, antiphase domain boundaries must almost certainly be avoided, which calls for a rigorous suppression of one of

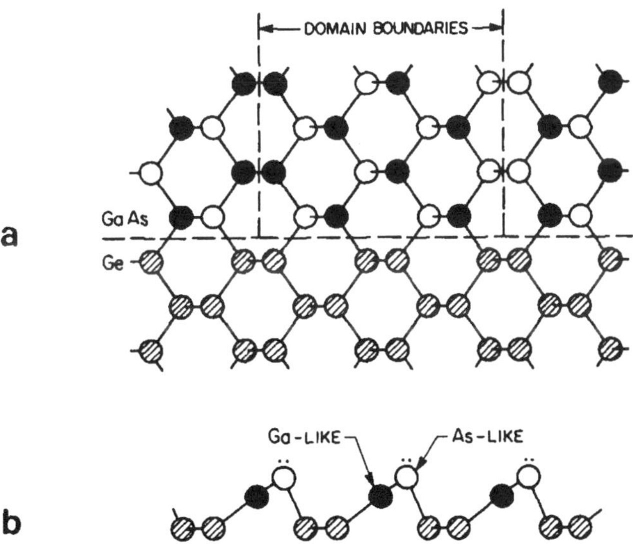

Fig. 14. (a) Occurrence of antiphase domain disorder in the growth of GaAs on an unreconstructed Ge {110} surface, due to the absence of a built-in bonding difference for the as-yet unoccupied surface sites belonging to the two sublattices. (b) Creation of Ga-like and As-like electronic configurations in the top Ge {110} atomic layer, due to reconstruction, aiding in the suppression of antiphase disorder inside the GaAs. From ref. [55].

the two nucleation modes. The problems in doing so depend very strongly on the exact atomic arrangement and on the dangling-bond configuration at the surface of the elemental semiconductor substrate. Unfortunately, they are particularly severe for the simplest nonpolar interface orientation, the {110} orientation. The situation is illustrated in fig. 14a, which shows that on an ideal and perfectly flat (= unreconstructed) Ge {110} surface the sites subsequently to be occupied by Ga and by As atoms have no built-in distinction between themselves. The relative Ga/As ordering at different nucleation sites should therefore be perfectly random, which in turn would lead to a high degree of antiphase domain disorder, with domain sizes of the order of the nucleation site separation, which is usually very small for good epitaxial growth.

The situation on the {112} surface is far more favorable. As fig. 13 shows, the unoccupied sites ahead of an ideal (112) surface are of two quite different kinds: Sites (labelled 1 in fig. 13) with two back bonds to the Si surface, and sites (Nos. 2 and 4) with only one back bond. One easily sees that the two kinds of sites belong to the two different sublattices. Now it is well known that the column-V elements P, As, and Sb, form chemical compounds with Ge and Si, whereas the column-III elements Al, Ga and In do not. One might therefore expect that the strongly-bonding column-V atoms might displace any column-III atoms from the doubly back-bonded sites (No. 1). But once site No. 1 has been occupied by a column-V atoms, site No. 2 becomes more favorable for occupancy by a column-III atom than by a column-V atom. This, in turn, favors occupancy of site No. 3 by another P atom, followed by another Ga atom on site No. 4. Apparently, this is indeed that happens: We have grown GaP on Si {112} by MBE [54], and tests show that the observed sublattice ordering is as described here, with no evidence of antiphase domains. Furthermore, although the electrical properties of these first GaP-on-Si {112} interfaces are still far from ideal, we were able to build bipolar n–p–n transistors with an n-type GaP emitter on a Si p–n base/collector structure, with emitter injection efficiencies up to 90%. This is still far below what would be desirable for practically useful devices (> 99%), but is far better than anything else ever achieved in the very difficult GaP-on-Si system. It raises the hope that device-quality polar-on-nonpolar hetero-interfaces might in fact be achievable.

Our above theoretical speculation was oversimplified in that the reconstruction of the free Ge or Si surface, which is unquestionably present, was ignored. because of the strong bonding difference present already in the unreconstructed {112} surface, any reconstruction on that surface [56] should be little more than a quantiative complication, unless the reconstruction somehow destroys the strong inherent surface site inequivalence, which is extremely unlikely. The situation on the {110} surface is entirely different. Here any reconstruction would *create* a site inequivalence (see fig. 14b), and if this inequivalence is of the right kind, it might convert a hopeless orientation into a promising one. As we have pointed out elsewhere [55], the simplest possible reconstruction, a bond rotation similar to that on GaAs {001}, and postulated by Harrison [57] to occur on Si {110}, is exactly of the most desirable kind. In fact, growth of GaAs on Ge {110} apparently free from antiphase disorder can be achieved under certain growth conditions [55], which unfortunately however do not appear to lead to device-quality electrical properties. The {112} surface, which has a built-in strong site inequivalence, is therefore preferable over the reconstructed {110} surface, which must rely on a tenuous surface reconstruction to achieve site selection. Our experimental experience [54] strongly confirms this expectation. We therefore consider our own former advocacy [55] of the reconstructed {110} surface as having been superseded by the subsequent realization of the inherently greater promise of the {112} orientation.

5.4. Small misorientations: nuisance or design parameter?

There is no such thing as a perfectly-oriented crystallographic interface. Any real interface will have deviations from perfect flatness and perfect orientation, as a result of which the $\langle 111 \rangle$ bonds are rotated out of the true hetero-interface plane by a small but non-zero angle θ. At apolar/nonpolar interface this will cause a finite built-in interface charge to appear, and even for small misorientations the resulting charge may be large by device standards. For the {112} interface, the charge density is easily shown to be

$$\sigma = \left(q\sqrt{3} / a^2 \right) \sin \theta.$$

If the tilt angle is small enough, this charge is not likely to be removed by the HKWG atomic re-arrangement, but is likely to act as a permanent *tilt doping*. A wafer orientation to within $\pm 0.5°$ ($\cong 10$ milliradian) is roughly the practical limit of current *routine* wafer orientation techniques. Assuming the lattice constant of GaAs, such a misorientation corresponds to an interface charge density of 4.7×10^{12} elementary charges per cm^2. This is a large charge, and much more accurate wafer orientation techniques than are in current use will be necessary. This is of course possible, but is a major nuisance. A highly (112)-selective etch would certainly help. However, one man's nuisance is often the next man's design parameter. *If* the orientation could be controlled to significantly better than 10^{-3} radian, a deliberate misorientation might become a practical means of introducing desirable interface charges into devices such as HEMT's. Because the interface charges would not be randomly distributed, but be located on quasiperiodic interface steps, they would scatter less, and even new superlattice effects might arise. Finally, by deliberately creating a controlled local variation in the interface tilt, one might even introduce lateral "doping" variations into device structures. It is a fitting notion on which to close a paper that addresses itself to the role of interfaces in submicron structures, more specifically, to the role of the interface nanostructure in determining the properties of devices containing those interfaces.

Acknowledgments

It is a pleasure to thank Dr. R.S. Bauer for inviting me to present this paper at this Symposium, and thereby providing the stimulus to order my thoughts on the topics discussed and to put them down on paper, something that otherwise would have been unlikely to occur. Many thanks are due to Drs. E.A. Kraut, J.R. Waldrop, R.W. Grant, D.L. Miller and S.P. Kowalczyk, for uncounted discussions. Last, but not least, I wish to acknowledge the profound influence that Professor W.A. Harrison has had on my thinking.

References

[1] See, for example, H. Kroemer, Japan. J. Appl. Phys. 20, Suppl. 20-1 (1981) 39.

[2] H. Kroemer, Proc. IEEE 70 (1982) 13.

[3] W. Shockley, US Patent 2,569,347, issued 25 Sept. 1951.

[4] R.L. Anderson, Solid-State Electron. 5 (1962) 341.

[5] For a general review, see A.G. Milnes and D.L. Feucht, Heterojunctions and Metal–Semiconductor Junctions (Academic Press, New York, 1972).

[6] An excellent recent review is contained in chs. 4 and 5 of H.C. Casey and M.B. Panish, Heterostructure Lasers (Academic Press, New York, 1978).

[7] H. Kroemer, Proc. IEEE 51 (1963) 1782.

[8] R. Dingle, in: Festkörperprobleme/Advances in Solid State Physics, Vol. 15, Ed. H.J. Queisser (Vieweg, Braunschweig, 1975) p. 21.

[9] R. Dingle, H.L. Störmer, A.C. Gossard and W. Wiegmann, Appl. Phys. Letters 33 (1978) 665.

[10] For a recent review, see T. Mimura, Surface Sci. 113 (1982) 454.

[11] For a review, see N. Holonyak, R.M. Kolbas, R.D. Dupuis, and D.D. Dapkus, IEEE J. Quantum Electron. 16 (1980) 170.

[12] H. Sakaki, L.l. Chang, R. Ludeke, C.-A. Chang, G.A. Sai-Halasz and L. Esaki, Appl. Phys. Letters 31 (1977) 211;
see also L.L. Chang and L. Esaki, Surface Sci. 98 (1980) 70.

[13] H. Kroemer, W.-Y. Chien, H.C. Casey and A.Y. Cho, Appl. Phys. Letters 33 (1978) 749.

[14] H. Kroemer, W.-Y. Chien, J.S. Harris, Jr. and D.D. Edwall, Appl. Phys. Letters 36 (1980) 295.

[15] Y.Z. Liu, R.J. Anderson, R.A. Milano and M.J. Cohen, Appl. Phys. Letters 40 (1982) 967.

[16] See, for example, J.R. Waldrop, S.P. Kowalczyk, R.W. Grant, E.A. Kraut and D.L. Miller, J. Vacuum Sci. Technol. 19 (1981) 573.

[17] G.F. Williams, F. Capasso and W.T. Tsang, IEEE Electron Devices Letters 3 (1982) 71;
see also F. Capasso, Surface Sci. 132 (1983) 527.

[18] H. Kroemer, Critical Rev. Solid State Sci. 5 (1975) 555.

[19] W.A. Harrison, J. Vacuum Sci. Technol. 14 (1977) 1016;
see also ref. [24] below.

[20] G. Margaritondo, A.D. Katnani, N.G. Stoffel, R.R. Daniel and T.-X. Zhao, Solid State Commun. 43 (1982) 163;
see also G. Margaritondo, Surface Sci. 132 (1983) 469.

[21] J.L. Shay, S. Wagner and J.C. Phillips, Appl. Phys. Letters 28 (1976) 31.

[22] J.C. Phillips, J. Vacuum Sci. Technol. 19 (1981) 545.

[23] W.R. Frensley and H. Kroemer, Phys. Rev. B16 (1977) 2642.

[24] W.A. Harrison, Electronic Structure and the Properties of Solids: The Physics of the Chemical Bond (Freeman, San Francisco, 1980); see especially section 10F.

H. Kroemer / Heterostructure devices   575

[25] J.O. McCaldin, T.C. McGill and C.A. Mead, Phys. Rev. Letters 36 (1976) 56. These authors expressed the correlation between valence band lineup and anion electronegativity for Schottky barriers; the approximate applicability of their result to heterojunctions appears to have been discussed first by W.R. Frensley and H. Kroemer, J. Vacuum Sci. Technol. 13 (1976) 810; see also ref. [23].

[26] For a very "physical" discussion of this theoretical foundation, see Harrison, ref. [24], especially chs. 1–3 and ch. 6.

[27] S.J. Anderson, F. Scholl and J.S. Harris, in: Proc. 6th Intern. Symp. on GaAs and Related Compounds, Edinburgh, 1976, Inst. Phys. Conf. Ser. 33b (Inst. Phys., London and Bristol, 1977) p. 346.

[28] The numerical values are based on Harrison's table 10-1 on p. 253 of ref. [24], except that we use the values from ref. [27] for the energy gaps of GaSb and AlSb.

[29] J.A. Van Vechten, Phys. Rev. 87 (1969) 1007. Van Vechten gives an extensive table of theoretical ionization energies, from which electron affinities are easily obtained by subtracting the energy gaps.

[30] This broken-gap lineup is, in fact, predicted by all three major predictive theories: The electron affinity rule, the Frensley–Kroemer theory, and the Harrison theory.

[31] G.C. Osbourn, J. Appl. Phys. Letters 53 (1982) 1536; J. Vacuum Sci. Technol. 21 (1982) 469; see also ref. [34] below.

[32] G.H. Döhler, Phys. Status Solidi (b) 52 (1972) 79,553;
G.H. Döhler, H. Künzel and K. Ploog, Phys. Rev. B25 (1982) 2365.

[33] Our calculation is to illustrate the basic idea only. The quoted composition falls into a solid solubility gap of uncertain width the existence of which has been reported. It may therefore be difficult or impossible to prepare. For a discussion and further references on this point see ch. 5 of ref. [6].

[34] P.L. Gourley and R.M. Biefeld, J. Vacuum Sci. Technol. 21 (1982) 473;
G.C. Osbourn, R.M. Biefeld and P.L. Gourley, Appl. Phys. Letters 41 (1982) 172.

[35] M.E. Davis, G. Zeidenbergs and R.L. Anderson, Phys. Status Solidi 34 (1969) 385.

[36] G.M. Guichar, C.A. Sébenne and C.D. Thuault, Surface Sci. 86 (1979) 789.

[37] R.S. Bauer and H.W. Sang, Jr., Surface Sci. 132 (1983) 479.

[38] H. Morkoc, L.C. Witkowski, T.J. Drummond, C.M. Stanchak, A.Y. Cho and J.E. Greene, Electron. Letters 17 (1981) 126;
see also H.L. Störmer, Surface Sci. 132 (1983) 519.

[39] It has been suggested by W.I. Wang (personal communication) that the ⟨110⟩ sequence dependence might be related to an as yet unexplained instability of ⟨110⟩-oriented (Al, Ga)As growth observed by him. For another report of a different kind of ⟨110⟩ growth instability see P. Petroff, A.Y. Cho, F.K. Reinhart, A.C. Gossard and W. Wiegmann, Phys. Rev. Letters 48 (1982) 190.

[40] R.C. Miller, W.T. Tsang and O. Munteanu, Appl. Phys. Letters 41 (1982) 374.

[41] E.A. Kraut, R.W. Grant, J.R. Waldrop and S.P. Kowalczyk, Phys. Rev. Letters 44 (1980) 1620.

[42] H. Kroemer and W.-Y. Chien, Solid-State Electron. 24 (1981) 655.

[43] H.K. Gummel and D.L. Scharfetter, J. Appl. Phys. 38 (1967) 2148;
see also C. Kittel and H. Kroemer, Thermal Physics, 2nd ed. (Freeman, San Francisco, 1980) ch. 13. For very unsymmetrically doped junctions, the GS correction is between 1 kT/q and 2 kT/q.

[44] H. Kroemer, RCA Rev. 17 (1956) 515.

[45] M. Weinstein, R.O. Bell and A.A. Menna, J. Electrochem. Soc. 111 (1964) 674.

[46] Zh.I. Alferov, V.I. Korolkov and M.K. Trukan, Soviet Phys.-Solid State 8 (1967) 2813.

[47] G. Zeidenbergs and R.L. Anderson, Solid-State Electron. 10 (1967) 113.

[48] N.N. Gerasimenko, L.V. Lezheiko, E.V. Lyubopytova, L.V. Sharanova, A.Ya. Shik and V. Shmartsev, Soviet Phys.-Semicond. 15 (1981) 626.

576   *H. Kroemer / Heterostructure devices*

[49] S.L. Wright, PhD Thesis, University of California, Santa Barbara, CA (1982).
[50] See, for example, C.M. Garner, C.Y. Su, Y.D. Shen, C.S. Lee, G.L. Pearson, W.E. Spicer, D.D. Edwall, D. Miller and J.S. Harris, Jr., J. Appl. Phys. 50 (1979) 3383; see also the references quoted there.
[51] W.A. Harrison, E.A. Kraut, J.R. Waldrop and R.W. Grant, Phys. Rev. B18 (1978) 4402.
[52] W.A. Harrison, in: Festkörperprobleme/Advances in Solid State Physics, Vol. 17, Ed. H.J. Queisser (Vieweg, Braunschweig, 1977) p. 135.
[53] W.A. Harrison, J. Vacuum Sci. Technol. 46 (1979) 1492.
[54] S.L. Wright, M. Inada and H. Kroemer, J. Vacuum Sci. Technol. 21 (1982) 534.
[55] H. Kroemer, K.J. Polasko and S.L. Wright, Appl. Phys. Letters 36 (1980) 763.
[56] R. Kaplan, Surface Sci. 116 (1982) 104.
[57] W.A. Harrison, Surface Sci. 55 (1976) 1.

Reprinted with permission from

H. Kroemer, ``Barrier Control and Measurements: Abrupt Semiconductor Heterojunctions,'' J. Vac. Sci. Technol. B, Vol. 2(3), pp. 433-439, 1984.

Barrier control and measurements: Abrupt semiconductor heterojunctions

Herbert Kroemer

Department of Electrical & Computer Engineering, University of California, Santa Barbara, California 93106

(Received 13 February 1984; accepted 18 March 1984)

A brief critical review is given of diverse techniques used to measure heterojunction band lineups; they range from very reliable to worthless. Another problem pertains to the heterosystems themselves: Data on systems in which two semiconductors from a different pair of columns of the periodic table are combined, should be reviewed with suspicion, although some selected pairs are probably trustworthy—but none in which a compound semiconductor was grown on an elemental one. Technologies that do not lead to device-quality interfaces also probably do not yield device-quality lineup data. A list of the most trustworthy experimental data is given. The simplest possible theoretical framework for a theory of band lineups is a model of linear superpositon of atomiclike bulk potentials. Such a model automatically leads to a theory that is linear and transitive, in which the band lineups are orientation independent, and in which a technology dependence of the band lineups requires a technology-dependent deviation of the atomic arrangement from the ideal one. The Harrison theory is both the simplest and the most successful theory of band lineups, although it still does not meet the needs of the device physicist. The set of most reliable data selected earlier agree very well with this theory, with a largest deviation of 0.18 eV and a standard deviation of 0.13 eV.

PACS numbers: 73.40.Lq, 73.30. + y, 68.48. + f

I. INTRODUCTION

From a device physics point of view the most important aspect of a semiconductor heterointerface, and the point of departure for all subsequent considerations, is the lineup of the bands at the interface. These band lineups may vary over a wide range, from the most common *straddling* lineup of Fig. 1(a) via the less common *staggered* lineup of Fig. 1(b), to the rare *broken-gap* lineup in Fig. 1(c).

The purpose of the present paper is to review the present status of our knowledge of those band lineups for the common diamond- and zinc-blende-type semiconductors, from both the experimental and the theoretical point of view. The paper draws heavily on two more extensive recent papers,[1,2] to which frequent reference will be made. The first of these[1] contains an extensive critique of various experimental methods that have been employed to determine band offsets, and of some of the ways in which nuisance effects such as spurious interface charges might falsify the apparent offsets. It also contains a discussion of technological problems that can make the (apparent or real) band offsets poorly reproducible, especially in mixed-column heterosystems like Ge/GaAs. The second paper[2] selects from the large amount of experimental data those that are most likely to be correct, followed by a detailed review of various lineup theories. The selected experimental data are compared with the theoretical predictions, especially those of the Harrison atomic orbital (HAO) theory.[3,4]

II. EXPERIMENTAL BAND LINEUPS

A. The problem

To assess the validity of any theory of band lineups, it is necessary to compare its predictions with band offsets that are already known experimentally with a degree of reliability sufficient to permit a meaningful test. Although the literature contains a very large number of lineup data for many different semiconductor pairs, few can be considered reliable enough to permit a meaningful test of lineup theories.[1] For example, for the widely studied Ge/GaAs system, conduction band offsets ranging from 0.09 to 0.54 eV have been claimed in the literature, a range corresponding to 68% of the energy gap of Ge. Many of those values *must* be wrong, and this makes all data suspect.

Ignoring ordinary measurements inaccuracies, one can identify four problem areas.

1. Indirect measurement techniques

Many techniques that have been employed determine the band offsets only very indirectly, by projecting the results of whatever measurement is employed, upon a preconceived model of the heterojunction. If the model is not valid, the resulting offset values may be invalid, too. In particular,

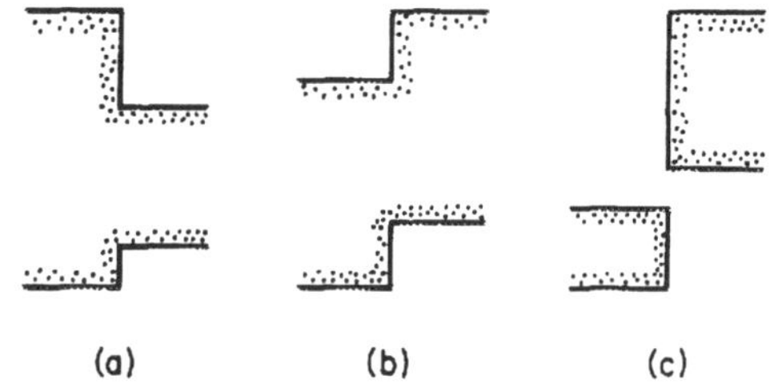

FIG. 1. Band discontinuities at abrupt semiconductor heterojunctions. Three different possible band lineups are shown: (a) "straddling" lineup, (b) "staggered" lineup, and (c) "broken-gap" lineup. From Ref. 2.

small residual interface charges tend to distort grossly the results of some measurement techniques. A critical assessment of various measurement techniques was given in Ref. 1 the results of which may be summarized as follows.

Probably the most reliable data are those obtained from sufficiently carefully performed UPS or XPS photoemission experiments[5] on very thin heterojunctions, *provided* the heterojuncion itself was prepared by a technology yielding high structural perfection. A close second to UPS/XPS measurements are optical *absorption* (not emission!) measurements on multiquantum well structures.[6] Capacitance–voltage (C–V) measurements on heterojunctions may or may not be reliable, depending on the exact nature of the measurement.[1] In fact, C–V profiling *through* an isotype heterojunction from an adjacent Schottky barrier[7] is potentially one of the most reliable techniques. Least reliable are I–V measurements; many—but not all—are essentially worthless.[1]

This assessment is strikingly different from the situation with Schottky barriers,[8] where I–V and C–V measurements are among the most reliable (and most widely used) techniques to determine Schottky barrier band lineups, ranking along with UPS/XPS techniques. The principal reason for this difference is the following: Heterointerfaces often contain non-negligible interface charges, which can grossly deform the band diagrams of the entire heterojunction and hence change most electrical properties of the structure.[1] At a Schottky barrier, such charges are located right at the metal surface, where they are unable to deform the bands far away from the interface. On the other hand, two of the best heterojunction techniques, Dingle's quantum well absorption technique,[6] and C–V profiling *through* an isotype heterointerface,[7] are fundamentally unusable to determine the band lineup at Schottky barriers.

2. Technology problems

The band offset data appear to depend somewhat on technological details of how the heterojunction is prepared. This dependence *must* reflect technology-dependent differences in the exact atomic arrangements near the interface. The exact nature of these differences (and their origin) is at present not understood. But it is clear that one cannot fully trust data that were taken on structures prepared under conditions significantly different from those employed for the high-quality device structures whose lineups are the real object of the theory.

3. Chemically induced interface dipoles

Many heterojunctions that have been studied involve two semiconductors from different columns or column pairs of the periodic table, such as Ge/GaAs, Ge/ZnSe, GaAs/ZnSe, InP/CdS, and many others. In all such systems, any interchanges of atoms across the interface will introduce atomic dipole moments that change the band offsets. Such atom interchange effects can, in general, not be prevented. In fact, it has been shown[9] that, for most crystallographic orientations, atom interchanges across the interface are *necessary* to prevent the accumulation of a huge interface-destabilizing net interface charge. The final result will be an interface with both a residual interface dipole and a residual

interface charge, the magnitudes of which depend sensitively on technology. These effects can be minimized by working with the electrically neutral (110) cleavage planes of the compounds, and by growing the junction at a low temperature. But the latter is only a compromise, because the low-temperature growth tends to lead to poor bulk properties not representative of a device-quality semiconductor.

4. Antiphase disorder

In heterojunctions between one of the column-IV elements and a III/V or a II/VI compound, severe antiphase disorder is likely to occur when the compound semiconductor is grown on the elemental semiconductor substrate, rather than in the opposite order.[1,2] There probably does not exist a heterosystem more ill suited to a test of band lineup theories than GaAs grown on (001)-oriented Ge; yet this combination has been one of the most widely studied—with predictably irreproducible results. In fact, recent work by Neave et al.[10] has demonstrated that these structures do indeed suffer from heavy antiphase disorder. In my judgement, such compound-on-element systems should be expected *not* to satisfy any simple lineup theory; only systems in which the element was grown upon the compound should be considered for testing such theories.

B. Reference systems for theory testing

When all these problems are taken into consideration, only two heterosystems remain that can truly serve as standards of comparison for lineup theories: The $Al_x Ga_{1-x}As/$ GaAs system and the InAs/GaSb system.

For the first of these, both superlattice absorption data[6] and XPS data[11] for (100)-oriented abrupt heterojunctions show that the valence band offset is 15% $\pm$ 3% of the direct energy gap at $k = 0$, both for x in the range 0.2–0.3, and for $x \cong 1$. If one assumes a linear relation with x, the data can be described by

$$\Delta\epsilon_v [Al_x Ga_{1-x} As \rightarrow GaAs] = (0.19 \pm 0.04)x \text{ eV}, \quad (1)$$

where we have adopted the convention that $\Delta\epsilon_v [A \rightarrow B]$ shall be positive if the band edge step is an upward step in going from A to B. Recent C–V profiling data[12] have confirmed the earlier band offset data.

For InAs/GaSb, various data[13] show beyond any doubt that this system is of the broken-gap variety [Fig. 1(c)], with a break in the gap of about 150 $\pm$ 50 meV. Combined with the 300 K energy gap of InAs (0.36 eV) this yields

$$\Delta\epsilon_v [InAs \rightarrow GaSb] = (0.51 \pm 0.05) \text{ eV}. \quad (2)$$

The very unusual nature of this broken-gap lineup makes this system a severe test of any theory of heterojunction lineups.

Compared to the (Al,Ga)As/GaAs and InAs/GaSb data, all other lineup data suffer from one uncertainty or another. The most likely to be reliable are the XPS/UPS data for InAs on GaAs[14] and for Ge on Si,[15] for which the following lineups have been reported:

$$\Delta\epsilon_v [InAs \rightarrow GaAs] = -0.17 \text{ eV}, \quad (3)$$

$$\Delta\epsilon_v [Ge \rightarrow Si] = -0.2 \text{ eV}. \quad (4)$$

The trouble with both systems is that they are badly lattice mismatched (7% and 4%). One must expect that the exact lineups depend on how exactly this mismatch is accommodated at the interface; hence they might be technology dependent. The Ge-on-Si value is subject to the additional criticism that the data were obtained on samples in which the Ge was grown at an unrepresentatively low temperature. As a result of these reservations, it is not clear how exactly the band offsets for both systems *should* agree with any theory that has made idealizing assumptions (even if implicitly) about the atomic structure of the interface and of the crystal adjacent to it.

There exist large numbers of lineup data on heterojunctions in which the two semiconductors come from different columns of the periodic table. As was mentioned earlier, and discussed extensively in Refs. 1 and 2, all such systems are prone to exhibit technology-dependent interface charges and interface dipoles. These effects depend very strongly on the crystallographic orientation of the interface: The two least-suspect orientations are the (110) and (112) orientations.[1,2,9,16,17] The widely used (001) and (111) orientations are highly nonideal for such systems, no matter how ideal they may be for III/V-only heterojunctions.

Of all mixed-column lineup data in the literature the ones I consider least likely to suffer from complications are the XPS data of Kowalczyk *et al.*[18] for heterojunctions of ZnSe grown on GaAs (110) at 300 °C (not their 23 °C growth data):

$$\Delta\epsilon_v[\text{ZnSe on GaAs(110)}] = 0.96 \pm 0.03 \text{ eV}. \quad (5)$$

Finally, there exist numerous data in which elemental Si or Ge was grown on a compound semiconductor—not the other way around—which because of their growth sequence are not subject to the exclusion on the grounds of likely antiphase disorder, discussed earlier. Probably the least-suspect data on these element-on-compound systems are the XPS data, again of Kowalczyk *et al.*,[18] for Ge grown on ZnSe(110):

$$\Delta\epsilon_v[\text{Ge on ZnSe(110)}] = -(1.52 \pm 0.03)\text{eV}. \quad (6)$$

It is only with considerable reluctance that I include among the reference systems what is one of the most widely studied heterosystem, Ge on GaAs. The lineup data on this system scatter so widely[19] that it appears difficult to decide which of the data are least unreliable, and the strong chemical interaction of Ge with As makes the system prone to chemical interface reactions.[20] However, recent data on MBE-grown Ge-on-GaAs(110) heterojunctions have tended to converge towards what appears to be the most carefully determined value, that of the Rockwell group.[21]

$$\Delta\epsilon_v[\text{Ge on GaAs(110)}] = -(0.53 \pm 0.03)\text{eV}, \quad (7)$$

obtained again by XPS, on junctions grown at 425 °C.

C. Anion correlation rule

There is strong independent evidence that in all systems such as (Al,Ga)As/GaAs and InAs/GaAs, in which the anion atom species (As) on both sides of the heterojunction is the same, the valence band offsets *should* be much smaller than the conduction band offsets.[22] This *common anion rule* arises from the theoretically well-established fact[23] that the valence band wave functions derive largely from the anion

atomic wave function. Taken together with the fact that the valence band wave functions tend to be more localized than the conduction band wave functions, this yields valence band energies that correlate strongly with the anion species. For semiconductor pairs with a common anion, the valence band offsets should therefore always be small compared to the conduction band offsets.

For semiconductor pairs with the common cation X, the valence band energies at the interface should correlate with the different anion electronegativities. For the III/V compounds this implies

$$\epsilon_v(\text{XP}) < \epsilon_v(\text{XAs}) < \epsilon_v(\text{XSb}). \quad (8)$$

If the correlation is strong enough, one might expect Eq. (8) to persist even if both anion and cation are different.

D. Linearity and transitivity

Many theories assume that there is a specific *absolute* energy associated with the various band edges of every individual semiconductor, and that the band offsets are simply the differences between the respective absolute band energies of the two semiconductors. With our sign convention

$$\Delta\epsilon_v[\text{A}\rightarrow\text{B}] = \epsilon_v(\text{B}) - \epsilon_v(\text{A}). \quad (9)$$

Such theories may be called *linear* theories. It is possible to test experimentally whether or not the experimental data can possibly satisfy a linear theory, without actually invoking a specific theory: Given the band lineups of two different semiconductor pairs A/B and B/C, having one semiconductor (B) in common, the lineup for the third possible pair A/C should follow by simple addition. With our sign convention,

$$\Delta\epsilon_v[\text{A}\rightarrow\text{B}] + \Delta\epsilon_v[\text{B}\rightarrow\text{C}] + \Delta\epsilon_v[\text{C}\rightarrow\text{A}] = 0. \quad (10)$$

This property has been referred to as *transitivity* by Frensley and Kroemer.[24] This prediction is independent of whatever theory one might wish to invoke for the band lineups, so long as it is a theory of this linear class. Because transitivity should be a common feature of entire classes of theories, tests for transitivity are useful tests of great generality, which do not involve fitting experimental data to any particular lineup theory, but which may rule out entire classes of theories if unsuccessful. Among our reference systems, the three pairs Ge on ZnSe, ZnSe on GaAs, and Ge on GaAs, form a nearly lattice-matched triplet suitable for testing transitivity under difficult conditions. When the lineups given earlier are combined, one finds the closure sum

$$\Delta\epsilon_v[\text{Ge on ZnSe}] + \Delta\epsilon_v[\text{ZnSe on GaAs}]$$
$$\quad - \Delta\epsilon_v[\text{Ge on GaAs}]$$
$$= -1.52 \text{ eV} + 0.96 \text{ eV} + 0.53 \text{ eV} = -0.03 \text{ eV} \quad (11)$$

an extremely small remainder, below the accuracy of the measurements themselves. Based on this limited evidence it would appear that transitivity is indeed an excellent assumption for well-prepared heterojunctions.

III. THEORY

A. The problem

The problem of the band offsets at an abrupt heterojunction can be divided into two independent parts[25]: (i) The problem of the lineup of the energy bands in each semiconductor relative to the periodic potential in the same semiconductor, and (ii) the problem of the alignment of the two periodic potentials relative to each other. The first problem is not itself a heterojunction problem but one in bulk band theory. The central problem of band lineup theory is that of the lineup of the two periodic potentials relative to each other.

Given the two periodic potentials within the two individual bulk semiconductors, we may always view each as a linear superposition of overlapping atomiclike potentials. Near the atomic nuclei, the atomiclike potentials resemble the potentials inside the free atoms. But in the regions between the atoms, especially within the interstices in the diamond and zinc blende structure, these potentials will be different from free-atom potentials; hence, our designation *atomiclike* potentials. For any given periodic potential, such atomiclike potentials can always be defined, and the bulk band structure may be viewed as being known relative to the atomiclike potentials in the crystals.

The simplest possible atomic theory of band lineups is obtained if the potential *throughout the entire structure* is approximated as a superposition of *unmodified* overlapping atomiclike potentials (Fig. 2). In the immediate vicinity of the interface itself, the potential would contain contributions from atoms on both sides of the interface, but with each atomiclike potential still being the same as deep inside the bulk of the particular semiconductor. In such a model, the relative lineup of the two bulk potentials is well defined. The band lineups are then also well defined, and the only problems are those of calculational technique.

In such a model there is no place for any crystallographic orientation dependence or technology dependence of the band lineups. All those must be due to deviations from the model.

In the vicinity of the interface one must expect charge and potential readjustments relative to the predictions of a simple linear superposition of atomiclike bulk charges and potentials. The charges will readjust in response to various forces, such as image forces, quantum-mechanical exchange forces, tunneling, etc. The overall result of these readjust-

TABLE I(A)–I(C). Band edge energies in eV of various semiconductors relative to the top of the valence band of GaAs, from the Harrison atomic orbital (HAO) theory (Refs. 3 and 4). The bottom entry in each box represents the valence band edge, the top the conduction band edge. The designations X, Δ, or L preceding the conduction band value indicates an indirect gap; the direct gap is then given in parentheses. From Ref. 2.

(A) III/V compounds

	P	As	Sb
Al	X: 1.95 (3.1) − 0.50	X: 2.17 (2.52) − 0.04	X: 2.44 (3.08) + 0.86
Ga	X: 1.79 (2.31) − 0.47	+ 1.42 0.00	+ 1.57 + 0.84
In	+ 1.24 − 0.11	+ 0.68 + 0.32	+ 1.29 + 1.12

(B) II/VI compounds

	S	Se	Te
Zn	+ 1.93 − 1.87	+ 1.82 − 1.05	+ 2.42 + 0.03
Cd	+ 0.97 − 1.59	+ 1.02 − 0.82	+ 1.81 + 0.21

(C) Si and Ge

Si	Ge
Δ: + 1.15 (4.21) + 0.03	L: + 1.08 (1.22) + 0.41

ments is an *electronic* interface readjustment dipole, which shifts the bands relative to one another, and relative to the linear superposition model.

In addition to these electronic dipoles we must expect *atomic* (or ionic) interface dipoles to occur if, for whatever reason, atoms from one semiconductor cross over the ideal interface into the other semiconductor. At least in mixed-column systems this can produce large net dipoles.[1,2]

These dipoles—of either origin—do not appear to be large (usually at most a few tenths of 1 eV). But they are the major bottleneck in the use of lineup theories for accurate predictions on the level of accuracy desired for device applications. For example, for the GaAs/(AlGa)As heterosystem, variations of band lineups with technology and with crystallographic orientation have been reported.[11] Such variations are inherently outside the possibility of the simple linear superposition model, and are therefore necessarily consequences of dipole shifts, presumably atomic ones.[2]

B. The Harrison atomic orbital (HAO) theory of band lineups

If one ignores theories that are in hopeless disagreement with experimentally observed band lineups of the conceptually and technologically simplest heterojunctions, between two nearly lattice-matched III/V compounds, the simplest remaining theory is Harrison's atomic orbital (HAO) theory.[3,4] It also appears to give the best agreement with experiment, a combination that makes it the standard of comparison against which all others must be measured.

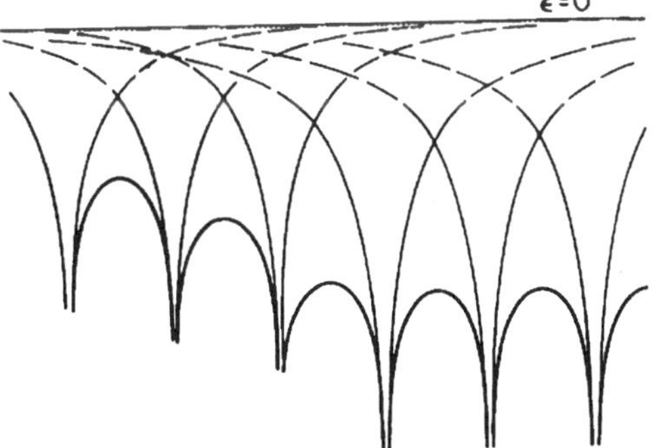

FIG. 2. Simple model of the potential energy within a few atoms of a heterointerface, as a linear superposition of overlapping atomiclike potentials. Within each semiconductor the individual atomiclike potential atomic species is the same for all atoms of that species. Near the interface the atomic potentials from the two sides overlap. From Ref. 2.

The HAO theory is conceptually a member of that class of theories that are based on a linear superposition of atomiclike potentials, unmodified by any dipole corrections at the interface. In fact, Harrison goes one step further by using neutral-free-atom potentials, and he expresses all band energies relative to the potential at infinity of the neutral free atoms. The fact that the actual *atomiclike* potentials inside the crystal are quite different in the space between the atoms, is accounted for only partially and quite indirectly, by adjusting the overlap matrix elements involving the wave functions on adjacent atoms, in such a way that the empirical energy band structures come out as accurately as is possible with this scheme. As we shall see, this works surprisingly well, but it would be an exaggeration to claim that we really understand *why* it works as well as it does. For more on this point the reader is directed to Ref. 2.

Because the valence band structures of the zinc-blende-type semiconductors are much simpler, and fit theoretical models much more accurately than the conduction band structures, Harrison expresses the band lineups in terms of the valence band offsets. The conduction band offsets are then obtained indirectly, from the predicted valence band offsets, by adding the difference between the accurately known *experimental* energy gaps.

We express here the results of the Harrison theory as a set of three simple tables, Tables I(A)–I(C), one each for the III/V compounds, the II/VI compounds, and the elemental semiconductors. For convenience, we have reexpressed all energies relative to the top of the valence band of GaAs rather than using Harrison's pseudovacuum level. The columns in Tables I(A) and I(B) represent equal anions, the rows equal cations. The bottom entry in each box is the energy of the valence band edge, the top entry the conduction band edge, obtained by adding the experimental 300 K energy gaps. For the III/V compounds the gaps were taken from the excellent compilation in Casey and Panish's book,[26] which are believed to be more accurate than the values used by Harrison. In those cases where the conduction band edge is not at $k = 0$, the notation X, Δ, or L has been added to indicate the location of the band edge, and the value at the Γ point has been given in parentheses.

These tables represent the best theoretical estimates that can currently be given for heterojunction band lineups.

Figure 3 shows a comparison of the experimental data for those systems selected above as having the most reliable experimental data. The fit speaks for itself. Table II gives the actual data. If one weighs all data points equally, one finds a mean error of only -0.016 eV, with a standard deviation of 0.13 eV. Also included are the data for the CdS-on-InP system, which has been invoked by Shay *et al.*[27] as a reference system, and which fits the HAO theory well, but which I have deliberately not included among the reference system, for reasons discussed in Ref. 2.

There can be no doubt that the HAO theory looks like it might serve as a useful predictive tool, better than any other we have available, although still not on the level of accuracy demanded by the device physicist. Yet this unexpectedly good fit by no means ends the search for a better understanding. To the contrary: having achieved something that ap-

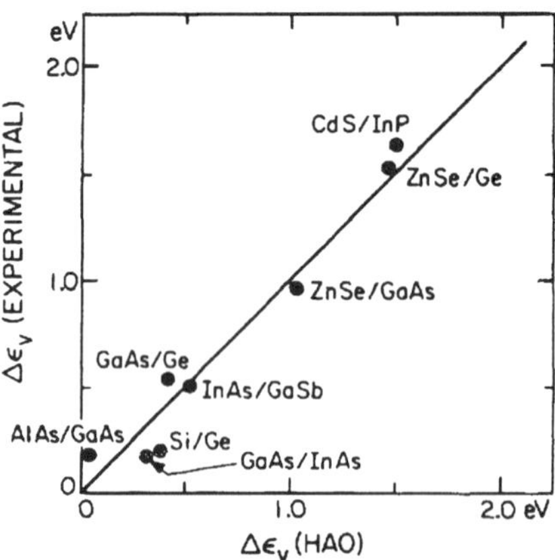

FIG. 3. The *experimental* valence band offsets in eV, for the seven reference systems selected in Sec. II C, plotted as a function of the *theoretical* valence band offsets predicted by the Harrison theory. The value for CdS/InP is also shown. The AlAs/GaAs value is extrapolated from $Al_xGa_{1-x}As/$ GaAs for $x \approx 0.3$. A numerical tabulation, with references, is given in Table II. From Ref. 2.

pears to be "approximately right" only doubles the incentive to answer the question[2] "why is it right—or is it?"

It should perhaps be specifically pointed out that the good fit contained in Fig. 3 and in Table II was not obtained by deliberately selecting from the experimental data those that would fit the HAO theory. The data were selected strictly on the basis of the sample preparation technique and the measurement technique employed, independent of whether the data did or did not fit the HAO theory or any other theory. In fact, the comparison with various theories was deliberately avoided until after the selection had been made.

I believe that the use of selected data is a more meaningful way to test the applicability of various theories than the use of unselected data, which tends to blame on the theories what are really faults of the experiment, and which tends to give an excessively pessimistic picture of *all* theories. Worse, it hurts the better theories proportionally more, by obscuring their superiority. For example, the recent comparison by Katnani *et al.*[28] obscures the very real differences between

TABLE II. Comparison of the valence band discontinuities in eV predicted by the Harrison atomic orbital (HAO) theory for selected reference systems, with experimental values. The values for CdS/InP are also given. The AlAs/GaAs value is the value extrapolated from $Al_xGa_{1-x}As/GaAs$ value with $x \approx 0.3$.

Heterojunction	HAO	Experimental	Ref.	Error
AlAs/GaAs	0.04	0.19	6, 11, 12	+ 0.15
InAs/GaSb	0.52	0.51	13	− 0.01
GaAs/InAs	0.32	0.17	14	− 0.15
Si/Ge	0.38	0.20	15	− 0.18
ZnSe/GaAs	1.05	0.96	18	− 0.10
ZnSe/Ge	1.46	1.52	18	+ 0.06
GaAs/Ge	0.41	0.53	5	+ 0.12
CdS/InP	1.48	1.63	27	+ 0.15

the quality of fit of the data to the HAO theory and to its competitors. Conversely, the electron affinity rule (see below), although definitely inferior to HAO, is really not as bad as the unselected-data comparison by Bauer *et al.*[29] makes it appear. In fact, I do not believe that the accuracy of the currently available data justifies the pessimistic rejection of *all* linear theories by those authors.

C. Alternate theoretical approaches

1. The Frensley–Kroemer pseudopotential (FKP) theory

The HAO theory was not the first attempt to predict heterojunction band offsets automatically, from the periodic potentials actually present inside the semiconductors, from the energetic position of the bands within those periodic potentials, and from the alignment of the periodic potentials relative to one another. It was preceded by the Frensley–Kroemer pseudopotential (FKP) theory,[24] which differed from HAO principally in calculational methodology. Although quite successful for some heterojunction pairs, especially the tricky broken-gap pair InAs/GaSb, its success was not as consistent as that of HAO, and the required calculations were more complicated. Never in widespread use, it has *de facto* been superseded by HAO. The interested reader is referred to Ref. 2, where a critical review and comparison with HAO is given.

2. The electron affinity rule (EAR)

Until the emergence of HAO the most widely used tool for the prediction of heterojunction band lineups was the Electron Affinity Rule (EAR),[26,30–32] which asserts that the conduction band offset is equal to the difference between the electron affinities χ_1 and χ_2 of the two semiconductors, with signs such that

$$\Delta\epsilon_c [1\rightarrow2] = \epsilon_{c2} - \epsilon_{c1} = -(\chi_2 - \chi_1). \tag{12}$$

In fact, the EAR continues to be widely used, although this continued use is probably due largely to the fact that the HAO theory and its demonstrably better agreement with existing experimental data are still not widely known. A critical review of the EAR and a comparison with HAO has recently been given in Ref. 2; it reassesses specifically the theoretical foundations of the EAR and the criticism of those foundations made earlier by the present writer,[25] as well as the defense of the EAR raised by others in return.[27,33]

3. Self-consistent interface potential theories

All the theories discussed so far achieve their successes—whatever it may be—by ignoring the interface itself, including any redistribution of charge at the interface, and with it any electronic interface dipole. To the extent that the theories, especially the HAO theory, are able to account for the experimentally observed band lineups, this disregard may be *empirically* justified, but conceptually it hardly represents a satisfactory state of affairs.

One possible approach towards the problem of heterojunction band lineups consists of a self-consistent quantum-mechanical treatment of the interface region itself. Following the pioneering work of Baraff *et al.*,[34] several efforts have

been made to determine heterojunction band lineups in this fashion, especially by Cohen and his co-workers.[35–38] Unfortunately, these calculations have so far not fulfilled the hope that they would provide a reliable predictive tool for accurate values of the band lineups, free of the questionable simplifications of the simpler theories, such as HAO, FKP, and EAR: Except for the (Al,Ga)As/GaAs heterojunction, the band lineups predicted by these self-consistent interface potential calculations have been in much poorer agreement than HAO with reliable observations. The discrepancy is especially strong for ZnSe/GaAs(110), where Ihm and Cohen[38] predict a staggered lineup quite incompatible with the experimental data. For quantitative details see, once again, Ref. 2.

Considering the outstanding success of pseudopotential theories in simpler problems, the reasons for this unsatisfactory state of affairs are probably not fundamental, but are purely matters of calculational accuracy. Presumably, this *could* be improved, at least to the point of *retrodictively* understanding the already observed band lineups. Whether such theories are likely to become accurate *predictive* tools remains doubtful, for reasons whose discussion goes beyond the scope of this article.[2]

[1] H. Kroemer, Surf. Sci. **132**, 543 (1983).

[2] H. Kroemer, in *Proceedings of the NATO Advanced Study Institute on Molecular Beam Epitaxy and Heterostructures, Erice, Sicily, 1983*, edited by L. L. Chang and K. Ploog (Martinus Nijhoff, The Netherlands, 1984).

[3] W. A. Harrison, J. Vac. Sci. Technol. **14**, 1016 (1977).

[4] W. A. Harrison, *Electronic Structure and the Properties of Solids: The Physics of the Chemical Bond* (Freeman, San Francisco, 1980), Sec. 10F.

[5] For an excellent review, see E. A. Kraut, R. W. Grant, J. R. Waldrop, and S. P. Kowalczyk, Phys. Rev. Lett. **44**, 1620 (1980).

[6] R. Dingle, in *Festkörperprobleme/Advances in Solid State Physics*, edited by H. J. Queisser (Vieweg, Braunschweig, 1975), Vol. 15, p. 21.

[7] H. Kroemer, W. -Y. Chien, J. S. Harris, and D. D. Edwall, Appl. Phys. Lett. **40**, 967 (1980).

[8] J. W. Waldrop, J. Vac. Sci. Technol. B (these proceedings).

[9] W. A. Harrison, E. A. Kraut, J. R. Waldrop, and R. W. Grant, Phys. Rev. B **18**, 4402 (1978). See also Ref. a.

[10] J. H. Neave, P. K. Larsen, B. A. Joyce, J. P. Gowers, and J. F. van der Veen, J. Vac. Sci. Technol. B **1**, 668 (1983).

[11] J. R. Waldrop, S. P. Kowalczyk, R. W. Grant, E. A. Kraut, and D. L. Miller, J. Vac. Sci. Technol. **19**, 573 (1981).

[12] R. People, K. W. Knecht, K. Alavi, and A. Y. Cho, Appl. Phys. Lett. **43**, 118 (1983).

[13] J. Sakaki, L. L. Chang, R. Ludeke, C. -A. Chang, G. A. Sai-Halasz, and L. Esaki, Appl. Phys. Lett. **31**, 211 (1977). See also L. L. Chang and L. Esaki, Surf. Sci. **98**, 70 (1980).

[14] S. P. Kowalczyk, W. J. Schaeffer, E. A. Kraut, and R. W. Grant, J. Vac. Sci. Technol. **20**, 705 (1982).

[15] G. Margaritondo, A. D. Katnani, N. G. Stoffel, R. R. Daniels, and T. X. Zhao, Solid State Commun. **43**, 163 (1982).

[16] H. Kroemer, K. J. Polasko, and S. L. Wright, Appl. Phys. Lett. **36**, 763 (1980).

[17] S. L. Wright, M. Inada, and H. Kroemer, J. Vac. Sci. Technol. **21**, 534 (1982).

[18] S. P. Kowalczyk, E. A. Kraut, J. R. Waldrop, and R. W. Grant, J. Vac. Sci. Technol. **21**, 482 (1982).

[19] See, for example, W. Mönch, R. S. Bauer, H. Gant, and R. Murschall, J. Vac. Sci, Technol. **21**, 498 (1982), and the references given there.

[20] See, for example, R. S. Bauer and J. C. Mikkelsen, J. Vac. Sci. Technol. **21**, 491 (1982) which contains extensive references to earlier work.

[21] R. W. Grant, J. R. Waldrop, and E. A. Kraut, Phys. Rev. Lett. **40**, 656

(1978); J. Vac. Sci. Technol. **15**, 1451 (1978).

[22]J. O. McCaldin, T. C. McGill, and C. A. Mead, Phys. Rev. Lett. **36**, 56 (1976).

[23]For a very "physical" discussion see Chaps. 1–3 and Chap. 6 of Harrison, Ref. 4.

[24]W. R. Frensley and H. Kroemer, Phys. Rev. B **16**, 2642 (1977). See also, J. Vac. Sci. Technol. **13**, 810 (1976).

[25]H. Kroemer, CRC Crit. Rev. Solid State Sci. **5**, 555 (1975).

[26]H. C. Casey, Jr. and M. B. Panish, *Heterostructure Lasers* (Academic, New York, 1978).

[27]J. L. Shay, S. Wagner, and J. C. Phillips, Appl. Phys. Lett. **28**, 31 (1976); see also S. Wagner, J. L. Shay, K. J. Bachmann, and E. Buehler, *ibid.* **26**, 229 (1975).

[28]A. D. Katnani and G. Margaritondo, Phys. Rev. B **28**, 1944 (1983).

[29]R. S. Bauer, P. Zurcher, and H. W. Sang, Jr., Appl. Phys. Lett. **43**, 663 (1983). The apparently good fit of the AlAs/GaAs lineup with the electron affinity rule, shown in Fig. 2 of that paper, is fictitious: It is based on an electron affinity literature value for AlAs that was calculated backwards from the lineup. To my knowledge, the true electron affinity of AlAs has never been measured independently.

[30]R. L. Anderson, Solid State Electron. **5**, 341 (1962).

[31]B. L. Sharma and R. K. Purohit, *Semiconductor Heterojunctions* (Pergamon, London, 1974), Chap. 2.

[32]A. G. Milnes and D. L. Feucht, *Heterojunctions and Metal–Semiconductor Junctions* (Academic, New York, 1972).

[33]J. C. Phillips, J. Vac. Sci. Technol. **19**, 545 (1981).

[34]G. A. Baraff, J. A. Applebaum, and D. R. Hamann, Phys. Rev. Lett. **38**, 237 (1977); J. Vac. Sci. Technol. **14**, 999 (1977).

[35]For an excellent review, see M. L. Cohen, Adv. Electron. Electron Phys. **51**, 1 (1980). See also J. Pollmann and A. Mazur, Thin Solid Films **104**, 257 (1983).

[36]W. E. Pickett, S. G. Louie, and M. L. Cohen, Phys. Rev. B **17**, 815 (1978).

[37]W. E. Pickett and M. L. Cohen, Phys. Rev. B **18**, 939 (1978).

[38]J. Ihm and M. L. Cohen, Phys. Rev. B **20**, 729 (1979).

Reprinted from

H. Kroemer and H. Okamoto, ``Some Design Considerations for Multi-Quantum-Well Lasers,'' Japan. J. Appl. Phys., Vol. 23, pp. 970-974, 1984.

JAPANESE JOURNAL OF APPLIED PHYSICS
VOL. 23, NO. 8, AUGUST, 1984 pp. 970-974

Some Design Considerations for Multi-Quantum-Well Lasers

Herbert KROEMER* and Hiroshi OKAMOTO

*Musashino Electrical Communication Laboratory, Nippon Telegraph and Telephone Public Corporation,
Musashino-shi, Tokyo 180*

(Received February 16, 1984; accepted for publication May 26, 1984)

This paper addresses itself to two different problems in the design of multi-quantum-well (MQW) lasers: (a) The problem of inter-well coupling by electron tunneling is investigated, and it is shown that the tunneling can be made sufficiently strong to permit efficient electron equilibration between wells, without destroying the advantages of the step-like distribution of states of a single well or of fully decoupled wells. (b) A modification of the quantum well array itself is proposed, in which additional narrower quantum wells are added outside the lasing well array itself. These subsidiary wells contain only a negligible concentration of electrons and hence do not participate in the laser action, but they should improve the quality of the epitaxial growth, and assist in both the electron capture and the optical confinement.

§1. Introduction

Quantum well lasers have a staircase-like distribution of states that is more favorable for laser action than the parabolic distribution in conventional double heterostructure (DH) lasers.[1,2] This advantage is fully retained in multi-quantum-well (MQW) lasers *if* the barriers between the wells are sufficiently thick and high so as to decouple the wells, with negligible tunneling between wells. But in this case the electron flow necessary to equilibrate the electron population amongst the wells is greatly impeded, because it must proceed by thermal emission over the barriers. As a result the electron population will be different in different wells, which means that some wells will not exhibit population inversion and hence gain, and the remaining wells will have to be pumped harder to achieve laser action. The overall result is a higher threshold current than would otherwise be necessary. This case has been studied by Dutta,[3] who has determined the very restrictive conditions that must be satisfied to achieve efficient electron equilibration amongst the wells in the absence of tunneling. If the barriers are made sufficiently thin that at least some tunneling occurs, the electron equilibration is greatly facilitated. But if the barriers are made too thin, the benefits of the staircase-like distribution of states are lost. One of the purposes of the present paper is to analyze the compromise between a favorable distribution of states on the one hand, and ease of electron flow by tunneling on the other, if the barriers are made somewhat penetrable: It is shown in §2 that an excellent compromise can be readily achieved, even if thermal emission over the barriers is neglected altogether.

A second objective of the present paper is to introduce refinements into the design of the quantum well array itself. In the design introduced here, the structure contains, in addition to the "normal" electron-confining lasing quantum wells, two additional kinds of narrower quantum wells: (a) "Padding wells" of about one-half the width of the lasing wells, separated from the lasing

well array by thin barriers of about one-half the width of the barriers inside the array. (b) Very narrow ($\cong 1$–2 nm) non-binding cavity surface wells, at the interfaces between the optical cavity and the external confinement regions that define the optical cavity separately from the lasing well array itself.

These subsidiary quantum wells do not themselves contribute to the laser action, but provide other benefits that facilitate the achievement of laser action in the lasing wells: (a) The wells on the substrate side improve the quality of the epitaxial growth; (b) the padding wells facilitate the collection of electrons into the lasing wells, and (c) all additional wells collectively improve the optical confinement somewhat. These benefits, discussed in detail in §3, come at negligible technological cost; they simply require additional shutter operations in the MBE system, of the same kind as those necessary for incorporating the lasing wells anyway.

§2. Well Coupling in Multi-Quantum-Well Structures

2.1 *Simple two-well structure*

2.1.1 *Distribution of states*

To understand the role of inter-well coupling, it is useful first to consider the case of a simple symmetrical two-well system (Fig. 1(a)); more complicated systems can be understood by simple generalization of this limiting case.

The two lowest bound states of such a system (assuming there are at least two bound states) are an even-parity state (energy E_e) for which the wave function has a minimum at the center of the barrier, followed by an odd-parity state (energy $E_o > E_e$) for which it has a null (Fig. 1(b)). The effect of these two states on the two-dimensional distribution of states is a double step, with two step edges separated by a small energy $\Delta E = E_e - E_o$ (Fig. 2) which, for weak inter-well coupling, remains small compared to the energies E_e and E_o themselves. Such a distribution will retain the advantages of the staircase-like distribution of states of a single well, but at a doubling of the number of states available for the laser action, *if* the step separation ΔE remains sufficiently smaller than the thermal energy kT. The condition

$$\Delta E < kT \cdot \ln 2 \tag{1}$$

---

*On leave from Department of Electrical and Computer Engineering, University of California, Santa Barbara, California 93106, USA.

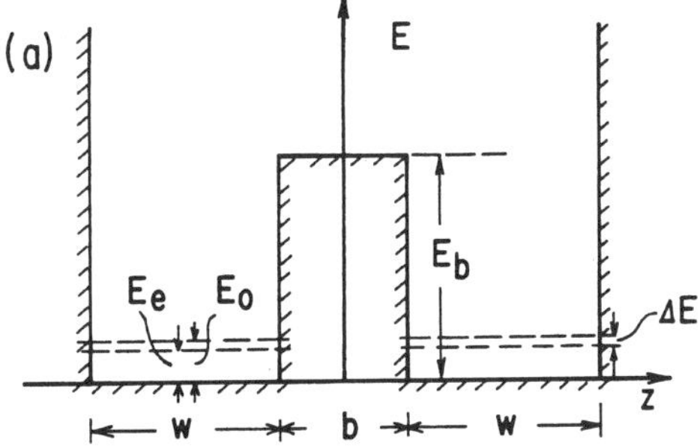

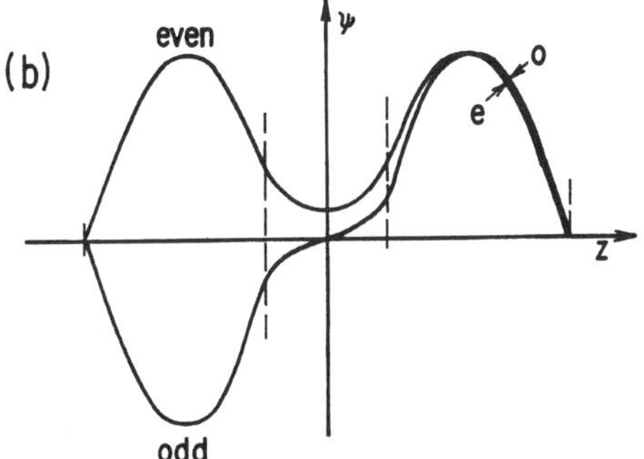

Fig. 1. (a) Symmetrical square well with infinitely high side walls, as simplest model for inter-well coupling in multi-quantum-well lasers. (b) Wave function of the two lowest-energy bound states, an even-parity state and an odd-parity state. The corresponding energies are E_e and E_o, with $E_e < E_o$.

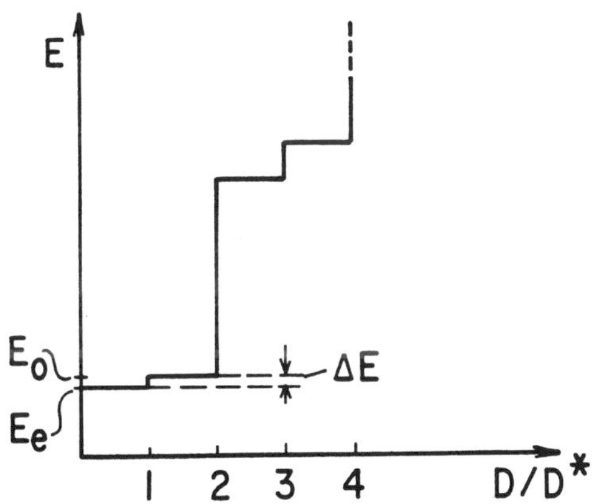

Fig. 2. Double-step distribution of density of states for a weakly-coupled double quantum well.

is a *sufficient* condition that the product (density of states) × (occupation probability) is larger at the energy E_o than at the energy E_e. If (1) is satisfied by a sufficiently wide margin, there is nothing to be gained by reducing ΔE further. If the electron quasi Fermi level approaches or exceeds E_e, the condition (1) may be relaxed.

We estimate ΔE here for the simple limiting case of the double well shown in Fig. 1(a), containing a barrier of width b and height E_b separating two wells of width

w, terminated by infinitely high walls. We use the effective mass approximation with conventional connection rules, that is, we assume that the logarithmic derivative ψ'/ψ of the effective mass wave function is continuous at the interface. Zhu and Kroemer[4] have recently criticized the conventional connection rules; they have shown that in general a delta function potential must be added at the interface. However, for the currently most important system, $Al_xGa_{1-x}As/GaAs$, the delta function correction is probably unimportant,[4] and we have neglected it, along with the effective mass difference between the two semiconductors. The calculations can easily be generalized to include corrections such as these.

We define three wave number k, K and G, according to

$$E = \frac{\hbar^2 k^2}{2m^*}, \quad E_b - E = \frac{\hbar^2 K^2}{2m^*}, \quad E_b = \frac{\hbar^2 G^2}{2m^*}. \quad \text{(2a,b,c)}$$

Note that G is energy-independent, and that k and K are interrelated according to

$$K^2 = G^2 - k^2. \quad (3)$$

One finds easily that the conventional wave function connection rules at $x = \pm b/2$ may be written in the form

$$-\cot kw = \frac{Kw}{kw} \frac{1 \mp \exp(-Kb)}{1 \pm \exp(-Kb)}, \quad (4)$$

where on the right-hand side the upper signs in ($\mp$) and ($\pm$) refer to the even-parity state, the lower signs to the odd-parity state. Because K in (4) depends on k through (3), eq. (4) may be viewed as a transcendental equation for kw, which cannot be solved in closed form. It is easily solved numerically by first inserting, on the right-hand side, a trial value for kw in the range $\pi/2 < kw < \pi$, together with the associated value of K. The left-hand side then yields a new value for kw in that range, which is re-inserted on the right-hand side. The process is repeated until convergence has been obtained, which tends to occur within a few iteration cycles. As an example, consider the following parameters: $E_b = 0.2$ eV, $m^* = 0.07\ m_e$, $w = 10$ nm, and $b = 3$ nm, corresponding to $Gw = 6.062$. The numerical iteration converges rapidly to $k_e w = 2.512$ and $k_o w = 2.819$ for the even- and odd-parity states, corresponding to $E_e = 0.0344$ eV and $E_o = 0.0433$ eV, implying a level splitting $\Delta E = 0.0089$ eV, well within the limit imposed by (1). The associated values of $\exp(-Kb)$ are 0.1911 and 0.1999, indicating that the attenuation of the wave function inside the barrier is appreciable, but far from complete.

Inasmuch as the energy splitting is caused by the $\exp(-Kb)$ terms in (4), it is clear that in the limit of weak coupling the energy splitting will scale linearly with this parameter. It is not difficult to show that in this limit, to the first order

$$\Delta E \cong \frac{8 E_\infty K_\infty^2}{G^2 (1 + K_\infty w)} \exp(-K_\infty b), \quad (5)$$

where E_∞ and K_∞ are the values of E and K in the limit $b \to \infty$. If the coupling is not negligibly small, this first-order approximation tends to underestimate the energy splitting somewhat. For our above example, in which coupling is far from negligible, eq. (5) yields 0.0078 eV, about 12 % *below* the exact iterated value.

The relation (5) still requires a determination of E_∞ and K_∞ by numerical iteration. If the coupling is sufficiently weak, one may go one step further and replace K_∞ by G, and k in the calculation of E_∞ by π/w, yielding

$$\Delta E = \frac{4\hbar^2\pi^2}{m^*w^2}\frac{\exp(-Gb)}{1+Gw}. \tag{6}$$

For our example, $\Delta E = 0.0099$ eV, about 11 % *above* the exact iterated value, slightly *more* accurate than the value obtained from (5). Although the error involved in going from (5) to (6) is quite large, it has the opposite sign of that involved in obtaining the approximation (5) in the first place, and accidental error cancellation leads to final results of similar accuracy.

2.1.2 Tunneling rate

The time necessary for tunneling through the barrier is easily estimated as follows. If an electron is initially in one of the two wells, its wave function will be a linear superposition of the even- and odd-parity state, with approximately equal amplitudes. Each state is associated with a time-dependent pure phase factor of the form $\exp(-iEt/\hbar)$. Because the two states have slightly different energies, their phase factors advance at slightly different rates, and the overall linear superposition wave function corresponds to an electron that oscillates back and forth between the two wells with a frequency

$$f = \Delta E/h. \tag{7}$$

If at $t=0$ the electron was in one of the two wells, it will be found in the other well after one-half period of this oscillation, that is, after the time

$$t_{\mathrm{T}} = h/2\Delta E, \tag{8}$$

which may be viewed as the time required for an electron to tunnel through the barrier. For $\Delta E = 10$ meV, roughly the upper limit of the desirable inter-well coupling energies, one finds $t_{\mathrm{T}} = 0.2$ ps, a tunneling time sufficiently short to permit efficient equilibration of the electrons between the wells.

In order for a net current to flow between two wells, the well occupancies must differ. If ΔN is the difference in the areal density of occupancy (i.e. the number of electrons per unit area), an electric current density

$$J = q\,\Delta N/t_{\mathrm{T}} \tag{9}$$

will flow between the two wells. We are interested in current densities of the order 1000 A·cm^{-2} or less; to achieve such a current density for $t_{\mathrm{T}} = 0.2$ ps requires an occupation density difference $\Delta N \cong 1.25\times10^9$ cm^{-2}. Such a value is only a very small fraction of the total electron density required to achieve laser action in the first place: The two-dimensional density of states for a single well is

$$D^* = \frac{m^*}{\pi\hbar^2}. \tag{10}$$

This is also the density of states per well for a double well above the energy E_0. If the electron quasi Fermi level coincided with the energy E_0, the occupation density per well would be

$$N = D^*\cdot kT\cdot\ln 2. \tag{11}$$

If we assume $m^* = 0.07\,m_{\mathrm{e}}$ and $T = 300$ K, we find $N = 5.24\times10^{11}$ cm^{-2}, roughly 400-times the density difference required to carry a tunneling current of 1000 A·cm^{-2}. Evidently, a population difference of a fraction of 1 percent is sufficient to carry the desired current.

2.2 Generalization to multiple quantum wells

When $n>2$ wells are present, but the wells remain weakly coupled, most of the above results to a simple double well with infinitely high end walls remain valid. Each of the states of a single well will then split up into a closely-spaced cluster of n steps. The distribution of states then becomes a distribution with n sub-steps being clustered together within a narrow energy interval ΔE. If the width of this interval remains sufficiently small compared to both its separation from the higher states, and to kT, this structure will again behave roughly as if the sub-step energies coincided, and it will thus retain the advantages of the staircase-like distribution of states of a single quantum well.

In the limit of weak inter-well coupling, the overall width of the cluster of n states will be governed by the nearest-neighbor coupling of two adjacent wells only, independently of the total number of wells. Furthermore, so long as the step height of the terminal potential walls is not much less than that of the barriers separating the wells, the nonzero penetration of the wave function into the end walls will have only a small effect on the energy level distribution. As a result, the energies of the two states E_e and E_0 of a simple double well remain a good approximation to the lower and upper boundary of the band of n states of an n-well laser structure, if the inter-well coupling in the latter remains weak.

2.3 Valence band wells, and doping considerations

We have concentrated exclusively on the conduction band wells, ignoring the valence band wells. In those semiconductor heterosystems that are of largest current interest for MQW lasers, especially in the (Al, Ga)As and (Al, Ga)Sb systems, the conduction band wells are much deeper than the valence band wells, hence they dominate the design. In particular, the shallow valence band wells mean that hole transport between these wells can easily take place by thermal emission over the hole barrier; tunneling is not necessary for efficient equilibration of the hole concentration: The bottleneck in the carrier equilibration process is the electron equilibration.

The hole equilibration process is further facilitated by the fact that the density-of-states considerations, which recommend designs with a higher hole than electron concentration in conventional DH lasers,[5] and hence strong p-type doping, also apply to MQW lasers.

We wish to point out here that there is an additional reason for achieving a high hole conductivity throughout the entire quantum well array, through deliberate p-type doping of the quantum well region, namely, to achieve efficient *electron* equilibration by tunneling: Any significant electrostatic potential differences between adjacent electron wells would cut off the tunneling current, similarly to the current decrease with increasing voltage in an Esaki tunnel diode. Ideally, the end-to-end electrostatic potential difference throughout the quantum well array should be kept small compared to the energy difference ΔE calculated above. A high hole conductivity throughout the entire quantum well array will assist in ahieving this goal. Under no circumstances should the transition from p-type to n-type take place within that array.

§3. Subsidiary Quantum Wells

3.1 *Padding wells*

There exists strong experimental evidence showing that the structural and electronic perfection of GaAs/(Al, Ga)As interfaces is much poorer if the GaAs is grown on the (Al, Ga)As than for (Al, Ga)As–on–GaAs growth.[6,7] The effect appears to be due to the accumulation of a background impurity,[7] probably carbon, which degrades the morphology of the growing (Al, Ga)As surface, thus leading to an atomically rough and probably defective interface. (For the purposes of the present paper, the exact mechanism of this deterioration is of only secondary importance.) The degree of interface deterioration depnds on the thickness of the last (Al, Ga)As layer grown preceding the interface. In superlattice structures that have been grown on a relatively thick initial layer of (Al, Ga)As, the first interface tends to be a poor interface, leading to poor luminescence properties of the first quantum well.[6,7] The growth morphology recovers quickly if a thin layer ($\cong 5$ nm) of straight GaAs is subsequently grown, and the inner GaAs–on–(Al, Ga)As interfaces of a true GaAs/(Al, Ga)As supperlattices, having been grown on comparatively thin (Al, Ga)As barrier layers following straight GaAs layers, tend to have a degree of perfection similar to that of (Al, Ga)As–on–GaAs interfaces.[6,8]

The above considerations suggest that it should be useful to precede the array of lasing wells in a quantum well laser by a "padding well" (Fig. 3(a)) sufficiently wide to permit recovery of the surface morphology, but sufficiently narrow to keep the electron wave function at the lasing energy level very small inside that padding well. This goal can be accomplished by a padding well approximately half as wide as the lasing wells, separated from the first lasing well by a barrier half as wide as the barriers separating the lasing wells. This is readily seen by considering the special case of the lowest-energy state of a superlattice containing an infinite number of wells (Fig. 3(b)). If the superlattice were suitably terminated, the lowest-energy wave function would be an even function about the centers of the internal potential barriers, with a minimum at the center of each barrier. If this wave function is now extended through the half-width padding well barrier, it will have zero slope at the outer barrier surface, with a small amplitude, equal to the amplitude at the center of the internal barriers. The amplitude will further decrease inside the padding well itself, and because the width of the padding well is approximately one-quarter wavelength, the wave function will be near one of its nulls at the outer edge of the well. For a certain height of the outside wall (slightly less than the barrier height between wells), the wave function will then decay exponentially to zero. In general, the height of the

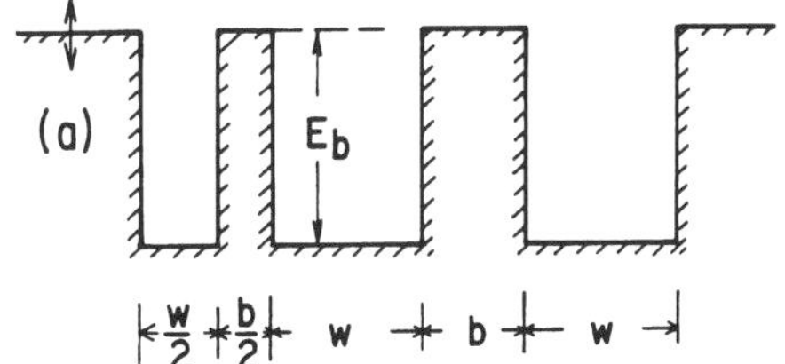

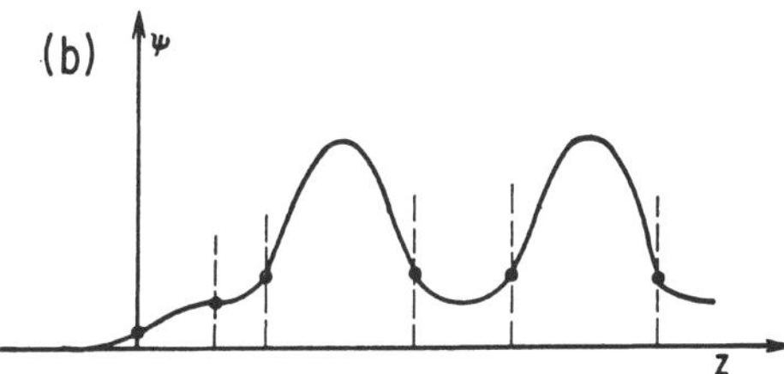

Fig. 3. (a) Padding well of width/w/2, separated from the lasing array by a barrier of width b/2. The barrier height outside the padding well need not be the same as that of the inner barriers. (b) Wave function for lowest-energy state of a multi-quantum-well array terminated by a padding well. The wave function amplitude inside the padding well is less than inside the internal barriers.

outside wall will be larger than this specific value, but the effect of this tends to be negligible: The true lowest-energy state of the terminated superlattice will be very close in energy to that for this special case, with very nearly the same wave function. In fact, all states of the lowest miniband of the superlattice will have a similarly low amplitude inside the padding well.

The padding well, being half as wide as the lasing wells, has its lowest-energy binding resonance at an energy very close to that of the second miniband of the lasing array. This is desirable, because it enables electrons captured by the padding well to tunnel into the lasing array. Because the barrier is thinner, and the lasing probability increases steeply with increasing energy, this tunneling is even more rapid than the already-rapid tunneling between the lasing wells themselves. This not only prevents any accumulation of electrons in the padding wells, it also assists somewhat in the capturing of electrons by the lasing array, especially for a small number of lasing wells.

An additional minor benefit of the padding well is that it contributes to an increase of the average refractive index inside the optical cavity of the laser structure, and hence to an improvement in the optical confinement.

The contributions of the padding well to both the electron capture and the optical confinement suggest that the use of padding wells at *both* ends of the lasing array should be beneficial, even though the initial reason for introducing such wells, the improvement in epitaxial growth, applies only on the substrate side.

3.2 *Cavity surface wells*

The padding well concept can be extended in such a way that it may be applied to the interface between the

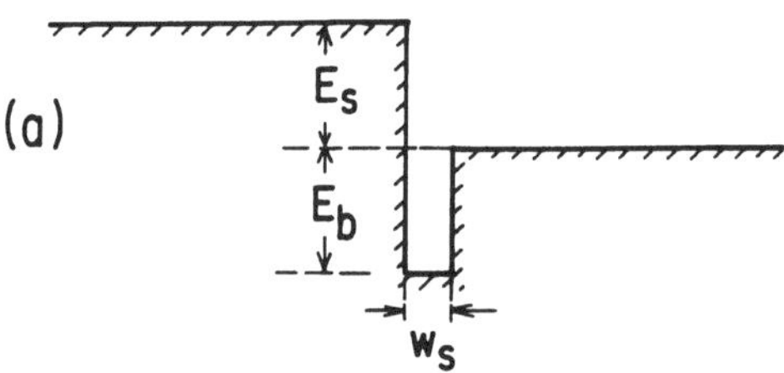

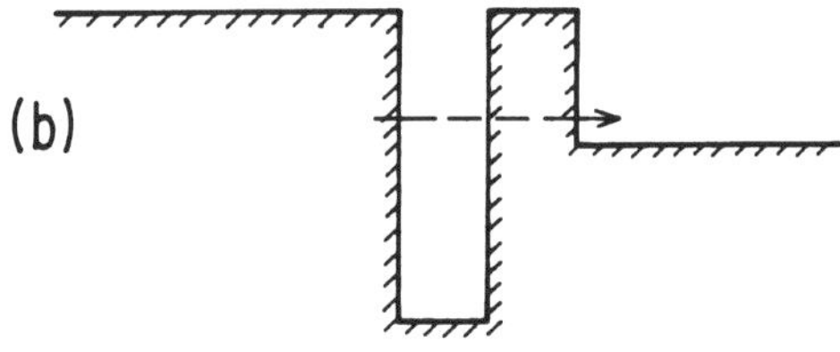

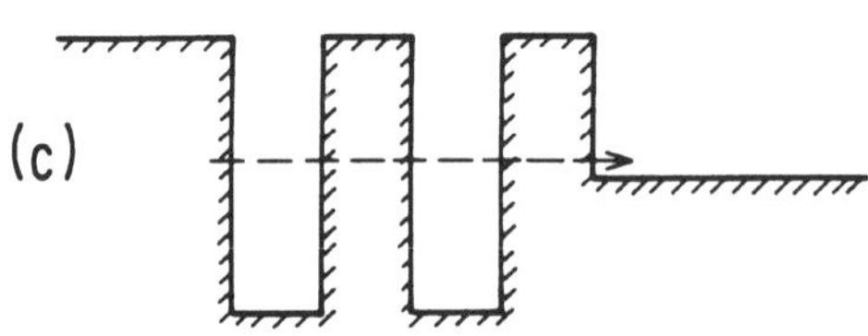

Fig. 4. (a) Simplest form of cavity surface well at the interface between the optical cavity region (lower gap, to the right) and the external optical confinement region (higher gap, to the left). The width of the well is chosen such that the lowest bound state has just been pushed out of the well. (b) Modified cavity surface well. The addition of a barrier on the cavity side pushes the energy of all states up, thus permitting a wider cavity before a truly bound state will occur. There will, however, now be a quasi-bound state above E_o, whose lifetime must be kept short by making the barrier sufficiently thin to permit tunneling through it. (c) Possible further modification into a multi-quantum-well structure.

optical cavity and the external optical confinement layers. It is known from elementary quantum mechanics that an *un*symmetrical quantum well need not have any bound states at all. For the simplest case of a square well with barrier heights E_b and $E_b + E_s$ (Fig. 4(a)), a simple calculations shows that there will be no bound states if the well width satisfies

$$w_s < \hbar \cdot (2m^* E_b)^{-1/2} \cdot \arctan \left[(E_s/E_b)^{1/2} \right]. \quad (12)$$

This calculation assumes equal effective masses in all three regions, and conventional connection rules at the interfaces, simplifications that are easily removed.

Such a non-binding surface well would not be able to capture electrons (whcich could not readily tunnel out of the well if the lasing well array is far removed from the edge of the optical cavity), but it would provide benefits similar to those of the inner padding wells that terminate the lasing well array. Quantitatively, the benefits of a configuration as simple as in Fig. 4(a) tend to be limited, however, because the condition (12) tends to require quite narrow wells. For example, assuming $E_s = E_b = 0.2$ eV and $m^* = 0.07\, m_e$, one obtains $w_s < 1.3$ nm, far narrower than the inner padding wells.

The width of the surface wells may be increased somewhat, to almost twice the limit (12), by going to a slightly more advanced configuration like that in Fig. 4(b), with quasi-bound states for energies above E_b. The confining barrier must then be kept sufficiently thin, so that any electrons captured will quickly leak out by tunneling.

Further refinements may be made by going to multiple surface wells, as shown for a two-well system in Fig. 4(c). The number of such wells is limited only by the need to retain a sufficiently rapid rate of tunneling from this array into the optical cavity. A quantitative analysis of these designs has not been carried out.

Acknowledgements

The research described in this paper was performed during a sabbatical visit of one of the authors (H. K.) to the Musashino ECL of NTT, and he wishes to express his deep appreciation to Dr. N. Kuroyanagi for his hospitality. The research itself grew out of discussions with Dr. Y. Horikoshi and Dr. H. Sugiura, and it benefitted greatly from discussions with them and others at ECL.

References

1) J. P. van der Ziel, R. Dingle, R. C. Miller, W. Wiegmann and W. A. Nordland, Jr.: Appl. Phys. Lett. **26** (1975) 463.
2) For a review see N. Holonyak, R. M. Kolbas, R. D. Dupuis and D. D. Dapkus: IEEE J. Quantum Electron. **16** (1980) 170.
3) N. K. Dutta: IEEE J. Quantum Electron. **19** (1983) 794.
4) Q.-G. Zhu and H. Kroemer: Phys. Rev. B27 (1983) 3519.
5) See, for example, H. C. Casey and M. B. Panish: *Heterostructure Lasers* (Academic Press, New York, 1978).
6) A. C. Gossard, W. Wiegmann, R. C. Miller, P. Petroff and W. T. Tsang: Collected Papers 2nd Int. Symp. Molecular Beam Epitaxy & Related Clean Surface Techniques, Japan Soc. of Appl. Phys. (1982) p. 39.
7) R. C. Miller, W. T. Tsang and O. Munteanu: Appl. Phys. Lett. **38** (1982) 372.
8) See, for example, T. J. Drummond, J. Klem, D. Arnold, R. Fischer, R. E. Thorne, W. G. Lyons and H. Morkoc: Appl. Phys. Lett. **42** (1983) 615.

Reprinted with permission from

**E. J. Caine, S. Subbanna, H. Kroemer, J. L. Merz, and A. Y. Cho,
``Staggered-Lineup Heterojunctions as Sources of Tunable
Below-Gap Radiation: Experimental Verification,'' Appl. Phys. Lett.,
Vol. 45(10), pp. 1123-1125, 1984.**

Staggered-lineup heterojunctions as sources of tunable below-gap radiation: Experimental verification

E. J. Caine,[a] S. Subbanna, H. Kroemer, and J. L. Merz
Department of Electrical & Computer Engineering, University of California, Santa Barbara, California 93106

A. Y. Cho
AT&T Bell Laboratories, Murray Hill, New Jersey 07974

(Received 4 June 1984; accepted for publication 7 September 1984)

We report experimental verification of the prediction of widely bias-tunable below-gap luminescence, from lattice-matched (p) (Al,In)As/ (n) InP heterojunctions, a system that has been predicted to have staggered lineup. The diodes, grown by molecular beam epitaxy, exhibit strong luminescence at 1.4 K, with a peak energy that shifts from 0.97 to 1.04 eV as the (pulsed) current density is increased from 4.5 to 40 A/cm^2. Nonshifting injection luminescence at 1.4 eV, due to hole injection into the n-InP substrate, was also present, but appreciably weaker ($<25\%$) than the interface luminescence. The spectra indicate that the band lineup in the (Al,In)As/InP system is indeed staggered, with a residual gap at the interface close to 0.96 eV. The corresponding conduction and valence-band offsets are 0.52 and 0.40 eV.

It has been predicted [1,2] that abrupt staggered-lineup heterojunctions offer the potential of light emission with a quantum energy below the gap of either constituent semiconductor. The postulated radiation mechanism is tunneling-assisted radiative recombination between the adjacent electron and hole accumulation wells in the conduction and valence bands at the junction interface (Fig.1), somewhat similar to that seen in "nipi" doping superlattices.[3] Because the electron and hole collection wells are extremely narrow, they exhibit strong voltage-dependent spatial quantization effects, and as a result, the photon energy of the tunneling-assisted interface luminescence should get pushed to higher energies,[2] making this luminescence strongly bias tunable.

In this letter we present results of the first experiments on a (p) (Al,In)As/ (n)InP heterojunction that verify the occurrence of tunable below-gap electroluminescence (EL).

The sample was grown by molecular beam epitaxy at 600 °C on (100) S-doped (2E18/cm^3) InP substrate material using a modified Riber model 2300 machine.[4,5] Approxi-

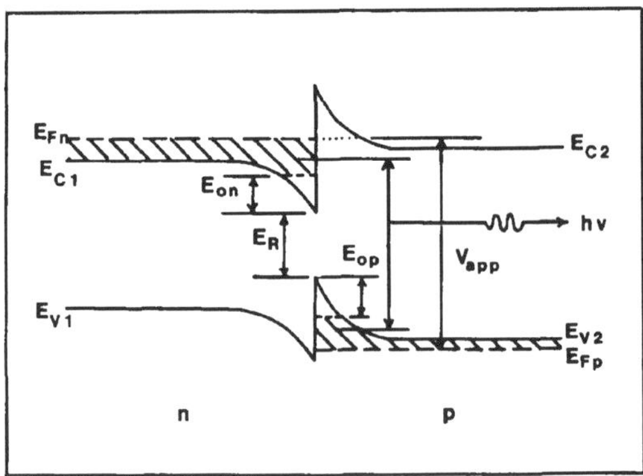

FIG. 1. Band diagram of p-Al$_{0.48}$In$_{0.52}$As/n-InP staggered-lineup heterojunction with sufficient forward bias applied to cause accumulation at the interface. Both semiconductors are assumed to be degenerately doped at the temperatures used. The splitting of the quasi-Fermi levels indicates an applied voltage (V_{app}) across the junction. The below-gap photon emission energy is given as the sum of E_R (interface residual gap) and the energies of the confined-particle states in the corresponding accumulation wells.

[a] On leave from Santa Barbara Research Center, Goleta, CA 93117

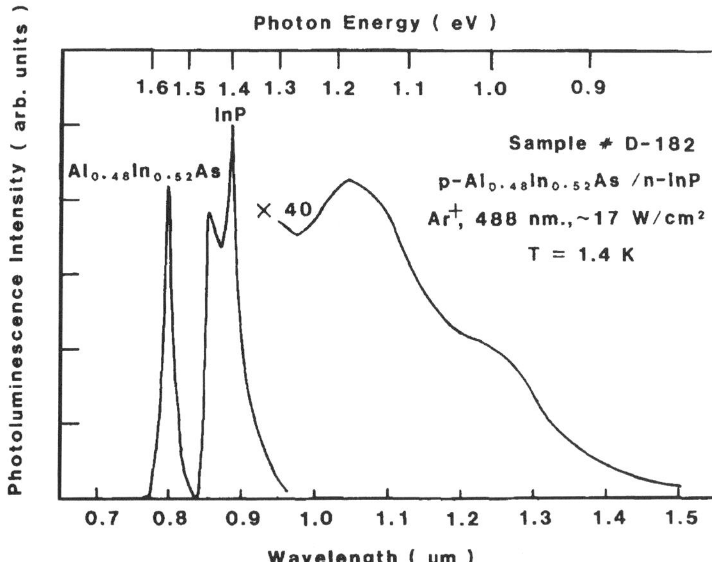

FIG. 2. Photoluminescence spectrum of a p-Al$_{0.48}$In$_{0.52}$As/n-InP sample.

mately 1.86 μm of lattice-matched Al$_{0.48}$In$_{0.52}$As, Be doped to 1.6E18/cm^3, was grown directly on the substrate after conventional surface oxide desorption.

Photoluminescence (PL) measurements were taken using 488-nm argon laser excitation radiation at ~ 17 W/cm^2. Figure 2 shows the PL spectrum indicating InP peaks at 0.886 and 0.87 μm and Al$_{0.48}$In$_{0.52}$As at 0.796 μm. (The curve has not been corrected for detector response.) Wakefield et al.[6] have recently reported the cathodoluminescence, and Ohno et al.[7] the photoluminescence, of similar material, showing the intensity peak of lattice-matched (Al,In)As to occur at 0.829μ ($T = 4$ K) and 0.804μ ($T = 4$ K), respectively. In addition, we saw a weak broad peak at wavelengths between 0.95 and 1.2 μm, of unknown origin, perhaps caused by interface defects.

To perform EL measurements, Au-Zn nonalloyed metal stripe top contacts were evaporated. Electroluminescence samples (~ 0.2 cm $\times 0.1$ cm) were then cleaved, and mounted onto small printed-circuit board strips, using silver conductive epoxy and gold wires. This holder was immersed into pumped liquid helium ($T \sim 1.4$ K) for EL measurements. Current pulses of 300-μs width at a repetition rate of 60 Hz were applied, and the spectrum of the light emerging from one cleavage facet was analyzed. A North Coast Scientific liquid nitrogen cooled germanium detector was used; for the wavelengths of interest, the detector responsivity is 1.46 times greater at 1.1 eV than 1.4 eV.

Six EL spectra were recorded, with injection current densities up to 40 A/cm^2, corresponding to input powers up to 2.4 W. Strong luminescence attributable to the postulated staggered-lineup luminescence (SLL) mechanism was observed, starting at current densities just below 4.5A/cm^2, where the luminescence peak occurred at 0.97 eV, increasing in intensity and shifting to higher quantum energies at higher current densities, reaching 1.04 eV at 40 A/cm^2. Also visible in all spectra was a weaker nonshifting InP peak at $\cong 1.4$ eV, attributable to hole injection from the wider gap

(Al,In)As layer into the InP. Figure 3 shows the actual spectra, plotted renormalized relative to the InP peak, shown at a fixed height; its actual intensity increased by about a factor 36 over the current range employed. The tunable SLL radiation is seen to shift by 70 meV, corresponding to 860 Å. The absence of any shifting (and broadening) of the InP peak indicates that sample heating was small, and it rules out any major contribution to the shift of the SLL by any unspecified thermal mechanism. The photon energy of the emission peak was found to vary almost exactly linearly with the logarithm of the current density (Fig. 4). For most models of forward current transport across semiconductor junctions

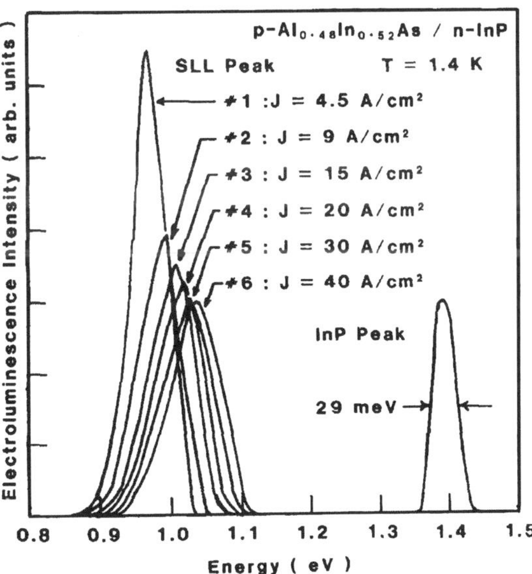

FIG. 3. Observed electroluminescence spectra for the various current densities applied. All spectra are normalized relative to a constant height of the InP peak. Sample area = 0.02 cm^2. Pulses of 300-μs duration and 60-Hz repetition rate were used.

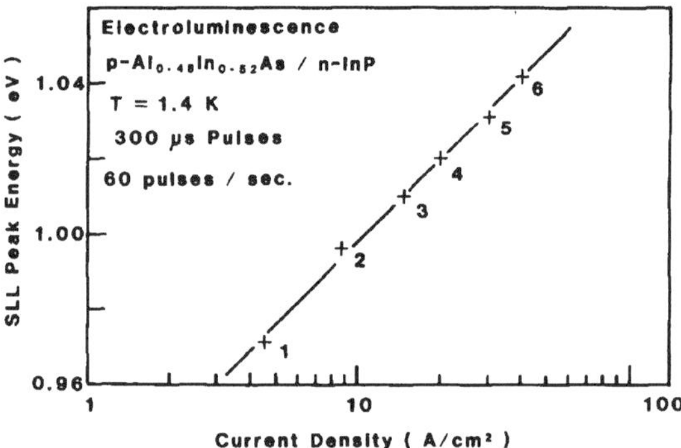

FIG. 4. Photon energy of the below-gap electroluminescence peak as a function of current density. The straight line is a least squares fit, with a slope of 72 meV/decade.

the current flow increases roughly exponentially with voltage. The behavior of Fig. 4 thus suggests a spectral shift close to linear with junction voltage—a reasonable result. The series resistance of the bulk semiconductor body, and the contact resistance, were too large ($\sim 1.5\,\Omega$) and insufficiently constant to permit an accurate extraction of the actual junction voltage, except in the range of current onset; a 1.4-K turn-on voltage of about 1.45 V was found.

It is evident that the SLL is both broader and stronger than the InP luminescence, indicating that the tunneling-assisted recombination process is more efficient than hole injection into the InP over or through the valence-band barrier. With increasing current density, the height of the SLL peak decreases to that of the InP peak, but at the same time the SLL line broadens from 51 (FWHM value) to 94 meV, compared to 29 meV for the InP luminescence. The integrated intensities for both signals increase slightly more rapidly than linearly with increasing current, with a SLL:InP ratio that drops slowly from about 6:1 to about 4.3:1. The broadening of the SLL signal is too large and of the wrong shape to

be readily explained by simple band filling; we suspect that residual interface defects play an important role.

No EL signal near 1.5 eV, attributable to electron injection into the (Al,In)As, was observed. This is as expected from the fact that (Al,In)As has a significantly wider energy gap than InP, and that almost all injection current should flow from the wider gap semiconductor into the narrower gap one.

The low-bias limit of the SLL quantum energy should coincide with the residual energy gap E_R at the heterojunction interface. Our data suggest a value $E_R \sim 0.96$ eV. This is about 0.14 eV less than the value 1.1 eV estimated in Ref. 2 by linear interpolation of the valence-band offsets predicted for AlAs/InP and InAs/InP by Harrison's theory.[8] The present value falls within the established range of deviations of "good" experimental data from the predictions of this theory.[9]

The authors wish to acknowledge Dr. K. Mohammed and Dr. I. Banerjee for help with the electroluminescence setup and D. Zak for technical assistance. The Santa Barbara part of this work was supported by the Office of Naval Research.

[1] Yu S. Mel'nikova, Sov. Phys. Semicond. **14**, 357 (1980).
[2] H. Kroemer and G. Griffiths, Electron. Devices Lett. **4**, 20 (1983).
[3] H. Künzel, G. H. Döhler, P. Ruden, and K. Ploog, Appl. Phys. Lett. **41**, 852 (1982).
[4] A. Y. Cho and K. Y. Cheng, Appl. Phys. Lett. **38**, 360 (1981).
[5] K. Y. Cheng, A. Y. Cho, and W. R. Wagner, Appl. Phys. Lett. **39**, 607 (1981).
[6] B. Wakefield, M. A. G. Halliwell, T. Kerr, D. A. Andrews, G. J. Davies, and D. R. Wood, Appl. Phys. Lett. **44**, 341 (1984).
[7] H. Ohno, C. E. C. Wood, L. Rathbun, D. V. Morgan, G. W. Wicks, and L. F. Eastman, J. Appl. Phys. **52**, 4033 (1981).
[8] W. A. Harrison, *Electronic Structure and the Properties of Solids: The Physics of the Chemical Bond* (Freeman, San Francisco, 1980), see especially Sec. 10F; See also W. A. Harrison, J. Vac. Sci. Technol. **14**, 1016, July/Aug. (1977).
[9] H. Kroemer, J. Vac. Sci. Technol. B. **2**, 433, July–Sept. (1984).

IEEE ELECTRON DEVICE LETTERS, VOL. EDL-6, NO. 4, APRIL 1985

Heterojunction Bipolar Transistor Using a (Ga,In)P Emitter on a GaAs Base, Grown by Molecular Beam Epitaxy

M. J. MONDRY AND H. KROEMER, FELLOW, IEEE

Abstract—We report the first N-p-n heterojunction bipolar transistor (HBT) using a (Ga,In)P/GaAs heterojunction emitter on a GaAs base. This combination is of interest as a potential alternate to (Al,Ga)As/GaAs, because of theoretical predictions of a larger valence band discontinuity and a smaller conduction band discontinuity, thus eliminating the need for grading of the emitter/base junction. The structure was grown by molecular beam epitaxy, with the base doping ($\sim 10^{19}$cm^{-3}) far exceeding the n-type doping ($\sim 5 \cdot {}^{17}$cm^{-3}) of the (Ga,In)P wide gap emitter ($E_g = 1.88$ eV). Common-emitter current gains of 30 were attained at a current density of 3000 A/cm^2, the highest current density achieved without burnout.

I. INTRODUCTION

HETEROJUNCTION BIPOLAR TRANSISTORS (HBT's) have stimulated much interest due to their predicted potential for high current gain and superior high-frequency performance [1], [2]. By utilizing an emitter with a wider energy gap than that of the base, the majority carriers of the base can be more strongly confined to this region. This confinement can be utilized to achieve high emitter efficiencies even in the presence of a (desirable) very high base-to-emitter doping ratio. To achieve this end it is, however important to minimize any electron blocking "spike barrier" in the conduction band, as it may occur at an abrupt emitter/base heterojunction if the band lineup at the latter is such that a large fraction of the total energy gap discontiuity occurs in the conduction band rather than in the valence band. It has been pointed out [1] that the overall effect of a fully developed spike is roughly the same as if the energy gap in the emitter were increased only by the amount of the valence band discontinuity ΔE_v rather than by the full energy gap discontinuity. An additional drawback of the conduction band spike is an undesirable increase in the emitter turnon voltage by about $\Delta E_v/q$.

The Ga-As based HBT work up to now has concentrated solely on GaAs paired with (Al,Ga)As. This dominance is due to that material system's inherent lattice matching, which yields interfaces of low defect densities. The energy band lineup in the (Al,Ga)As/GaAs system is, however, less than ideal because of the unfavorable band lineup: about 62 percent of the energy gap difference occurs in the conduction band [4], causing an undesirable potential barrier to electron injection from the emitter. Although compositional grading of the emitter–base junction can alleviate this drawback [5], a heterojunction with a majority of its energy gap discontinuity in the valence band would be much more desirable.

The (Ga,In)P/GaAs system has been proposed as such an alternate for GaAs-based HBT's [6]. A linear interpolation between the theoretical band offsets predicted by the Harrison theory [7] of band lineups for GaP/GaAs and InP/GaAs heterojunctions yields a valence-band offset of 0.29 eV and a conduction-band offset of 0.16 eV for the lattice-matched Ga$_{0.51}$In$_{0.49}$P/GaAs heterojunction. It is not clear exactly how reliable these specific numerical values are, but the Harrison theory tends to describe at least the general trends of band lineups very well, and we therefore believe that the overall prediction of a substantially larger valence-band offset than conduction-band offset is probably reliable. In this letter, we present what we believe to be the first report on the fabrication of a (Ga,In)P/GaAs HBT, and the demonstration of the wide-gap emitter effect in this material system.

II. MBE GROWTH AND DEVICE FABRICATION

The epitaxial layers of the HBT reported here were grown by MBE in a Varian MBE 360 system. A (100)-oriented GaAs substrate, Si-doped to 10^{18} cm^{-3}, was used. At a substrate temperature of 560°C, a 0.25-μm-thick GaAs collector, non-intentionally doped n-type to 10^{16} cm^{-3}, was grown directly on the substrate, followed by a 0.15-μm-thick GaAs base, p-type doped with Be to 10^{19} cm^{-3}. This was followed by the growth of the (Ga,In)P emitter.

The MBE technology of the (Ga,In)P/GaAs system is relatively undeveloped [8][1], [9], with lasers being the only kind of device reported so far [10]. In our procedure, separate Ga and In sources and a novel P$_2$ source were utilized to grow the (Ga,In)P emitter layer. The P$_2$ source consisted of a GaP decomposition source [11], modified with a baffle to condense out the undesired Ga flux component [12]. Once proper Ga and In fluxes were obtained, as determined by ion gauge measurements at the substrate

Manuscript received November 5, 1984; revised January 21, 1985. This work was supported by a Grant from Hewlett-Packard, Rockwell, and Xerox, under the University of California MICRO program.

M. J. Mondry was with the Department of Electrical and Computer Engineering, University of California, Santa Barbara, CA 93106. He is now with the McDonnell Douglas Microelectronics Center, Huntington Beach, CA 92647.

H. Kroemer is with the Department of Electrical and Computer Engineering, University of California, Santa Barbara, CA 93106.

[1] This paper [8] contains references to most of the earlier work in this system.

IEEE ELECTRON DEVICE LETTERS, VOL. EDL-6, NO. 4, APRIL 1985

176

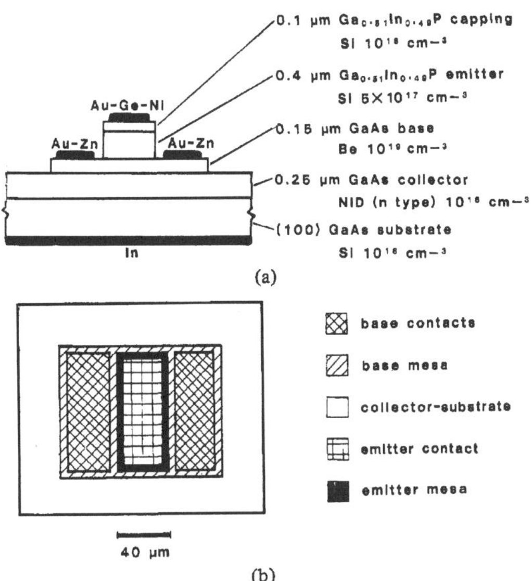

Fig. 1. Heterojunction bipolar transistor. (a) Schematic cross section. (b) Scaled top view.

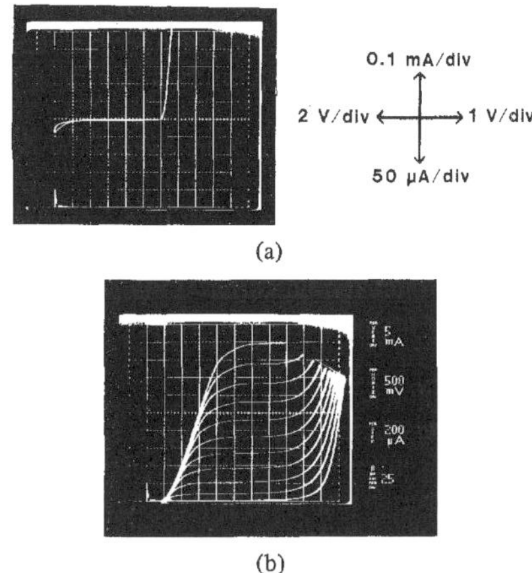

(a)

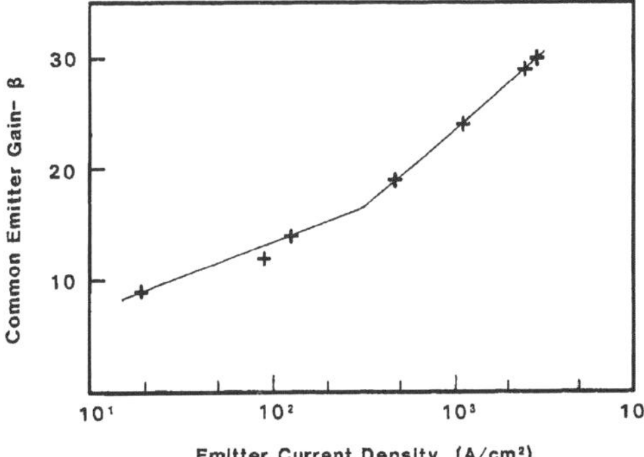

(b)

Fig. 2. (a) $I\text{-}V$ characteristics of GaAs/(Ga,In)P p-N base/emitter heterojunction. (b) Common-emitter characteristics of (Ga,In)P/GaAs heterojunction bipolar transistor.

position, the substrate temperature was reset to 510°C. Using a P_2 beam of 3×10^{-6} torr, a 0.4-μm-thick $Ga_{0.51}In_{0.49}P$ layer, n-type doped with Si to 5×10^{17} cm^{-3}, was grown at a rate of 0.5 μm/h. The Si doping was then increased to 10^{18} cm^{-3} for another 0.1 μm, for ohmic contact to the emitter. Note that the base doping exceeded that of the emitter by over an order of magnitude.

Discrete transistors with relatively large lateral dimensions were fabricated from the grown wafer using the mesa structure in Fig. 1(a) and (b). The reason for the large dimensions was to keep the processing of the nonstandard wafers as simple as possible. The emitter was contacted with evaporated Au-Ge-Ni. The emitter mesa was then defined by selectively etching down to the GaAs base with HCl. Au–Zn base contacts were evaporated to the exposed p-type GaAs, and the base mesa was etched using H_2SO_4:H_2O_2:H_2O = 10:3:87. The contacts were then alloyed at 450°C for 1 min. in a H_2:N_2 = 15:85 gas mixture. The collector was contacted with In on the backside of the substrate, which thus served as a portion of the device. The emitter–base and collector–base junction areas were 3.2×10^{-5} cm^2 and 9.6×10^{-5} cm^2.

III. RESULTS AND DISCUSSION

A typical $I\text{-}V$ characteristic of the p-GaAs/N-(Ga,In)P base/emitter heterojunction is shown in Fig. 2(a). The diodes have ~1.0-V forward current turn-on voltages, with hard breakdown between 12- and 14-V reverse bias. The low current turn-on voltage supports our speculation that there is a much lower conduction-band spike barrier in this system than in ungraded (Al,Ga)As/GaAs emitter/base junctions.

In Fig. 2(b), the common-emitter characteristics of a typical HBT are shown, and Fig. 3 gives the current gain β as a function of emitter current density. It can be seen that the

Fig. 3. Common-emitter current gain versus emitter current density. The two straight lines drawn through the data and the kink represent an ad hoc fit; they have no theoretical basis.

current gains are low at low current densities, but increase strongly with increasing density. This behavior suggests that appreciable space-charge layer recombination takes place at the emitter/base junction, presumably due to defects at the still-imperfect heterojunction interface. At high current densities this defect current is swamped by the injection current, which increases more strongly with forward bias—like exp (qV/kT)—than the defect current. Due to burnout problems, we were unable to achieve current densities above a relatively low 3000 A/cm^2 in these first and still excessively large devices. At these densities, β-values of 30 were obtained, still increasing with current, and suggesting that much higher β-values could be achieved if the current densities could be raised, even for the present defect current densities.

The current densities were limited by what appeared to be "forward secondary breakdown" [14], leading to a burning out of the device. Device failure was characterized by the collector shorting to the emitter. The shorted emitter/collector combination showed degraded but still rectifying $I\text{-}V$ characteristics against the base. The formation of the short circuit was often preceded by gain instability, symptomatic of thermal effects. Power calculations assuming uniform dissipation suggest a junction temperature less than 65°C at room temperature ambient. Hence, the heating is attributed to localized thermal runaway, perhaps caused by localized defects at the (Ga,In)P/GaAs interface.

The relatively large geometries of the devices prevented the determination of the high-frequency characteristics. The dc performance was also degraded by the large dimensions of the device, which caused emitter current crowding effects.

The current gain of this nonoptimized HBT is at present far below the gains achieved with the more highly developed (Al,Ga)As/GaAs material system [3], [5], [13], but is comparable to the first (Al,Ga)As/GaAs HBT reported [15]. As indicated, the current gain characteristics of Fig. 3 imply that higher gains than 30 should be achievable, but the failure mechanism indicates that materials problems still exist. Further improvement of the growth parameters, particularly an increase of the substrate temperature, is expected to increase the injection efficiency and current handling capabilities of the (Ga,In)P/GaAs heterojunctions.

ACKNOWLEDGMENT

The authors are grateful to B. R. Hancock for numerous discussions and for supplying the mask set, to S. Subbanna for performing photoluminescence measurements, to W. E. Gardner for providing X-ray analysis, to Dr. E. J. Caine for many useful discussions, and to D. Zak for his technical assistance.

REFERENCES

[1] H. Kroemer, "Heterostructure bipolar transistors and integrated circuits," *Proc. IEEE*, vol. 70, no. 1, pp. 13–25, Jan. 1982.

[2] For recent progress in this area see, for example: P. M. Asbeck, D. L. Miller, R. J. Anderson, R. N. Demling, R. T. Chen, C. A. Liechti, and F. H. Eisen, "Application of heterojunction bipolar transistors to high speed, small-scale digital integrated circuits," in *1984 GaAs IC Symp. Tech. Dig.*, pp. 133–136; H. Ito, T. Ishibashi, and T. Sugeta, "High current gain AlGaAs/GaAs hetero-

junction bipolar transistors with heavily doped base," in *Proc. 16th Conf. Solid State Dev. Mater.*, (Kobe, Japan), Extended Abs., 1984, pp. 351–354.

[3] D. Ankri, A. Scavennec, C. Besombes, C. Courbet, F. Heliot, and J. Riou, "Diffused epitaxial GaAlAs-GaAs heterojunction bipolar transistor for high frequency operation," *Appl. Phys. Lett.*, vol. 40, no. 9, pp. 816–818, May 1982.

[4] M. Watanabe, J. Yoshida, M. Mashita, T. Nakanisi, and A. Hojo, in *Proc. 16th Conf. Solid State Dev. and Mater.*, (Kobe, Japan), Extended Abs., 1984, pp. 181–183.

[5] P. M. Asbeck, D. L. Miller, R. A. Milano, J. S. Harris, Jr., G. R. Kaelin, and R. Zucca, "(Ga,Al)As/GaAs bipolar transistors for digital integrated circuits," in *IEDM Tech. Dig.*, 1981, pp. 629–632. See also, J. R. Hayes, F. Capasso, R. J. Malik, A. C. Gossard, and W. Wiegmann, "Optimum emitter grading for heterojunction bipolar transistors," Appl. Phys. Lett., vol. 43, no. 10, pp. 949–951, Nov. 1983.

[6] H. Kroemer, "Heterostructure bipolar transistors: what should we build?," *J. Vac. Sci. Technol. B*, vol. 1, no. 2, pp. 126–130, Apr.–June 1983.

[7] W. A. Harrison, "Elementary theory of heterojunctions," *J. Vac. Sci. Technol.*, vol. 14, no. 4, pp. 1016–1021, July/Aug. 1977.

[8] See, for example: P. Blood, J. S. Roberts, and J. P. Stagg, "GaInP grown by molecular beam epitaxy doped with Be and Sn," *J. Appl. Phys.*, vol. 53, no. 4, pp. 3145–3149, Apr. 1982.

[9] Y. Kawamura, H. Asahi, and H. Nagai, "Molecular beam epitaxial growth of undoped low-resistivity $In_xGa_{1-x}P$ on GaAs at high substrate temperatures (500–580°C)," *Japn. J. Appl. Phys.*, vol. 20, no. 11, pp. L807–L810, Nov. 1981.

[10] G. B. Scott, J. S. Roberts, and R. F. Lee, "Optically pumped laser action at 77 K in GaAs/GaInP double heterostructures grown by molecular beam epitaxy," *Appl. Phys. Lett.*, vol. 37, no. 1, pp. 30–32, July, 1980.

[11] S. L. Wright and H. Kroemer, "Operational aspects of a gallium phosphide source of P_2 vapor in molecular beam epitaxy," *J. Vac. Sci. Technol.*, vol. 20, no. 2, pp. 143–148, Feb. 1982.

[12] M. J. Mondry, E. J. Caine, and H. Kroemer, "A GaP decomposition source for producing a dimer phosphorus molecular beam free of gallium and tetramer phosphorus," *J. Vac. Sci. Technol.*, to be published.

[13] See, e.g., P. M. Asbeck, D. L. Miller, W. C. Peterson, and C. G. Kirkpatrick, "GaAs/GaAlAs heterojunction bipolar transistors with cutoff frequencies above 10 GHz," *IEEE Electron Dev. Lett.*, vol. EDL-3, pp. 366–368, Dec. 1982.

[14] A. Blicher, *Field-Effect and Bipolar Power Transistor Physics.* New York: Academic, 1981 Sec. 10.4.3 and 10.5.1.

[15] W. P. Dumke, J. M. Woodall, and V. L. Rideout, "GaAs-GaAlAs heterojunction transistor for high frequency operation," *Solid-State Electron.*, vol. 15, pp. 1339–1343, 1972.

Reprinted from

D. I. Babic and H. Kroemer, ``The Role of Nonuniform Dielectric Permittivity in the Determination of Heterojunction Band Offsets by C-V Profiling Through Isotype Heterojunctions,'' Solid-State Electron., Vol. 28(10), pp. 1015-1017, 1985.

Solid-State Electronics Vol. 28, No. 10, pp. 1015–1017, 1985
Printed in Great Britain.

0038–1101/85 $3.00 + .00
© 1985 Pergamon Press Ltd.

THE ROLE OF NONUNIFORM DIELECTRIC PERMITTIVITY IN THE DETERMINATION OF HETEROJUNCTION BAND OFFSETS BY C–V PROFILING THROUGH ISOTYPE HETEROJUNCTIONS

DUBRAVKO I. BABIC and HERBERT KROEMER

Department of Electrical and Computer Engineering, University of California, Santa Barbara,
CA 93106, U.S.A.

(*Received* 17 *December* 1984; *in revised form* 29 *January* 1985)

Abstract—Up to now, C–V profiling through isotype heterojunctions has been performed assuming a uniform dielectric permittivity throughout the heterostructure. We extend the interpretation of C–V data to the case of a semiconductor with position-dependent dielectric permittivity, and we show that the variation of the dielectric permittivity across an isotype heterojunction interface has no effect on the determination of the heterojunction band discontinuity and the interface charge density.

Capacitance–voltage (C–V) carrier concentration profiling is potentially a good method for the determination of heterojunction band offsets[1,2,3]. One of the questions that has so far been left open is the following: To what extent does any variation of the dielectric permittivity across the heterojunction affect the band offset determination, and what is the interpretation of the carrier concentration profile obtained by C–V profiling if the dielectric permittivity is position dependent? We extend here the interpretation of the C–V data to the case of a nonuniform dielectric permittivity, and we show that the dielectric permittivity variation cancels out of the band offsets measurement.

We consider an n-type semiconductor with an arbitrary impurity profile $N(x)$ and—for generality—a continuously varying dielectric permittivity $\epsilon(x)$, as shown in Fig. 1. At the surface ($x = 0$) of the semiconductor a Schottky barrier (SB) is assumed to be present, to which a dc reverse bias has been applied. The free carrier distribution $n(x)$ is assumed to be completely depleted at the position of the Schottky barrier, $n(0) = 0$. In the capacitance measurement a small voltage variation is superimposed on the dc bias at the SB. A small change ΔV in the voltage will deplete a charge distribution $\Delta n(x)$ (Fig. 1), referred to as the incremental displaced electron distribution (IDED) in [4]. The total amount of charge depleted and the magnitude of the voltage increment are then given in terms of $\Delta n(x)$ by the two relations

$$\Delta Q = q \int_0^\infty \Delta n(x)\, dx, \qquad (1)$$

$$\Delta V = q \int_0^\infty \frac{dx}{\epsilon(x)} \int_0^x \Delta n(y)\, dy. \qquad (2)$$

The capacitance is $C = \Delta Q / \Delta V$.

We reduce the problem of capacitance measurement on a semiconductor with nonuniform dielectric permittivity to the much simpler and already-covered case of capacitance measurement on a semiconductor with a uniform dielectric permittivity of some value ϵ_u. To this end we define an *equivalent* position variable χ via

$$\chi = \epsilon_u \int_0^x \frac{dy}{\epsilon(y)}, \qquad (3)$$

which implies

$$\frac{d\chi}{\epsilon_u} = \frac{dx}{\epsilon(x)}. \qquad (4)$$

We also define an *equivalent* incremental displaced electron distribution $\Delta\eta(\chi)$ via

$$\Delta\eta(\chi) = \frac{1}{\epsilon_u} \epsilon(x)\, \Delta n(x). \qquad (5)$$

The overall transformation evidently conserves charge:

$$\Delta\eta(\chi)\, d\chi = \Delta n(x)\, dx, \qquad (6)$$

which implies

$$\Delta Q = q \int_0^\infty \Delta\eta(\chi)\, d\chi. \qquad (7)$$

It also conserves the voltage increment ΔV associated with each depletion charge increment: insertion of (4) and (5) into (2), and execution of an integration by parts, yields

$$\Delta V = \frac{q}{\epsilon_u} \int_0^\infty \Delta\eta(\chi)\, \chi\, d\chi. \qquad (8)$$

1016   D. I. BABIC and H. KROEMER

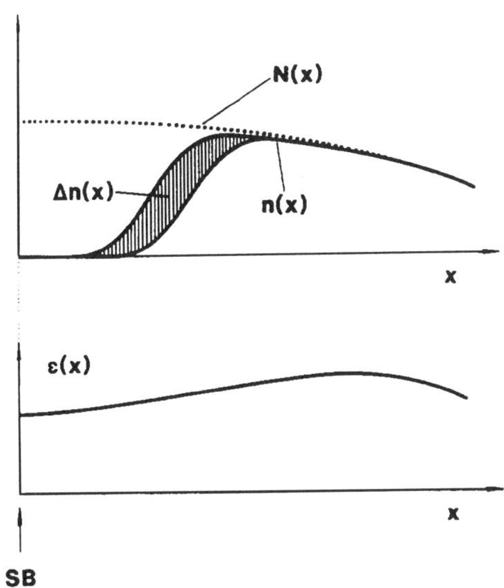

Fig. 1. Nonuniform doping profile and dielectric permittivity of an *n*-type semiconductor. The Schottky barrier is placed at $x = 0$.

But this is exactly the relation for the voltage increment caused by an IDED $\Delta\eta(\chi)$ in a uniform-ϵ semiconductor.

The central point is now the following: If $\Delta n(x)$ is the exact IDED when the bias voltage is changed from V to $V + \Delta V$ in a nonuniform-ϵ semiconductor with the donor distribution $N(x)$ and an effective density of states $N_C(x)$, then $\Delta\eta(\chi)$ is the exact IDED for the same bias voltage increment in an *equivalent* uniform-ϵ semiconductor with the *equivalent* doping distribution

$$\mu(\chi) = \frac{1}{\epsilon_u}\epsilon(x)N(x), \qquad (9)$$

and with the equivalent effective density of states distribution

$$\eta_C(\chi) = \frac{1}{\epsilon_u}\epsilon(x)N_C(x). \qquad (10)$$

Furthermore, at any given bias V the associated equivalent total electron distribution is given by

$$\eta(\chi) = \frac{1}{\epsilon_u}\epsilon(x)n(x). \qquad (11)$$

To prove this equivalence claim, we consider the Poisson–Boltzmann equation in the original semiconductor,

$$\frac{d}{dx}\left(\epsilon(x)\frac{d\Phi(x)}{dx}\right) = -\rho(x) = q[n(x) - N(x)]. \qquad (12)$$

For simplicity, we assume here that the semiconductor is nondegenerate, in which case

$$n(x) = N_C(x)\exp\left(-\frac{E_C(x) - E_F}{kT}\right). \qquad (13)$$

However, this assumption is actually unnecessary, and the proof is easily extended to the degenerate case. In (12) and (13) $\Phi(x)$ is the local electrostatic potential, while $E_C(x)$ and E_F are the conduction band edge and the Fermi level. Insertion of (4) in (9)–(11) transforms the Poisson–Boltzmann equation into

$$\frac{d^2\Phi^*(\chi)}{d\chi^2} = \frac{q}{\epsilon_u}(\eta(\chi) - \mu(\chi)), \qquad (14)$$

with

$$\eta(\chi) = \eta_C(\chi)\exp\left(-\frac{E_C^*(\chi) - E_F}{kT}\right). \qquad (15)$$

Here $\Phi^*(\chi)$ and $E_C^*(\chi)$ are the original electrostatic potential and conduction band edge distributions, transformed into functions of χ rather than x, via

$$\Phi^*(\chi) = \Phi^*[\chi(x)] \equiv \Phi(x), \qquad (16)$$

$$E_C^*(\chi) = E_C^*[\chi(x)] \equiv E_C(x). \qquad (17)$$

Equations (14) and (15) represent the true Poisson–Boltzmann equation of a semiconductor with a uniform dielectric permittivity equal to ϵ_u, with the doping distribution $\mu(\chi)$ and with the effective density of states distribution $\eta_C(\chi)$. The transformation from (12) and (13) to (14) and (15) shows that with every solution of the Poisson–Boltzmann equation in the original nonuniform-ϵ semiconductor, it is possible to associate a solution in the equivalent uniform-ϵ semiconductor. The electrostatic potential distribution in the equivalent semiconductor differs from that in the original semiconductor; the difference is simply that each point on the original $\Phi(x)$ curve has been moved "horizontally" along the abscissa, to the transformed abscissa position $\chi = \chi(x)$, with an unchanged ordinate value Φ. The same is true for the transformation of $E_C(x)$. In particular, any abrupt band variations present in the original semiconductor will occur with the unchanged height in the equivalent semiconductor, although in general at a different position. Inasmuch as the abscissa end points are not changed by the transformation, $\chi(0) = 0$ and $\chi(\infty) = \infty$, the equivalent solution exhibits the same *overall* variation in both Φ and E_C, between 0 and ∞:

$$\Phi^*(\infty) - \Phi^*(0) = \Phi(\infty) - \Phi(0), \qquad (18)$$

$$E_C^*(\infty) - E_C^*(0) = E_C(\infty) - E_C(0). \qquad (19)$$

This means that the original and the equivalent solutions correspond to the same bias voltage V.

Because the transformation also conserves charge, the equivalent pairs of solutions exhibit the same dependence of total depleted charge on bias voltage $Q = Q(V)$, and hence the same capacitance–voltage relation $C(V)$. This means that it is possible to interpret C–V measurements on nonuniform-ϵ semiconductor in terms of an *apparent* (i.e. Debye-averaged[4]) electron distribution $\bar{\eta}(\chi)$ in the equivalent uniform-ϵ semiconductor, via the familiar relation

$$\bar{\eta}(\bar{x}) = \frac{2}{q\epsilon_u}\left\{\frac{d}{dV}\left(\frac{1}{C^2}\right)\right\}^{-1}, \qquad (20)$$

where

$$\bar{x} = \frac{\epsilon_u}{C}. \qquad (21)$$

Because of Debye averaging, this apparent profile will, in general, differ significantly from the true electron distribution $\eta(\chi)$ even in the equivalent semiconductor especially in the presence of an abrupt heterojunction.

However, for the purpose of principal interest here, the extraction of heterojunction band offsets, this difference is immaterial: it was shown earlier[1, 4] that such offsets can be extracted even from the Debye-averaged profile, if the actual donor distribution is known. Hence, it is possible to extract the band offsets of the equivalent semiconductor structure from such a measurement. But as we stated earlier, the band offsets in the actual and the equivalent semiconductor are of the same height; hence the extraction of the band offsets from the Debye-averaged apparent profile of the equivalent uniform-ϵ semiconductor automatically yields the correct band offset for the original nonuniform-ϵ semiconductor structure. Similarly, any interface charge σ_i is preserved. Therefore, σ_i and the conduction band discontinuity ΔE_C [1] are given by

$$\sigma_i = \int_0^\infty (\bar{\eta}(\chi) - \mu(\chi))\,d\chi, \qquad (22)$$

and

$$\Delta E_C = \frac{q^2}{\epsilon_u}\int_0^\infty (\bar{\eta}(\chi) - \mu(\chi))(\chi - \chi_i)\,d\chi$$

$$+ kT\ln\left[\frac{\mu_2\eta_{C1}}{\mu_1\eta_{C2}}\right]. \qquad (23)$$

Here χ_i is the *equivalent* position of the interface, $\chi_i = \chi(x_i)$, given by

$$\chi_i = \epsilon_u\int_0^{x_i}\frac{dy}{\epsilon(y)}, \qquad (24)$$

$$\chi_i = \frac{\epsilon_u x_i}{\epsilon_1}, \qquad (25)$$

where x_i is the true position of the interface. The asymptotic values of the effective doping profile on

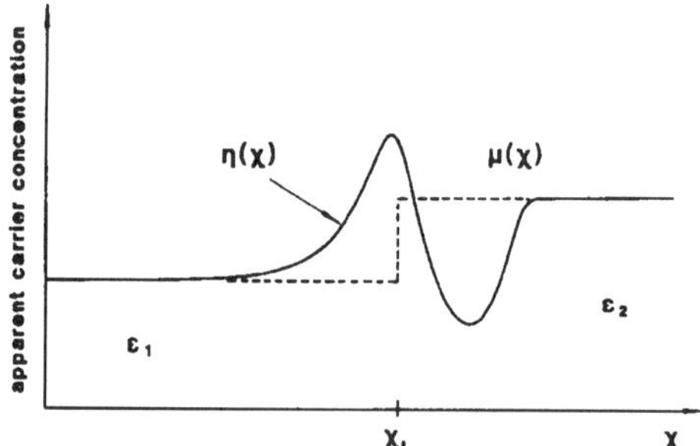

Fig. 2. A typical apparent carrier concentration profile through an isotype heterojunction. The position of the interface is denoted with χ_i and the SB is placed at $\chi = 0$.

the two sides of the HJ are denoted with μ_1 and μ_2. The equivalent conduction-band densities of states of the two semiconductors are denoted with η_{C1} and η_{C2}.

The values of μ_1 and μ_2 are given by the flat portions at the ends of the apparent carrier profile (Fig. 2). It is important to note that the flat portions in the apparent profile do not give the true doping levels, but their transformed values (see eqn (9)).

There are, in principle, no constraints on the value of the reference dielectric permittivity ϵ_u. There is a significant advantage in chosing ϵ_u equal to ϵ_1, that is, to the dielectric permittivity of the layer closer to the SB: in this case χ_i will be equal to the true interface position x_i (see eqn (25)).

C–V carrier concentration profiling through heterojunctions has up to now been performed neglecting the dielectric permittivity variation. Even though the absolute values of the uniform doping levels around the heterojunctions obtained in this way are not accurate the errors cancel out in the calculation of ΔE_C and σ_i. Note that the formalism presented here applies to compositionally graded heterojunctions as well[5].

Acknowledgements—We wish to thank Bruce Hancock for many useful discussions, and an anonymous reviewer for pointing out that our proof can be extended without making the assumption (13) of nondegeneracy. This work was supported by the Office of Naval Research.

REFERENCES

1. H. Kroemer, Wu-Yi Chien, J. S. Harris Jr., and D. D. Edwall, *Appl. Phys. Lett.* **36**(4), 295–297 (1980).
2. R. People, K. W. Wecht, K. Alavi, and A. Y. Cho, *Appl. Phys. Lett.* **43**(1), 118–120 (1983).
3. M. O. Watanabe, J. Yoshida, M. Mashita, T. Nakanisi, and A. Hojo, *Extended Abstracts of the 16th Conf. on Solid State Devices and Materials*, 181–184 Kobe, 1984.
4. H. Kroemer and W.-Y. Chien, *Solid-State Electron.* **24**, 655–660 (1981).
5. H. Kroemer, *Appl. Phys. Lett.* **46**(5), 504–505 (1985).

Reprinted from

H. Kroemer, ``Band Offsets at Heterointerfaces: Theoretical Basis, and Review of Recent Experimental Work,'' Surf. Sci., Vol. 174, pp. 299-306, 1986.

Surface Science 174 (1986) 299–306
North-Holland, Amsterdam

299

BAND OFFSETS AT HETEROINTERFACES: THEORETICAL BASIS, AND REVIEW, OF RECENT EXPERIMENTAL WORK

Herbert KROEMER

*Department of Electrical and Computer Engineering, University of California,
Santa Barbara, California 93106, USA*

Received 2 September 1985; accepted for publication 15 September 1985

New data obtained since early-1984 have forced a drastic re-assessment of the energy band lineups for the most important of all heterointerfaces, the GaAs/(Al, Ga)As interface. The paper concentrates on the most trustworthy of the new data, and on the three new measurement techniques used to obtain them. An attempt is made to reconcile the remaining discrepancies between the new data, and to select a set of most probable lineup rules.

1. Introduction

In 1983 the writer presented a detailed review [1] of our understanding of the problem of heterostructure band offsets, as it existed at the time. The picture then available suggested a rapid convergence of this understanding. Experimentally, for the most important of all heterosystems, the (Al, Ga)As system, Dingle's 85:15 rule [2] for the ratio of conduction to valence band offsets in the direct-gap range of (Al, Ga)As/GaAs heterojunctions, had become accepted by a large majority of those working in the field. Theoretically, the Harrison atomic orbital (HAO) theory [3] appeared to give a surprisingly good account of all those heterostructure band offsets for which the data appeared trustworthy on other grounds. There were some unresolved problems, to be sure, but few individuals thought that they were precursors of major difficulties. Around mid-1984, this complacent picture collapsed almost abruptly: Dingle's rule was suddenly refuted by a whole string of new data [4,5], and with it collapsed the faith in the HAO theory, which was unable to account for the band offsets in the most important of all heterosystems, the (Al, Ga)As system.

As it stands right now, *experimentally*, the band offset ratio in the direct-gap range of (Al, Ga)As/GaAs heterojunctions appears to be close to a 62:38 ratio. *Theoretically*, we have no widely accepted *predictive* theory at all, although there is a new contender, Tersoff's theory [6], but it has not received widespread acceptance amongst theorists, even though experimentalists seem to like it: its predictions are in quite good agreement with many of the experimental data.

The present paper is an abbreviated update on the earlier review, concentrating on those *selected* aspects of the band offset problem where some consensus appears to be re-emerging. A more complete review is intended for another time and place.

More specifically, the paper will discuss briefly three techniques for determining heterostructure band offsets that have moved to the forefront since early-1984: (a) charge transfer techniques [7], (b) C/V profiling techniques [8–10], and (c) thermionic current flow over square heterobarriers [11,12]. An important aspect of all three techniques is that they were applied successfully to p–p interfaces as well as n–n interfaces, and it is precisely the consistency of the valence band offset data with the conduction band offset data that gives the new lineup data a much greater trustworthiness than the older data ever possessed.

A second purpose of this paper is to give a review of what this writer believes to be the most trustworthy of the new data, and to synthesize from these data – still somewhat contradictory – a set of band offset values for the (Al, Ga)As system which, while possibly still controversial, should at least have the virtue that it should be hard to predict the *direction* in which any single value in this set will change during the inevitable next revision.

Time and space limitations do not permit a discussion of many other points, such as: (a) the current status of *predictive* theories of band offsets, like Tersoff's theory [6]; (b) the origins of difficulties in the earlier techniques; (c) various "minor" new techniques. Nor has an atempt been made to give complete references to those techniques that are discussed; often only the most recent reference is given.

2. New experimental techniques and selected results

2.1. Preliminaries

As the reader will recall, Dingle's 85:15 rule was originally established [2] through optical absorption measurements on periodic multi-quantum-well (MQW) structures, or *superlattices*. Chronologically, the first challenge to the validity of this rule that could not be easily explained away, arose from another set of optical absorption measurements on MQW structures, with just a different well shape, which called for more symmetrical band offsets [4]. It next emerged that Dingle's data could also be interpreted just as well in terms of more symmetrical band offsets, *if* one was willing to treat the hole effective masses as adjustable parameters. Inasmuch as the actual hole effective masses were quite well known, such an approach appeared purely ad-hoc, and although it could in principle be readily justified by band mixing of the light- and heavy-hole band at the interface, this approach appeared too artificial to cause by itself a wholesale rejection of Dingle's rule. The deathblow came from another direction.

2.2. Charge transfer techniques

In the early years of (Al, Ga)As heterostructures, especially of heterostructure field effect transistors (HFTs), essentially all work was done on n–n structures. p–p structures appeared to be of little interest, for two separate reasons: (a) The hole mobilities were known to be so much lower than the electron mobilities, that there appeared to be little interest in p-type HFTs. (b) The valence band offsets predicted by Dingle's rule were too low to permit the accumulation of a two-dimensional hole gas of useful density at the interface, except possibly at low temperatures. Eventually, p-type structures were studied, however, and it became clear immediately that the 2D hole concentrations were far larger than expected, completely incompatible with the small valence band offsets predicted by Dingle's rule. A technique to determine band offsets from the density of the quasi-two-dimensional carrier gas at a heterojunction was developed by Stern and his co-workers [7], who showed that the 2D hole gas data on binary–binary p–p GaAs/AlAs heterojunctions required a valence band offset of 0.45 ± 0.05 eV. Some supplementary data on n–n junctions placed the valence band offsets at the upper limit of this uncertainty range, at 0.5 eV. According to the widely accepted data quoted in Casey and Panish's book [13], the total energy gap of (Al, Ga)As for less than 45%Al increases linearly with the Al concentration, with a slope of 12.5 meV/%Al. Assuming a valence band offset slope of 4.5 meV/%Al, this leaves only 8.0 meV/%Al for the conduction band offset slope, or a $\Delta E_{\mathrm{C}}:\Delta E_{\mathrm{V}}$ ratio of 64:36, far below Dingle's 85:15 ratio.

2.3. C/V profiling

In 1980, the present writer and his co-workers pointed out [8] that it is possible to determine the band offset at an isotype heterojunction by C/V profiling through that junction from an adjacent Schottky barrier. Performed on some LPE-grown n-type heterojunctions, the measurements showed a conduction band offset of only about 66% of the total energy gap difference, significantly below Dingle's rule. Rather than taking our lower number at face value, we argued that it was due to grading effects at the LPE-grown junctions. What we failed to realize at the time was that the effect of grading would cancel out again in the final calculation of the band offset from the raw data [14]. Actually, an inspection of our 1980 paper shows that we recognized that there was a problem about the extent to which grading would lead to a

  H. Kroemer / Band offsets at heterointerfaces

reduction of the band offset value obtained by our technique, but we did not realize until late-1984 that the C/V profiling technique is truly grading-independent.

By that time the C/V technique had been used with stunning success by Watanabe et al. [9], who studied both n–n and p–p junctions with it, and obtained offset slope values of 7.7 and 4.7 meV/%Al, adding up to 12.4 meV/%Al, almost exactly Casey and Panish's value for the overall energy gap slope. This was the first time that really precise data for *both* conduction and valence band offsets had been determined *separately and independently of each other*, by a common technique. The excellent self-consistency of the data with the known gap slope contributes tremendously to the trustworthiness of these data. Another important self-consistency check in that work involved the use of computer reconstruction from the band offset data of what *should have been* the experimental C/V profile, and comparison of that reconstructed profile with the experimental one. The agreement was excellent.

Shortly after the work of Watanabe et al., Okumura et al. [10] performed another set of careful C/V measurements of conduction band offsets, obtaining a somewhat higher conduction band offset slope, 8.3 meV/%Al. Unfortunately, Okumura et al. did not perform the two kinds of self-consistency checks of Watanabe et al., hence it is somewhat more difficult to assess the accuracy of their data.

The C/V profiling technique is beginning to be applied to other heterosystems as well. The most impressive work so far is that of Forrest et al. [15], who studied the important lattice-matched (Ga, In)(P, As)/InP system, finding that the conduction band offset is 39% of the energy gap difference, over the entire lattice-matched composition range.

2.4. Thermionic current flow over square heterobarriers

Isotype heterojunctions somewhat resemble Schottky barriers, showing qualitatively similar rectification effects. From an analysis of the temperature dependence of the I/V characteristics it should, in principle, be possible to obtain the barrier heights, and from those the band offsets. The trouble with this idea is that the effective barrier heights are easily distorted by relatively small uncontrolled defect charges near the interface, leading to grossly incorrect band offset values. For example, many early (Al, Ga)As/GaAs n–n heterojunctions did not rectify at all, even at reduced temperatures [16], a result totally imcompatible with the already-known work of Dingle. This space-charge uncertainty was finally brought under control, by Hickmott et al. [11], through the use of square heterostructure barriers with two heavily-doped low GaAs regions enclosing a not intentionally doped (Al, Ga)As barrier region. Such structures are basically capacitors with semiconductor electrodes, similar to MOS-type capacitors, and MOS-type C/V measurements make it possible to determine the magnitude and the centroid of any parasitic charges in the barrier region, and thereby to correct for the effects of these charges. Measurements on n-type structures [11] yielded conduction band offsets consistent with an offset ratio in the 60:40 to 65:35 range, but the technique proved particularly adaptable to p-type structures, where parasitic charges were essentially absent. Batey and Wright [12] utilized this fact to determine the valence band offsets for (Al, Ga)As/GaAs heterojunctions over the entire composition range, and they found a purely linear relation with surprisingly small data scatter, with an offset slope of 5.5 meV/%Al.

2.5. Band-to-band consistency, and the energy gap uncertainty

In the direct-gap range of the (Al, Ga)As system, the conduction and valence band offsets should add up to the presumably known energy gap difference between the two semiconductors. The data of Watanabe et al. [9] met this demand perfectly, but if one adds, say, the largest of the recent conduction band offset values (by Okumura et al. [10]) and the largest of the recent valence band offset data (by Batey and Wright [12]), one obtains 13.8 meV/%Al, clearly far above the Casey/Panish [13] value of 12.5 meV/%Al. This raises the question as to the accuracy of the E_g versus x calibrations in *all* (Al, Ga)As work, especially of those calibrations that were obtained through electron microprobe measurements. A re-determination of this calibration has recently been reported by Miller et al. [17], who claim a significantly steeper slope, 14.5 meV/%Al. This value is larger than the abovementioned largest plausible band offset value of 13.8 meV/%Al, and is hence suspect. But the very existence of a large 2 meV/%Al discrepancy between these new data of Miller et al. and the earlier Casey and Panish data suggests that the accuracy of the latter cannot be taken for granted. Any critical assessment of the band offsets in the (Al, Ga)As system thereby necessarily requires the inclusion of a re-assessment of the E_g versus x calibrations.

3. Re-conciliation of experimental lineups

3.1. The low-aluminum range

It is clear that there are still residual discrepancies between the various new data cited in this paper, and the discrepancies could be easily increased by including other recent data that I have omitted from this brief review, for various reasons. We must somehow choose, and preferably by a more rational process than simply averaging over all data. I believe that the most trustworthy data for the band offset *ratio* $\Delta E_C : \Delta E_V$ are the data of Watanabe et al. [9], who found a 62:38 ratio: Theirs is the only work that determined both conduction and valence band offsets separately, by exactly the same technique. Also, their data passed with flying colors the important self-consistency check of computer reconstruction from the band offset data, and comparison of that reconstructed profile with the experimental one.

Next, I propose to accept Batey and Wright's claim [12] of a perfectly linear ΔE_V versus x relation throughout the entire composition range, but with a somewhat smaller slope, 5.0 meV/%Al, at the upper end of Wang and Stern's uncertainty range [7]. My reasons for this preference go beyond a desire not to exceed the uncertainty range of Wang and Stern: If an activation energy is itself temperature-dependent, a determination of this energy from an Arrhenius plot, as in the work of Batey and Wright, leads not to the *actual* value in the temperature range of the measurements, but to the value *extrapolated* to $T = 0$. Inasmuch as the energy gaps decrease with increasing temperature, one should expect the same for the valence band offset, and the value of Batey and Wright *should* be somewhat larger than the true room-temperature value. Also, any residual quantum-mechanical reflection effects at the barrier would probably mimic a slightly higher barrier.

If one combines the 5.0 meV/%Al valence band offset slope with the 62:38 offset ratio, one finds a conduction band offset slope of 8.13 meV/%Al, significantly above the 7.7 meV/%Al value of Watanabe et al. To reconcile this discrepancy, I make the ad-hoc assumption that the Al composition parameter x in the work of Watanabe et al. is too high by a factor $8.13/7.70 = 1.056$. Together with the proposed valence band slope his would imply an energy gap slope of 13.1 meV/%Al, larger than the Casey and Panish value, but still significantly smaller than the value of Miller et al. [17], which is simply incompatible with today's best heterostructure band offset data. Only the future can tell what will come of this ad-hoc assumption made here.

3.2. The crossover point and the high-aluminum range

According to Casey and Panish, the direct/indirect crossover takes place at $x = 0.45$, at which point the energy gap has increased by $0.45 \times 1.247\,\text{eV} = 0.561$ eV, of which 62%($= 0.347$ eV) should occur in the conduction band. I see no reason to challenge the energy gap value at the crossover point. But with a steeper slope parameter, the crossover presumably takes place already at $x = 0.45/1.056 = 0.43$.

The predicted conduction band offset at the crossover point of 0.347 eV should be the highest conduction band offset available in the (Al,Ga)As/GaAs system; for higher Al concentrations the conduction band offset declines again (fig. 1), to 0.239 eV in the AlAs/GaAs binary/binary limit. This behavior has the interesting consequence that (Al, Ga)As/AlAs heterojunctions should have a *staggered lineup* for $x > 0.239/0.813 = 0.29$, with the largest conduction band offset, 0.347 eV $-$ 0.239 eV $= 0.108$ eV, occurring at the crossover com-

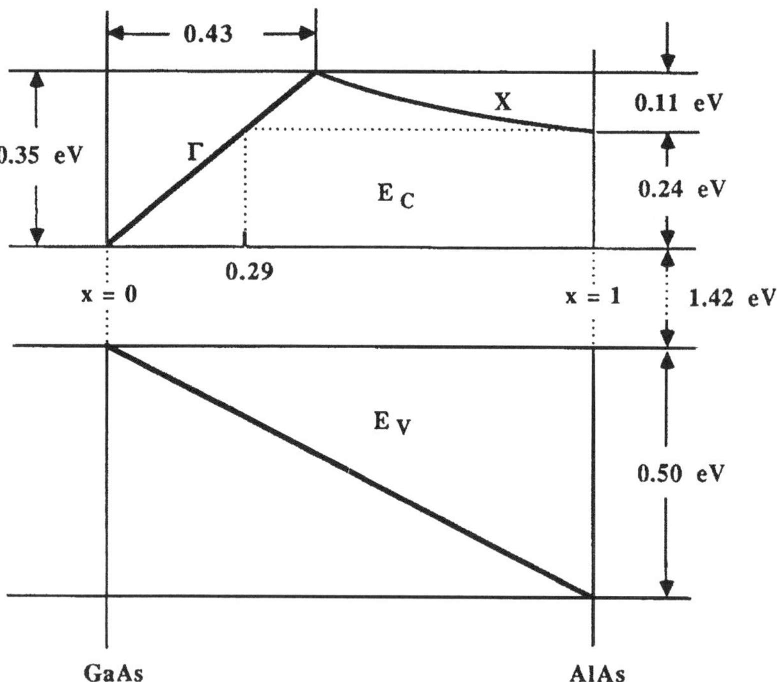

Fig. 1. Proposed reconciliation of experimental conduction and valence band energies in the (Al, Ga)As heterosystem.

position. Experimental evidence for such a staggered lineup at $x = 0.4$ has already been reported [18]. Such a staggered lineup would have numerous far-reaching consequences, a discussion of which would go beyond the scope of this paper.

Acknowledgements

It is a pleasure to acknowledge extensive and very useful discussions with several individuals, especially Dr. S.L. Wright of IBM (who also made a valuable preprint available), and Mr. J.R. Waldrop and Drs. R.W. Grant and E.A. Kraut of Rockwell. This work was supported by the Office of Naval Research.

References

[1] H. Kroemer, in: Molecular Beam Epitaxy and Heterostructures, Eds. L.L. Chang and K. Ploog (Nijhoff, The Hague, 1985) p. 331; see also: J. Vacuum Sci. Technol. B2 (1984) 433.

306 *H. Kroemer / Band offsets at heterointerfaces*

[2] R. Dingle, in: Advances in Solid-State Physics, Vol. 15. Festkörperprobleme, Ed. H.J. Queisser (Pergamon/Vieweg, London/Braunschweig, 1975) p. 21.

[3] W.A. Harrison, J. Vacuum Sci. Technol. 14 (1977) 1016.

[4] R.C. Miller, A.C. Gossard, D.A. Kleinman and O. Munteanu, Phys. Rev. B29 (1984) 3740.

[5] For detailed references, see, for example: G. Duggan, J. Vacuum Sci. Technol. B3 (1985) 1224.

[6] J. Tersoff, Phys. Rev. B30 (1985) 4874.

[7] W.I. Wang and F. Stern, J. Vacuum Sci. Technol. B3 (1985) 1280.

[8] H. Kroemer, W.-Y. Chien, J.S. Harris and D.D. Edwall, Appl. Phys. Letters 36 (1980) 295.

[9] M.O. Watanabe, J. Yoshida, M. Mashita, T. Nakanisi and A. Hojo, Appl. Phys. 57 (1985) 5340.

[10] H. Okumura, S. Misawa, S. Yoshida and S. Gonda, Appl. Phys. Letters 46 (1985) 377.

[11] T.W. Hickmott, P.M. Solomon, R. Fischer and H. Morkoç, J. Appl. Phys. 57 (1985) 2844.

[12] J. Batey and S.L. Wright, to be published; see also Surface Sci. 174 (1986) 320.

[13] H.C. Casey and M.B. Panish, Heterostructure Lasers (Academic Press, New York, 1978).

[14] H. Kroemer, Appl. Phys. Letters 46 (1985) 504.

[15] S.R. Forrest, P.H. Smith, R.B. Wilson and M.L. Kaplan, Appl. Phys. Letters 45 (1984) 1190.

[16] See, for example: C.M. Garner, Y.D. Shen, C.Y. Su, G.L. Pearson and W.E. Spicer, J. Vacuum Sci. Technol. 15 (1978) 1480.

[17] N.C. Miller, S. Zemon, G.P. Werber and W. Powazinik, J. Appl. Phys. 57 (1985) 512.

[18] T.J. Drummond and I.J. Fritz, Appl. Phys. Letters 47 (1985) 284.

IEEE ELECTRON DEVICE LETTERS, VOL. EDL-8, NO. 1, JANUARY 1987

An (Al,Ga)As/GaAs Heterostructure Bipolar Transistor with Nonalloyed Graded-Gap Ohmic Contacts to the Base and Emitter

M. A. RAO, E. J. CAINE, STEPHEN I. LONG, SENIOR MEMBER, IEEE, AND HERBERT KROEMER, FELLOW, IEEE

Abstract—Graded regions of n-(Ga,In)As and p-Ga(As,Sb) were incorporated side-by-side as emitter and base contacts, respectively, into an n-p-n (Al,Ga)As/GaAs heterostructure bipolar transistor (HBT). The process involved two separate molecular beam epitaxy (MBE) growths, leading to base contact regions that were self-aligned to the emitter mesas. The devices could be easily probed with pressure contacts even prior to any metallization, and excellent characteristics were obtained after final metallization. Contact resistivities of 5×10^{-7} and 3×10^{-6} $\Omega \cdot cm^2$ were measured for n- and p-type graded-gap ohmic contact structures, respectively.

I. INTRODUCTION

IN 1981, Woodall *et al.* [1] proposed and demonstrated a nonalloyed graded-gap scheme for obtaining ohmic contacts to n-type GaAs, by first growing a graded transition from GaAs to InAs and then making a nonalloyed metallic contact to the InAs. The underlying idea was as follows. It is well known that at a metal-to-InAs interface the Fermi level is pinned inside the InAs conduction band [2], hence this interface by itself acts as an ideal negative-barrier ohmic contact. However, if the GaAs-to-InAs transition were not graded, it would act as a quasi-Schottky barrier with a barrier height close to the conduction-band offset ΔE_c of the GaAs/InAs heterojunction, about 0.9 eV [3], and the contact would be poor overall. Sufficient grading flattens out the heterojunction barrier, and leads to an excellent ohmic contact with properties that make it an attractive alternative to the widely used Au/Ge/Ni/Au alloyed system [4]–[6]. For p-type GaAs, the Ga(As,Sb) system could be similarly used, as proposed by Chang and Freeouf [7].

In the present work, we report the side-by-side incorporation of such n- and p-type graded-gap contacts as emitter and base contacts of an n-p-n (Al,Ga)As/GaAs heterostructure bipolar transistor (HBT). Not only are the contact resistivities

Manuscript received October 7, 1986. This work was supported by the Air Force Office of Scientific Research under Contract AFOSR-82-0344 and by the Semiconductor Research Corporation.

M. A. Rao, S. I. Long, and H. Kroemer are with the Department of Electrical and Computer Engineering, University of California, Santa Barbara, CA 93106.

E. J. Caine was with the Department of Electrical and Computer Engineering, University of California, Santa Barbara, CA 93106. He is now with GEC Hirst Research Center, Wembley, Middlesex, England.

IEEE Log Number 8612481.

competitive with those of alloyed contacts, but the nonalloyed contacts are noninvasive. They are, therefore, particularly attractive for bipolar transistors, because emitter regions can be employed that are much thinner than those possible with alloyed contacts, leading to reduced emitter resistances. Also, the possibility of alloying through the thin base region is eliminated with nonalloyed base contacts.

II. GRADED-GAP CONTACT GROWTH PROCEDURE

As a preliminary to transistors, n- and p-type graded-gap contacts grown by molecular beam epitaxy (MBE) on (100)-oriented semi-insulating GaAs substrates were investigated separately. To keep the series path resistance of the graded region low, the doping in the graded region should be as high as possible and the graded region should be as narrow as possible, limited only by the requirement to flatten the quasi-Schottky barrier mentioned in the introduction. Theoretical investigations [8] show that graded regions as narrow as 30 nm should be permissible. For the n-type graded-gap contact structures, approximately 200 nm of 1×10^{18} cm^{-3}(Si-doped) n$^+$-GaAs was grown at 600°C, followed by approximately 30 nm of n$^+$-(Ga,In)As compositionally graded from GaAs to InAs. The grading was achieved by ramping the temperatures of the Ga and In furnaces. In addition, during the growth of the graded region the substrate temperature was ramped down from 600°C to 500°C because the temperature of congruent sublimation for (Ga,In)As decreases with increasing indium fraction [9]. Approximately 30 nm of n$^+$-InAs was grown above the graded region. The doping level in the graded region and the InAs was about 3×10^{18} cm^{-3}.

For the p-type graded-gap contact structures, a 200-nm p$^+$-GaAs buffer layer was grown first, followed by 30-nm p$^+$-Ga(As,Sb) compositionally graded from GaAs to GaSb. The growth of the graded region was initiated by opening the shutter of the Sb source. Both the antimony-to-gallium and the *initial* arsenic-to-gallium *atomic* flux ratios were approximately 3:1. In the presence of such an arsenic flux at a substrate temperature of 600°C, very little (< 1 percent) antimony gets incorporated into the growing material [10]. During the growth of the graded region the substrate temperature was then ramped from 600°C to 470°C, for two reasons.

1) The temperature of congruent sublimation of GaSb is approximately 455°C [9] which is much lower than that of GaAs. Consequently GaSb requires a lower growth temperature than GaAs. 2) The substitution of arsenic by antimony is enhanced at lower growth temperatures [10], thus facilitating the growth of the graded layer. In addition, to accomplish a complete transition to GaSb, the power to the arsenic furnace was turned off 3 min into the growth of the graded region. The arsenic flux dropped to a tenth of its initial value at the end of the growth of the graded region, while the antimony flux remained constant. About 30 nm of p^+-GaSb was grown above the graded layer. The p^+-layers were doped with Be to about 5×10^{18} cm^{-3}.

III. CONTACT RESISTIVITY MEASUREMENTS

The specific contact resistivities of the n- and p-type graded-gap contacts were measured with a four-point Kelvin cross-resistor structure [11]. For nonalloyed (Ga,In)As graded-gap contacts to n-GaAs, contact resistivities down to 5×10^{-7} $\Omega \cdot$cm^2 were obtained. For p-type Ga(As,Sb) graded-gap contacts, resistivities down to 3×10^{-6} $\Omega \cdot$cm^2 were measured. However, the Kelvin measurement technique is extremely sensitive to misalignment between the mesa and the ohmic metal [12]. In general, the measurement overestimates the contact resistivity and it is this pessimistic upper limit that is reported here. The true value of the resistivity could be much lower than the measured value.

For comparison, the typical values for Au/Ge/Ni n-type and Au/Zn p-type alloyed contacts found in the literature are 1×10^{-6} $\Omega \cdot$cm^2 [5] and 7×10^{-6} $\Omega \cdot$cm^2 [13]. It can be seen that our measured values are competitive with the typical values reported in the literature.

IV. TRANSISTORS

The HBT structure was grown by MBE on a (100)-oriented n^--GaAs substrate. The doping levels, compositions, and thicknesses of the initial HBT layers are shown in Fig. 1. The region between the base and the emitter was digitally graded with a narrow-well (Al,Ga)As/GaAs superlattice [14]. The Al mole fraction in the (Al,Ga)As emitter was 0.25. The GaAs was grown at 600°C, and the (Al,Ga)As at 650°C, with arsenic-to-group III *atomic* flux ratios of approximately 2:1. The growth temperature was reduced to 600°C at the end of the emitter layer, and 30 nm of (Al,Ga,In)As compositionally graded from (Al,Ga)As to InAs was grown, followed by 30 nm of InAs, using a procedure similar to the one described for the growth of the (Ga,In)As graded-gap contact in Section II. The graded region and the InAs were heavily doped n-type with Si to about 3×10^{18} cm^{-3}.

The sample was then removed from the MBE system and SiO$_2$ was deposited using plasma-enhanced chemical vapor deposition (PCVD) at 300°C. The SiO$_2$ was patterned into the emitter mesa regions using standard photoresist techniques. The SiO$_2$ was used as a mask for etching the emitter mesas, with H$_3$PO$_4$:H$_2$O$_2$:H$_2$O = 4:1:50. The sample was then rinsed in solvents, rinsed in deionized water for 10 min, and reloaded into the MBE system for the base contact regrowth.

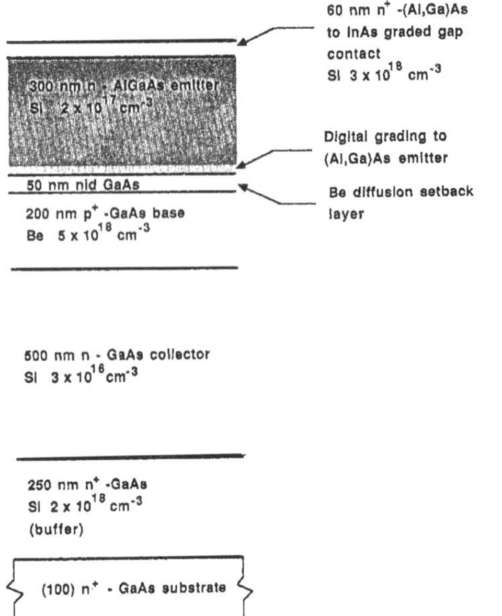

Fig. 1.   Heterostructure bipolar transistor layer diagram. The structure was grown by MBE.

The oxide was desorbed at approximately 620°C in an arsenic ambient. A small carbon peak was detected by Auger electron spectroscopy after the oxide desorption. Approximately 100 nm of p^+-GaAs was grown to smooth out the restart interface. Then 30 nm of p^+-Ga(As,Sb) compositionally graded from GaAs to GaSb was grown, followed by 30 nm of p^+-GaSb, as described in Section II. The doping in the p^+-layers was about 5×10^{18} cm^{-3}. Because of the undercut due to the emitter mesa etch, the MBE regrowth of the p-type contacts leads to base contact regions that are self-aligned with respect to the emitter mesas, a very desirable feature.

One of the potential advantages of graded-gap contacts is that they are capable of withstanding high-temperature processing. Our growth procedure showed that the (Ga,In)As capped with SiO$_2$ could clearly withstand high temperatures of 620°C during the regrowth of the p-type graded-gap contact.

After the sample was removed from the MBE system the material grown on top of the SiO$_2$ was found to be polycrystalline. Furthermore, it appeared to be penetrated by an HF etch; the SiO$_2$ was easily etched off with HF, along with the material on top of the SiO$_2$. Base mesas were then etched, first using HF:H$_2$O$_2$:H$_2$O = 10:1:100 to etch the Ga(As,Sb) and then using the phosphoric acid-based etch mentioned above to etch the GaAs. At this point the device *I–V* characteristics could be obtained even prior to any final metallization, using tungsten probe tips to directly contact the semiconductor surfaces of the base and the emitter. Indium alloyed on the backside of the n^+-wafer formed the collector contact. Examples of *I–V* curves obtained in this manner are shown in Fig. 2(a).

Finally, a dielectric (SiO$_2$) was deposited by PCVD, contact

32

IEEE ELECTRON DEVICE LETTERS, VOL. EDL-8, NO. 1, JANUARY 1987

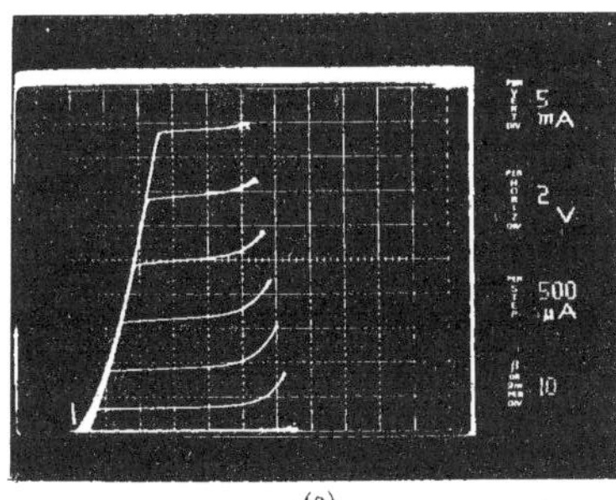

(a)

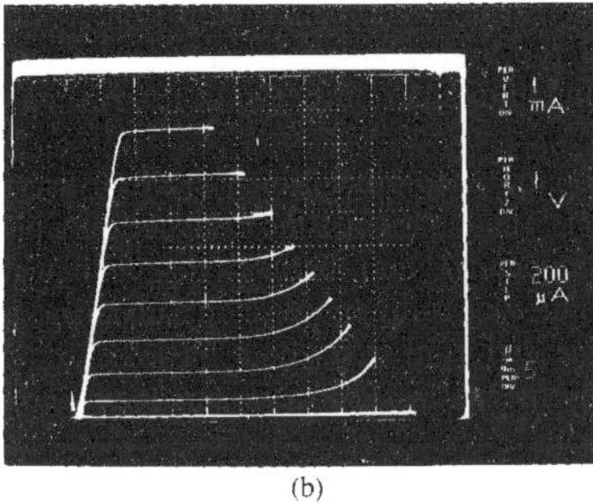

(b)

Fig. 2. Transistor I–V curves obtained (a) prior to metallization. Emitter area is $150 \times 150~\mu m^2$. (b) After final metallization. Emitter area is $20 \times 20~\mu m^2$. These transistors were from a different run than those corresponding to (a).

windows were cut, and Cr/Au metallization was deposited by evaporation. Both the base and the emitter metallization were done in one step. No alloying was done. The device I–V curves obtained after final metallization are shown in Fig. 2(b). It should be noted that the I–V curves in Fig. 2(b) were obtained on devices fabricated in a different run than those corresponding to Fig. 2(a). The I–V curves were very similar to those obtained on conventional alloyed contact transistors which were fabricated in a different run.

ACKNOWLEDGMENT

The authors would like to thank A. Yuen, T. Hausken, and S. Subbanna for useful discussions, and D. Zak for technical assistance.

REFERENCES

[1] J. M. Woodall, J. L. Freeouf, G. D. Pettit, T. Jackson, and P. Kirchner, "Ohmic contacts to n-GaAs using graded band gap layers of $Ga_{1-x}In_xAs$ grown by molecular beam epitaxy," *J. Vac. Sci. Technol.*, vol. 19, p. 626, 1981.

[2] C. A. Mead and W. G. Spitzer, "Fermi level position at metal-semiconductor interfaces," *Phys. Rev.*, vol. 134, p. 713, 1964.

[3] S. P. Kowalczyk, W. J. Schaeffer, E. A. Kraut, and R. W. Grant, "Determination of the InAs-GaAs (100) heterojunction band discontinuities by x-ray photoelectron spectroscopy (XPS)," *J. Vac. Sci. Technol.*, vol. 20, p. 705, 1982.

[4] T. Hara and T. Inada, "Trends in ion implantation in gallium arsenide," *Soild-State Technol.*, vol. 22, p. 69, 1979.

[5] T. S. Kuan, P. E. Batson, T. N. Jackson, H. Rupprecht, and E. L. Wilkie, "Electron microscope studies of an alloyed Au/Ni/Au-Ge ohmic contact to GaAs," *J. Appl. Phys.*, vol. 54, p. 6952, 1983.

[6] A. A. Ketterson, F. Ponse, T. Henderson, J. Klem, C. -K. Peng, and H. Morkoç, "Characterization of extremely low contact resistances on modulation-doped FET's," *IEEE Trans. Electron Devices*, vol. ED-32, p. 2257, 1985.

[7] L. L. Chang and J. L. Freeouf, "Ohmic contacts to p-type semiconductors," *IBM Tech. Discl. Bull.*, vol. 24, p. 4065, 1982.

[8] H. Kroemer, unpublished results.

[9] C. E. C. Wood, K. Singer, T. Ohashi, L. R. Dawson, and A. J. Noreika, "A pragmatic approach to adatom-induced surface reconstruction of III-V compounds," *J. Appl. Phys.*, vol. 54, p. 2732, 1983.

[10] C.-A. Chang, R. Ludeke, L. L. Chang, and L. Esaki, "Molecular-beam epitaxy (MBE) of $In_{1-x}Ga_xAs$ and $GaSb_{1-y}As_y$," *Appl. Phys. Lett.*, vol. 31, p. 759, 1977.

[11] S. J. Proctor and L. W. Linholm, "A direct measurement of interfacial contact resistance," *IEEE Electron Device Lett.*, vol. EDL-3, p. 294, 1982.

[12] W. M. Loh, K. Saraswat, and R. W. Dutton, "Analysis and scaling of Kelvin resistors for extraction of specific contact resistivity," *IEEE Electron Device Lett.*, vol. EDL-6, p. 105, 1985.

[13] T. Sanada and O. Wada, "Ohmic contacts to p-GaAs with Au/Zn/Au structure," *Japan. J. Appl. Phys.*, vol. 19, p. L491, 1980.

[14] S. L. Su, R. Fischer, W. G. Lyons, O. Tejayadi, D. Arnold, J. Klem, and H. Morkoç, "Double heterojunction GaAs/$Al_xGa_{1-x}As$ bipolar transistors prepared by molecular beam epitaxy," *J. Appl. Phys.*, vol. 54, p. 6725, 1983.

Electron concentrations and mobilities in AlSb/InAs/AlSb quantum wells

Gary Tuttle, Herbert Kroemer, and John H. English
Department of Electrical and Computer Engineering, University of California, Santa Barbara,
California 93106

(Received 5 December 1988; accepted for publication 24 February 1989)

We present data on the electron concentrations and mobilities in deep (≈ 1.3 eV) AlSb/InAs/
AlSb quantum wells grown by molecular-beam epitaxy. High electron sheet concentrations of
the order 10^{12} cm^{-2}, found in the not-intentionally doped wells, indicate the presence of a deep
donor in the AlSb barriers. Typical mobilities are between 22 000 and 28 000 cm^2/V s at room
temperature, increasing with decreasing temperature, and leveling out below 50 K at values
between 175 000 and 330 000 cm^2/V s. The temperature-independent low-temperature
mobilities indicate a nonthermal scattering mechanism, possibly interface roughness scattering.
Under illumination the wells exhibit a strong negative photoconductivity, which is explained as
a natural consequence of the band structure of the wells.

The InAs/AlSb heterojunction system, first studied by
Chang *et al.* in 1984,[1] is a potentially very interesting system.
From the known valence-band offset of the GaSb-InAs
system, ΔE_v [InAs$\rightarrow$GaSb] $= 0.5$ eV,[2,3] and from the
recently determined valence-band offset in AlSb/GaSb,
ΔE_v [AlSb$\rightarrow$GaSb] $= 0.4$ eV,[4] one estimates a slightly
staggered valence-band offset for InAs/AlSb,
ΔE_v [InAs$\rightarrow$AlSb] $= 0.1$ eV. Together with the large energy gap of AlSb (1.55 eV), this implies a very large conduction-band offset of about 1.3 eV [Fig. 1(a)]. Capacitance
measurements by Nakagawa[5] on rectifying InAs/AlSb junctions (both *n-N* and *n-P* junctions) support the predicted
1.3-eV conduction-band offset.

Such a large conduction-band offset implies the possibil-

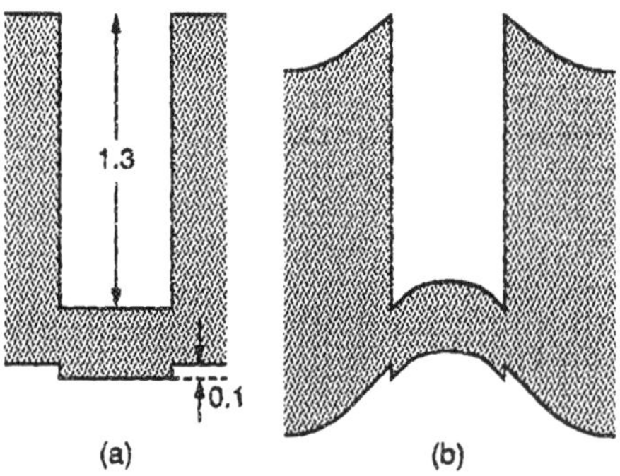

FIG. 1. (a) Predicted band lineup for an InAs/AlSb quantum well. (b) Band bending in the presence of an excess charge stored inside the well.

ity of very deep InAs/AlSb quantum wells (Fig. 1), with the very high electron mobility of InAs inside the well, a combination of considerable interest for quantum-well field-effect transistors[6] and possibly for other devices. Mobilities greater than 10^5 cm^2/V s have already been reported for InAs quantum wells with Ga-rich *ternary* (Al,Ga)Sb and Ga(As,Sb) barriers,[7] with such barrier compositions that the broken gap of InAs/GaSb was closed to a staggered gap by adding either Al or As to GaSb. In this communication, we show that similarly high and even higher mobilities can also be achieved with Ga-free *binary* AlSb barriers.

The samples reported here were grown by molecular-beam epitaxy (MBE) in a Varian modular GEN II system, equipped with elemental group III and group V sources, the latter producing As$_4$ and Sb$_4$ beams. Bulk InAs layers grown in this machine show n-type background concentrations on the order of 1×10^{15} cm^{-3}, and 77-K mobilities near 60 000 cm^2/V s. Not-intentionally doped bulk layers GaSb are p-type at a level of 1×10^{16} cm^{-3}. Not-intentionally doped bulk AlSb layers have an extremely high resistivity, and we have been unable to measure their background carrier concentration or even the residual conductivity type.

All growths were performed on semi-insulating GaAs substrates. They have an 8% smaller lattice constant than AlSb, and to obtain good electrical properties, relatively thick GaSb or AlSb buffer layers were found important. In our two best growths, reported here, we employed a 1-μm GaSb buffer layer, followed by a 2-μm AlSb buffer, which in turn was followed by a ten-period (25 Å + 25 Å) GaSb/AlSb superlattice. The latter aids considerably in obtaining a smooth morphology,[8] but transmission electron microscopy (TEM) measurements have shown that it fails in its original objective of suppressing misfit-induced threading dislocations. Based on TEM measurements on other lattice-mismatched structures investigated in our laboratory,[9] we would expect our samples to have at least 10^7 threading dislocations per cm^2.

The GaSb layer was doped with Te at a level of $\approx1\times10^{16}$ cm^{-3} in an effort to compensate the holes normally found in undoped GaSb.[10] Following this first GaSb layer, the Te source was shuttered and cooled, and the rest of the structure was grown without intentional doping. The

GaSb and AlSb buffer layers were grown at a substrate temperature of 530 °C as measured by infrared pyrometry. The temperature was then lowered to 500 °C just prior to the growth of the GaSb/AlSb superlattice, and held there for the remainder of the growth.

The quantum well itself consisted of a 120-Å InAs well sandwiched between two 200-Å AlSb barriers. The As flux was preset to a level to keep the InAs growth just barely on the As-stable side, as indicated by RHEED reconstruction pattern. Our experience, and that of others,[11-13] has shown that minimizing the As excess is necessary for obtaining high-mobility InAs. No substrate rotation was employed during the growth, so that RHEED pattern changes could be observed at the interfaces. In the growths reported here, neither the AlSb barriers nor the InAs quantum wells were intentionally doped.

To protect the AlSb from reaction with the water vapor in the air, a 50-Å GaSb cap layer was grown on top of the structure. More details of the growth of the quantum-well structures will be presented elsewhere.

Electron sheet concentrations and mobilities were measured using conventional Hall bridges and van der Pauw cloverleaf patterns. Small dots of indium were alloyed to the samples to form electrical contacts. These contacts almost certainly alloyed through to the underlying buffer layers. However, the low mobilities and carrier concentrations in these layers would have a small, if not negligible, effect on the measurements. The Hall measurements were done at temperatures ranging from 300 to 15 K using a closed-cycle cryostat.

In Fig. 2 we show the electron concentration and mobility of our most extensively characterized sample (a conventional Hall bar sample), as it was cooled in the cryostat in the absence of any illumination. As can be seen, the room-temperature mobility is $\approx25\,000$ cm^2/V s, and the mobility rises monotonically as the temperature is lowered down to 50 K. Below this temperature, the mobility remains essentially constant at 230 000 cm^2/V s. The carrier sheet concentration decreases slightly from its room-temperature value of 1.25×10^{12} cm^{-2} to 8.0×10^{11} cm^{-2} at 15 K. Other sam-

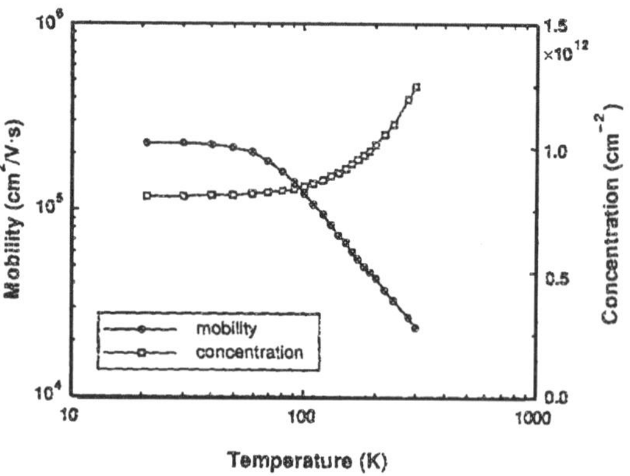

FIG. 2. Electron sheet concentration and mobility as functions of temperature for a 120-Å-wide InAs/AlSb quantum well.

ples from this growth and from an essentially identical second growth behaved similarly, but with some quantitative variations: The room-temperature mobilities ranged from 22 000 to 28 000 cm²/V s, and the low-temperature mobilities from 175 000 to 330 000 cm²/V s.

The high electron concentrations in these not-intentionally doped structures—found in all our growths—cannot not be due to any background donor in the InAs itself: (i) the electron concentration is three decades higher than the known background; (ii) the high mobilities are incompatible with the amount of impurity scattering such a high donor concentration would introduce.

We can also rule out that the electrons are neutralized by holes in the valence band of the AlSb, as might be the case in a broken-gap InAs/GaSb structure: The Fermi level in the quantum well can be estimated to be almost 0.2 eV above the bottom of the well (including quantization effects), far too high to create a sufficiently high hole concentration. The electrons must arise by transfer from some kind of donor on the AlSb side of the interface. Inasmuch as not-intentionally doped AlSb contains an insufficient concentration of shallow donors, we conclude that the electrons in the quantum well are supplied by a deep donor in the AlSb. If sufficiently deep, such donors would not manifest themselves in bulk AlSb except to the extent that they would compensate intentionally added shallow acceptors. But even a donor as deep as 1 eV below the conduction band could readily drain into the deep InAs quantum well, and cause the Fermi level in the well to rise by more than 0.2 eV, if present in sufficiently large concentrations. Whatever the nature of the deep donors, their concentration must be high: Simple band bending arguments show that the neutralizing donor charge must reside within at most 2×10^{-6} cm of the heterointerface, which calls for donor concentrations of at least 5×10^{17} cm^{-3}, and probably more. The high mobilities speak against a true interface defect, but they do not rule out a defect caused by the presence of the interface, but residing some distance away, inside AlSb. An arsenic antisite defect on an aluminum site would be a purely speculative possibility. This picture of deep donors near the interface is consistent with the arguments used to explain the excess concentrations of electrons found in InAs/GaSb quantum wells.[14]

The idea that the responsible deep donor is absent far away from the interface is supported by the weak decrease of the electron concentration with decreasing temperature: This decrease implies a decrease in the band bending outside the well, which in turn implies that the Fermi level inside the AlSb bulk moves lower with decreasing temperature, as would be the case for low hole concentration p-type AlSb.

An unexpected property of these wells is a pronounced negative persistent photoconductivity at temperatures below 130 K. By flashing a green light-emitting diode (LED) mounted adjacent to the sample, the carrier concentration in the well is substantially reduced. This is in striking contrast to the persistent positive photoconductivity caused by D-X centers in GaAs/(Al,Ga)As structures. Hand-in-hand with the reduction in electron concentration goes a reduction in mobility. At temperatures below 100 K, the changes persist for hours after the light is turned off, but they disappear if the

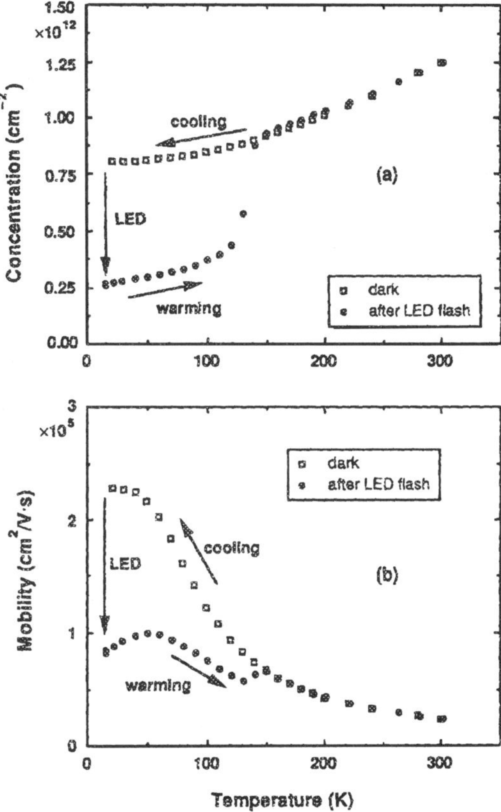

FIG. 3. (a) Electron concentration and (b) mobility in an InAs/AlSb quantum well measured during cooldown in the dark, and during warmup after temporary illumination at 15 K with a green LED.

sample is permitted to warm above about 100 K, having completely disappeared at 130 K. In Fig. 3, the concentration and mobility taken in the dark are plotted along with the concentration and mobility obtained as the sample was warmed after the LED had been flashed.

We believe that the negative photoconductivity has a straightforward explanation: The illumination with green light is bound to create electron-hole pairs in the AlSb. The holes are driven by the band bending towards the well, where they can recombine with the electrons in the well, either directly, or across the staggered-lineup interface, similar to the case of InP/(Al,In)As.[15] This depletes the well. At sufficiently low temperatures, the remaining repulsive band bending barrier would prevent the well from being refilled, thus creating the persistence of the effect. Additional contributions to the initial electron removal might be direct photoemission of electrons from the well, or deep-level effects, but in the absence of any experimental evidence, we see no compelling need to invoke such additional mechanisms.

The reduction in mobility that goes hand in hand with the reduction in electron concentration is very likely due to the same screening effect that was already observed by Störmer et al.[16] in GaAs/(Al,Ga)As structures. Using the cumulative illumination method employed by these authors, we found that the low-temperature mobility decreases roughly linearly with decreasing electron concentration.

The leveling out of the dark mobilities below 50 K implies a nonphonon scattering mechanism at low temperatures, meaning either impurity scattering or interface roughness scattering. Our highest mobilities are sufficiently high that they approach the interface roughness limit predicted by Sakaki *et al.*[17] for GaAs/AlAs quantum wells of the same thickness. As was shown by these authors, in that limit the mobility will vary with the sixth power of the well width. In our samples there was a considerable variation in the measured mobilities of samples taken from the same substrate. Since no substrate rotation was used in the growth of our AlSb/InAs samples, a 10% variation in growth rate across the substrate would be expected, corresponding to more than 60% variation in mobility, close to the variations observed. Furthermore, although the mobility was found to vary widely, the carrier concentrations in the various samples were nearly identical, and the manner in which carrier concentration depended on temperature and illumination was consistent from sample to sample. This suggests that roughness scattering might play a major role in determining the low-temperature mobility of InAs/AlSb quantum wells.

We wish to express our thanks for many discussions to Seshadri Subbanna, Atsushi Nakagawa, and Professor Art Gossard. This work was supported by the Semiconductor Research Corporation and by the Office of Naval Research.

One of us (G. T.) gratefully acknowledges the fellowship support of Hewlett-Packard.

[1] C. A. Chang, L. L. Chang, E. E. Mendez, M. S. Christie, and L. Esaki, J. Vac. Sci. Technol. B **2**, 214 (1984).
[2] H. Sakaki, L. L. Chang, R. Ludeke, C. A. Chang, G. A. Sai-Halasz, and L. Esaki, Appl. Phys. Lett. **31**, 211 (1977).
[3] L. L. Chang and L. Esaki, Surf. Sci. **98**, 70 (1980).
[4] G. J. Gualtieri, G. P. Schwartz, R. G. Nuzzo, and W. A. Nuzzo, Appl. Phys. Lett. **49**, 1037 (1986).
[5] A. Nakagawa, H. Kroemer, and J. H. English, Appl. Phys. Lett. (in press).
[6] G. Tuttle and H. Kroemer, IEEE Trans. Electron Devices **ED-34**, 2358 (1987).
[7] H. Munekata, L. Esaki, and L. L. Chang, J. Vac. Sci. Technol. B **5**, 809 (1987).
[8] G. Griffiths, K. Mohammed, S. Subbanna, H. Kroemer, and J. L. Merz, Appl. Phys. Lett. **43**, 1059 (1983).
[9] T. Y. Liu and P. M. Petroff (unpublished).
[10] S. Subbanna, G. Tuttle, and H. Kroemer, J. Electron. Mater. **17**, 297 (1988).
[11] W. J. Schaffer, M. D. Lind, S. P. Kowalczyk, and R. W. Grant, J. Vac. Sci. Technol. B **1**, 688 (1983).
[12] B. R. Hancock and H. Kroemer, J. Appl. Phys. **55**, 4239 (1984).
[13] S. M. Newstead, R. A. A. Kubiak, and E. H. C. Parker, J. Cryst. Growth **81**, 49 (1988).
[14] H. Munekata, T. P. Smith, and L. L. Chang, J. Cryst. Growth **95**, 235 (1989).
[15] E. J. Caine, S. Subbanna, H. Kroemer, J. L. Merz, and A. Y. Cho, Appl. Phys. Lett. **45**, 1123 (1984).
[16] H. L. Störmer, A. C. Gossard, W. Wiegmann, and K. Baldwin, Appl. Phys. Lett. **39**, 912 (1981).
[17] H. Sakaki, T. Noda, K. Hirakawa, M. Tanaka, and T. Matsusue, Appl. Phys. Lett. **51**, 1934 (1987).

Reprinted with permission from

**P. F. Hopkins, A. J. Rimberg, R. M. Westervelt, G. Tuttle, and H. Kroemer,
``Quantum Hall effect in InAs/AlSb quantum wells,''
Appl. Phys. Lett., Vol. 58(13), pp. 1428-1430, 1991.**

Quantum Hall effect in InAs/AlSb quantum wells

P. F. Hopkins,[a] A. J. Rimberg, and R. M. Westervelt
*Division of Applied Sciences and Department of Physics, Harvard University, Cambridge,
Massachusetts 02138*

G. Tuttle and H. Kroemer
*Department of Electrical and Computer Engineering, University of California, Santa Barbara,
California 93106*

(Received 10 December 1990; accepted for publication 24 January 1991)

We demonstrate via low-temperature electron transport measurements the realization of a
high-mobility ($> 300\,000$ cm^2/V s) two-dimensional electron gas in unintentionally
doped InAs/AlSb single 120 Å quantum wells grown on GaAs substrates by molecular beam
epitaxy. Magnetoresistance and Hall measurements at $T \sim 0.4$ K show a well-formed
quantum Hall effect, with effects due to spin splitting observed at filling factors as high as
$\nu = 17$. The electron densities of these wells could be reduced by a factor ~ 5 by using
the negative persistent photoconductivity of these samples.

Quantum well structure made from InAs, with barriers
made from AlSb, GaSb, or (Al,Ga)Sb alloys, possibly with
the addition of As to the barrier, are natural candidates for
the study of low-temperature magnetotransport effects
such as the quantum Hall effect and the Shubnikov–de
Haas effect. Of all binary III/V compounds, InAs has the
second highest *intrinsic* electron mobility. Only InSb has a
higher mobility, but it suffers from the absence of an approximately lattice-matched barrier material that would
permit the construction of quantum wells with low-defect
barrier interfaces, necessary for highest mobility transport.
In the present work, we report on InAs quantum wells
with straight (unalloyed) AlSb barriers, of the kind reported earlier by Tuttle *et al.*[1,2] Although slightly less-well
lattice matched and technologically slightly more difficult
to grow than GaSb or (Al,Ga)Sb barriers, AlSb barriers
have the advantage that they eliminate complications due
to a broken-gap band structure at the interface. These complications are of interest in their own right, and were in fact
recently studied by Munekata *et al.*[3] but in the present
work it was desired to avoid them.

Quantum wells of InAs/AlSb were first studied by
Chang *et al.*,[4] who also reported the first observation of the
quantum Hall effect in such structures. However, those
early samples still suffered from relatively low mobilities
($\sim 15\,000$ cm^2/V s), and the transport measurements were
made only down to $T \sim 4.2$ K, so that much detail remained unresolved. In the years since the pioneering work
of Chang *et al.*, great improvements in the technology of
the InAs/AlSb have been made. Low-temperature ($\leqslant 10$
K) mobilities up to 330 000 cm^2/V s were reported by
Tuttle *et al.*[2] in 1989, and more recently, a value as high as
613 000 cm^2/V s has been reported by Chalmers *et al.*[5] in
a quantum well with a modified interface structure. The
work reported here is a continuation of the earlier work of
Tuttle *et al.*, drawing on the same high-mobility samples,
and extending the magnetotransport measurements down

to $T \sim 0.4$ K.

More specifically, we report on low-temperature magnetoresistance and Hall effect data on two 120 Å InAs/
AlSb quantum well samples, with the magnetic field applied perpendicularly and parallel to the plane of the
electron layer. These data provide evidence for a high quality, high-mobility, two-dimensional electron gas layer in
these wells.

The unintentionally doped 120 Å InAs/AlSb samples
discussed in the present letter were grown by molecular
beam epitaxy on GaAs substrates, with InSb-like quantum
well interfaces, as described by Tuttle *et al.*[2] Because of the
lattice mismatch (7%) between the InAs/AlSb system and
GaAs, we expect the samples to have threading dislocation
densities of 10^7 cm^{-2} or higher,[1] and one of the objectives
of this study was to see if these dislocations posed a serious
obstacle in obtaining high-quality two-dimensional (2D)
transport. Hall bars (7×3 mm^2) with three pairs of Hall
voltage probes were photolithographically defined, and
contacts were made to the electron gas by alloying dots of
indium at 300 °C for 5–10 min. The samples were immersed in He-3 to temperatures $T \sim 0.4$ K, and measured
in magnetic fields to 23 T at the Francis Bitter National
Magnet Laboratory. Transverse magnetoresistance and
Hall effect measurements were taken with the applied magnetic field B perpendicular to the plane of the electron
layer. The two distinct magnetoresistance measurements
with B in the plane of the electron layer were taken, with B
parallel and perpendicular to the current. All resistance
measurements were taken using low-frequency (typically
11 Hz) ac lock-in techniques; the current bias levels (< 0.2
μA) and magnetic field ramp rates (< 1 T/min) were kept
low to avoid sample heating. Illumination of the samples to
reduce the carrier density via the negative persistent photoconductivity effect[2] was done at < 77 K with a red (640
nm measured at 77 K) or green (560 nm at 77 K) light-
emitting diode (LED) mounted near the sample. Several
sheets of filter paper were used as a diffuser to insure more
uniform illumination of the sample; this was checked by
comparing data from different contact pairs. In addition,
care was taken to insure that no room light reached the

---

[a] Current address: Materials Department, University of California, Santa
Barbara, CA 93106.

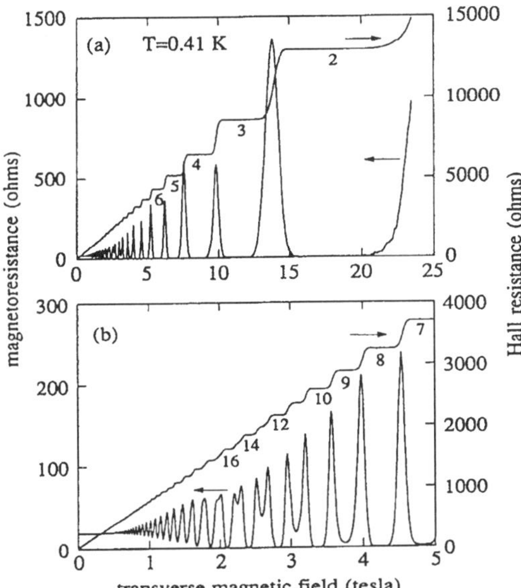

FIG. 1. Transverse magnetoresistance and Hall resistance at $T = 0.41$ K for the unilluminated sample with $n_s = 8.3 \times 10^{11}$ cm^{-2}: (a) 0–23 T; (b) 0–5 T.

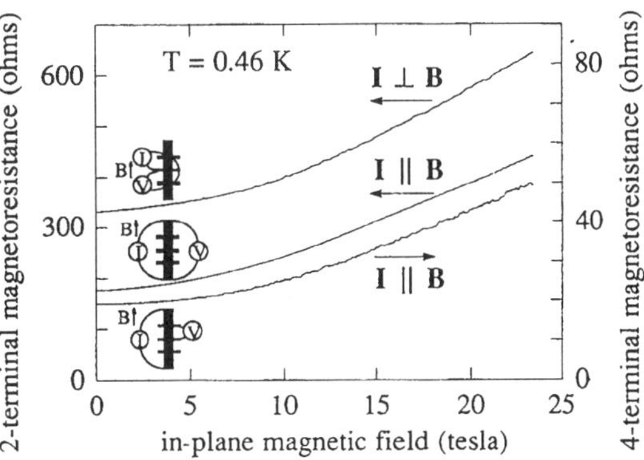

FIG. 2. Magnetoresistance and Hall resistance for the unilluminated sample at $T = 0.46$ K and 0–23 T with the magnetic field applied in the plane of the electron layer. Data were taken both for the current I along B, $I\|B$, and perpendicular to B, $I{\perp}B$. The two-terminal and four-terminal resistances for $I\|B$ and the two-terminal resistance for $I{\perp}B$ are shown.

sample when cooled down; this was verified by the reproducibility of the unilluminated magnetoresistance and Hall data over several cooldowns.

The low-temperature magnetoresistance and Hall resistance for the unilluminated sample with the magnetic field applied perpendicular to the electron layer is shown in Figs. 1(a) and 1(b). Figure 1(a) shows the $T = 0.41$ K data up to 23 T for the sample, which has a low-temperature carrier density $n_s = 8.3 \times 10^{11}$ cm^{-2} and a zero magnetic field mobility $\mu = 320\,000$ cm^2/V s. Figure 1(b) shows the 0–5 T range from Fig. 1(a) in more detail. At this temperature, well-formed quantized Hall plateaus and zeroes in the magnetoresistance can be observed; resistance minima touch zero as low as $B = 2.2$ T, corresponding to the filling factor $\nu = 16$. The quantized Hall resistance plateaus occur at the values $R_{xy} = h/\nu e^2$, to the accuracy of the measurements. Spin splitting is apparent in Fig. 1(b) up to a filling factor $\nu = 17$ at $B \sim 2.0$ T, indicating a spin splitting large compared to the thermal broadening of the Landau levels. This is what one might expect from the large effective bulk g factor[6] for InAs, $g = -15$ (compared to $g = -0.44$ for GaAs). Magnetoresistance measurements in tilted magnetic fields could be used to check the magnitude of the g factor quantitatively, and to determine whatever exchange enhancement it might exhibit.[7] Note that the data in Figs. 1(a) and 1(b) show no evidence for the fractional quantum Hall effect.

The Fourier transform of the low-field (<2 T) magnetoresistance versus $1/B$ data shows a strong peak at a single frequency f_i, indicating a single occupied subband; this frequency $f_i = (h/2e)n_s$ corresponds to the same total density as that given by the slope of the Hall resistance.

This agreement and the presence of zero-resistance minima both argue against the existence of parallel conducting paths in these samples under these experimental conditions. In addition, the constant average slope of the Hall resistance and the observed field positions of the zero resistance minima indicate that in the range 0–23 T no apparent loss of carriers is occurring with increasing magnetic field.

Figure 2 shows the magnetoresistances up to 23 T for two orientations with of magnetic field applied in the plane of the electron layer: (a) current I parallel to the field, $I\|B$, and (b) current perpendicular to the field, $I{\perp}B$. For the orientation $I\|B$, the Hall bar was oriented lengthwise along B, with the current biased along the sample and the voltage measured across the current probes (two-terminal resistance) and across two voltage probes (four-terminal resistance). To obtain the $I{\perp}B$ measurement, the sample was left in the same orientation with respect to B but current biased through the middle Hall voltage probe pair.

The data in Fig. 2 show a weak monotonic increase, and suggest that the magnetoresistance with applied field in the plane is isotropic with respect to the current direction. The four-terminal resistance increases roughly as B^2. The absence of oscillations in the magnetoresistances for B in the plane demonstrate the two-dimensional behavior of the electron gas. Using the measured sheet density and effective mass $m^*/m_e = 0.023$, we obtain a Fermi energy of 90 meV, much less than that needed to occupy a second subband in these wells in zero magnetic field,[8] in agreement with the data. We note that the magnetic length $l_m = \sqrt{\hbar/eB}$ is equal to one-half the well width of 120 Å at $B = 18$ T.

Using the negative persistent photoconductivity of these samples, we were able to reduce the carrier density of the wells. With increasing illumination the zero-field resistance increases along with the slope of the Hall resistance and the period of the Shubnikov–de Haas oscillations, all corresponding to the decrease in electron density in the

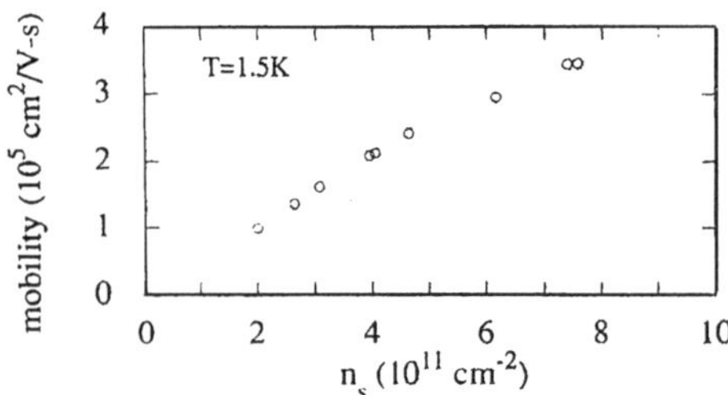

FIG. 3. Hall mobility vs sheet density measured at $T = 1.5$ K for a series of illuminations with both the red and green LEDs.

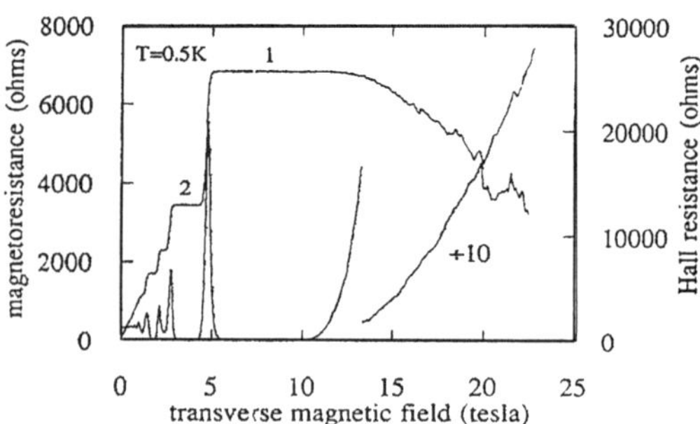

FIG. 4. Transverse magnetoresistance and Hall resistance at $T = 0.5$ K for the illuminated sample with $n_s = 1.7 \times 10^{11}$ cm^{-2}.

wells. Magnetoresistance and Hall resistance data for two pairs of contacts were compared and were found to agree for all illumination levels.

Figure 3 shows a composite plot of the $T = 1.5$ K measured Hall mobilities versus sheet densities, for illuminations at low temperatures with both the red and green LEDs separately; the sample was warmed to room temperature between the red illuminations and the green illuminations. With the red LED we were able to reduce n_s by a factor of ~ 2, and with green by a factor of ~ 5. The mobility has an sublinear dependence on sheet density; such a sublinear dependence is expected in GaAs/(Al,Ga)As heterostructures when the scattering is dominated by the background doping in the well.[9] We found that the mobility does not depend on the illumination photon energy, but that the $\mu(n_s)$ carrier density after saturation with illumination does.

The data of Fig. 3 are consistent with the model of the negative persistent photoconductivity effect given by Tuttle et al.,[1] according to which the illumination creates electron-hole pairs in the AlSb barriers [E_g(300 K) = 1.55 eV → 800 nm]. The holes are attracted and captured by the well, where they recombine with electrons inside the well to reduce n_s. The electrons in the AlSb fail to overcome the repulsive barrier at the edges of the well, and are captured by ionized deep donors in the AlSb. However, this model does not readily explain why the green LED has a larger effect on the carrier density than the red LED when both are used to saturation. No attempt was made to resolve this question.

To study the extreme quantum limit in these samples at obtainable magnetic fields, one sample was illuminated with the green LED until saturation to obtain the lowest carrier density $n_s = 1.7 \times 10^{11}$ cm^{-2} ($\mu = 88\,000$ cm^2/V s). Figure 4 shows the $T \sim 0.5$ K, 0–23 T magnetoresistance and Hall resistance data for this carrier density. Data were taken in both transverse magnetic field directions (normal and reversed) and subtracted to remove any re-

sidual magnetoresistance pickup in the Hall voltage probes due to misalignment. For fields above the $v = 1$ plateau, this pickup was substantial ($\sim 50\%$ of the Hall signal) because of the sharply increasing magnetoresistance. Beyond the $v = 1$ Hall plateau, both the magnetoresistance and the Hall resistance become "noisy", with the magnetoresistance increasing sharply and the Hall resistance dropping. No evidence of the fractional quantum Hall effect can be seen in these data.

In conclusion, we have investigated the low-temperature magnetoresistance and Hall effect of 120 Å InAs/AlSb single quantum wells grown on GaAs substrates by Tuttle et al.[1,2] We find that the wells contain a high-mobility 2D electron gas which shows quantum Hall effect behavior.

The authors wish to express their appreciation to Professor Art Gossard for several discussions. The work at Harvard was supported by the National Science Foundation (DMR-89-20490), that at Santa Barbara by the Office of Naval Research.

[1] G. Tuttle, H. Kroemer, and J. English, in *III-V Heterostructures for Electronic/Photonic Devices*, edited by C. Tu, V. D. Mattera, and A. C. Gossard, MRS Symposia Proceedings, San Diego, Aug. 1990, Vol. 145 (Materials Research Society, Pittsburgh, 1989), p. 393.
[2] G. Tuttle, H. Kroemer, and J. H. English, J. Appl. Phys. **65**, 5239 (1989).
[3] H. Munekata, T. P. Smith, and L. L. Chang, J. Cryst. Growth **95**, 235 (1989).
[4] C. A. Chang, L. L. Chang, E. E. Mendez, M. S. Christie, and L. Esaki, J. Vac. Sci. Technol. B **2**, 214 (1984).
[5] S. A. Chalmers, H. Kroemer, and A. C. Gossard, Proceedings of the 7th International Conference on Molecular Beam Epitaxy, J. Cryst. Growth (to be published).
[6] J. Konopka, Phys. Lett. A **26**, 29 (1967).
[7] See, for example, T. Ando, A. Fowler, and F. Stern, Rev. Mod. Phys. **54**, 551 (1982).
[8] For an empty 120 Å well with infinite barriers, the energy level spacing is $E_2 - E_1 \sim 260$ meV, ignoring nonparabolicity.
[9] A. Gold, Appl. Phys. Lett. **54**, 2100 (1989).

Reprinted with permission from

I. Sela, D. E. Watkins, B. K. Laurich, D. L. Smith, S. Subbanna, and H. Kroemer, ``Modulated photoabsorption in strained $Ga_{1-x}In_xAs/GaAs$ multiple quantum wells,'' Phys. Rev. B, Vol. 43(14), pp. 11884-11892, 1991.

PHYSICAL REVIEW B   VOLUME 43, NUMBER 14   15 MAY 1991-I

Modulated photoabsorption in strained $Ga_{1-x}In_xAs/GaAs$ multiple quantum wells

I. Sela, D. E. Watkins, B. K. Laurich, and D. L. Smith
Los Alamos National Laboratory, Los Alamos, New Mexico 87545

S. Subbanna* and H. Kroemer
Department of Electrical and Computer Engineering, University of California at Santa Barbara, Santa Barbara, California 93106
(Received 19 November 1990)

Modulated photoabsorption measurements in strained $Ga_{1-x}In_xAs/GaAs$ multiple quantum wells are presented. The modulating intensities vary from a few to about 10^5 W/cm^2. The absorption near the first heavy-hole exciton is probed with a tunable Ti:sapphire laser. The modulating beam is either from the same Ti:sapphire laser as the test beam or from an Ar^+-ion laser whose photon energy is much larger than the first heavy-hole exciton transition energy. A dramatic difference is observed in the modulated transmission spectra for the two modulating wavelengths. This difference in behavior can be explained as arising from screening of the residual surface electric field by Ar^+-ion-laser excitation but not by Ti:sapphire laser excitation. The Ar^+-ion laser creates high-energy carriers that are initially free to drift in the surface field before they are captured in the quantum wells. Carriers excited by the low-photon-energy Ti:sapphire laser are created in the quantum wells and therefore cannot effectively screen the surface field. We present a model based on surface-field screening and exciton saturation for Ar^+-ion-laser modulation and exciton saturation alone for Ti:sapphire laser modulation that describes the observed results.

I. INTRODUCTION

The optical properties of the strained-layer $Ga_{1-x}In_xAs/GaAs$ materials system have received considerable attention.[1-10] Strain splits the heavy- and light-hole bands and, therefore, excitonic transitions involving these bands are usually well separated spectrally. In one sense, this spectral separation of corresponding heavy- and light-hole transitions simplifies optical studies of $Ga_{1-x}In_xAs/GaAs$ compared to optical studies of $GaAs/Ga_{1-x}Al_xAs$ in which corresponding heavy- and light-hole transitions are spectrally close. In this paper we present an experimental study of modulated photoabsorption near the first heavy-hole exciton in $Ga_{1-x}In_xAs/GaAs$ multiple quantum wells (MQW). The modulating intensities used in the study vary from a few to about 10^5 W/cm^2. The absorption is probed with a spectrally tunable beam from a Ti:sapphire laser. The modulating beam is either from the same Ti:sapphire laser as the test beam (and therefore has the same wavelength as the test beam) or from an Ar^+-ion laser whose photon energy is much larger than the first heavy-hole exciton transition energy. For the case in which the modulating beam is from the Ti:sapphire laser, an electron-hole pair created by absorption of the modulating beam is confined in a $Ga_{1-x}In_xAs$ quantum well. For the case in which the modulating beam is from the Ar^+-ion laser, electrons and holes created by absorption of the modulating beam have a higher energy than the GaAs barriers so that they are initially free to move throughout the quantum structure. Subsequently, these electrons and holes lose energy and become confined in the quantum wells.

We observe a dramatic difference between the modulated transmission spectra using the two different modulating wavelengths. Modulation with the Ti:sapphire laser leads to increased transmission (reduced absorption) at the exciton resonance for all modulation intensities. These results are well described by saturation of the excitonic transition[11-14] and can be parametrized by a single saturation intensity over a wide intensity range. Very different behavior is observed for modulation with the Ar^+-ion laser. At modulation intensities below a few kW/cm^2 (the precise value is sample dependent), modulation with the Ar^+-ion laser leads to reduced transmission (increased absorption), whereas at higher modulating intensities, it leads to increased transmission. At modulation intensities greater than several kW/cm^2, modulation with the Ar^+-ion laser shows only the same exciton saturation as does modulation with the Ti:sapphire laser.

We attribute the different behavior observed at the two modulating wavelengths to screening of surface electric fields by the electrons and holes generated by the Ar^+-ion laser. These surface electric fields arise because the energy bands at the surface are at a different position, relative to the Fermi energy, than they are deeper in the structure. This band bending corresponds to the surface electric fields. They are the same fields responsible for modulated photoreflectance.[15-18] The surface fields shift the excitonic transition to the red and decrease the strength of the excitonic transitions by increasing the spatial separation of the electron and hole wave functions in the quantum well. The electrons and holes created by Ar^+-ion-laser modulation screen the surface fields because these carriers are created at energies higher than the GaAs barriers. This allows them to move freely in

43   11 884   ©1991 The American Physical Society

response to the surface fields until they lose energy and are captured in the quantum wells. The fields are not effectively screened by electron-hole pairs generated by Ti:sapphire laser modulation because each electron-hole pair is generated in a quantum well and is therefore not free to move in response to the field. Screening of the surface fields by Ar^+-ion-laser modulation increases the strength of the excitonic transition and shifts the transition energy to the blue, as compared to the case with unscreened fields. Thus, screening tends to reduce transmission (increase absorption) at the exciton resonance. Surface-field screening saturates as the carrier density increases, so the usual exciton saturation effect dominates the modulated transmission spectra for both Ar^+-ion and Ti:sapphire laser modulation at high intensity. We present modulated transmission spectra of $Ga_{1-x}In_xAs/GaAs$ multiple quantum wells using both Ar^+-ion and Ti:sapphire laser modulation over a wide intensity range. A model based on surface-field screening and exciton saturation for Ar^+-ion-laser modulation and exciton saturation alone for Ti:sapphire laser modulation that describes the observed results is presented.

The paper is organized as follows: Our experimental approach is described in Sec. II, results of the optical measurements are presented in Sec. III, Sec. IV contains a discussion of the theoretical model, and our results are summarized in Sec. V.

II. EXPERIMENTAL APPROACH

MQW samples were grown by molecular-beam epitaxy on semi-insulating GaAs substrates with 0.1-μm undoped GaAs buffer layers. There were 30 undoped periods in the MQW's, each consisted of a 7-nm $Ga_{1-x}In_xAs$ well and a 14-nm GaAs barrier. A 30-nm undoped GaAs cap layer completed the structure. Four samples were studied: two with indium concentration given by $x = 0.12$ (samples 1) and two with indium concentration $x = 0.17$ (samples 2). For each concentration one sample was grown along the [100] crystal axis and the other was grown along the [211]B crystal axis.[19] Corresponding samples were grown simultaneously to obtain similar structures. Rutherford backscattering measurements were used to determine the In concentrations. Photoluminescence measurements taken at very low optical intensity show a strong, sharp peak associated with the 1hh exciton. No peaks associated with extrinsic bound excitons were observed. We measured the free carrier recombination time by using a picosecond optical (532-nm) excitation source to generate the carriers while monitoring the change in sample transmission at the exciton resonance using a cw Ti:sapphire laser beam and a fast photodiode. In all four samples the change in transmission induced by the picosecond laser pulse had a decay constant of about 2 ns. All transmission measurements were made after the samples had been mechanically polished with a 5° wedge between the surfaces to eliminate interference effects.

Three types of measurements were made: transmission as a function of wavelength, differential transmission as a function of test-beam wavelength and both test- and pump-beam intensity, and differential photoreflectance. All of the measurements were made with the samples at low temperature (~ 5 K) to avoid heating effects by the cw laser beams. Two experimental approaches were used (Fig. 1). Figure 1(a) shows the usual pump-probe approach in which the Ti:sapphire laser beam is split into a pump and a probe (or test) beam. Figure 1(b) shows a variation on this approach where the Ar^+-ion laser which is used to pump the Ti:sapphire laser is also used for the pump beam. (In another variation a HeNe laser was used to pump the sample.) This figure also shows the setup for the differential photoreflectance measurement. In all cases the two laser beams were focused onto the same spot on the sample and spatial overlap was verified visually using a microscope and an infrared viewer. The spot size of the pump beam was kept bigger than that of the test beam. Spot size was determined using a razor blade mounted on a differential micrometer stage with 1-μm resolution. The razor-blade edge was located at the focal plane of the lens and in front of a power meter that collected the whole laser beam. The diameter for 10% to 90% of full power transmission was 40 μm for the Ti:sapphire pump beam and 20 μm for the test beam. The two laser beams were chopped at different frequencies. The change in test-beam transmission due to the pump-beam modulation was measured using a Si photodiode at D_3 and a lock-in amplifier at the sum of the two chopper frequencies. This approach suppresses scattered light and photoluminescence. Photodiodes at positions D_1 and D_2 were used to monitor input signal intensities after the laser beams were attenuated by neutral density filters. The additional detector D_4 in Fig. 1(b) was used

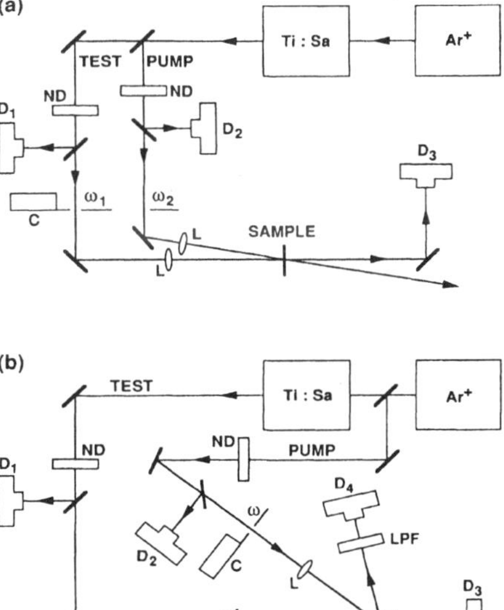

FIG. 1. Schematic of experimental setup for (a) Ti:sapphire laser modulation and (b) Ar^+-ion-laser modulation.

11 886

I. SELA *et al.*

43

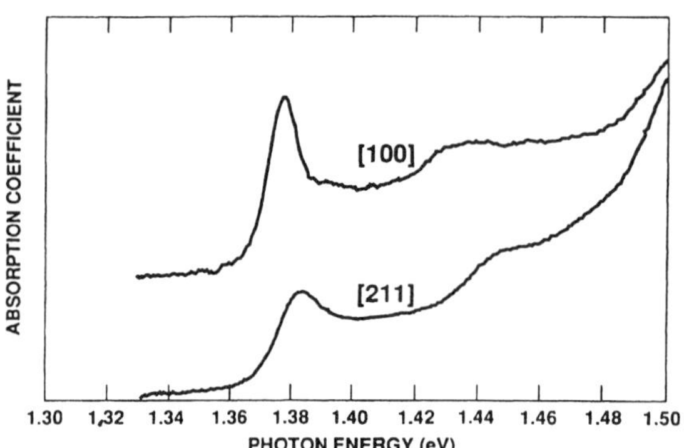

FIG. 2. Measured absorption spectra for the [100] and [211] oriented samples.

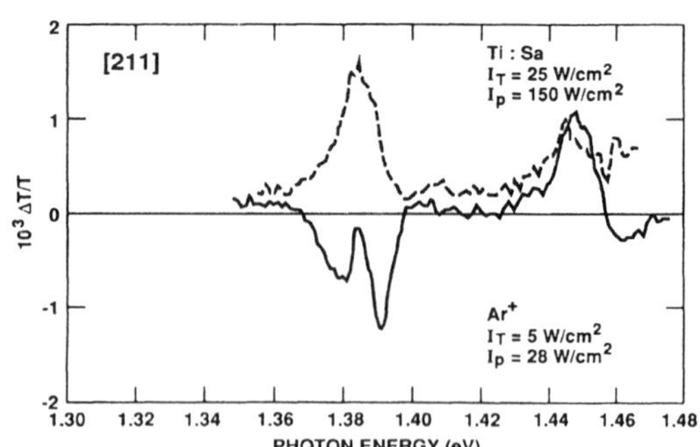

FIG. 4. Low-intensity modulated transmission spectra for the [211] sample with Ti:sapphire laser modulation (dotted line) and Ar$^+$-ion-laser modulation (solid line).

for photoreflectance measurements. A low-pass optical filter was placed in front of D_4 to block the scattered light from the Ar$^+$-ion-laser beam. Quoted incident intensities were corrected for reflection at the cryostat windows and the sample front surface.

III. EXPERIMENTAL RESULTS

Figure 2 shows the results of absorption measurements on samples 2. The peaks near 1.38 eV are caused by the 1hh exciton, and the shoulders near 1.44 eV are related to the 1lh exciton. The absorption increases sharply above 1.5 eV because of the GaAs substrate. Results from samples 1 are qualitatively similar to samples 2 for all the measurements and will not be shown here.

Normalized differential transmission measurements for the [100] and [211] samples are shown in Figs. 3 and 4, respectively. These measurements were made with low pump- and test-beam intensities. In both figures, the dashed curve was measured using the Ti:sapphire laser for the pump [Fig. 1(a)] and the solid curve was measured using the Ar$^+$-ion laser for the pump beam [Fig. 1(b)].

Measurements made with a HeNe pump laser gave similar results to the Ar$^+$-ion pump laser. The peaks near 1.38 eV in Figs. 3 and 4 are clearly due to the 1hh exciton (see Fig. 2). The change in signal phase for the two pump wavelengths indicates the existence of more than one mechanism for inducing a change in transmission. The

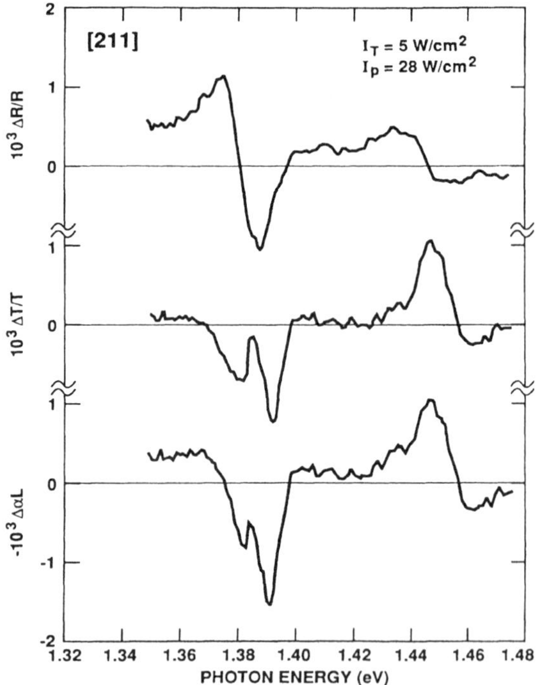

FIG. 5. Low-intensity modulated reflectivity spectrum (upper panel) and modulated transmission spectrum (central panel) for the [211] sample with Ar$^+$-ion-laser modulation. The modulated absorption spectrum (lower panel) is calculated from the measured reflectivity and transmission spectra.

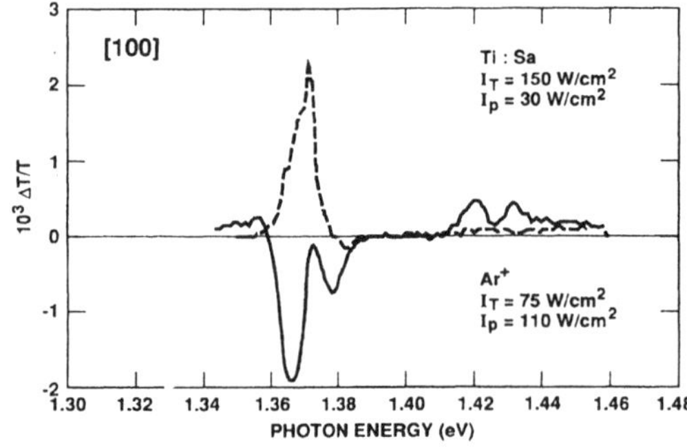

FIG. 3. Low-intensity modulated transmission spectra for the [100] sample with Ti:sapphire laser modulation (dotted line) and Ar$^+$-ion-laser modulation (solid line).

positive signal indicates a decrease in exciton absorption and is the usual observation in similar experiments.[11] In this case the mechanism is exciton saturation caused by phase-space filling (PSF).[11–14] The negative signal indicates an increase in exciton absorption in the presence of the pump beam. This additional effect is the result of surface-field screening.

Figure 5 shows the results of simultaneous measurements of the differential photoreflectance (upper trace) and the differential transmission (middle trace) for the [211] sample. These measurements were made at low intensity with the Ar^+-ion-laser pump beam. The corrected change in differential absorption signal is shown in the lower trace (see the Appendix). The correction tends to reduce the double negative peaks to a single negative peak, giving a clearer indication of the increase in absorption when the pump beam is incident on the sample. similar results were obtained for the [100] sample.

The change in transmission for the [100] sample at high pump- and test-beam intensities is shown in Fig. 6. Both the Ti:sapphire and Ar^+-ion pump-beam results are displayed in the figure. The signal for the high-intensity resonant pumping case (lower part of the figure) is qualitatively the same as for low-intensity resonant pumping case. The signal for the high-intensity Ar^+-ion pump beam (upper part of Fig. 6) has changed phase relative to

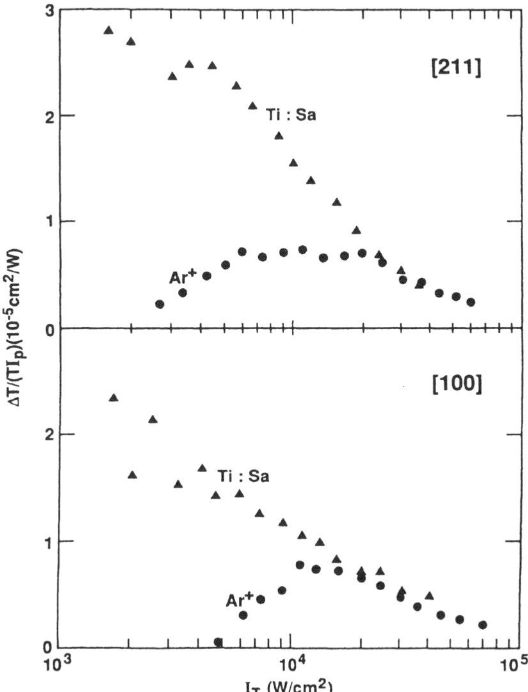

FIG. 7. Normalized modulated transmission for the [211] sample (upper panel) and the [100] sample (lower panel) over a large intensity range with a Ti:sapphire laser (triangle) and an Ar^+-ion laser (circles). The test-beam wavelength is at the peak of the first heavy-hole exciton transition.

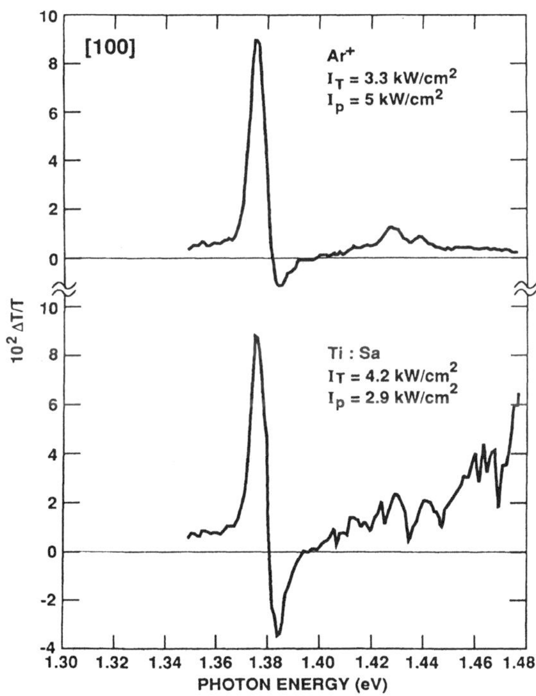

FIG. 6. High-intensity modulated transmission spectra of the [100] sample with Ar^+-ion-laser modulation (upper panel) and Ti:sapphire laser modulation (lower panel).

the low-intensity result in Fig. 3. The behavior shown in Fig. 6 is consistent with the exciton saturation mechanism independent of pump-beam wavelength.

The difference between the two pump-beam wavelengths was further examined by measuring the magnitude of the differential transmission signal as a function of the pump intensity for a test-beam wavelength fixed at the peak of the 1hh transition. In this experiment we maintained a ratio of pump-beam intensity to test-beam intensity of 0.1. Figure 7 presents the normalized differential transmission divided by the pump-beam intensity as a function of the test-beam intensity. A clear difference is observed in the results for the two pump-beam wavelengths at low intensities. The data associated with Ti:sapphire pumping are the anticipated results for saturation of a two-level excitonic system. The data associated with Ar^+-ion-laser pumping show a second modulation mechanism that dominates at low intensity.

IV. THEORETICAL MODEL

The absorption spectrum for the [100] multiple-quantum-well structure shown in Fig. 2 has two major features: the 1hh exciton transition at about 1.378 eV

11 888   I. SELA *et al.*   <u>43</u>

and the 1lh exciton at about 1.435 eV. The [211] sample behaves qualitatively similar to the [100] sample (Ref. 19). The principal modulation effects occur at the 1hh exciton absorption, so we concentrate on describing that transition.

At low intensity and without the presence of an electric field, the multiple-quantum-well absorption coefficient can be written as

$$\alpha(\omega) = \frac{4\pi\omega}{C\sqrt{\epsilon_\infty}} \operatorname{Im}[\chi(\omega)] , \tag{1a}$$

where

$$\operatorname{Im}[\chi(\omega)] = \sum_B \frac{e^2}{m^2\omega^2(b+d)} \sum_{e,h} |\langle U_e|\hat{\eta}\cdot\mathbf{P}|U_h\rangle|^2 \left[\sum_{n=0}^{\infty} \frac{1}{\pi a^2(n+\frac{1}{2})^3} \left[\frac{\hbar/T_2}{(\varepsilon_n - \hbar\omega)^2 + (\hbar/T_2)^2} \right] \right.$$
$$\left. + \frac{\mu_\|}{2\pi\hbar^2} \left[\frac{2}{1+e^{-\beta/\gamma}} \right] \{\pi/2 + \tan^{-1}[(\hbar\omega - \varepsilon_B)T_2/\hbar]\} \right] , \tag{1b}$$

$$\beta = \pi \left[\frac{\hbar^2}{2\mu_\| a^2 \varepsilon_B} \right] , \tag{1c}$$

and

$$\gamma = \begin{cases} (\hbar\omega - \varepsilon_B)/\varepsilon_B, & \hbar\omega > \varepsilon_B \\ 0, & \hbar\omega < \varepsilon_B . \end{cases} \tag{1d}$$

Here, ϵ_∞ is the high-frequency dielectric constant; B sums over the band-to-band transitions with transition energy ε_B; β and γ are terms that appear in the Sommerfeld enhancement factor; b and d are the thickness of the quantum well and barrier, respectively; U_e and U_h are the zone-center wave function for the electron and hole, respectively; the sum on e,h accounts for band-edge degeneracies; $\hat{\eta}$ is the photon polarization vector; $\mathbf{P}$ is the momentum operator; the sum on n is over bound exciton ground and excited states with transition energy ε_n; a is the exciton Bohr radius; T_2 is a lifetime broadening factor; and $\mu_\|$ is the reduced effective mass for carrier motion in the quantum well. Broadening of multiple-quantum-well optical spectra is usually caused by inho-

mogeneity, not by lifetime effects. Thus, in principle, a small lifetime broadening should appear in Eq. (1b) and the result should be convolved with a Gaussian to describe this inhomogeneous broadening. However, we are not interested in a detailed line-shape analysis and we simply chose T_2 to give the observed full width at half maximum of the transition. The calculated absorption spectrum is shown in Fig. 8. The parameters used in the calculation are given in Table I.

The effect of exciton saturation can be described by a two-level model.[11,12] We divide the bound-state contribution in Eq. (1b) by $(1 + I/I_s)$, where I is the incident optical intensity and I_s is a saturation parameter. We use a single value of I_s for all the ground- and excited-state bound exciton transitions. The excited bound exciton states for a given band-to-band transition make a very weak contribution to the absorption spectrum and a more careful treatment of their saturation properties is not warranted unless a very detailed line-shape analysis were attempted. We concentrate on describing the modulation of the 1hh transition and choose I_s by comparing with the observed results for this transition. We do not separately fit a saturation intensity to the 1lh transition and, as a result, we do not describe the modulation of that transition as well as the 1hh transition.

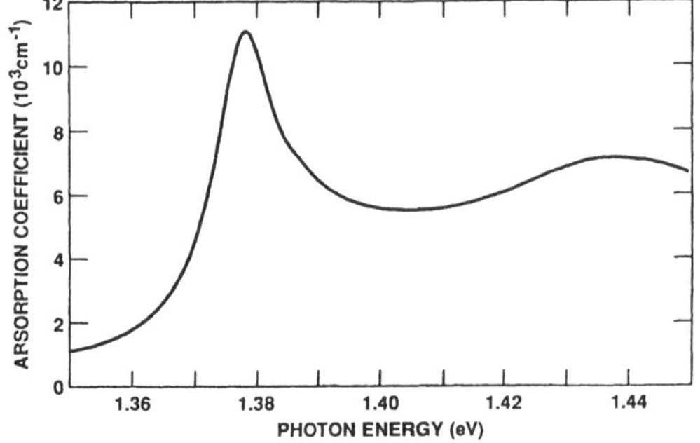

FIG. 8. Calculated absorption coefficient as a function of photon energy at low intensity.

TABLE I. Parameters used in the calculation.

ε_B (eV)	1.386
$|P|^2$ (eV)	18.0
$\mu_\|(m_0)$	0.07
$\hbar/T_2$ (meV)	6.0
ε_{ex} (meV)	8.0
a_{ex} (Å)	175.0
c (eV cm^2/V^2)	2×10^{-12}
d (eV cm^2/V^2)	3×10^{-9}
I_s (W/cm^2)	2.5×10^4
E_0 (V/cm)	7500
τ_R (ns)	2
τ_c (ps)	2
μ (cm^2/V s)	400

The surface fields are caused by surface charge, which results from Fermi-level pinning, and compensating space charge in the quantum-well structure. From the growth conditions, we expect that the quantum-well structure is lightly p type (low 10^{15} cm^{-3}) because of carbon incorporation. As a simple model, we take a uniform surface field, resulting from a positive sheet of surface charge and a compensating negative sheet of charge, representing the charged-carbon acceptors, whose distances from the surface are the thickness of the quantum-well structure L (about 0.7 μm in this case). If σ_0 is the unscreened sheet charge density, the unscreened surface field is

$$E_0 = \frac{4\pi}{\epsilon}\sigma_0 , \tag{2a}$$

where ϵ is the static dielectric constant. If σ is the screening charge, the screened surface field is

$$E = \frac{4\pi}{\epsilon}(\sigma_0 - \sigma) . \tag{2b}$$

To model σ, we note that the carriers must be generated by the modulating Ar$^+$-ion laser and they must be separated by the surface field. We set

$$\sigma = eNl , \tag{3a}$$

where N is the average volume density of electron-hole pairs generated by the laser in the quantum-well structure

$$N = \frac{I_m \tau_R}{\hbar\omega_m L} . \tag{3b}$$

Here I_m is the intensity of the modulating Ar$^+$-ion laser, $\hbar\omega_m$ is the photon energy from this laser, and τ_R is the carrier recombination lifetime. (The Ar$^+$-ion laser is completely absorbed in the quantum-well structure.) In Eq. (3a), l is the average distance by which the electron-hole pair is separated by the surface field (Ref. 20):

$$\frac{1}{l} = \frac{1}{\mu E \tau_c} + \frac{1}{L} , \tag{3c}$$

where τ_c is the time for carrier capture in the quantum well and μ is an untrapped carrier mobility for motion along the growth axis of the quantum-well structure. (Because of scattering from the interface potential discontinuities, we expect that μ should be smaller than the usual low-field carrier mobilities in GaAs.) In the experiment, the Ar$^+$-ion-laser light is absorbed very near the surface. Holes must be transported to neutralize the charged acceptors in the quantum-well structure. Thus, μ and τ_c refer to hole transport and trapping. Solving for E gives

$$E = \frac{[E_0 - (x+y)] + \{[E_0 - (x+y)]^2 + 4yE_0\}^{1/2}}{2} , \tag{4a}$$

where

$$x = \frac{e4\pi}{\epsilon}\frac{I_m}{\hbar\omega_m}\tau_R \tag{4b}$$

and

$$y = \frac{L}{\mu\tau_c} . \tag{4c}$$

The screening model described above is clearly simplified, but it contains the essential physical features of the surface-field screening process.

There are several parameters in Eq. (1b) which change their value when an external field is applied, including the exciton binding energies and Bohr radii, the band-to-band transition energies ε_B, and the squared optical matrix elements $|\langle U_e|\hat{\eta}\cdot\mathbf{P}|U_h\rangle|^2$. For the magnitude of fields considered here ($E \leq E_0 \sim 10^4$ V/cm), the change in exciton parameters is quite small[21] and the band-to-band transition energy and squared transition matrix elements are quadratically decreasing functions of the applied field.[22] The surface field is not spatially uniform and its nonuniformity leads to a broadening of the optical spectra. We interpret the solution of Eq. (4) as the maximum value of the surface field and take half this solution as the average value of the surface field. We then have

$$\varepsilon_B = \varepsilon_B^0 - c(E/2)^2 , \tag{5a}$$

$$\frac{\hbar}{T_2} = \left[\frac{\hbar}{T_2}\right]^0 + \frac{c}{2}(E)^2 , \tag{5b}$$

and

$$|\mathbf{P}|^2 = (|\mathbf{P}|^2)^0 - d(E/2)^2 , \tag{5c}$$

where E is the solution to Eq. (4), $|\mathbf{P}|^2$ is the squared transition matrix element[23] in Eq. (1b), $\hbar/T_2$ is the half-width-at-half-maximum broadening of the transition, and the superscript 0 indicates the value at zero electric field. The coefficients c and d are determined by numerical calculation[22] of the effect of an electric field on quantum-well optical transitions. These values of c and d are listed in Table I. To calculate a modulated absorption spectrum, we choose a test-beam intensity I_t and a pump-beam intensity I_p. Both I_t and I_p contribute to the exciton saturation. When describing Ar$^+$-ion-laser modulation, I_p contributes to surface-field screening [Eq. (4)], but when describing Ti:sapphire laser modulation, there is no surface-field screening (i.e., $E = E_0$). The absorption coefficient is calculated both with the pump laser on and with it off. Subtracting gives $\Delta\alpha$, the absorption coefficient with the pump beam on minus that with the pump beam off. The parameters used in the calculations are given in Table I. We have chosen parameters to give a qualitative overall description of the data discussed above rather than attempting a detailed fit of the data from a particular sample.

In Fig. 9 we show calculated results for low-intensity modulated absorption ($-\Delta\alpha L$), both including the effect of surface-field screening to describe Ar$^+$-laser modulation and neglecting the effect of surface-field screening to describe Ti:sapphire laser modulation. The intensities used in the calculation correspond to those in Fig. 4. After the reflection corrections shown in Fig. 5 are taken into account, the calculation gives a reasonable description of the low-intensity modulated absorption spectra.

In Fig. 10 we show calculated results for high-intensity

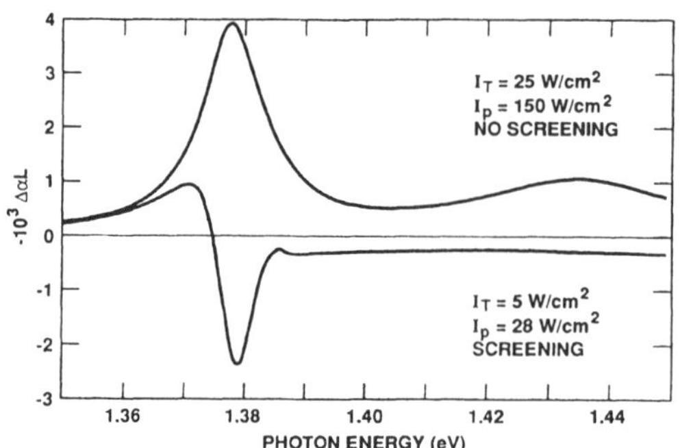

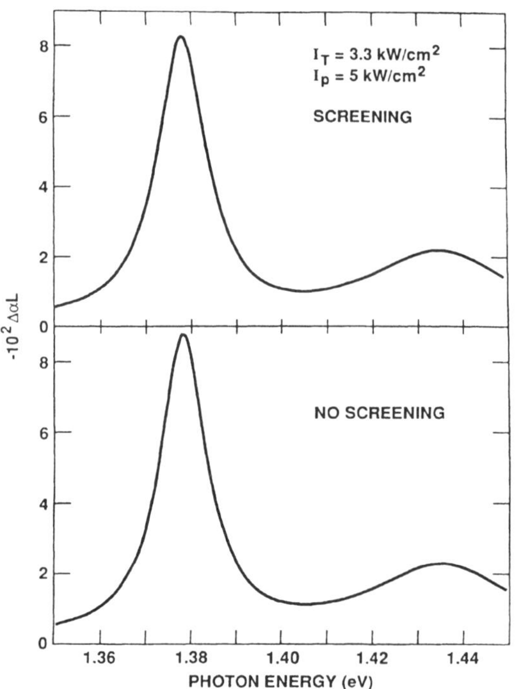

FIG. 9. Calculated absorption modulation spectra, at low modulation intensity, including surface-field screening (lower curve) and neglecting surface-field screening (upper curve).

modulated absorption ($-\Delta\alpha L$), both including and neglecting the effect of surface-field screening. The intensities used in the calculation correspond to those in Fig. 6. At these high intensities, the effect of surface-field screening has saturated and exciton saturation dominates the modulated photoabsorption spectra. The calculations, including and neglecting the surface screening

effect, are essentially the same. There is good correspondence between the calculation shown in Fig. 10 and the experimental spectra shown in Fig. 6.

In Fig. 11 we show calculated results for the normalized modulated absorption over a large intensity range both including and neglecting the effect of surface-field screening. The results in Fig. 11 are calculated at the peak of the 1hh exciton transition (1.378 eV). The pump intensity is 0.1 times the test intensity. The curve calculated neglecting surface-field screening is essentially that describing saturation of a two-level system. The curve calculated including surface-field screening deviates strongly from that neglecting it at low intensities but the two curves largely coincide at high intensities. There is a good correspondence between the calculated results shown in Fig. 11 and the experimental results shown in Fig. 7.

The model described in this section is simplified, but it provides a good description of all the observed results. The parameters used in the calculation are physically reasonable. The transition energies and linewidths were taken from the observed results. The transition matrix element, exciton parameters, effective masses, and electric-field dependence of the transition energy and matrix element were taken from quantum-well electronic-structure calculations.[22] (These parameters were not adjusted to match the experimental results.) The recombination lifetime was measured. The saturation intensity I_s was chosen to fit the exciton saturation observed in the Ti:sapphire modulation experiments. The unscreened surface field E_0 capture time τ_c, and untrapped carrier mobility μ (only the product $\tau_c\mu$ actually enters the model) were chosen to describe the Ar^+-ion modulation experiments. A surface field of 7.5×10^3 V/cm is a typical value. The value 2 ps used for τ_c is consistent with calculations for this trapping time in $GaAs/Ga_{1-x}Al_xAs$ QW.[24] The value 400 cm^2/V s for μ implies a scattering mean free path of about 170 Å (Ref. 25). This is a reasonable value (about one period) for a hole whose energy is above the GaAs valence-band edge (i.e., the hole is not confined by the GaAs barriers) in these structures.

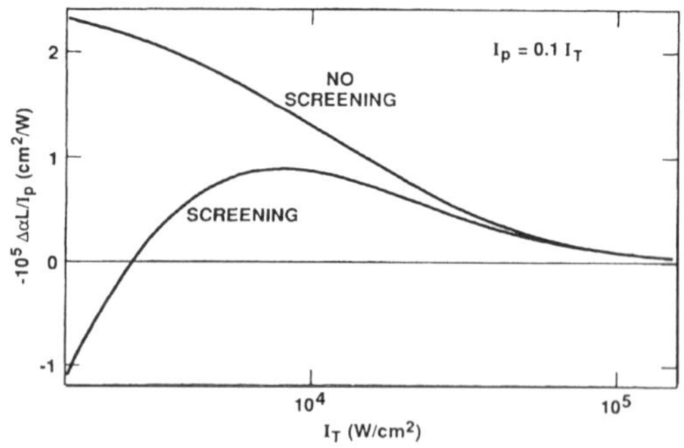

FIG. 10. Calculated absorption modulation spectra, at high modulation intensity, including surface-field screening (upper panel) and neglecting surface-field screening (lower panel).

FIG. 11. Calculated normalized absorption modulation including surface-field screening (lower curve) and neglecting surface-field screening (upper curve). The absorption modulation is calculated at the peak of the first heavy-hole exciton.

V. SUMMARY AND CONCLUSIONS

We have presented the results of an experimental study of modulated photoabsorption near the first heavy-hole exciton in $Ga_{1-x}In_xAs/GaAs$ multiple quantum wells in which the modulating intensity was varied from a few to about 10^5 W/cm^2. The absorption was probed with a spectrally tunable beam from a Ti:sapphire laser. The modulating beam was either from the Ti:sapphire laser or an Ar$^+$ laser. A dramatic difference is observed in modulated transmission spectra using the two different modulating wavelengths. Modulation with the Ti:sapphire laser is well described by saturation of the excitonic transition. Modulation with the Ar$^+$-ion laser also shows the effect of screening of surface fields in the quantum structure. (Such screening does not occur in Ti:sapphire laser modulation because the electron-hole pairs created by this laser are trapped in the quantum wells.) The surface-field screening significantly changes the modulated absorption spectra. We present a model based on surface-field screening and exciton saturation for Ar$^+$-ion-laser modulation and exciton saturation alone for Ti:sapphire laser modulation that describes the observed results.

ACKNOWLEDGMENTS

We thank T. E. Mitchell and O. Unal for transmission electron measurements on these samples. The work of two of us (S.S. and H.K.) was supported by the Office of Naval Research.

APPENDIX: REFLECTIVITY CORRECTIONS

In this appendix we describe how the modulated reflectivity and modulated transmission results were combined to give the modulated absorption spectrum shown in Fig. 5. The modulated reflectivity from the front surface of the sample was measured. (Reflections from the cryostat windows and the back surface of the sample were spatially separated from the reflection at the sample front.) The reflected intensity I_R and transmitted intensity I_T can be written as

$$I_R = R_F I_I \qquad \text{(A1a)}$$

and

$$I_T = (1 - R_F)(1 - R_B)e^{-\alpha L}I_I \ , \qquad \text{(A1b)}$$

where I_I is the incident intensity, R_F and R_B are the reflectivities of the front and back surfaces of the sample, and α and L are the absorption coefficient and thickness of the quantum-well structure. The modulated laser changes R_F and α. Expanding to lowest order gives

$$\frac{\Delta I_T}{I_T} = \frac{-\Delta R_F}{(1 - R_F)} - \Delta\alpha L \qquad \text{(A2a)}$$

and

$$\frac{\Delta I_R}{I_R} = \frac{\Delta R_F}{R_F} \ . \qquad \text{(A2b)}$$

Therefore we have

$$-\Delta\alpha L = \frac{\Delta I_T}{I_T} + \frac{\Delta I_R}{I_R}\left[\frac{R_F}{1 - R_F}\right] \ . \qquad \text{(A2c)}$$

We used $R_F = 0.3$ in constructing Fig. 5.

*Present address: IBM, East Fishkill, Hopewell Junction, NY 12533.

[1]J. Y. Marzin, M. N. Charasse, and B. Sermage, Phys. Rev. B **31**, 8298 (1985).

[2]I. J. Fritz, B. L. Doyle, T. J. Drummond, R. M. Biefeld, and G. C. Osbourn, Appl. Phys. Lett. **48**, 1606 (1986).

[3]J. Menendez, A. Pinczuk, D. J. Werder, S. K. Sputz, R. C. Miller, D. L. Sivco, and A. Y. Cho, Phys. Rev. B **36**, 8165 (1987).

[4]T. G. Andersson, Z. G. Chen, V. C. Kulakovskii, A. Uddin, and J. T. Vallin, Phys. Rev. B **37**, 4032 (1988).

[5]N. G. Anderson, W. D. Laidig, R. M. Kolbas, and Y. C. Lo, J. Appl. Phys. **60**, 2361 (1986).

[6]G. Ji, D. Huang, U. K. Reddy, T. S. Henderson, R. Houdre, and H. Morkoc, J. Appl. Phys. **62**, 3366 (1987).

[7]S. H. Pan, H. Shen, F. H. Pollak, W. Zhuang, Q. Xu, A. P. Roth, R. A. Masut, C. Lacelle, and D. Morris, Phys. Rev. B **38**, 3375 (1988).

[8]B. K. Laurich, K. Elcess, C. G. Fonstad, J. G. Beery, C. Mailhiot, and D. L. Smith, Phys. Rev. Lett. **62**, 649 (1989).

[9]K. J. Moore, G. Duggan, K. Woodbridge, and C. Roberts, Phys. Rev. B **41**, 1090 (1990).

[10]K. J. Moore, G. Duggan, K. Woodbridge, and C. Roberts, Phys. Rev. B **41**, 1095 (1990).

[11]D. S. Chemla, D. A. B. Miller, P. W. Smith, A. C. Gossard, and W. Wiegmann, IEEE J. Quantum Electron. QE-20, 265 (1984).

[12]S. H. Park, J. F. Morhange, A. D. Jeffrey, F. A. Morgan, A. Chavez-Pirson, H. M. Gibbs, S. W. Koch, N. Peyghambarian, M. Derstine, A. C. Gossard, J. H. English, and W. Wiegmann, Appl. Phys. Lett. **52**, 1201 (1988).

[13]R. Zimmermann, Phys. Status Solidi B **146**, 371 (1988).

[14]M. Wegener, I. Bar-Joseph, G. Sucha, M. N. Islam, N. Sauer, T. Y. Chang, and D. S. Chemla, Phys. Rev. B **39**, 12 794 (1989).

[15]H. Shen, P. Parayanthal, F. H. Pollak, M. Tomkiewicz, T. J. Drummond, and J. N. Schulman, Appl. Phys. Lett. **48**, 653 (1986).

[16]B. V. Shanabrook, O. J. Glembocki, and W. T. Beard, Phys. Rev. B **35**, 2540 (1987).

[17]H. Shen, S. H. Pan, F. H. Pollak, and R. N. Sacks, Phys. Rev. B **37**, 10 919 (1988).

11 892   I. SELA *et al.*   <u>43</u>

[18]O. J. Glembocki and B. V. Shanabrook, Superlatt. Microstruct. **5**, 603 (1989).

[19]Transmission electron microscopy studies of these samples suggest that considerable strain relaxation occurred. Similar samples consisting of 10 MQW periods showed much lower crystal defect densities; see O. Unal, B. K. Laurich, and T. E. Mitchell, *MRS Symposium Proceedings* (Materials Research Society, Pittsburgh, 1990), Vol. 183, pp. 183–186.

[20]The average rate for capture ($1/\tau_c$) and the average rate to traverse the sample ($\mu E / L$) add to give the average rate at which holes are separated from the electrons. Dividing by the average velocity (μE) gives the reciprocal of the average separation ($1/l$).

[21]J. A. Brum and G. Bastard, Phys. Rev. B **31**, 3893 (1985).

[22]See, for example, D. L. Smith and C. Mailhiot, Phys. Rev. Lett. **58**, 1264 (1987).

[23]We define

$$|\mathbf{P}|^2 = \frac{2}{m} \sum_{e,h} |\langle U_e | P_x | U_h \rangle|^2 \ .$$

[24]J. A. Brum and G. Bastard, Phys. Rev. B **33**, 1420 (1986).

[25]We use an effective mass of $0.45m_0$ and an average hole energy of 35 meV (one optical phonon energy) above the GaAs barriers to make this estimate.

Reprinted from

S. A. Chalmers, H. Kroemer, and A. C. Gossard, ``The growth of (Al,Ga)Sb tilted superlattices and their heteroepitaxy with InAs to form corrugated-barrier quantum wells," J. Cryst. Growth, Vol. 111, pp. 647-650, 1991.

Journal of Crystal Growth 111 (1991) 647–650
North-Holland

647

The growth of (Al,Ga)Sb tilted superlattices and their heteroepitaxy with InAs to form corrugated-barrier quantum wells

S.A. Chalmers, H. Kroemer and A.C. Gossard

Electrical and Computer Engineering Department and Materials Department, University of California, Santa Barbara, California 93106, USA

We have demonstrated the molecular beam epitaxial growth of (Al,Ga)Sb tilted superlattices (TSLs) on 2° vicinal (100) GaSb and GaAs substrates. The existence of (Al,Ga)Sb TSLs proves that step-flow growth can occur in this material system, and in the presence of strain. Lateral fluctuations in the tilt angle of the superlattice are observed and are found to be caused by a non-uniform distribution of incorporated adatoms which is correlated with the surface step density. The (Al,Ga)Sb TSL growth was also combined with InAs growth to form an InAs quantum well with "corrugated" barriers consisting of (Al,Ga)Sb TSLs. The electron mobilities in this structure exceeded 6×10^5 cm^2/V·s.

1. Introduction

Aluminum antimonide, gallium antimonide, and indium arsenide form an interesting family of compounds for molecular beam epitaxial (MBE) growth because they are nearly lattice-matched and they can be combined to form heterojunctions with straddling, staggered, and broken gap line-ups. To learn more about the MBE growth of these compounds we have investigated their step-flow growth on vicinal surfaces, which gives information on the surface diffusion of the deposited adatoms and on the morphology of the growing surface. Successful step-flow growth of these compounds also raises the possibility of fabricating new structures in this material system, such as tilted superlattices (TSLs) [1,2] The growth of more conventional structures may also be improved by step-flow growth on vicinal surfaces, which may be more ordered and better controlled than layer-by-layer growth on singular surfaces, because we can choose the orientation and therefore the bonding nature of the surface steps. This may be particularly important when forming interfaces across which the relatively volatile anion atoms are changed.

To investigate the step-flow growth of AlSb and GaSb we attempted to grow (Al,Ga)Sb TSLs, and found that in spite of the 0.65% lattice mismatch between these two materials, they can be grown together by step-flow to form uniform TSLs. We also demonstrated that AlSb and GaSb step-flow growth can be successfully combined with InAs growth by growing an InAs quantum well with "corrugated" (Al,Ga)Sb TSL barriers, which exhibits electron mobilities inside the well in excess of 6×10^5 cm^2/V · s.

2. Experimental

All growths were performed in a Varian Gen II MBE system with solid sources and an arsenic cracker. Both not-intentionally-doped GaSb and semi-insulating GaAs substrates were used, with the surfaces misoriented from (100) by 2° toward [011], which produces terraces with an average width of 20 atoms. The GaSb substrates were cleaned with a standard de-grease sequence and etched in 1 HF : 1 H$_2$O$_2$: 100 H$_2$O for 2 min, while the GaAs substrates were used as received. All growth rates were approximately 0.3 monolayers/s. The Sb_4 beam equivalent pressure (BEP) was approximately 1×10^{-6} Torr, and the As_2 BEP was approximately 2×10^{-6} Torr.

3. Results and discussion

We grew the (Al,Ga)Sb TSLs by sequentially depositing approximately 0.4 monolayers (ML) of AlSb and 0.6 ML of GaSb. To lower the temperature for AlSb step-flow during TSL growth, we deposited the AlSb by migration-enhanced epitaxy (MEE) [3,4] (i.e., by alternating the Al and Sb beams), while we grew the GaSb by concurrent Ga and Sb deposition. We found that in this growth mode we were able to grow well-defined (Al,Ga)Sb TSLs at substrate temperatures from 450 to 510°C, on both GaSb and GaAs substrates. Fig. 1 shows a transmission electron microscopic (TEM) cross-section of four 12 nm thick TSLs grown on a GaAs substrate at 490°C after buffer layers of 100 nm of GaAs (600°C), 50 nm of AlSb (560°C), and 300 nm of GaSb (510°C) were deposited, in that order. We can see from the figure that even these relatively thin buffer layers resulted in a fairly uniform surface, as manifested in the uniform periodicity of the TSLs. This is surprising considering the 7% lattice mismatch between the GaAs substrate and the (Al,Ga)Sb layers which results in a high density of threading dislocations, and can be attributed to the smoothing properties of

Fig. 1. Dark-field TEM micrograph of 12 nm thick (Al,Ga)Sb TSLs grown on 2° A-type GaAs substrate, with $p = 1.00$, 1.01, 1.03, and 1.05 (top to bottom). The steps descend left to right. The diagonal feature of the right side is a threading dislocation, resulting from the lattice mismatch due to growing on a GaAs substrate.

Fig. 2. Dark-field TEM micrograph of a 12 nm thick (Al,Ga)Sb TSL grown on a 2° A-type GaSb substrate, which shows variations in the TSL tilt angle.

the AlSb and GaSb buffer layers. TSLs grown on the GaSb substrates, with a 300 nm GaSb buffer layer grown at 510°C, were similar but exhibited better uniformity than those grown on GaAs substrates. The exact number of monolayers of AlSb and GaSb deposited per cycle, p, for the different TSLs in fig. 1 was 1.00, 1.01, 1.03, and 1.05 from top to bottom, as determined from the resulting tilt angles. The TSLs exhibit a high contrast between the AlSb-rich and GaSb-rich regions, indicating that these components are well separated, but there is also contrast between the GaSb-rich regions and the GaSb spacer layers, indicating that the separation is not complete. Although strain in the region of the dislocation seen in fig. 1 makes TEM imaging in that area difficult, we can see that TSLs formed, and therefore step-flow growth occurred, even very close to the dislocation.

An interesting feature of the (Al,Ga)Sb TSLs is that their tilt angle can vary with position, as seen in fig. 2. This TEM cross-section shows a 12 nm thick TSL grown on a GaSb substrate after a 300 nm GaSb buffer was grown at 510°C. Careful analysis of the local TSL thickness shows that the changing tilt angles are due to a variation in the local growth rate across the TSL, resulting in varying p-values. The resulting variations in p are much too great to be explained by inhomogeneous strain or non-uniform adatom deposition, but must rather be caused by a non-uniform redistribution of the group III adatoms that occurs *after* deposition. Because step edges play a dominant role in determining adatom distribution during step-flow growth, we might expect there to be a relation between an area's growth rate and its step density. Fig. 3 plots the p-value versus average step density for various locations of the TSL in fig. 2. The strong correlation between the two indicates that p is greater in regions that contain more steps. This sort of correlation, to differing degrees, was

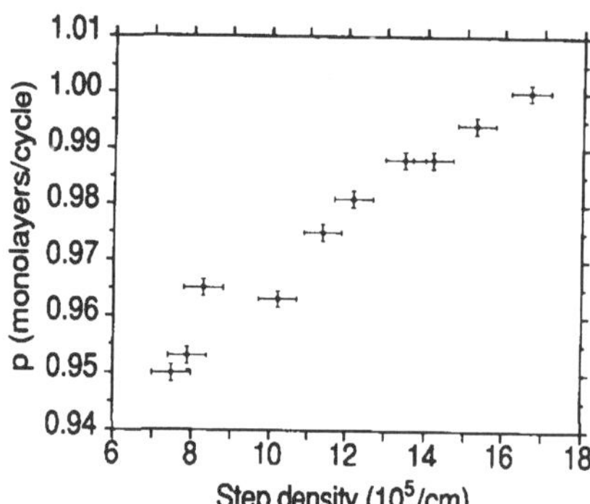

Fig. 3. Plot showing the correlation between the average local step density at various locations of the TSL in fig. 2 and the local p-value calculated from the TSL tilt.

seen in all the (Al,Ga)Sb TSLs examined. This behavior and therefore the tilt variations can be explained if we consider two different models of step-flow growth.

In one extreme limit where diffusing adatoms stick to the nearest step edge, p would be independent of terrace width, and the terrace widths would eventually equalize, in contradiction both to fig. 3 and to the actual appearance of the TSL pattern.

In the opposite limit, a model that assumes that the adatoms are *completely* free to diffuse over distances large compared to the terrace widths and that the sticking probability of an adatom at a step is the same at all step edges, would lead to a local growth rate that is simply proportional to the local step density, a correlation far stronger than what is observed.

Our data can be explained by a model between these two extremes, where some of the adatoms diffuse over distances equal to several terrace widths. Considering the bonding energies of AlSb and GaSb, and the analogous situation in the (Al,Ga)As system, the adatoms diffusing over these long distances are probably Ga. This result suggests that to grow TSLs with less tilt angle variation, we need to grow them on a surface with a more uniform step distribution, which may be accomplished by simply growing a thicker buffer layer.

To demonstrate that step-flow growth in the (Al,Ga)Sb system can be successfully combined

with InAs growth, we grew a corrugated-barrier quantum well (CBQW) that consisted of a 20 nm wide InAs quantum well with 12 nm (Al,Ga)Sb TSL barriers, as diagrammed in fig.4a. The average barrier composition was chosen to be 30% Al (i.e. 0.3 ML of AlSb and 0.7 ML of GaSb deposited sequentially per cycle), so that any TSL formation in the barriers would result in the well/barrier interface being alternately broken gap (GaSb-rich barrier regions) and staggered gap (AlSb-rich barrier regions), which we felt might result in interesting transport effects. The structure was grown on a 2° A-type GaAs substrate, preceded by 50 nm of AlSb grown at 560°C and a 2.5 μm thick GaSb buffer layer grown at 500°C. The InAs well was grown at 500°C and the TSL barriers at 490°C.

It has been shown that to achieve high electron mobilities in an InAs quantum well with AlSb barriers, it is critical that the bottom barrier/well interface be InSb-like [5]. To achieve this, we grew the bottom interface in the sequence 0.3 ML of Al, 6s of Sb, 0.5 ML of In, 3 s of As, and then the

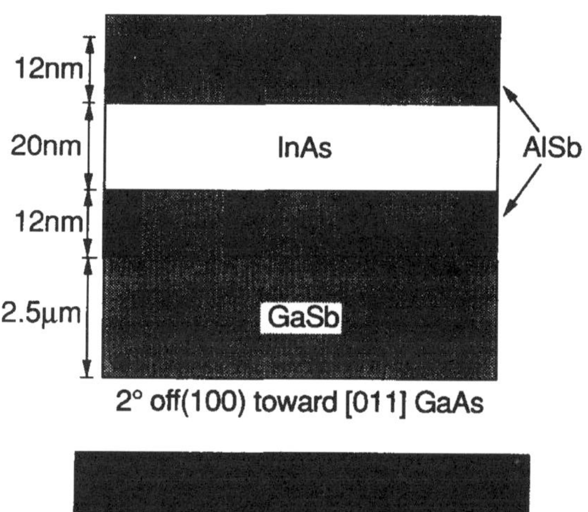

Fig. 4. (a) Schematic diagram of the corrugated-barrier quantum well (CBQW) that we attempted. (b) Dark-field TEM micrograph of the corrugated-barrier quantum well diagrammed in (a).

 S.A. Chalmers et al. / Growth of (Al,Ga) Sb TSLs

remainder of the InAs well by concurrent In and As deposition. After the InAs well was complete, we had a growth interruption of 50 s with As on to smooth the InAs surface, and then we started the top barrier with a 0.7 ML GaSb deposition. A TEM cross-section of the resulting structure is shown in fig. 4b. We can see that TSLs formed in both barriers, but they appear to be of lower quality than TSLs shown in fig. 1. Aside from a thicker buffer layer and a slightly lower Al content, the only difference in the growth of the bottom TSL of the CBQW and the TSLs shown in fig. 1 was the presence of a background As_2 pressure of approximately 3×10^{-7} Torr BEP during the CBQW growth, which was unavoidable since As_2 was needed for the InAs well growth. We suspect that the presence of As_2 on the growing surface may have interfered with the step-flow growth and therefore caused the poor TSL quality. In any case, a survey over the entire TEM sample revealed no apparent degradation of the top TSL with respect to the bottom one, indicating that the InAs layer itself had little adverse effect on the TSL formation. A puzzling feature of the micrograph in fig. 4b is the appearance of dark regions at both InAs/(Al,Ga)Sb TSL interfaces, which are visible over the entire TEM sample. We do not know whether they are caused by strain, or are due to a different compound, such as InSb, having formed at the interfaces, but it should be noted that the two dark regions appear to be very similar in spite of the fact that the interfaces were grown in different sequences.

Shubnikov–De Haas and Hall effect measurements revealed a very high electron mobility of 6.1×10^5 cm^2/V · s with a carrier density of 9.8×10^{11} cm^{-2} at 2 K, which we believe is the highest reported mobility for an InAs quantum well to date. Another sample, which was grown simultaneously with this sample but on a semi-insulating (100) GaAs substrate, had an electron mobility of only 3.6×10^5 cm^2/V · s with a carrier density of 1.1×10^{12} cm^{-2} at 2 K. Neither sample exhibited negative persistent photoconductivity, in contrast to high mobility AlSb/InAs wells [5]. Surprisingly, the measurements showed no significant anisotropy of the transport properties, relative to the direction of the TSL stripes. We speculate that

the absence of any anisotropy might be related to the presence of the intervening dark layers in fig. 4b, which may somehow shield the electrons in the well from the corrugations in the barrier. This point clearly needs more research, as does the understanding of the mechanism responsible for the high electron mobility. Of particular interest will be the effects of different substrate orientations and different interface growth sequences.

4. Conclusions

We have shown that AlSb and GaSb can be grown by step-flow and can be combined to grow tilted superlattices on 2° A-type GaAs and GaSb substrates, in spite of a 0.65% lattice mismatch. Lateral fluctuations in the tilt angle of the superlattice are observed and are found to be caused by a non-uniform adatom redistribution which is correlated with the surface step density. (Al,Ga)Sb TSL growth was also combined with InAs to form a corrugated-barrier quantum well (CBQW) with electron mobilities greater than 6×10^5 cm^2/V · s, and which showed no anisotropy relative to the stripe direction of the corrugated barriers.

Acknowledgements

This work was supported by the National Science Foundation Science and Technology Center for Quantized Electronic Structures (S.A.C. and A.C.G.) and by the Office of Naval Research (H.K.).

References

[1] P.M. Petroff, A.C. Gossard and W. Wiegmann, Appl. Phys. Letters 45 (1984) 620.
[2] J.M. Gaines, P.M. Petroff, H. Kroemer, R.J. Simes, R.S. Geels and J.H. English, J. Vacuum Sci. Technol. B6 (1988) 1378.
[3] F. Briones, D. Golmayo, L. Gonzalez and A. Ruiz, J. Crystal Growth 81 (1987) 19.
[4] Y. Horikoshi, M. Kawashima and H. Yamaguchi, Japan. Appl. Phys. 27 (1988) 169.
[5] G. Tuttle, H. Kroemer and J.H. English, J. Appl. Phys. 67 (1990) 3032.

Reprinted with permission from

H. Kroemer, C. Nguyen, and B. Brar, ``Are there Tamm-state donors at the InAs-AlSb quantum well interface?'' J. Vac. Sci. Technol. B, Vol. 10(4), pp. 1769-1772, 1992.

Are there Tamm-state donors at the InAs–AlSb quantum well interface?

Herbert Kroemer, Chanh Nguyen, and Berinder Brar
*Department of Electrical and Computer Engineering, University of California, Santa Barbara,
California 93106*

(Received 28 January 1992; accepted 20 March 1992)

An explanation of the temperature dependence of the electron concentration in InAs/AlSb
quantum wells, along with the high electron mobilities found, calls for a model involving a high
concentration of interface donors whose electron scattering cross section is much less than that
of conventional point defect donors. We postulate that the interface donors are in fact not point
defects, but a Tamm-state-like band of de-localized interface states not associated with defects,
but with the strong discontinuity in the periodic potentials on the two sides of the interface.
Heuristic arguments supporting this hypothesis are given.

I. INTRODUCTION

The InAs–AlSb quantum well (QW) system is characterized by unusually deep wells (≈ 1.35 eV) combined with
the high electron mobilities of high-purity InAs.[1,2] One
characteristic feature of these wells has always been the
occurrence of high electron sheet concentrations, usually
of the order 10^{12} electrons per cm^2, even though these wells
were usually not deliberately doped. In their original work,
Tuttle *et al.*[1] attributed those electrons to "transfer from ...
a deep donor in the AlSb ... into the deep InAs quantum
well." However, difficulties with this explanation were apparent almost from the outset: The wells invariably show
an *increase* in sheet concentration with increasing temperature, by an *additive* amount that is typically between 2
$\times 10^{11}$ and 3×10^{11} cm^{-2}, in going from low temperatures (≈ 10 K) to room temperature. The magnitude and
shape of this increase varies remarkably little from sample
to sample, even though the base level on which this increase is superimposed may vary by up to an order of
magnitude.

It was noticed already by Tuttle[3] that such a temperature dependence is opposite to what one would expect from
a simple model involving a spatially uniform concentration
of donors in the barriers, draining their electrons into the
deep wells. Tuttle showed that one *could* explain the temperature dependence of the carrier concentrations if one
was willing to make implausible *ad hoc* assumptions, such
that the (not intentionally doped) AlSb barriers are
strongly *p* type ($> 10^{16}$ cm^{-3}), with a thin region of a high
concentration ($> 10^{17}$ cm^{-3}) of fully ionized donors adjacent to the well. Apart from the question of how such a
distribution might come about, very strong remote ionized
impurity scattering should be present in such model, incompatible with the high 10 K mobilities observed
($> 3 \times 10^5$ cm^2/V s), even if allowance is made for screening. In fact, the compatibility with the high observed mobilities is one of the central problems of *any* model of the
origin of the high electron concentrations, made more
complicated by the observation that the mobilities increase
with increasing electron concentration.[1,4]

A recent quantitative systematic study of electron concentrations versus barrier parameters[5,6] has shed more
light on this problem. That study showed that the electron

sheet concentration per well could be cleanly separated
into three different contributions.

(i) A bulk contribution proportional to the barrier
width between the wells, identified as being due to a deep
bulk donor in the AlSb barriers draining into the even
deeper QWs. Typical bulk concentrations found were of
the order 10^{16} cm^{-3}, the exact value depending on
molecular-beam epitaxy (MBE) machine status. This bulk
contribution was essentially temperature-independent, indicating that the donor level is far above the Fermi level in
the well, which is itself above the lowest quantum state in
the well.

(ii) A surface contribution correlating strongly with the
thickness of the top barrier, in a way that quantitatively
matches a model involving a *very high concentration* of
surface states on top of the entire structure, at an energy
about 0.85 eV below the bottom of the conduction band of
AlSb. In structures with thin top barriers (<100 nm) this
was the *dominant contribution*. The model predicts an essentially temperature-independent electron concentration
resulting from this surface state mechanism, and the data
were in excellent agreement with this prediction.

(iii) A substantial "remainder" that does not correlate
with the two kinds of barrier thicknesses, and which contains essentially the entire temperature dependence. At low
temperatures, this remainder was approximately 3
$\times 10^{11}$ cm^{-2}, increasing by about 3×10^{11} cm^{-2} in going
from 10 to 300 K. It could not be attributed to donors
either inside the barrier or at the surface. By default, this
contribution must be attributed to donors in the well itself
or at the InAs–AlSb interface(s).

The understanding of this remainder is the topic of the
present article.

II. PHENOMENOLOGICAL DONOR MODEL

We analyze here our earlier single-well data,[6] shown in
Fig. 1, for the three wells with the thickest top barriers,
and hence, the smallest electron transfer from surface donors. We note first that an increase in electron concentration with increasing temperature, in a well already containing a degenerate carrier concentration, calls for a donor
below the Fermi level, which in turn implies a donor sheet
concentration even higher than the added electron sheet

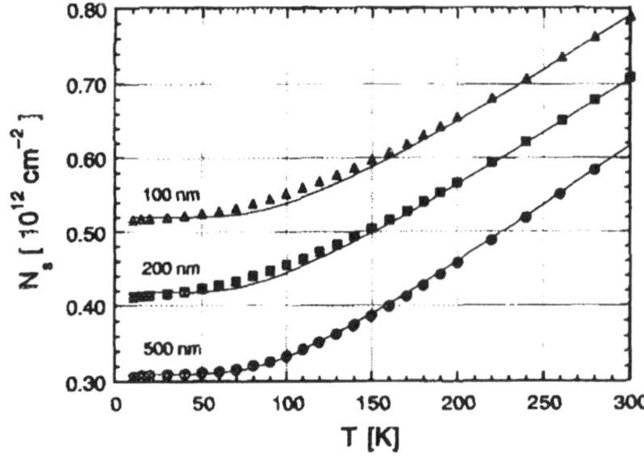

FIG. 1. Electron sheet concentration vs temperature, for three 15 nm wide InAs–AlSb single QWs with top barrier thicknesses of 100, 200, and 500 nm (not counting a 10 nm GaSb cap). The solid curves represent a fit to the theoretical model of Fig. 2(b).

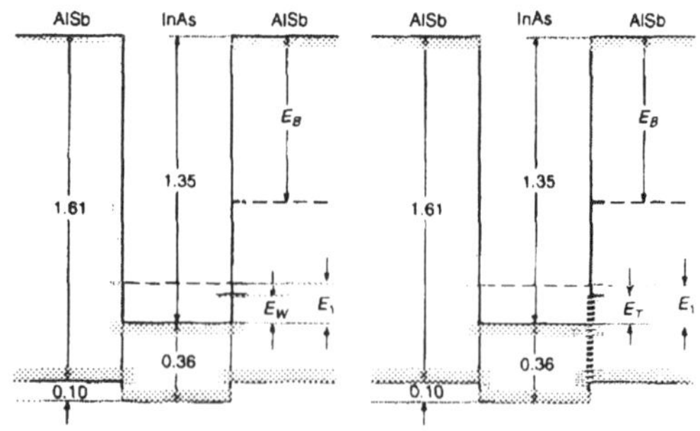

FIG. 2. Proposed energy level diagrams for interface and barrier donors. [(a) left] Discrete donor level model. [(b) right] Interface band model.

concentration. As we shall see in Table I, the data require a donor with a sheet concentration above 3×10^{12} cm^{-2}. The volume equivalent ($> 2 \times 10^{18}$ cm^{-3}) of such a sheet concentration is about three decades above the known background doping of InAs in our MBE system. We see no way to reconcile this discrepancy, and hence conclude that the donor associated with the well is *somehow* associated with the InAs–AlSb interface.

We fit the temperature dependence to two slightly different phenomenological models for this postulated interface donor, both illustrated in Fig. 2. The first model assumes a *discrete* donor at an energy E_W above the bottom of the InAs well, with a sheet concentration N_W. The second model assumed that the donor is broadened into a conventional two-dimensional (2D) band the *top* of which is at an energy E_T above the bottom of the well and with a 2D density of states that is treated as energy-independent over the range of interest, a few kT from the top of the band. Such a constant density may be expressed in terms of a density-of-states effective mass m_T^* associated with motion *along* the heterointerface, via the familiar relation

$$D_2 = \frac{m_T^*}{\pi \hbar^2}. \tag{1}$$

The ionization of the bulk donors causes substantial temperature-dependent band bending in the thick barriers

both above and below the single wells. This band bending depends on the bulk donor ionization energy; hence, the latter must be included as a fitting parameter in both models. The surface donor energy was not included as a fitting parameter, but was assumed to be fixed at 0.85 eV below the AlSb conduction band. Nonparabolicity effects in the InAs well were included.

Both models produced acceptable fits; the fits shown in Fig. 1 are for the interface band model. Note especially the excellent fit to the band model for the 500 nm sample, which has the smallest contribution to the electron concentration from sources other than the interface donor. The discrete-donor model fits (not shown) are only slightly inferior, indicating a low sensitivity to the exact model employed. Tables I and II give the fitting parameters for both models.

III. TAMM-STATE INTERPRETATION

The high donor sheet concentrations demanded by a fit to the experimental data, if interpreted in terms of conventional spatially localized donors, are completely incompatible with the high electron mobilities observed in these wells: When localized interface donors are *deliberately* introduced at much lower concentrations ($\approx 1 \times 10^{12}$ cm^{-2}), the mobilities are found to plunge to between 3000 and 5000 cm^2/V s,[7] almost two decades below what is observed in the not deliberately doped wells. Yet we typically observe low-temperature (10 K) mobilities between 100 000 and 300 000 cm^2/V s,[8] and in unpublished recent work we have observed 4.2 K mobilities as high as 850 000 cm^2/V s.

TABLE I. Optimal fitting parameters to the discrete interface donor model of Fig. 2(a), for the three 15 nm wide single wells whose data are shown in Fig. 1. The quantities n_B and N_W are the volume concentration of the barrier donors and the sheet concentration (per well, not per interface) of the well donors.

W	500	200	100	nm
E_B	0.90	0.85	0.90	eV
n_B	0.42	0.52	0.63	10^{16} cm^{-3}
E_W	0.020	0.045	0.041	eV
N_W	5.3	3.5	3.14	10^{12} cm^{-2}

TABLE II. Optimal fitting parameters for the same samples as in Table I, but fitted to the interface *band* model. The 2D density of states is expressed in terms of a corresponding density-of-states mass m_T^*, via Eq. (1), counting both surfaces as a single band.

W	500	200	100	nm
E_B	0.96	0.95	0.97	eV
n_B	0.47	0.68	0.71	10^{16} cm^{-3}
E_T	0.039	0.052	0.057	eV
m_T^*	0.28	0.21	0.21	m_e

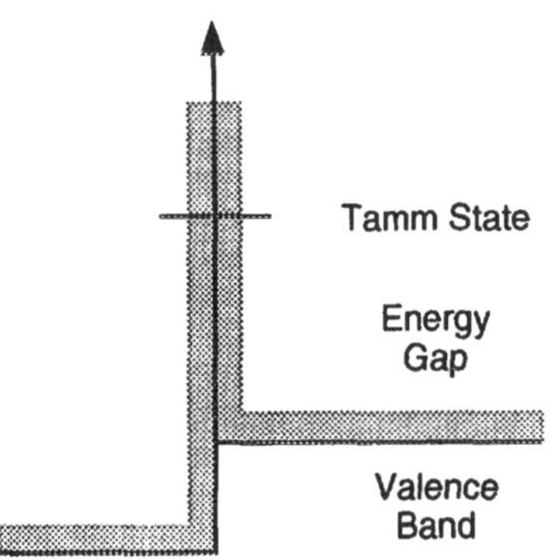

FIG. 3. Delta function correction to the simple abrupt-step model of energy band offsets, after Zhu and Kroemer.[13] The case shown represents a hole-attracting delta function attached to a valence band offset. Its bound states would act as donorlike interface Tamm states.

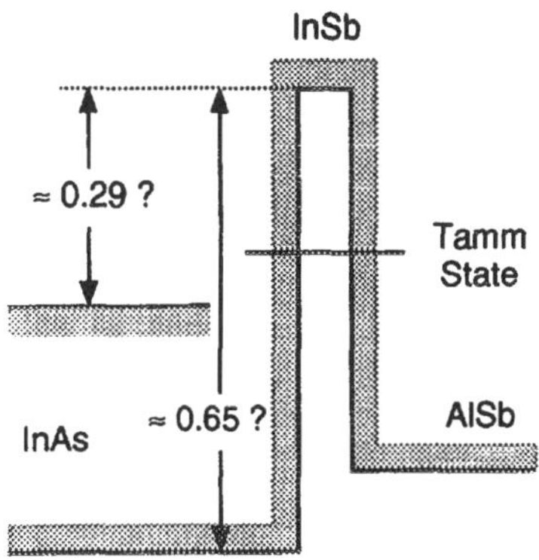

FIG. 4. Heuristic interface model of the valence band at an InAs/AlSb heterojunction in terms of a delta-functionlike narrow InSb QW inserted between the InAs and AlSb barriers.

To resolve the discrepancy, we propose that the interface donor is not a conventional point defect donor, but a band of interface Tamm states[9–11] associated with the drastic discontinuity between the periodic potentials on the two sides of the interface, and split off from the valence band of the structure. Being neutral when occupied, these interface states would act as ionizable deep donors. The key point is that such Tamm states should be essentially plane wavelike in the direction parallel to the interface. If ionized, their charge distribution should be de-localized and essentially uniform along the interface, leading to a potential not capable of scattering the electrons in the well. This would resolve the conflict between the high density of the states and the lack of any significant mobility reduction in the wells.

Although Tamm's original 1932 work[9] addressed itself to *surface* states, it was recognized as early as 1949 by James[11] that such states might also occur at what is today called a heterojunction, the interface between two crystals. More recently, and starting from a different point of view, Kroemer and Zhu[12,13] pointed out that the common simple model of an abrupt heterojunction as a simple step in the band edges is oversimplified, and that for interfaces between semiconductors with sufficiently different band structure, delta function spikes should be added to the band diagram. Such delta functions may be either attractive or repulsive. An attractive delta function would be capable of splitting off states from "its" band. This concept is illustrated in Fig. 3 for the case of a "hole-attracting" delta spike added to a valence band offset. Although Kroemer and Zhu did not use the term "Tamm states" for such states, the latter are in fact exactly the kind of interface Tamm states discussed by James. (The terminology has recently been resurrected in the original context of surface states, but applied to semiconductor superlattices rather than bulk semiconductors.[14,15])

We believe that there are strong reasons to expect that a phenomenon such as that discussed by Kroemer and Zhu

takes place in the valence band of an InAs/AlSb heterojunction of the kind employed in our InAs/AlSb QWs. In all our work, the MBE shutter sequence was designed to drive the bond arrangement across the interfaces towards an InSb-like rather than AlAs-like arrangement.[7] Recent Raman scattering measurements by Sela[16] on some of our wells have indeed shown a strong Raman signal corresponding to In–Sb bonds. A (weaker) In–Sb signal was found even at interfaces in which a shutter sequencing encouraged AlAs-like interfaces, suggesting that the In–Sb bond configuration across the interface is indeed the preferred one.

To the first order, taking the limit of a nonatomistic continuum model, it should be legitimate to view an InSb-like interface as a case of a minimum-width ($\approx$0.3 nm) InSb QW inserted between an InAs and an AlSb barrier (Fig. 4). Given such a model, there can be no doubt that the band lineup at the InAs–InSb interface is a broken-gap lineup, with the top of the valence band of InSb at an energy substantially higher than the bottom of the conduction band of InAs. The band lineup at the InAs–GaSb already has a 150 meV break in the gap. Going from GaSb to InSb should raise the valence band on the Sb-side further, and widen the break. Exact lineup data appear to be unavailable, but we may estimate the lineup from the data of Kurtz *et al.*[17] on the band lineups at InSb–InAs$_x$Sb$_{1-x}$ heterojunctions. Near the Sb-rich end of the composition range, these authors find a shift of the heavy-hole band by about 6.54 meV for every 1% change in the alloy composition, including strain effects. Assuming that this rate persists throughout the composition range (equivalent to assuming that the strong energy gap bowing in that system is pure conduction band bowing), leads to a prediction of an InSb heavy hole band about 654 meV above the InAs valence band, that is, about 294 meV above the InAs conduction band.

For a *wide* inserted InSb hole QW, we would therefore expect the top of that well to project almost 300 meV above the InAs conduction band edge. Due to quantization effects in the extremely narrow (≈ 0.3 nm) residual well, the highest hole state would of course be lower. In a square well with "ordinary" barriers, this quantization would push the highest hole state very close to the AlSb valence band edge. But because of the broken gap, we would expect much weaker quantization, and the hole state postulated by our data fit, 40–50 meV above the InAs conduction band, near the bottom of the break in the gap, appears plausible.

Our heuristic model naturally leads to a 2D band model rather than a discrete donor level model. Such a preference is also suggested by other considerations. It was pointed out already by Shockley[10] that the *spatial* density of Tamm states should be equal to the surface density of atoms (or, more precisely, of primitive cells), of the order 10^{15} cm^{-2}. But states of such a high density are bound to broaden into a 2D band.

The idea that the Tamm states are plane wavelike suggests further that the states near the top of the Tamm-state band should be characterized by an effective mass, with a 2D density of states appropriate for that mass, of the form (1). In the absence of band mixing effects, one might expect a density-of-states mass equal to twice the heavy-hole mass for InSb ($m_{hh} \approx 0.45m_e$), the factor 2 arising because there are two interfaces. However, for the kind of broken-gap interfaces present, band mixing effects are almost certainly very strong, leading to a lower density-of-states mass. Our fitting values of 0.21–0.28m_e fall within the range of this expectation.

For energies deeper inside the Tamm-state band, the simple effective-mass model with its constant density of states will of course break down, and the density of states will increase. However, what matters for purposes of occupancy statistics is the density of states within a few kT of the top of the Tamm-state band; hence, our simple model postulated earlier.

IV. CONCLUSIONS

We have advanced what at this point must still be viewed as a speculative hypothesis, which clearly calls for additional research, both theoretical and experimental.

On the theoretical end, suitable calculations of the energy level structure at the InAs–AlSb interface are required, going beyond our heuristic arguments. We have at this point not conducted such calculations.

On the experimental end, a first task should be studies of the effects of systematic modifications of the interface growth procedures, going beyond the original work of Tuttle.[7] Such studies have been initiated and will be reported in due course.

Furthermore, the postulated interface Tamm states should manifest themselves in numerous other ways. For example, they would probably provide an efficient electron-hole recombination path. Recent work by Pekarik on AlSb/InAs/AlSb P-n-P transistors[18] shows strong evidence of rapid electron-hole recombination at the emitter-base heterointerface. The postulated Tamm states might also explain our persistent failure to observe any recombination radiation from optically pumped InAs QWs.

ACKNOWLEDGMENTS

The authors wish to acknowledge fruitful and encouraging discussions with Professor Lu Sham (UCSD), Professor Andrew Briggs (Oxford), and Professor Jörg Kotthaus (München). This work was supported by the Office of Naval Research.

[1] G. Tuttle, H. Kroemer, and J. H. English, J. Appl. Phys. **65**, 5239 (1989).
[2] A. Nakagawa, H. Kroemer, and J. H. English, Appl. Phys. Lett. **54**, 1893 (1989).
[3] G. Tuttle, Ph. D. thesis, University of California at Santa Barbara, 1991.
[4] P. F. Hopkins, A. J. Rimberg, R. M. Westervelt, G. Tuttle, and H. Kroemer, Appl. Phys. Lett. **58**, 1428 (1991).
[5] C. Nguyen, B. Brar, H. Kroemer, and J. H. English, J. Vac. Sci. Technol. B **10**, 898 (1992).
[6] C. Nguyen, B. Brar, H. Kroemer, and J. H. English, Appl. Phys. Lett. **60**, 1854 (1992).
[7] G. Tuttle, H. Kroemer, and J. H. English, J. Appl. Phys. **67**, 3032 (1990).
[8] C. R. Bolognesi, H. Kroemer, and J. H. English, J. Vac. Sci. Technol. B **10**, 877 (1992).
[9] I. Tamm, Phys. Zeitschrift Sowjetunion **1**, 733 (1932).
[10] W. Shockley, Phys. Rev. **56**, 317 (1939).
[11] H. M. James, Phys. Rev. **76**, 1611 (1949).
[12] H. Kroemer and Q.-G. Zhu, J. Vac. Sci. Technol. **19**, 143 (1982).
[13] Q.-G. Zhu and H. Kroemer, Phys. Rev. B **27**, 3519 (1983).
[14] H. Ohno, E. E. Mendez, J. A. Brum, J. M. Hong, F. Agulló-Rueda, L. L. Chang, and L. Esaki, Phys. Rev. Lett. **64**, 2555 (1990).
[15] F. Y. Huang and H. Morkoç, J. Appl. Phys. **71**, 524 (1992).
[16] I. Sela, C. Bolognesi, and H. Kroemer, Appl. Phys. Lett. **60**, 3283 (1992).
[17] S. R. Kurtz, G. C. Osbourn, R. M. Biefeld, and S. R. Lee, Appl. Phys. Lett. **53**, 216 (1988).
[18] J. Pekarik, H. Kroemer, and J. H. English, J. Vac. Sci. Technol. B **10**, 1032 (1992).

Reprinted from

B. Brar, H. Kroemer, and J. H. English, ```Quasi-direct' narrow GaSb/AlSb (100) quantum wells,'' J. Cryst. Growth, Vol. 127, pp. 752-754, 1993.

Journal of Crystal Growth 127 (1993) 752–754
North-Holland

JOURNAL OF **CRYSTAL GROWTH**

"Quasi-direct" narrow GaSb–AlSb (100) quantum wells

Berinder Brar, Herbert Kroemer and John English

Department of Electrical and Computer Engineering, University of California, Santa Barbara, California 93106, USA

In GaSb–AlSb quantum wells, GaSb is expected to make a transition to an X-valley semiconductor for well widths less than 2 nm. For narrow X-valley quantum wells the unassisted radiative transitions are no longer forbidden, allowing for the possibility of a "quasi-direct" transition from the X conduction band to the Γ valence band. Photoluminescence characterization of such narrow well GaSb–AlSb multi-quantum well structures has been performed. Spectra from wells as thin as a few monolayers were observed. A comparison of the measured transition energy with a simple calculation supports the idea that the observed transitions are indeed quasi-direct transitions.

1. Introduction

When the well width w_z of (100)-oriented GaSb quantum wells with (wide) AlSb barriers is reduced, quantization effects cause two successive changes in the band structure inside the well: (i) Around $w_z \approx 4$ nm, the low-effective-mass Γ-valley is pushed above the set of four equivalent L-valley pairs, leading to a Ge-like indirect-gap band structure inside the well, a phenomenon we have extensively studied previously [1,2]. (ii) Below $w_z \approx 2$ nm, quantization pushes the L-valleys themselves above the *lowest* X-valley pair. The X-valleys are split into two sets with different orientations in k-space relative to the plane of the wells. The lowest set consists of that valley pair which has the largest effective mass in the quantization direction. The k-space displacement vectors from $k = 0$ for this valley pair are perpendicular to the plane of the well. Inside the well, this pair forms a 2D band with an energy minimum where the *in-plane* k-component is zero, similar to a 2D minimum derived from a direct-gap Γ-valley. The key point is now that unassisted radiative transitions from this 2D minimum to the valence band maximum at $k = 0$ are no longer forbidden: In sufficiently narrow wells, the k-conservation selection rules hold only for the *in-plane*

k-components, but break down for the k-component *in the quantization direction*. Consequently, "quasi-direct" interband transitions should be allowed in sufficiently narrow GaSb–AlSb wells.

In the present work we extend our earlier studies of GaSb–AlSb wells into this quasi-direct regime by studying the luminescence of quantum wells with widths chosen to approximate an integer number N of GaSb monolayers (= MLs; 0.305 nm per ML), with N ranging from 5 down to 1. All wells down to a single GaSb ML showed strong photoluminescence, comparable in intensity to the signal from wider, direct, wells.

2. Growth and structure

The structures were grown on (100) semi-insulating GaAs substrates, using a modular Varian GEN II MBE machine, equipped with elemental group III and V sources. The antimony source was uncracked molecular Sb_4. To accommodate the 7% lattice mismatch between the epilayer and the substrate, a 1 μm AlSb buffer layer was grown at 570°C. After the growth of the buffer layer, the temperature was lowered to 530°C for the GaSb–AlSb quantum well structures themselves. A growth rate of 1 μm/h and a group-

V: group-III beam flux ratio of 5:1 were employed.

The AlSb buffer layer was followed by five sets of GaSb–AlSb multi-quantum wells, separated by 20 nm of AlSb. *Within* each set there were fifteen quantum wells with a fixed GaSb well width, separated by 5 nm of AlSb barriers. The five sets corresponded to multi-quantum wells of 1, 2, 3, 4, and 5 MLs, and were ordered such that the narrower wells, with higher PL emission energy, were closest to the surface. A 5 s antimony flush was employed at both interfaces of the GaSb wells to smooth the growth surface.

3. Results and discussion

The PL spectra of the structures were measured at 1.4 K with a 1 m McPherson monochromator using a 1 μm blazed grating and a North Coast Ge detector cooled to 77 K. A 488 nm Ar$^+$ laser line was used as a pump source with an excitation intensity of 10 W/cm^2. The PL signal as a function of emission energy is shown in fig. 1.

The highest-energy peak corresponds to the 1 ML wells which lie closest to the surface, as described earlier. With the exception of the 1 ML peak, the wells further from the surface have a weaker PL signal, presumably from a decay in intensity of the pump beam due to absorption.

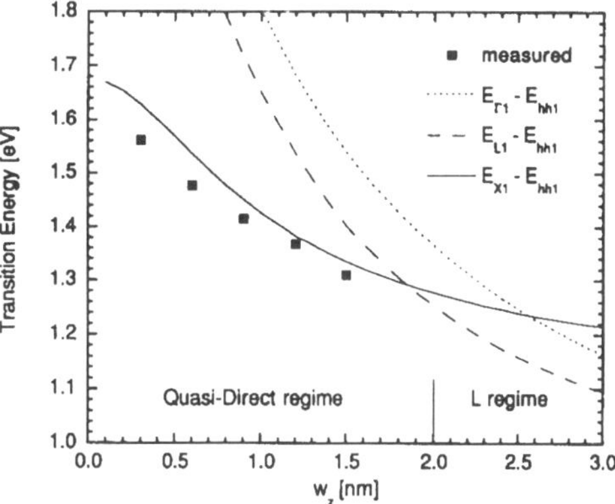

Fig. 2. Comparison between measured and calculated quasi-direct transition energies. For the X-valley, only the high-effective-mass curve is shown.

We speculate that the comparatively weak 1 ML peak and the comparatively strong 2 ML peak reflect an island-like growth mode, resulting in a shift of intensity from the 1 ML peak to the 2 ML peak [#1]. The uniformity effects over the wafer have not been systematically studied so far.

The observed transition energies in fig. 1 agree reasonably well with a simple square-well model for an X-valley-to-heavy-hole model, using appropriate values for band energies [3], band offsets [4], and effective masses (and their anisotropies) [5]. Fig. 2 shows a comparison between the measured and the calculated transition energies, indicating a well-width dependence that clearly reflects the weak quantization of the high-mass "perpendicular" pair of X-valleys.

4. Conclusions

Narrow well ($w_z < 2$ nm) GaSb–AlSb multi-quantum well samples with 1, 2, 3, 4, and 5 ML wells were grown using MBE. Low-temperature PL measurements show strong, clearly defined peaks corresponding to each of the well widths. A simple calculation suggests that below $w_z \approx 2$ nm,

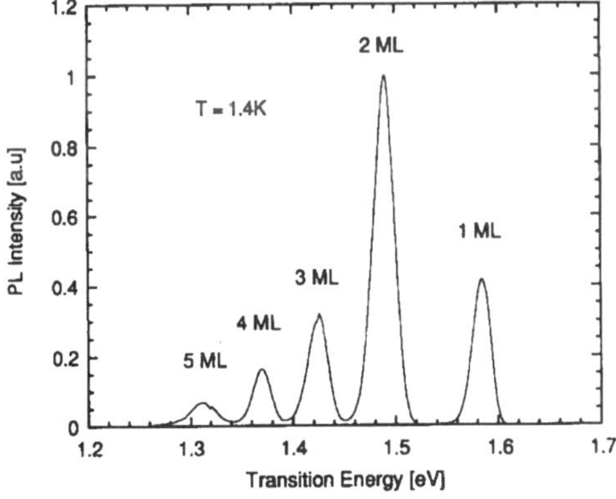

Fig. 1. Low-temperature PL spectra from GaSb–AlSb multi-quantum wells.

[#1] We have observed such a shift in narrow InAs–AlSb quantum wells also.

quantization effects cause the GaSb to be an X-valley semiconductor. A comparison of the measured transition energy with the calculated values supports the idea that the observed transitions are from X conduction band states to Γ valence band states. The strength of the signal is explained by introducing the new concept of a quasi-direct transition, whereby the interband selection rules for k-conservation are relaxed in narrow quantum wells. A quantitative analysis of the matrix element for the quasi-direct transition has not been attempted in the present paper.

We note that such a quasi-direct transition should also be realizable for other materials systems with a suitable band structure (InAs–AlSb?), and for other crystallographic orientations. For example, we would expect them in L-valley semiconductors if the quantum confinement is along the [111] direction, as in the Ge (111) quantum wells with (Si,Ge) alloy barriers.

Acknowledgments

We would like to thank Chanh Nguyen for many useful discussions. This work was supported by the Jet Propulsion Laboratory.

References

[1] G. Griffiths, K. Mohammed, S. Subbanna, H. Kroemer and J.L. Merz, Appl. Phys. Letters 43 (1983) 1059.
[2] U. Cebulla, A. Forchel, G. Tränkle, G. Griffiths, S. Subbanna and H. Kroemer, Superlattices Microstruct. 3 (1987) 429.
[3] C. Alibert, A. Joullie, A.M. Joullie and C. Ance, Phys. Rev. B 27 (1983) 4946.
[4] U. Cebulla, G. Tränkle, U. Ziem, A. Forchel, G. Griffiths, H. Kroemer and S. Subbanna, Phys. Rev. B 37 (1988) 6278.
[5] O. Madelung, in: Data in Science and Technology, Ed. R. Poerschke (Springer, Berlin, 1991).

Reprinted with permission from

H. Kroemer, ``Semiconductor Heterojunctions at the Conference on the Physics and Chemistry of Semiconductor Interfaces: A Device Physicists Perspective,'' J. Vac. Sci. Technol. B, Vol. 11(4), pp. 1354-1361, 1993.

Semiconductor heterojunctions at the Conference on the Physics and Chemistry of Semiconductor Interfaces: A device physicist's perspective

Herbert Kroemer
ECE Department, University of California, Santa Barbara, California 93106

(Received 25 January 1993; accepted 16 April 1993)

After a very slow start, heterojunctions have emerged as one of the central topics of the Conference on the Physics and Chemistry of Semiconductor Interfaces. The presentation describes this emergence, starting from such items as the electron affinity rule of conduction band offsets, Dingle's first determination of the GaAs–(Al,Ga)As band lineups, and the first lineup theories. Some of the blind alleys in this development (85:15 Rule, Common-Anion Rule, and others) are retold by one of the participants. The treatment then turns to a few of the most recent developments, such as the emerging role of *ab initio* computations as a quasiexperimental tool, plus a few developments this writer finds worth speculating about. The treatment is from the perspective of a device physicist, rather than a surface scientist, and some thoughts are offered on why there is not more commonality between heterojunctions and Schottky barriers.

I. INTRODUCTION

The field of heterojunctions is one where the interests of many different kinds of individuals intersect: surface scientists, device physicists, and crystal growers, to mention just three groups; the list could be extended. The Conference on the Physics and Chemistry of Semiconductor Interfaces (PCSI) has provided a forum for discussion that would bring us together as no other conference has done, to the benefit of all of us, and of the heterojunction field itself— even if those benefits were not always acquired without some pain. A glance at the PCSI Proceedings shows that heterojunctions, although essentially absent at PCSI-1, and coming into their own only slowly, are now rivaling Schottky barriers as a central topic of PCSI, the exact balance depending on where one draws the line on what constitutes a heterojunction paper and what does not. At an anniversary session like this one, it is useful to look back and to review how the present role of heterojunctions within PCSI developed, and this is how I see my task for this presentation.

A glance at a stack of Proceedings shows that this task is actually hopeless. There is no way I can talk about more than a small minority of the topics of interest; I must select beyond any rationally defensible criterion, to the point that the selection inevitably becomes a highly personal one, along the lines *"PCSI papers that I remember* (plus a few that I *should* have remembered)" But in this way I know at least what I am talking about, and I hope my review will provide a useful perspective despite the personal bias.

This bias is that of a physicist who came to heterojunctions via semiconductor devices—hence my subtitle. As a device physicist, I am interested in heterointerfaces with a sufficiently low defect density that they have a chance to be actually used in some device (at least in a future device), rather than just being a place where two different semiconductors happen to come together: In essence, the interface *is* the device. A lattice-mismatched interface between GaAs and Si, with a misfit dislocation every 10 nm, and a GaAs layer full of antiphase domains may be a fascinating

crystal growth challenge (to which I have devoted several years of work myself), but it is not a heterojunction in the sense in which I use the word here.

This unfriendly attitude towards interface defects is one of the aspects that distinguished heterostructure device physicists from their Schottky barrier friends. To the Schottky barrier scientist the interface defects that pin the barriers are an essential part of Schottky barrier physics; to the device physicist interface defects are usually just a nuisance, to be gotten rid of as quickly as possible, and understanding these defects is simply a tool to get rid of them faster.

Much of my discussion will necessarily involve band lineup questions, and hence overlap Margaritondo's presentation following mine. But with the two of us coming to heterojunctions from opposite ends, our presentation will be quite orthogonal. In fact, I hope that the reader will find the presentations complementary to each other, giving a fuller picture than what either of us could have presented by himself.

II. PCSI 1: PRECURSORS

The acronym PCSI originally meant *Physics of Compound Semiconductor Interfaces*, which implies first and foremost interfaces between compound semiconductors, and in those days, before the advent of Si–Ge heterojunctions, that was synonymous with heterojunctions. I have no reason to believe that the initiators of this conference intended such a narrow interpretation—it would not have made much sense, nor would it have led to the eminently successful PCSI that we know and cherish. The acronym was simply a misnomer, almost immediately recognized as such, even though it took eight years to rename the conference in a way that reflected its actual contents—while retaining the acronym.

I recall PCSI-1 at Fort Collins vividly, but if there were actually any formal talks about heterojunctions in the sense of interfaces between semiconductors, I do not remember them. It was a workshop on essentially *any* kind of interface between *one* semiconductor and almost anything else, ranging from semiconductor-vacuum interfaces, better known as surfaces, to the Si-oxide interface, or to metal-semiconductor interfaces, also known as Schottky barriers. The latter dominated PCSI from the beginning, and it was only very slowly that heterojunctions assumed the important role that they play today. I do not recall that at PCSI-1 there were any papers involving the interface between *two* semiconductors, and the *one* semiconductor was more often silicon than a compound—unless one classifies silicon as cubic silicon silicide.

The heterojunction people simply did not show up—despite very extensive and successful work on heterojunction devices: By 1974, the *idea* for the double-heterostructure laser was 11 years old, and its experimental demonstration was four years old. Successful heterostructure bipolar transistors (HBTs) had also been demonstrated. But most of this successful heterojunction work was done by people who were basically device oriented rather than physics oriented, and whose interests were well met by the annual Device Research Conference, and later, after the emergence of molecular-beam epitaxy (MBE), by the annual MBE Workshop, not to mention assorted other specialized conferences that sprang up subsequently.

Furthermore, the band lineup problem so dear to physicists like Margaritondo and myself was of little interest to the early device people: All the successful device structures were GaAs–(Al,Ga)As structures grown by liquid phase epitaxy (LPE), which usually led to graded structures, for which the band offsets have little effect on the device performance. Far from being a nuisance, this grading was actually desirable in the early devices, and an interest in band lineups did not develop until MBE became an important heterojunction technology in the early 70s.

For those who *were* interested in understanding and predicting band lineups, there was only the Electron Affinity Rule (EAR). The electron affinity is defined as the work required to remove an electron from the bottom of the conduction band to the space outside the material, just beyond the range of the image force at the surface (Fig. 1). The EAR asserted that the conduction band offset at an abrupt heterojunction is equal to the difference in the electron affinities of the two semiconductors

$$\Delta\epsilon_c = \chi_2 - \chi_1. \tag{1}$$

But this could not be true: Electron affinities contain highly variable electrostatic dipole contributions, for example from absorbed surface layers, such as water. If Eq. (1) were correct, the band offsets would change if a surface dipole layer far away from the junction changes, a nonsensical notion. Somehow, the dipoles had to be subtracted out. I later worked out the correction;[1] instead of Eq. (1), the correct rule may be written

$$\Delta\epsilon_c - \delta_i = (\chi_2 - \delta_2) - (\chi_1 - \delta_1), \tag{2}$$

where δ_1 and δ_2 are the electrostatic dipole contributions contained in the two electron affinities, and δ_i is the electrostatic dipole—if any—associated with the heterointerface itself. This relation evidently reduces to Eq. (1) only

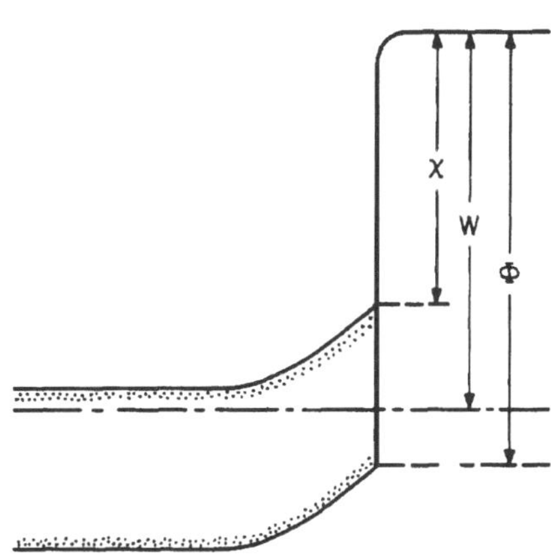

FIG. 1. Electron affinity χ, work function W, and ionization energy (or photothreshold) Φ, at the surface of a semiconductor.

if the three dipole contributions cancel, an unlikely event for three dipoles of unrelated physical origin.

In any case, the rule was academic, not only because the dipole contribution to the electron affinities were not known, but the electron affinities themselves for many semiconductors of interest had never been measured, including specifically those of AlAs and (Al,Ga)As alloys. The electron affinity values for these materials sometimes found in the literature are purely fictitious: They were calculated by using the incorrect rule (1) backward, from measured GaAs–(Al,Ga)As band offsets—which later turned out to be incorrect themselves.

Last, but not least, the interface defect problem that has been one of the dominant themes of PCSI, did not initially rear its ugly head with heterojunctions. Compared to all other kinds of interfaces, good LPE heterojunctions appeared to have a fabulous level of perfection, with defect densities sufficiently low that they were actually hard to measure by the then-existing *direct* techniques; they were most easily estimated indirectly, by their effect on *device* performance. This, too, was to change later, and when that time arrived, PCSI was ready, by providing a discussion forum that would bring people with different interests together in a way that none of the existing conferences did, nor any of the more recent conferences.

III. PCSI 2 THROUGH 4: THE STAGE GETS SET

It was not until PCSI-2 at the University of California at Los Angeles (UCLA) that the first papers dealing with "true" heterojunction problems appeared—true in the sense described on the Introduction. Initially, they represented a pitifully small fraction. The classification is somewhat arbitrary, but I would count only four out of the 39 papers in the Proceedings as true heterojunction papers. Two of those four dealt with heterojunction photocathodes, an important problem throughout the 1970s, but the other two dealt with the central band lineup problem of heterojunction physics: one was experimental, the other theoretical.

A. Dingle's rule

The central experimental heterojunction article of PCSI-2 was Dingle's classical paper[2] on the first GaAs–(Al,Ga)As quantum wells and their luminescence properties. It was a great paper, a "first" in more ways than one: It was the first PCSI paper to discuss quantum wells at all, the first based on MBE-grown and hence abrupt heterojunctions (where band lineups finally mattered). But more important from the PCSI perspective, it introduced quantum well luminescence as a powerful new tool for determining those band lineups—even if the interpretation of such data later turned out to be more tricky than was suspected in 1975. Although the work had already been published the year before (and the PCSI presentation was not its first conference presentation), it was new to many of the attendees, and a valuable contribution to the conference. This kind of willingness to include review-type presentations of already-published work helped in shaping PCSI as the great forum for discussion we know, rather than as just another place for as yet unpublished work.

What made Dingle's work so exciting to heterojunction physicists of all persuasions was that it provided the first good data for the band lineups in the most important of all heterojunction systems, the GaAs–(Al,Ga)As system. Using a very simple Kronig–Penney-type model, ignoring all possible complications, Dingle showed that the data provided an excellent fit to a model with an 85:15 conduction-to-valence band offset ratio. By hindsight, it would have been better if the fit had been less good: We would have retained some more skepticism about those numbers. As it was, the fit appeared so good that it "just had to be true." The possibility that an equally good (or even better) fit might be obtained to a significantly different more complicated model with different band offsets, was simply not considered: "Dingle's Rule," as it soon became known, had become gospel, and the few heretics who doubted it were ignored. Just how sacred Dingle's Rule had in fact become, is perhaps best illustrated by a painful personal reminiscence: By early 1980 we had developed the *C–V* profiling technique for measuring band offsets,[3] obtaining a much lower ratio for the two band offsets. Rather than viewing this as an indication of the incorrectness of Dingle's Rule, we interpreted it as a result of compositional grading in our own junctions, by an argument that subsequently turned out to be a fallacy.[4]

The first serious challenge to the validity of Dingle's Rule came only several years later, from within the optical-measurements community itself. Extracting band offsets from quantum well luminescence spectra is a rather indirect procedure that is highly model dependent, relying on assumptions of uncertain validity, such as the absence of valence band mixing, about which little was known. Measurements on parabolic rather than rectangular quantum wells, by Miller *et al.*,[5,6] which were actually somewhat less model dependent, could not be fitted to Dingle's 85:15 ra-

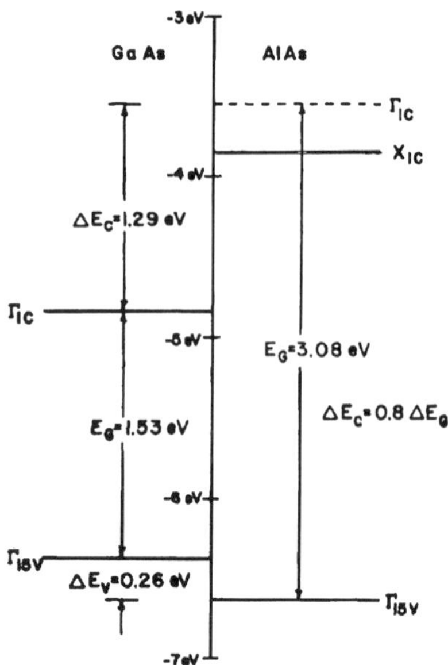

FIG. 2. The first theoretical prediction of the GaAs–AlAs energy band lineup, from Frensley and Kroemer (Ref. 8).

tio, requiring a revision to the ratio widely accepted today, somewhere between 62:38 and 60:40.

None of this distracts in any way from Dingle's achievement—nor was Dingle responsible for other people taking his results as gospel. But this excessive faith *did* eventually delay progress.

B. Theory

The theoretical heterojunction lineup paper at PCSI-2 was my own, on "*Problems in the Theory of Heterojunction Discontinuities.*" The first word, *Problems*, was deliberate: My objective was a programmatic one, to identify problems rather than to offer a solution to what was clearly a difficult problem. My point of departure was to show that the electron affinity rule was conceptually unsound, and the detour via the surface was in any event too indirect; a procedure that, in the later words of Walter Harrison, "*replaces one simple problem by two very complicated problems.*"[7] I suggested a methodology to calculate the band offsets from the bulk band structures of the participating semiconductors.

One year later, at PCSI-3, Bill Frensley and myself presented the first results of an attempt to predict heterojunction band offsets from bulk band structures, based on pseudopotentials.[8] Figure 2 shows our prediction of the GaAs–AlAs band offsets, which was remarkably close to Dingle's Rule—with absolutely no fudging! Bill and I were acutely aware of the crudeness of our pseudopotentials, and we proceeded to refine them, but the agreement with Dingle's Rule deteriorated. It was not until years later that it became clear that we had in fact been on the right track, but by then both Bill and myself had turned to other problems, and our theory had been superseded by later theories, starting with Walter Harrison's theory, also presented first at PCSI, one year later, at PCSI-4.[7]

While the pseudopotential method of F&K in essence approached the problem from the nearly free electron end, Harrison started from the opposite end, based on an atomic orbital tight-binding model. It was computationally much simpler than our pseudopotential approach, and gave slightly better agreement with the more reliable ones of the emerging data. These two early attempts were followed by numerous others, for many of which PCSI provided the first forum for discussion, and which tend to agree reasonably well with experimental data. The PCSI-15 article by Cardona and Christensen makes good reading on where we now stand on this issue.[9]

There were two problems with the early theories. The first was that the lineup criteria always contained an element of "*ad hocism.*" As I put in a rhetorical question in a review I gave at PCSI-11, referring to the Harrison theory (but just as applicable to our own theory): "*why is it right—or is it?*"[10]

The second problem was that all these theories were in the last analysis retrodictive rather than predictive theories: They tested our understanding of the basic physics, but they did not meet the device physicist's need for theoretical data that are accurate enough for practical device design. In fact, at least Walter Harrison quite frankly admitted that conceptual simplicity rather than device-level accuracy was his goal. Frensley and myself had hoped to be able to develop a theory of true predictive value, hence our choice of the less-intuitive and mathematically more involved pseudopotentials. Our purely local two-parameter pseudopotentials were far too crude to achieve that goal—and we knew that. We had intended to refine the procedure by going to more advanced nonlocal pseudopotentials, but for a variety of reasons—discouragement and a lack of cash being two of them—we never got around to that.

IV. "VIRTUAL REALITIES": *AB INITIO* COMPUTATIONS

I believe today that, if you want to have a truly predictive tool, the proper route is to go to purely numerical self-consistent first-principles calculations. The excellent paper of Dandrea *et al.*[11] at last year's PCSI-19, which yielded remarkably good agreement with many experiments, suggests that we are indeed getting close to that goal. This is a remarkable change from the situation in the 1970s, when some of the first such attempts gave even poorer agreement with good experimental data than the early crude theories. Some of those early results were even qualitatively wrong.[10]

One criticism that has been raised against self-consistent first-principles calculations has been that they are cumbersome and computationally expensive single-shot calculations that teach little systematics, and in which the physical insight is often lost.[9,10] However, it does not have to be that way: As Cardona and Christensen pointed out,[9] the insight *can* be regained by careful analysis of the numerical

data. There is no doubt that too little of the latter is currently being done, and I wish to urge my number-crunching friends to direct more of their attention to that important issue. However, in the last analysis, the computational approach is, in this regard, no worse than the experimental one, where a single measurement by itself also teaches no systematics and often yields little physical insight. The systematics and much of the physical insight come from the comparison of many experimental data, often acquired by different experimenters using different methods. In fact, given both a computer algorithm with true predictive power, and a strong attitude towards intensive postmortem "brain analysis," the systematics and at least some of the physical insight may be easier to achieve in this way than by real-world experimentation: both the "sample preparation" and the "measurement technique" are more reproducible under the virtual reality environment of a computer, than in a real laboratory.

Nor is the computational approach any more cumbersome than setting up a lab—certainly not to the computational physicist actually working in that environment. Both are one-time setup expenses, and once made, the "sample" turn-around time, and the operating expenses per unit result are almost certainly less in the virtual lab than in the real one. None of this should be misconstrued as a call for abolishing real experiments: I call for *supplementing* them, not replacing them. Even if we had full confidence in our results, we still need real-world technology to grow the devices that are the goal of much of the whole exercise. I do not expect to see, during my own lifetime, an environment where reality checks are no longer necessary.

Some may perhaps ask: if this is so, then why bother with all the computation? What is at stake is sometimes the decision on whether actually to go ahead with the hugely expensive new technology necessary to get the real data. Let me give here a recent real-world example, which illustrates that point.

It has been suggested that $(Ga,In)P$, lattice matched to GaAs, might be a better material than $(Al,Ga)As$ for some of the most important GaAs-based heterostructure devices,[12] because $(Ga,In)P$–GaAs heterojunctions have been predicted to offer highly desirable higher valence band offsets (290 meV) than $(Al,Ga)As$–GaAs with the usual low Al fraction ($<25\%$) that is necessary to minimize highly deleterious DX centers. In addition, the use of $(Ga,In)P$ would make it possible to get entirely rid of aluminum with its obnoxious affinity to oxygen. At the time this suggestion was made, phosphorous was hardly an attractive alternative to aluminum, but with the advent of gas-source MBE and especially organometallic vapor phase epitaxy (OMVPE), this is no longer a serious issue, and the $(Ga,In)P$–GaAs approach is now being actively pursued. However, there are major uncertainties in the band lineups themselves, which scatter over a range almost 200 meV wide, far beyond reasonable experimental error estimates of the measurement techniques employed. Initial *direct* band offset measurements by the C–V technique[13] yielded a valence band offset of 250 meV, somewhat less than the theoretical prediction, thereby reducing somewhat the incentive to develop the (Ga,In)P–GaAs device technology. Some recent *indirect* measurements on actual experimental devices confirm such values, whereas other data suggest much larger values,[14] up to 450 meV. The discrepancies have not gone away with time, and inspection of some of the data suggests that the differences are not due to errors in the measurement techniques, but reflect technology-induced differences in the atomic arrangements, caused by Ga-versus-In and P-versus-As site competition at the crossover from GaAs to (Ga,In)P. Apparently, some growth procedures happen to yield a "good" atomic arrangement whereas others do not. The problem is ideally suited for a computational approach that permits the investigator to actually explore different atomic arrangements, and the stakes are high. In fact, at PCSI-19, Dandrea *et al.*[11] took a first stab at this problem, and showed that significant band offset differences can be readily expected. Another article at PCSI-19, by Foulon *et al.*,[15] using tight-binding calculations, came to similar conclusions. But it appears to me that the computational approach is better suited for taking the next step, that of actually *guiding* the technology on how to achieve the best band lineups, rather than simply modeling whatever the technology yields.

V. THE COMMON-ANION RULE

Dingle's 85:15 Rule was not the only blind alley for heterojunction lineup theorists: No discussion of the history of this field is complete without at least mentioning (and chuckling over) the Common Anion Rule (CAR). In fact, the two debacles form a closely related "self-consistent pair."

Like Dingle's Rule, the CAR started out as a triumph: At PCSI-3, McCaldin, McGill, and Mead reported a remarkably accurate rule for the band lineup of gold Schottky barriers to a wide range of compound semiconductors.[16] The energetic separation between the valence band edge and the metal Fermi level was found to *"depend almost entirely on the anion alone i.e., the cation plays little, if any role."* The authors were able to give a very convincing explanation in terms of the valence band orbitals, and they wrote *"we expect that this concept will be useful in predicting the value of discontinuities in bands at semiconductor–semiconductor interfaces."* In support of their expectation they quoted Dingle's 85:15 ratio for the GaAs–(Al,Ga)As system. Additional support to the rule was provided by the early band lineup theories, which showed strong trends towards such a rule, trends that had not been explicitly built into the theory.

Once the 85:15 rule collapsed, the CAR collapsed with it, and the theoretical justifications for it vanished shortly afterwards: As the theories became refined, they yielded results with increasingly larger band offsets for common-anion systems, and there are numerous theoretical articles in the PCSI Proceedings of later years that explain what went wrong with the simple early theories, and why the CAR should in fact *not* hold for semiconductor heterojunctions, whatever its merits for Schottky barriers.

In defense of McCaldin *et al.*, it must be said that their *expectation* was eminently reasonable in the light of what was known in 1976. The role of defects in Schottky barrier heights was not understood yet, and it was reasonable to assume that Schottky barrier lineups and heterojunction lineups were much more closely related than they turned out to be.

VI. MEASUREMENT TECHNIQUES

Of all the techniques to measure band offsets, the most direct one, the ultraviolet photoelectron spectroscopy/x-ray photoelectron spectroscopy (UPS/XPS) technique, has been practically "owned" by PCSI. The technique does not depend on having "device-quality" interfaces, and as a result more heterojunction band offset pairs have been measured by it than by any other method. There have sometimes been questions as to the applicability of those data to true heterojunctions, but the XPS data have stood the test of time remarkably well. I will say no more about it here, leaving that topic to Margaritondo, one of the principal practitioners of that technique.

Despite their stunning impact at PCSI-2, band offset determinations by quantum-well photoluminescence (PL) remained under-represented at PCSI. In contrast to the XPS technique, they require device-quality heterojunction pairs for which a highly developed device technology already exists, which makes them too restrictive in their applicability to serve as general-purpose techniques. Although a large amount of work in this field has been done, very little of it found its way into PCSI. With few of the practitioners of the technique being members of the PCSI constituency, the technique stayed largely outside PCSI, being presented at device-oriented conferences providing a more sympathetic audience.

The same is true for band offset determinations by C–V profiling, one of the most widely used techniques,[3] which can be very accurate. Having been the originator of the C–V technique, I may be to blame at least in part of this neglect: Even though the idea for this technique sprang from a paper by Garner *et al.* at PCSI-5,[17] the "devicey" nature of the method caused me to present it at the DRC.

A number of "electrical" techniques other than C–V profiling had come into existence over the years, and by ~1984, a rich set of band lineups were available, taken by different techniques, sometimes agreeing, sometimes not. The PCSI Proceedings for PCSI-12 in Tempe provide a good overview over the state of the band lineup problem at that time.

One of the items of discussion at PCSI-12 concerned the persistent minor discrepancies between the optical and other techniques.[18] As the demise of Dingle's Rule had shown, the interpretation of the quantum-well PL data was unpleasantly model dependent, and it was not surprising that there remained discrepancies. The issue was not resolved at PCSI-12; a reasonable convergence was not achieved until one year later, at PCSI-13, where Wilson *et al.*[19] reported on optical determinations of the band lineups in the AlAs–(Al,Ga)As system, with one of the materials being AlAs rather than GaAs, and the other having an Al fraction of 37%. For this combination the hetero-junctions have a staggered lineup, in which conduction and valence bands on one side are shifted in the same direction relative to those on the other side, but without leading to a broken gap as in the InAs/GaSb system. It had been pointed out[20] that in such systems one should be able to observe below-gap radiation originating from the recombi-nation of electrons on one side with holes on the other (a "spatially indirect" system), with a photon energy given by the "residual gap" of the system. Obviously, here was an-other approach to measure band offsets optically, at least for heterojunctions with staggered lineups, and this was the technique Wilson *et al.* used. The results were in much better agreement with electrical lineups than had been the case for the quantum well luminescence data.

Although superficially similar to the quantum well lu-minescence technique, the staggered-lineup luminescence technique differs in one essential aspect. Not relying on quantization in a narrow well, it is almost free of the prob-lems, such as the complicated quantization-induced split-ting of the valence band degeneracy, that make the inter-pretation of quantum well luminescence data so model dependent. It is, in fact, one of the most "direct" tech-niques available. Although restricted in its applicability to staggered-lineup systems, the technique deserves more at-tention, because of its potential for yielding very accurate values. I am therefore happy to see that at this year's PCSI, Bimberg's Berlin group reports on the use of this technique to provide what I believe are the first really accurate band lineups for the staggered-lineup lattice-matched InP–(Al,In)As system.[21] Drawing on the Harrison theory, the likelihood of a staggered nature for this system had first been pointed out by Kroemer and Griffiths,[20] and the ex-istence of staggered-lineup radiation at such junctions was subsequently confirmed by Caine *et al.*,[22] but the samples of these authors suffered from severe interface defects caused by the absence of a good MBE technology for InP at that time, and the radiation probably was not band-to-band, but involved defect states. The Berlin data are there-fore a great advance, and a demonstration of the power of the technique. It is interesting to note that they are in excellent agreement with the original theoretical predictions—indicating how good the simple old Harrison theory is.

VII. WAVE FUNCTION CONNECTION RULES

In band diagrams, band offsets at abrupt heterojunction are almost invariably represented as simple steps. Such un-modified rectangular steps represent the *work* involved in transferring a band edge electron from one side to the other, but they imply that the effect of the discontinuity on the *phase* of the electron waves is the same as for a true free-space potential barrier (except for the changes inher-ent in the change from a free-electron mass to two effective masses). When a wave impinges on a heterointerface, both the reflected wave and any transmitted wave will have a certain phase relation with the incident wave, and there is no *a priori* reason that the phase relations for Bloch waves scattered by a band edge step should be the same as for free

electrons waves scattered by a normal potential step, even after allowance has been made for the transition to effective masses. The problem was briefly touched upon by myself and Zhu at PCSI-9, where we showed that deviations should indeed be expected, and that they could be repre-sented in the band diagram by adding a delta function scatterer to the step.[23] The problem remained dormant (at least within PCSI) because the effects appeared to be neg-ligibly weak at the all-important GaAs-(Al,Ga)As inter-face, and probably at other common-anion heterojunctions as well. But with an increasing interest in more exotic ma-terials systems, the issue can probably be no longer ig-nored: At last year's PCSI-19, we found it necessary to invoke precisely such delta function modifications at the interface of the InAs–AlSb interface, leading to interface Tamm states, to explain certain properties of these wells.[24] The jury is still out on this Tamm state hypothesis, but quite independently of whether or not this hypothesis is confirmed, I am convinced that the inclusion of such phase-correcting delta scatterers at the interface is an issue that will play a role at future PCSI conferences.

VIII. TRANSITIVITY AND SUCH

Since the earliest days of heterojunction band offset measurements, there have been occasional reports on an alleged technology dependence of the measured band off-sets, including specifically a dependence on the growth di-rection. Now it is clear that any such dependence must reflect differences in the atomic arrangement, either as a result of defects, or—more interesting—as a technology dependence of the bonding across the interface, even in the absence of defects. I find most such reports implausible, but there are two classes of exceptions to my skepticism. (a) Such dependences should be expected for heterojunc-tions where the semiconductors on the two sides of the interface involve different columns in the periodic table, such as, say, GaAs–Ge and GaAs–ZnSe interfaces. How-ever, in those cases, variations in band offsets probably get swamped by other problems, such as lack of electrical neu-trality, cross doping, antiphase disorder (for polar-on-nonpolar growth), and other headaches. (b) A more sub-tle case of great recent interest occurs when both the cation and the anion change in going across the interface, as at InAs–AlSb heterojunctions, which can be grown both with InSb-like and AlAs-like interfaces,[25] with drastically dif-ferent properties. Much of the difference is probably due to a high concentration of defects at the AlAs–like interfaces, but at last year's PCSI-19, Waldrop *et al.* reported a clear 70 meV difference in the band lineups.[26] If anything, this difference is surprisingly small.

IX. TRANSPORT PROPERTIES

There is probably no physical property more sensitive to the exact values of the band offsets than the electron trans-port across an isotype heterojunction. To the naive, the study of such properties should therefore be a powerful tool for an accurate determination of band offsets. Unfor-tunately, this transport is just as sensitive to all sorts of distortions of the band structure, such as spurious defect changes near the interface, compositional grading effects, and what have you.

Nothing shows the problem better than the current–voltage characteristics of the GaAs–(Al,Ga)As n-N heterojunctions grown by LPE in the early 1970s. Even though such structures have a substantial conduction band offset, and should therefore rectify similar to Schottky barriers, early investigators never observed any rectification. Prior to Dingle's work, this could be justified by assuming that the band offset was too small, but this excuse vanished once Dingle's work had appeared; the favored excuse then became that in those LPE-grown junctions the conduction band spike had been obliterated by the graded nature of the LPE interfaces. At PCSI-4, Garner et al.[17] demolished that hypothesis with a paper showing the complete absence of rectification effects, even at low temperatures, in LPE-grown GaAs–(Al,Ga)As heterojunctions that were grown as abruptly as possible by LPE, and for which Auger profiling showed that the grading was far too weak to explain the total lack of rectification down to 77 K.

The lack of rectification was never satisfactorily explained. The problem simply went away as technology improved, but it was to take many years before transport data were successfully used to determine band offsets, and even then only in systems with an already very highly developed technology. Garner himself suggested—with a distinct lack of enthusiasm—that maybe the absence of rectification might be explained by surface leakage, but he was evidently more concerned with pointing out the seriousness of the problem than with any specific explanation.

Although Garner's paper did not seem to draw the attention it deserved, I myself was very much influenced by it. While puzzling over the problem, I considered the possibility that maybe the LPE-grown heterojunctions were not so perfect after all, but contained an interface donor of some kind. That could convert the notch-and-spike band diagram into a pure notch with no remaining spike (Fig. 3), and hence with no rectification. But if there were interface donors, they should manifest themselves by C-V profiling *through* the junction, from a Schottky barrier placed on top of the epilayer. When I discussed this idea with Jim Harris, then at Rockwell, he showed me some strange-looking C-V data actually taken at Rockwell on some nonrectifying n-N junctions. I did not really understand C-V profiling at the time, and was puzzled that the data did not look at all like a donor sheet, but like a smeared-out dipolelike structure. I soon realized what some others had known all along, that C-V profiling does not measure the distribution of donors, but of electrons, and that we were seeing a partial electron depletion in the barrier region of the heterojunction, and an electron enhancement in the notch region. Eventually, I also realized that the smeared-out appearance of the profile did not indicate a smeared-out electron distribution, but was an artifact of the C-V profiling process itself. The clincher came when I realized that it was possible to extract the band offset directly from the smeared-out apparent electron distribution without having to reconstruct the true electron

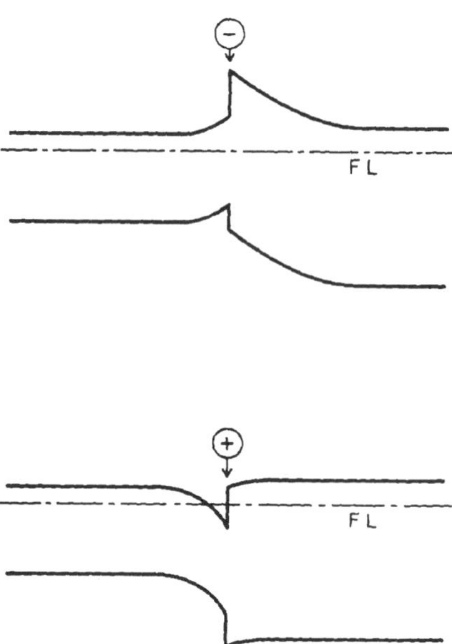

FIG. 3. Effect of interface charges on the net barrier height controlling the current flow across a heterojunction barrier.

distribution.[3] Thus, the C-V profiling technique actually grew out of a PCSI paper.

The story has an ironic ending: As I mentioned earlier, Dingle's Rule struck again. When we remeasured the Rockwell junctions and applied the new profiling theorem to the data, we obtained (in addition to the expected interface charge) a band offset significantly smaller than Dingle's rule demanded, and we explained the discrepancy in terms of a lowering of the barrier by compositional grading. It was only years later that it became clear that our data had in fact been correct, and that C-V band offset values were immune to the kind of grading we had postulated.[4] Oh, well...

Back to transport properties. With the important exception of regular contributions of articles from Tom McGill's group on tunneling through heterobarriers, articles on electron transport both across and along heterojunctions remained sparse. There were some excellent invited review articles keeping the PCSI audience informed about important transport development, but once again the key contributors were not members of the PCSI core constituency.

X. WHAT NEXT?

New problems continue to emerge, and I expect PCSI will continue to play a key role whenever the problems strongly benefit from that unique interplay of surface scientists, device physicist, crystal growers, etc., that I cited in the Introduction as being PCSIs great strength. One of the areas where I anticipate a large amount of activity is that of quantum wires, based both on the tilted-superlattice approach of Fukui and Saito[27] and of Gaines et al.,[28] first reported at PCSI-15, as well as on newer principles. As a teaser, let me predict for PCSI-20+ an emerging interest in semiconductor-superconductor "heterojunctions" (or are these Schottky barriers—or do we care?)

ACKNOWLEDGMENT

This work was supported by the Office of Naval Research.

[1] H. Kroemer, Crit. Rev. Solid State Sci. **5**, 555 (1975).
[2] R. Dingle, Crit. Rev. Solid State Sci. **5**, 585 (1975).
[3] H. Kroemer, W.-Y. Chien, J. S. Harris, and D. D. Edwall, Appl. Phys. Lett. **36**, 295 (1980).
[4] H. Kroemer, Appl. Phys. Lett. **46**, 494 (1985).
[5] R. C. Miller, D. A. Kleinman, and A. C. Gossard, Phys. Rev. B. **29**, 3740 (1984).
[6] R. C. Miller, D. A. Kleinman, and A. C. Gossard, Phys. Rev. B. **29**, 7085 (1984).
[7] W. A. Harrison, J. Vac. Sci. Technol. **14**, 1016 (1977).
[8] W. R. Frensley and H. Kroemer, J. Vac. Sci. Technol. **13**, 810 (1976).
[9] M. Cardona and N. E. Christensen, J. Vac. Sci. Technol. B **6**, 1285 (1988).
[10] H. Kroemer, J. Vac. Sci. Technol. B **2**, 433 (1984).
[11] R. G. Dandrea, C. B. Duke, and A. Zunger, J. Vac. Sci. Technol. B. **10**, 1744 (1992).
[12] H. Kroemer, J. Vac. Sci. Technol. B **1**, 126 (1983).
[13] M. A. Rao, E. J. Caine, H. Kroemer, S. I. Long, and D. I. Babic, J. Appl. Phys. **61**, 643 (1987).
[14] See, for example, W. T. Masselink, M. Zachau, W. T. Hickmott, and K. Hendrickson, J. Vac. Sci. Technol. B. **10**, 966 (1992).
[15] Y. Foulon, C. Priester, G. Allan, J. C. Garcia, and J. P. Landesman, J. Vac. Sci. Technol. B **10**, 1754 (1992).
[16] J. O. McCaldin, T. C. McGill, and C. A. Mead, J. Vac. Sci. Technol. **13**, 802 (1976).
[17] C. M. Garner, Y. D. Shen, C. Y. Yu, G. L. Pearson, and W. E. Spicer, J. Vac. Sci. Technol. **15**, 1480 (1978).
[18] G. Duggan, J. Vac. Sci. Technol. B **3**, 1224 (1985).
[19] B. A. Wilson, P. Dawson, C. W. Tu, and R. C. Miller, J. Vac. Sci. Technol. B **4**, 1037 (1986).
[20] H. Kroemer and G. Griffiths, IEEE Electron Dev.Lett. **4**, 20 (1983).
[21] J. Böhrer, A. Krost, and D. Bimberg, J. Vac. Sci. Technol. B **11**, 1642 (1993).
[22] E. J. Caine, S. Subbanna, H. Kroemer, J. L. Merz, and A. Y. Cho, Appl. Phys. Lett. **45**, 1123 (1984).
[23] H. Kroemer and Q.-G. Zhu, J. Vac. Sci. Technol. **21** 551 (1982).
[24] H. Kroemer, C. Nguyen, and B. Brar, J. Vac. Sci. Technol. B **10**, 1769 (1992).
[25] G. Tuttle, H. Kroemer, and J. H. English, J. Appl. Phys. **67**, 3032 (1990).
[26] J. R. Waldrop, G. J. Sullivan, R. W. Grant, E. A. Kraut, and W. A. Harrison, J. Vac. Sci. Technol. B **10**, 1773 (1992).
[27] T. Fukui and H. Saito, J. Vac. Sci. Technol. B **6**, 1373 (1988).
[28] J. M. Gaines, P. M. Petroff, H. Kroemer, R. J. Simes, R. S. Geels, and J. H. English, J. Vac. Sci. Technol. B **6**, 1378 (1988).

Reprinted with permission from

**H. Kroemer, ``Proposed Negative-Mass Microwave Amplifier,''
Phys. Rev., Vol. 109(5), p. 1856, 1958.**

PHYSICAL REVIEW VOLUME 109, NUMBER 5 MARCH 1, 1958

Letters to the Editor

PUBLICATION of brief reports of important discoveries in physics may be secured by addressing them to this department. The closing date for this department is five weeks prior to the date of issue. No proof will be sent to the authors. The Board of Editors does not hold itself responsible for the opinions expressed by the correspondents. Communications should not exceed 600 words in length and should be submitted in duplicate.

Proposed Negative-Mass Microwave Amplifier

HERBERT KRÖMER*

RCA Laboratories, Princeton, New Jersey
(Received November 11, 1957)

AT high kinetic energies, electrons and holes in semiconductors have negative effective masses. In an electric field the current contribution from negative-mass carriers is opposite to the electric field, and a semiconductor containing a sufficient number of such carriers would have a negative resistance. It could thus be used as an active element in an oscillator or amplifier,[1] up to frequencies of the order of the reciprocal collision time of the carriers, i.e., up to about 1000 kMc/sec.

Any attempt to accelerate normal carriers up into the negative-mass energy range faces the difficulty that optical-phonon scattering and avalanche ionization will, in most cases, prevent the carriers from reaching that range.

We want to point out that negative masses occur at energies close to the band edge if the energy contours are re-entrant there, as is the case, for example, for the heavy holes in germanium. There, the energy contours are re-entrant along the six [100] directions of k space[2] (Fig. 1). Along these directions, then, the *transverse* effective masses are negative, i.e., along the k_x axis,

$$m_y = \hbar^2 \left/ \frac{\partial^2 \epsilon}{\partial k_y^2} \right. = m_z = \hbar^2 \left/ \frac{\partial^2 \epsilon}{\partial k_z^2} \right. < 0.$$

From the data of reference 2, one calculates

$$m_y = m_z \approx -0.3 m_0$$

for germanium.

A microwave amplifier, therefore, is proposed, consisting of a p-type Ge crystal with a strong dc bias field, applied in the [100] direction, and the electrical rf field of a resonant cavity or a wave guide perpendicular to it; the bias field shifts the total hole population away from $k=0$ and toward the inside of the conical region of negative transverse mass. The rf field will then be amplified if the majority of the holes are brought inside this cone.

In order to make this possible, scattering of the holes out of the negative-mass cone must be avoided. A totally inelastic type of scattering is desirable where the holes after each collision return to nearly $k=0$. If the collision cross section is large enough, optical phonon scattering is of this type. Shockley's interpretation[3] of the high-field mobility data[4] in Ge suggests that this might be the case in Ge at biasing fields of a few thousand volts/cm.

This amplifier principle is not restricted to germanium, but should hold for all semiconductors with re-entrant energy contours and high optical-phonon cross sections, such as p-type Si and III–V compounds.

The detailed theory will be presented elsewhere shortly.

Note added in proof.—Preliminary experiments with Ge have not shown the effect. This is interpreted as being the result of Ge's not having a high enough scattering cross section, which is likely because the two atoms per unit cell are identical. Experiments on compounds have not been performed yet.

That space-charge instabilities caused by the negative dc conductivity can be avoided, and how, will be shown with the detailed theory.

* Now at Philips Laboratories, Hamburg, Germany.
[1] W. Shockley and W. P. Mason, J. Appl. Phys. **25**, 677 (1954).
[2] Dresselhaus, Kip, and Kittel, Phys. Rev. **98**, 368 (1955).
[3] W. Shockley, Bell. System Tech. J. **30**, 990 (1951).
[4] E. J. Ryder, Phys. Rev. **90**, 766 (1953).

Penetration of Magnetic Fields through Superconducting Films

A. L. SCHAWLOW

Bell Telephone Laboratories, Murray Hill, New Jersey
(Received January 17, 1958)

BARDEEN, Cooper, and Schrieffer[1] have recently formulated a molecular theory of superconductivity. Their analysis leads to a nonlocal relation between current and field not very different from that

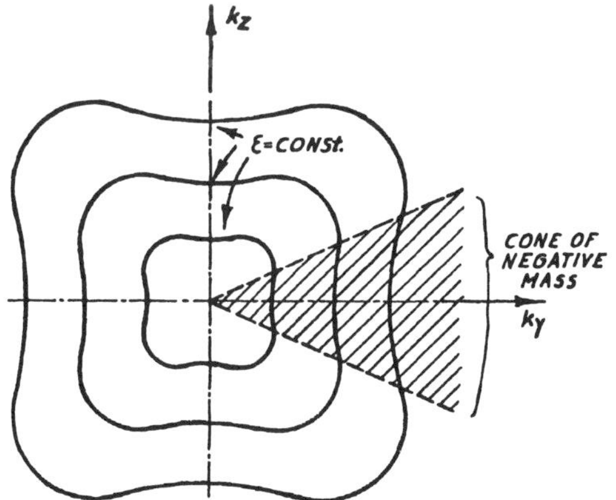

FIG. 1. Schematic energy contours for heavy holes in Ge.

Correspondence

Theory of the Gunn Effect

Gunn [1] has recently discovered a new kind of current oscillations at microwave frequencies, in n-type GaAs and InP. In his paper, Gunn discusses several possible explanations for these oscillations. Most of these explanations he rejects outright; about the remainder he has serious reservations.

The purpose of this correspondence is to point out that most, if not all, of the known properties of the Gunn effect can be explained, at least qualitatively, if it is assumed that these semiconductors have a negative differential bulk conductivity above the oscillation threshold field and that the current oscillations are due to the periodic nucleation and disappearance of traveling space-charge instability domains, of the kind discussed by Ridley [2].[1]

Assume that the drift velocity vs field behavior is characterized by a negative differential mobility range as in Fig. 1. Ridley [2] has shown that a crystal cannot be biased stably in that range but that it will break up into domains of lower and higher fields, corresponding to the points L and V in Fig. 1, and that these domains will travel with a velocity equal to the drift velocity v_L of the carriers. As one high-field domain moves out of the crystal at the positive electrode a new domain gets nucleated at or near the negative end. In sufficiently short samples only one nucleation center will be active, leading to coherent oscillations, with a frequency approximately equal to v_L/L, where L is the sample length. A detailed investigation [3] shows that the frequency is somewhat larger than this value and that it increases slowly with increasing voltage, in agreement with Gunn's observations.

The terminal current associated with such domain travel oscillates between the values corresponding to the valley velocity and to the threshold value, also in agreement with the observations, if one assigns Gunn's low current limits to the valley drift velocity.

Such a field-controlled bulk-type negative conductance cannot be stabilized by loading it with a sufficiently low impedance, in contrast to the interface-type negative conductance of a single Esaki tunnel diode. It behaves essentially like a large number of series-connected tunnel diodes. This is in agreement with Gunn's observation that it is impossible to stabilize the current at the low value.

According to Ridley's simple model the domains, and thereby the oscillations, should disappear if fields in excess of the valley field are applied. However, Ridley's model does not take into account the fact that, even in this case, the field must pass

Manuscript received October 14, 1964.
[1] Ridley himself hints at this possibility in "Electric bubbles and the quest for negative resistance," *New Scientist*, vol. 22, pp. 352–355; May, 1964.

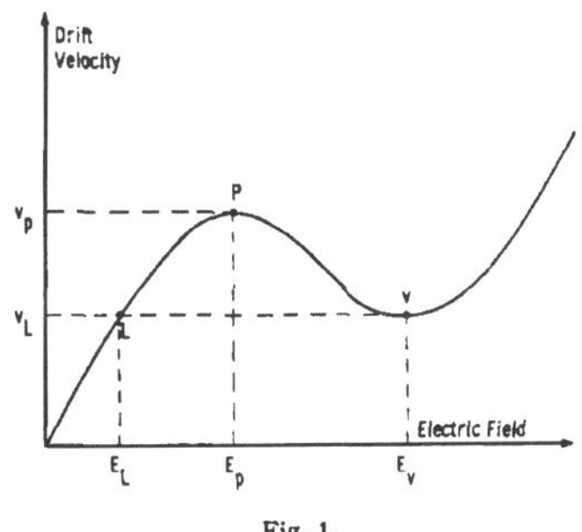

Fig. 1.

through the negative mobility range for a finite distance, near the electrodes. As Shockley [4] has pointed out, such an oversimplified treatment of the boundary conditions may lead to profound errors in problems of this sort. Some form of instability is, therefore, likely to remain in these high-field cases, although it is not likely to have the form of strong coherent transit-time oscillations. All of this is consistent with the work of Day [5] of this laboratory, who succeeded in strongly attenuating existing oscillations by increasing the field strength.

Another refinement necessary in Ridley's model is the fact that space-charge limitations prevent the domain walls from being arbitrarily thin, particularly at the electron-depleted positive end of the high-field domain, where the space-charge density cannot exceed the donor space charge. For a sufficiently low net donor density the domain wall thickness exceeds the sample length and no domains are possible. If one assumes a difference between low and high fields of a few thousand volts per cm, this occurs when the product of net donor density and sample length is less than a few times 10^{10} cm^{-2}. This was also observed by Day [5].

We wish to suggest that the origin of the negative differential mobility is Ridley and Watkins' mechanism [6] of electron transfer into the satellite valleys that occur in the conduction bands of both GaAs and InP. Gunn [1] has rejected this mechanism on the grounds that the energy separation of the satellite valleys would require electron temperatures of the order of 4000°K, while the experimental electron temperature at 80 per cent of the threshold field is only of the order of 400°K. However, this argument overlooks the fact that the combined density of states of the six (100) satellite valleys is about 400 times that of the main valley [7]. Hilsum [8] has considered this fact and has estimated a velocity peak field of 3000 v/cm, an electron temperature at this field of only about 670°K, and of only about 530°K at 80 per cent of the peak field, all values for the already high lattice temperature of 373°K, with lower values for lower temperatures.

Such drastic changes in electron distribution are, of course, not instantaneous in either time or space. This indicates a third

necessary modification of Ridley's model. The details of such rate limitations have not been worked out but one consequence that can be readily predicted is that the threshold field will increase with decreasing sample length, in order to accomplish the distribution change along the shorter available distance. This dependence, too, has been observed.

HERBERT KROEMER
Varian Associates
Palo Alto, Calif.

REFERENCES

[1] J. B. Gunn, "Instabilities of current in III–V semiconductors," *IBM J. Res. & Dev.*, vol. 8, pp. 141–159; April, 1964.
[2] B. K. Ridley, "Specific negative resistance in solids," *Proc. Phys. Soc.*, vol. 82, pp. 954–966; 1963.
[3] H. Kroemer, to be published.
[4] W. Shockley, "Negative resistance arising from transit time in semiconductor diodes," *Bell. Sys. Tech. J.*, vol. 33, pp. 799–826; July, 1954.
[5] G. F. Day, personal communication.
[6] B. K. Ridley and T. B. Watkins, "The possibility of negative resistance effects in semiconductors," *Proc. Phys. Soc.*, vol. 78, pp. 293–304; 1961.
[7] H. Ehrenreich, "Band structure and electron transport of GaAs," *Phys. Rev.*, vol. 120, pp. 1951–1963; December, 1960.
[8] C. Hilsum, "Transferred electron amplifiers and oscillators," PROC. IRE, vol. 50, pp. 185–189; February, 1962.

RF Characteristics of Thin Dipoles

In the above paper [1] Mack and Reiffen give an analysis of scattering by dipoles which is fundamentally in error, although their results are approximately correct. It is the purpose of this communication to call attention to the correct formulation, and to show to what extent the formulas of Mack and Reiffen apply.

The error lies in the identification of power "dissipated" in a Thevenin equivalent circuit with reradiated power. A little thought shows that this cannot generally be true, since an open-circuited antenna would then scatter no field. Also what justification would one have for using the Thevenin equivalent over the Norton equivalent? This would give diametrically opposite predictions as to reradiated power. In general, no identification of a power dissipated in an equivalent circuit can be made.

The correct formulation for scattering by antennas was first given by Y. Y. Hu for center-loaded dipoles [2], and this was later generalized to arbitrary antennas [3], [4]. The exact formulation for an antenna loaded by an impedance Z_L yields an echo area [4].

$$\frac{\sigma}{\lambda^2} = \frac{1}{\pi} \left| \Delta Z_o' - \frac{Z_{oa}' Z_{ao}'}{Z_a + Z_L} \right|^2 \tag{1}$$

Manuscript received June 23, 1964.

Reprinted from IEEE SPECTRUM
Vol. 5, No. 1, January 1968
pp. 47–56

Negative conductance in semiconductors

Although progress in semiconductor research has proceeded at an astonishing pace during the last two decades, two recent discoveries are on the verge of revolutionizing the fields of microwaves and solid-state physics

Herbert Kroemer Fairchild Semiconductor

Until recently, investigators have been frustrated in their attempts at applying microwave and milli-meter-wave frequencies to semiconductor devices. During the last few years, the discovery of avalanche transit-time and Gunn effects in bulk semiconductors has been met with overwhelming enthusiasm. The successful fabrication of models presently utilizing these negative-conductance phenomena has given these high-frequency devices an optimistic outlook for the future.

During the 19 years since the discovery of the transistor, the application of higher and higher operating frequencies has been one of the most persistent and frustrating objectives of semiconductor device research. Unfortunately, progress toward this elusive goal has been exceedingly slow, no matter how relentless the undertaking. The last few years, however, have introduced two discoveries that completely alter the complexity of this problem. It is possible for today's investigator to predict accurately semiconductor devices for the future that will not only operate over the entire conventional microwave frequency range, but the millimeter-wave frequency range as well.[1] The two discoveries referred to are embodied in the following events: (1) the realization of avalanche transit-time devices in silicon[2] (along the lines first proposed by Read[3] in the 50s), and (2) the discovery[4-6] and utilization[1] of the Gunn effect in gallium arsenide. It is the intention of this article to describe these two phenomena and their negative-conductance properties and applications in semiconductors.

Although they differ from each other in many respects, avalanche transit-time and the Gunn effect do have several aspects in common. For one, neither of them is related to the mechanisms of transistors. For another, both effects depend on the properties of "hot" electrons; that is, on the properties of electrons whose energy is large compared with kT, perhaps of the order of several tenths of an electronvolt or more. Finally, both phenomena involve the use of hot electrons in a two-terminal negative conductance of one form or another— the reason for the title of this article. The individual nature of negative conductance for these two cases is entirely separate, however. In avalanche transit-time devices, the negative conductance is caused by a phase shift (exceeding 90°, and ideally near 180°) between current and voltage. In devices that are related to the Gunn effect, the negative conductance is caused by a local negative conductivity within GaAs in strong electric fields; that is, at every point inside a GaAs crystal of such a device, the local current density decreases whenever the local electric field increases beyond a certain threshold.

The approach of any article attempting to deal with negative conductance in semiconductors should be two-fold. First, it should describe and perhaps explain the physical origins of the two known effects causing negative conductance. And second, it should describe the manner in which these negative conductance effects can lead to the experimentally observed device behavior. This second aspect is particularly important in devices that are related to the Gunn effect, since, as we shall observe, a local negative conductance in these devices is capable of an almost incredible variety of external appearances. Consequently, a greater proportion of this article places emphasis on the Gunn-effect type rather than on the avalanche transit-time type of negative conductance. Furthermore, we shall concentrate on the fairly well-established basic concepts as they are presently understood, rather than on the large amount of experimental data and theoretical details still in a state of rapid transition—restricting any comments concerning technology to a minimum.

This emphasis on basic concepts implies that the material will not be presented in its chronologic sequence of discovery. The Gunn effect, in fact, notoriously exemplifies the case where the sequence of events leading to a discovery did not at all follow the sequence in which one might want to order and present such material conceptually, once the discoveries had taken place. Those who are not familiar with the discovery chronology of the Gunn effect should find this conceptual presentation of the facts easier to understand. Those who are already familiar with the experimental details and the order in which they were uncovered may at least gain a new point of view– hopefully, discovering that there is more order in this phenomenon than they might have suspected.

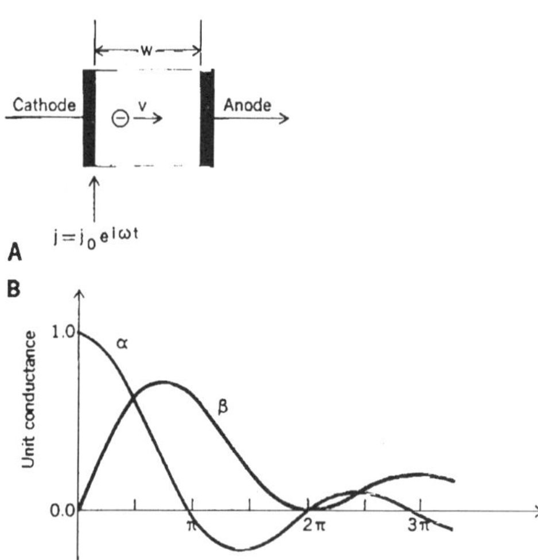

FIGURE 1. Negative conductance resulting from the combined phase shift of avalanche buildup and transit delay in silicon. A— p-n junction. B—Phase-shift characteristic.

FIGURE 2. Experimental (Ruch and Kino) and theoretical (Butcher and Fawcett) velocity-field characteristics of gallium arsenide.

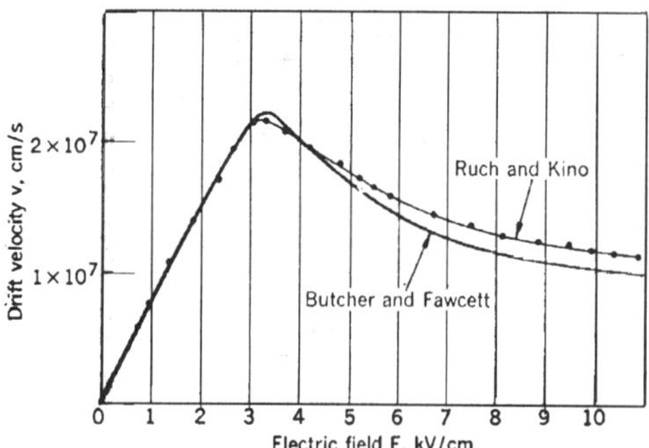

Phase-shift negative conductivity in silicon

Avalanche transit time. As previously stated, the negative conductance of avalanche transit-time devices is attributed to a phase shift between the current and voltage of p-n or p-i-n junctions that are biased into the avalanche breakdown range. This phase shift consists of two components. One of these is a phase delay caused by the finite transit time of electrons through the space-charge layer of the junction. The other is a phase delay caused by the avalanche multiplication process itself, resulting from the condition that the rate at which electron-hole pairs are generated inside the avalanching region of a reverse-biased p-n junction is proportional to the density of electron-hole pairs that are already present in that region. For fields sufficiently far into the avalanche breakdown range, this rate of generation will exceed the rate at which electron-hole pairs can leave the avalanche region; and, as a result, both density and current will exponentially grow with time. What ultimately limits the current is its own space charge, which weakens the field inside the avalanching region of the p-n junction to exactly the value necessary to sustain the avalanche without further growth.

Obviously, this space-charge buildup takes time. If one should now superimpose a small alternating voltage of sufficiently high frequency over such a sustained avalanche, the space-charge readjustment would not be able to follow that voltage, and one would obtain a high-frequency ac component of the avalanche current that would not be space-charge limited. This current will continue to build up during every cycle of the alternating voltage, even after that voltage has gone through its maximum; ideally, so long as the voltage is positive. However, this means that the alternating current will go through its maximum at that instant at which the alternating voltage declines and goes through zero. In other words, the avalanche current exhibits a 90° inductive phase delay relative to the driving alternating voltage. Any additional phase delay can then lead to a negative conductance.

This (ideally) 90° avalanche phase delay combines with the phase delay caused by the finite transit time of electrons through the space-charge region of the p-n junction in the way described by Fig. 1. If one assumes that the periodic ac density $j_0 e^{i\omega t}$ is only generated within a very thin layer at the cathode end of the space-charge region in this p-n junction,* and that the electrons travel with a uniform velocity v through this space-charge region, then one must sum over all the current contributions inside the p-n junction (generated at different times) in order to compute the total current that flows through the external leads. Calculations show that this current is given by

$$J(t) = (\alpha - i\beta)j_0 e^{i\omega t}$$

where

$$\alpha = \frac{\sin \omega\tau}{\omega\tau} \qquad \beta = \frac{1 - \cos \omega\tau}{\omega\tau}$$

* The assumption of only a very thin avalanche region all the way on the cathode side tremendously simplifies the analysis, without sacrificing any of the essential concepts. If the avalanche region is located at an intermediate position, and/or if it occupies a significant fraction of the space-charge layer thickness, both the mathematical analysis and– to some extent—the frequency behavior of the device, become more complicated, and do not introduce any newer physical concepts. The reader interested in these problems is directed to Ref. 1.

Here, $\omega = 2\pi f$; and τ is the transit time through the space-charge region. The two quantities α and β are shown in the curves depicted in Fig. 1. Both curves decay in an oscillatory fashion with increasing frequency, and although α oscillates between positive and negative values, β is always positive or zero.

If one now assumes that the original local current density that is generated at the cathode interface is a purely inductive current created by a strong avalanche multiplication, or

$$j_0 = \frac{E_0}{i\omega\lambda}$$

where λ is some proportionality factor that might be called an inductivity, then it is found that the overall conductance G of this device (the real part of the admittance) is proportional to $-\beta/\omega\lambda$. The idealized conductance, therefore, is either negative or zero for all frequencies, with the largest negative value occurring for a frequency approximately equal to one half of the inverse transit time, or a transit angle of $180°$. Its value is zero for the inverse transit-time frequency, and for all integer multiples of this frequency. It is this negative conductance brought about by the β term displayed in Fig. 1 that represents the underlying mechanism of avalanche transit-time devices.

In our idealized example, this negative conductance exists essentially for all frequencies. In practice, however, one is restricted to operate at a frequency close to the reciprocal transit time. Three reasons account for this. First, negative conductance attains its maximum at this point. Second, this is where the α term, which represents either an inductance or a capacitance, depending on its sign, goes through zero. Finally, our simplifying assumption that there exists a purely inductive current density j_0 is never entirely correct. This is particularly true at low avalanche multiplications or at low frequencies where space-charge effects become dominant; in these cases, the avalanche phase delay is always somewhat less than $90°$. In this case, the current density j_0 is not purely inductive as it flows through the cathode interface, but contains a conductive contribution, without a phase shift, and this conductive portion of the internal current density contributes to the overall conductance in proportion to the α term of Fig. 1.

This α term is, of course, always positive for low frequencies and, for sufficiently low frequencies, even a weak α term is much larger than the β term. It is, therefore, not possible to obtain an overall negative conductance with transit angles less than $10°$ to $30°$, and negative conductance has usually already disappeared for much larger transit angles.

For transit angles larger than $180°$ and approaching $270°$, the negative conductance of the device rapidly decreases and the capacitance increases. Therefore, for all practical purposes, one must consider that avalanche transit-time devices work well only in a frequency range around a $180°$ transit angle, although this range may be rather wide- -typically, as wide as $2:1$.

Device aspects. Having sufficiently described the physical mechanism of these devices, an evaluation of their merits, in comparison with such other devices as those based on the Gunn effect, must consider three primary aspects. One is available technology, the second is noise behavior, and the third is the relation of frequency to transit time, and therefore to the thickness of the device. Of these three, the existence of a sophisticated technology is the strongest asset of impact avalanche transit-time devices—an enormous advantage indeed. Since these devices can be fabricated from silicon, using essentially the same materials and device technology that are used for transistors and integrated circuits, the problems encountered in developing new materials or in formulating a new technology are virtually nonexistent. I believe this advantage to be so important that, in all applications where these devices can provide the required performance, their usage will be given priority.

Drawbacks, however, are encountered in the other two device considerations. Unfortunately, the avalanche multiplication process is a relatively noisy one[7,8]—similar to the noise in a photomultiplier—and it does not, at the present time, appear likely that these devices will ever incorporate low-noise characteristics. For those applications where minimum noise is a necessity, and there are many, one must look for some other class of devices, such as those utilizing the Gunn effect, which offer substantially less noise.

The second drawback, involving the relationship of frequency to device thickness, is a more indirect one. The implication is that, for a given cross-sectional area of a device, the capacitance increases in proprotion to the frequency, and the impedance decreases in inverse proportion to the square of the frequency. Since, in practice, one always has to work at some reasonable impedance level (whatever that level may be), an increase in frequency must also be followed by a reduction in area of a proportion equal to the inverse square of the frequency. This, of course, leads to a power falloff of 6 dB per frequency octave. In essence, this is only an ultimate limitation, since the situation is presently fairly academic and does not become an important consideration until one reaches millimeter-wave frequencies (unless one is interested in very high pulsed powers). Furthermore, because the high state of development for existing materials and technology is presently available in producing these devices, the realization of these frequencies is possible right now rather than, say, in five years. Indeed, Bowman and Burrus at Bell Telephone Laboratories, Inc., have achieved oscillations at frequencies ranging to 340 GHz, with power levels at 300 GHz still in the milliwatt range.[9] It is quite likely, however, because these investigations were based on an already highly developed technology, that the power levels obtained are already fairly close to the ultimate limits.

Local negative conductivity in gallium arsenide

The Gunn effect. Since the current density in a semiconductor is proportional to both the density of electrons and their drift velocity, a decrease in current density—with an increasing electric field—can be brought about by a decrease in either of these quantities. Both types of decrease have actually been observed. For example, a decrease of electron density, in the presence of an increasing field, can be brought about by field-enhanced trapping. This occurs in gold-doped germanium at cryogenic temperatures, and in high-resistivity GaAs at room temperatures. However, these field-enhanced trapping effects are, in general, very slow, and the device potential of this form of negative conductivity is, at best, very limited. Certainly, the microwave frequency range is beyond this potential.

The second possibility for reducing current density in an increasing field, that of decreasing the drift velocity, takes place in GaAs above a 3-kV/cm field strength. It is this mechanism which underlies the Gunn effect, although that was not recognized at the time the effect was discovered. This decrease in drift velocity, in turn, is a result of the conduction-band structure of GaAs. That the GaAs band structure could lead to a decrease in drift velocity with increasing field, i.e., a negative mobility, had already been recognized by Ridley and Watkins[10] and by Hilsum[11] before the discovery of the Gunn effect. Particularly, Hilsum's paper was quite specific and quantitative about this possibility. However, these predictions were promptly ignored, and it was not until after the discovery of the Gunn effect that they were taken seriously and invoked as a means of explaining the Gunn effect.[12]

Even then, direct measurement of the dependence of drift velocity on the electric field, and direct evidence for the existence of the negative differential mobility were not available. Measurements of this kind have finally been published during the last year by several groups working independently on this problem.[13-17] Figure 2 gives an example of perhaps the most beautiful and certainly the most direct result—that of Ruch and Kino[16] at Stanford University. Their experimental curve speaks for itself, and we merely wish to add that other evidence obtained from Gunn-effect models[6,18] has shown that the drift velocity maintains a value approximately equal to Ruch and Kino's minimum value in fields up to the order precipitating avalanche breakdown (about 200–300 kV/cm). Moreover, Fig. 2 shows a theoretical curve by Butcher and Fawcett[19] that represents a continuation and extension of the work Hilsum performed in 1961. The close agreement between both curves indicates the accuracy level theoretical understanding of this effect has attained.

As previously noted, negative-mobility behavior is a result of the conduction-band structure of GaAs (Fig. 3). GaAs is a direct-gap semiconductor with a conduction-band minimum occurring at $k = 0$, the center of the Brillouin zone. It is this fact that accounts for the low-minority carrier lifetimes that make GaAs a poor transistor material, as well as for the high-radiation recombination probability that makes it an excellent laser material. This property is also an essential ingredient of the Gunn effect. It is typical of direct-gap semiconductors that the conduction-band effective mass is relatively low, and GaAs is no exception to this rule. Its effective mass is of the order of 7 percent of the free-electron mass m_0.

In addition to this central valley, GaAs possesses what we shall call satellite valleys; that is, additional conduction-band minima with higher effective masses at higher energies energies that are large compared with kT. This, too, is a very common feature of the type of semiconductor under discussion. In the case of GaAs, the satellite valley energy is about 0.36 eV. These satellite valleys are located along the $\langle 100 \rangle$ directions of k space (probably on the surface of the Brillouin zone), and the combined density-of-states mass for all satellite valleys is of the same order as the free-electron mass itself.[20]

At room temperature, and in the absence of any strong electric fields, all electrons will occupy the central valley, since the energy separation between it and the satellite valleys is large compared with kT. Because of the very low effective mass in the central valley, electrons have the high mobility that is characteristic of these semiconductors; but, for the same reason, electrons are also very easily heated by an external electric field. When this heating does take place, an increasing fraction of electrons reaches energies approaching the magnitude of the

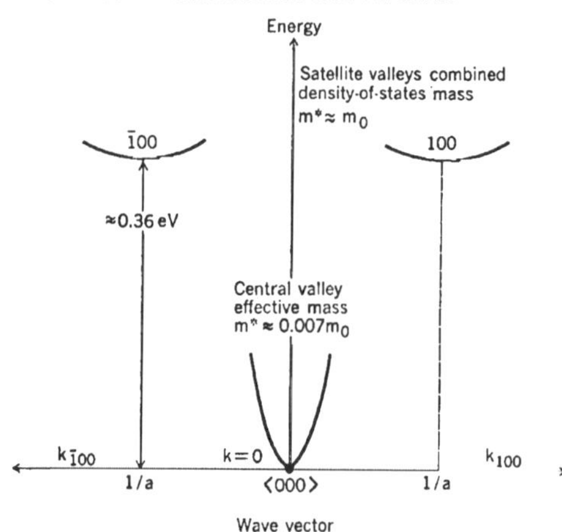

FIGURE 3. Conduction-band structure of GaAs.

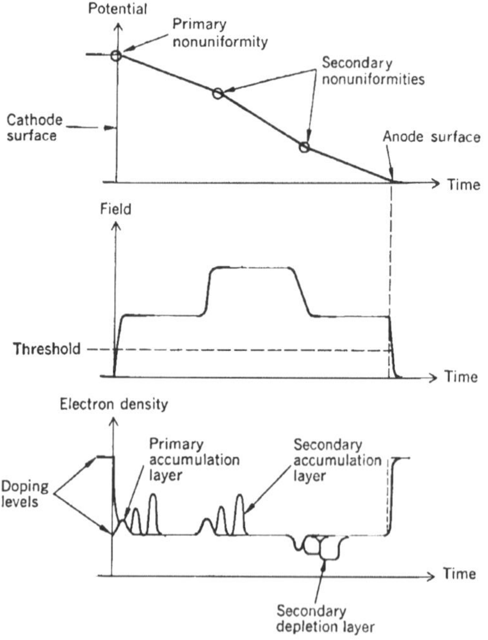

FIGURE 4. Traveling accumulation and depletion layer instabilities in a medium of negative conductivity.

satellite-valley energies, thereby becoming scattered into these valleys. The large effective mass of the satellite valleys, and the possibility of intervalley scattering between them, cause the mobility of these high-energy electrons to be much lower than the mobility of the central-valley electrons. As a result, the overall average electron mobility and conductivity of such a crystal will decrease as soon as the electric field reaches a value large enough to sustain electron transfer into the satellite valleys. This, in itself, does not imply the existence of a negative differential mobility or conductivity, but once the electron transfer has started, the fraction of high-energy electrons steeply increases with the increasing electric field. Even though the velocity of those electrons that stay behind in the central valley keeps increasing in proportion to the electric field, the rate of electron loss from the high-velocity central valley into the low-velocity satellite valleys is so rapid that, above a certain threshold field, the current, and, therefore, the average drift velocity, actually drops with further increase of the electric field, leading to the negative differential mobility that is observed. In GaAs, this effect takes place somewhere between 3.0 and 3.5 kV/cm, and the transfer appears to be essentially complete above 20 kV/cm.

Similar behavior could be expected for any other semiconductor with a similar band structure; namely, a low-mass central valley combined with a set of high-mass satellite valleys. Prerequisites also include an energy separation between high- and low-mass valleys that is large compared with kT but smaller than the energy gap of the semiconductor. The latter is necessary so that avalanche breakdown does not set in before the transfer of electrons into the satellite valleys. There are, indeed, several semiconductors that fulfill this requirement and do exhibit the Gunn effect. Of the III–V compounds, these are InP,[4] and InAs under hydrostatic pressure.[21] This hydrostatic pressure must be large enough to raise the InAs energy gap above its satellite valley energy, which for low pressures is greater than the energy gap, thus leading to avalanche breakdown rather than the Gunn effect. Of the II–VI compounds, both CdTe[22] and ZnSe[23] have exhibited the Gunn effect, and this list will probably expand with future developments.

Devices without internal space charge—the LSA mode of the Gunn effect. To continue our evaluation of devices that rely upon this negative mobility, it is obvious that the conceptually simplest possible model would consist of a uniformly doped semiconductor with a pair of parallel ohmic contacts, and completely devoid of any internal space charges. In such a case, the internal electric field would be uniform and simply proportional to the applied voltage. The current, in turn, would be proportional to the drift velocity at this field level. The entire device would then have a current–voltage characteristic of the same relative shape as the velocity–field characteristic (Fig. 2). That is, it would have a voltage-controlled negative conductance. Coupled to an external resonant circuit, this device could then excite oscillations at a frequency determined by the resonant frequency of the circuit (combined with the capacitance of the device), but independent of the transit time of the electrons. This concept of an extremely simple mode of oscillation for a negative-conductivity crystal has actually been materialized by Copeland[24,25]; but it has been given a rather obscure name—the LSA mode of the Gunn effect,

or simply the LSA diode. The abbreviation LSA originally stood for "large signal amplification"; more recently, for "limited space-charge accumulation." However, even the latter interpretation of these three letters obscures the fact that, conceptually, this is the simplest mode of oscillation for a negative-mobility crystal. For all practical purposes, it should be considered *the* fundamental mode of oscillation.

The term "limited space-charge accumulation" merely reflects the historical truth that, in earlier experiments, the Gunn effect had always been associated with a very strong formation of space-charge layers.[1-4] Our imperfect understanding of the effect had originally led us to believe that space-charge-free oscillations were not possible, until Copeland of Bell Telephone Laboratories, Inc., actually discovered such oscillations during computer simulations of the Gunn effect, establishing the criteria for their experimental realization. We shall discuss these criteria briefly in a subsequent section. As it turned out, Shuskus and Shaw of United Aircraft had apparently observed this mode of oscillation before,[26,27] but its nature was, at that time, not recognized.

Within a few months after Copeland's pioneering work, Copeland himself[28,29] and many others[30] utilized the LSA mode in drastically extending both the power and the frequency range of Gunn-effect devices. The highest frequencies achieved thus far with this mode are of the order of 150 GHz.* This does not yet approach the highest frequencies achieved with avalanche transit-time oscillators.[9] That is really not surprising, in view of the fact that both the materials and the device technology of gallium arsenide are much less developed than that of silicon. Nevertheless, for those frequencies at which LSA oscillators have operated, they have substantially outperformed avalanche transit-time devices with respect to both power and noise.

Since this area is undergoing extremely rapid deveopment, it would be rather pointless to quote any specific— and unpublished- achievements at this time, for they would almost certainly be obsolete by the time this article appeared in print.[46] I would like to state my belief that the chain of events, from the discovery of the Gunn effect to the discovery of the LSA mode, represents one of the greatest breakthroughs in semiconductor physics, probably the biggest since the discovery of the transistor itself. It is assuredly one that is likely to revolutionize the microwave field, particularly in the millimeter-wave region, as much as the gas laser is revolutionizing the field of optics.

Space-charge Instabilities

The critical nl product. Let us now turn our attention to those modes of the Gunn effect that do involve space-charge effects. Historically, they were discovered first, since they are, experimentally, considerably simpler to obtain than the LSA mode. Conceptually, however, they are definitely more complicated, a condition that has contributed to the slow initial understanding of the mechanics underlying the Gunn effect.

Space-charge complications associated with the Gunn effect arise in a medium of negative differential conductivity because any nonuniformity of the electric field, hence any space charge, tends to build up exponen-

* J. A. Copeland, personal communication.

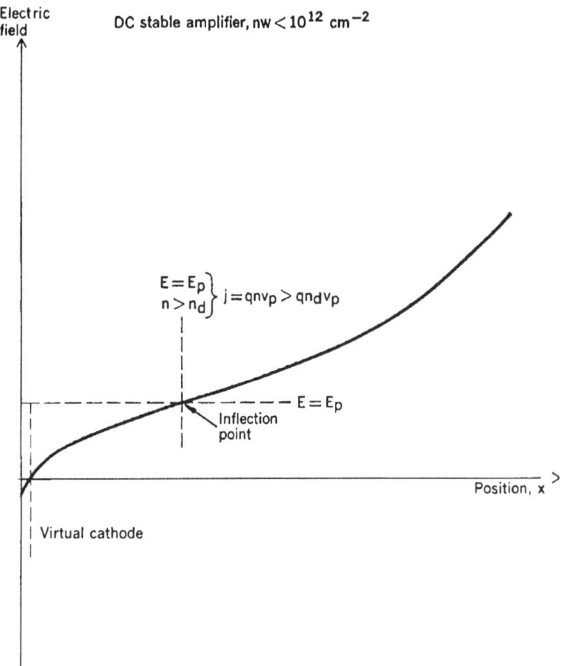

FIGURE 5. Field distribution in a subcritically doped negative-mobility crystal.

FIGURE 6. Theoretical conductance versus frequency of a 100-μm-thick crystal free from ionized impurities at a 4.8-kV/cm bias field.

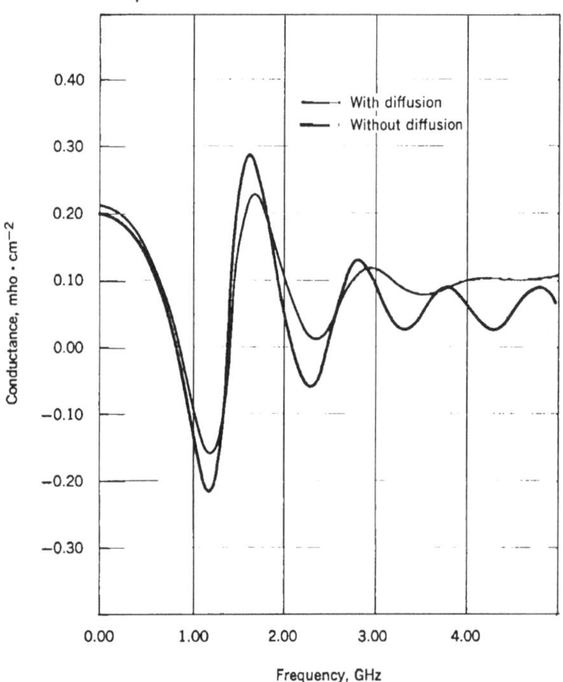

tially with time. This is demonstrated in Fig. 4. The top of this illustration shows an idealized potential distribution, such as one might find immediately after the application of a bias potential. Included are three different types of field nonuniformities. The first of these is simply caused by a field transition at the cathode interface, ranging from low fields inside the cathode to high fields inside the remainder of the crystal. This "primary" field nonuniformity is, of course, inevitable. The other two field nonuniformities indicated in Fig. 4 represent the two types that may arise from either imperfect doping or merely statistical fluctuations, e.g., noise. These will be referred to as "secondary" field nonuniformities. Since, at least in the one-dimensional case, each field nonuniformity is associated with a space charge, each of these nonuniformities must correspond to either an accumulation or a depletion of electrons (called "accumulation" or "depletion" layers). The primary field nonuniformity at the cathode is related to a primary accumulation layer; the two secondary field nonuniformities are related to an accumulation layer and a depletion layer, respectively. If one now assumes that the field in all three sections of the crystal is within the range of negative mobility, then it is apparent that both the secondary accumulation and depletion layers must increase in strength with time. This follows from reasoning that the middle region, which has the highest electric field, actually has the lowest drift velocity. As a result, electrons tend to pile up at the left edge of this high-field region, where an electron accumulation layer already exists; and the electron density continues to decrease at the right edge, where a depletion layer exists. This, in turn, further increases the field strength of the middle high-field region and further decreases the field strength of the outer low-field regions — thereby also increasing the rate of space-charge layer formation.

We are confronted with a runaway process. Given enough time, it would continue until the field strength in the low-field regions had dropped so far below the threshold field that the drift velocity in these regions would once again reach the same value as the drift velocity in the high-field region. In other words, the velocity along a low-field branch may reach the same value as the minimum velocity along the high-field branch.[31] It remains to be demonstrated that this limiting space-charge buildup is only developed for sufficiently high doping levels.

Since all of this space-charge buildup takes place in a moving electron stream, both the accumulation and depletion layers travel along with approximately the same velocity as that stream. This is not only true for the secondary accumulation and depletion layers, but also for part of the primary accumulation layer, which detaches itself from the cathode and moves as an accumulation layer through the crystal. In fact, during the initial stages, this is usually the strongest of the various space-charge layers; and, in weakly doped crystals, it may remain the dominant space-charge layer throughout the entire period of travel.

Ultimately, all of these space-charge layers will disappear into the anode. Naturally, the primary accumulation layer will disappear last, followed by a rebuilding of the field inside the crystal to values above the threshold field. At that time, a new set of space-charge layers, consisting of one primary and possibly several secondaries, will form. It is this periodic nucleation and dissipation of space-charge layers that give rise to the current oscillations of the Gunn effect.

A complete theory of this growth process, particularly in the presence of statistical fluctuations in the crystal, is difficult to expound, and requires the use of numerical techniques.[32,33] However, so long as the deviations from a uniform field are still weak– during the early stages of the space-charge buildup—the growth of these space-charge layers is given by

$$Q(x, t) = Q(x - vt, 0) \exp\left(\frac{t}{\tau_n}\right)$$

where

$$\tau_n = \frac{\epsilon\epsilon_0}{qn|\mu_n|}$$

is the absolute value of the negative dielectric relaxation time of a crystal with negative mobility μ_n, with ϵ the dielectric constant, q the electronic charge, and n the electron density. If this law remained valid throughout the entire growth of the space-charge layer, then the maximum growth during one transit time would be given by

$$G_n = \exp\left(\frac{l}{v\tau_n}\right) = \exp\left(\frac{qnl|\mu_n|}{\epsilon\epsilon_0 v}\right)$$

where l is the thickness of the crystal. Now, obviously, in order for space-charge instabilities to occur, this total growth factor must be large in comparison with unit value. This means that the product of doping (electron density) and thickness (sample length) must satisfy the inequality

$$nl > \frac{\epsilon\epsilon_0}{q}\frac{v}{|\mu_n|} \approx 10^{12} \text{ cm}^{-2}$$

When this requirement is not satisfied, the formation of strong space-charge instabilities should not be expected.[12] In fact, oscillations probably will not occur at all, any current generated through the crystal remaining stable.

Small-signal amplification for subcritical doping ($nl \ll 10^{12}$ cm^{-2}). At first, one might expect that a subcritically doped crystal would have a static negative conductance similar to that of the tunnel diode. This, however, is not the case. In 1954, Shockley[34] had already mathematically shown that a decrease of drift velocity in an increasing electric field will not, in general, lead to a decrease in static current as the static voltage increases. The underlying explanation for this, in the case of such weak doping, is that the primary accumulation layer extends essentially throughout the entire crystal. Hence, as the voltage increases, the amount of charge stored in this layer increases in such a way that the overall current through the device continues increasing, in spite of the decrease in drift velocity of the individual electrons.

This situation is described in more detail in the dc stable amplifier of Fig. 5, where electric field versus position is shown for such a crystal. This electric field, somewhere within the crystal, must increase from the low values at the cathode contact to the high values corresponding to the applied bias. In the process, it must also pass through a plane where it equals the threshold field E_p of the velocity–field characteristic. Within this plane, of course, the velocity is equal to the peak velocity, v_p, and the current is given by

$$j = qnv_p$$

Since this plane is in the region of an increasing field,

it must also be in a region of excessive electron density. Therefore, the following inequality must hold

$$j > qn_dv_p$$

where n_d is the doping density ($n_d < n$). Furthermore, this current must be constant throughout the entire thickness of the device. However, since the velocity is lower everywhere other than in the threshold plane, one must conclude that the electron density must be higher everywhere than in this plane. Since a space charge already existed in this plane, it follows that there must be an even larger

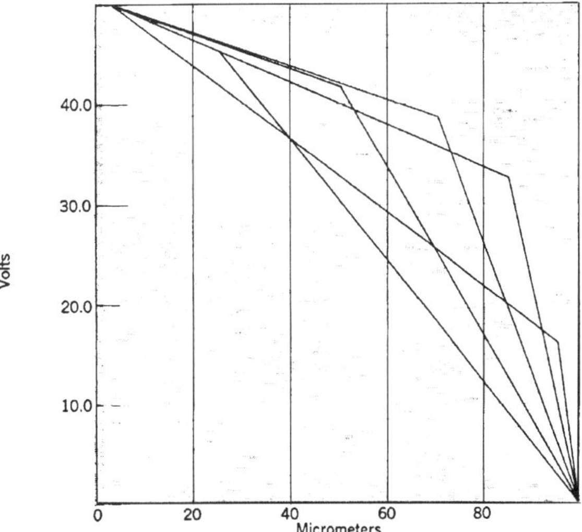

FIGURE 7. Dynamics of the pure accumulation mode, representing potential distribution for a doping level $n_d = 10^{14}$ cm^{-3} in intervals of 0.2 ns after voltage turn-on.

FIGURE 8. Electron density corresponding to Fig. 7.

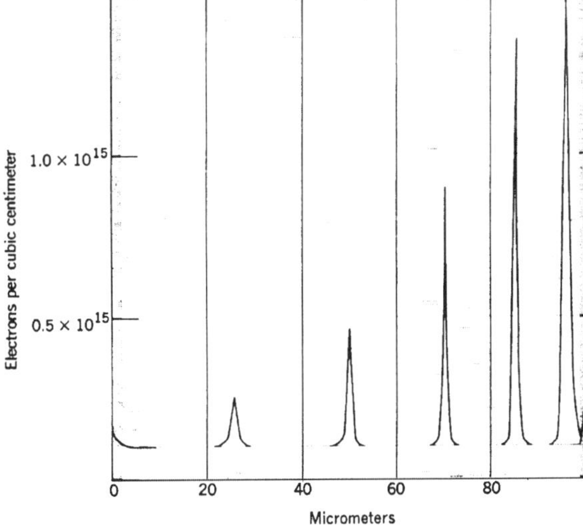

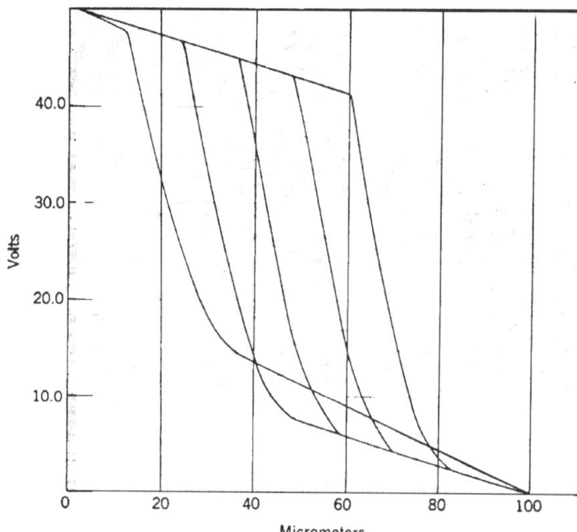

FIGURE 9. Dynamics of the mature dipole mode, representing potential distribution for the same doping level and time intervals as Fig. 7.

FIGURE 10. Quenching of accumulation and dipole modes in a resonant circuit that periodically drives the bias field below threshold.

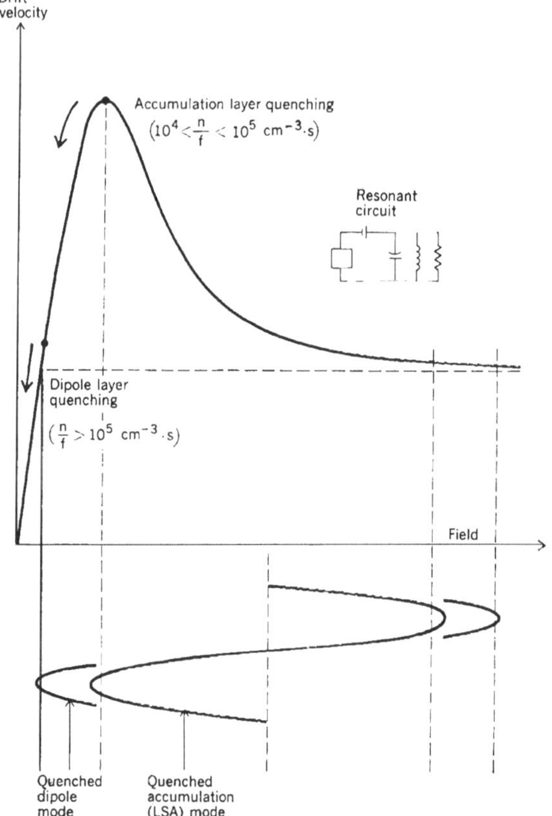

space charge throughout the entire crystal.

With increasing bias voltage, the entire field curve will get shifted upwards. As a result, the plane in which the electric field passes through the threshold value will shift to the left— closer to the cathode. While this happens, the curve also gets steeper, indicating, of course, that the excess amount of electron density above the doping density will increase. Because the velocity in the threshold plane is, by definition, constant and equal to the peak velocity, the current density must also increase with an increase in bias. This is, of course, intrinsically what is contained in Shockley's 1954 theorem.

It was discovered by Thim et al.[35] of Bell Telephone Laboratories, Inc., that such a subcritically doped crystal will exhibit a negative differential conductance at high frequencies, specifically at frequencies close to the transit-time frequency. Both Thim and others[36–41] have extensively used this property to construct experimental microwave amplifiers from bulk GaAs. Our own personal investigations have included extensive research on the theory of these amplifiers,[42] particularly within the limits of zero doping and in the absence of trapping effects. Figure 6 represents one of the results of these theoretical calculations for the specific case of a 100-micrometer-thick crystal under a bias field of about 4.8 kV/cm, calculated both with and without the inclusion of electron diffusion effects. It is discerned from this illustration that there is, indeed, a pronounced negative conductance around the transit-time frequency, even for zero doping. Experimentally, such a negative conductance has never been observed within the limit of very high resistivity. This almost certainly results because such GaAs always contains high densities of traps, and the theory is not applicable to crystals with trapping effects. It should be noted that numerous statements have appeared in the professional literature to the effect that, even in the absence of traps, negative conductance within the limit of zero doping should be nonexistent. However, these statements can be attributed to improper linearizations in the theory, and are certainly incorrect.[42]

The pure accumulation mode ($nl \approx 10^{12} \ \text{cm}^{-2}$).[32, 33] As one increases the doping level to around 10^{12} electrons per square centimeter, space-charge instabilities — and that means oscillations– do, indeed, set in. At that doping level, however, since the dielectric relaxation time is still sufficiently long and the resulting space-charge layer growth sufficiently weak, the space-charge dynamics are entirely dominated by the propagation and growth of the primary accumulation layer. The secondary accumulation and depletion layers play a very subordinate role in this effect.

Figure 7 exhibits the calculated dynamics of the internal potential distribution for such a crystal under the influence of a 5-kV/cm bias. The increasingly sharp bends in potential, caused by the growing accumulation layer, are easily recognized. Apparently, the field near the cathode first drops to low values, and then increases to higher values--the origin of the current oscillations. Figure 8 denotes the electron density of this crystal, and shows particularly well how the primary accumulation layer grows, and, incidentally, how its movement slows down as it approaches the anode.

The mature dipole (Gunn) mode ($nl \gg 10^{12} \ \text{cm}^{-2}$). Further increases in the doping level to values larger than 10^{12} electrons per square centimeter sufficiently shorten the dielectric relaxation time so that even relatively weak internal-field inhomogeneities or space-charge fluctuations can build up into fully developed space-charge layers. As it turns out, however, the simultaneous existence of more than one accumulation and more than one depletion layer is unstable. Furthermore, out of the multiplicity of accumulation and depletion layers that may initially form, only one dipole pair will remain.[32] This invariably leads to the type of propagating dipole domain that is embodied in Fig. 9.[32,33] Normally, the doping level which this illustration represents would not be sufficient to produce this mode; however, the depletion layer was created by initializing the computations with a weak electron depletion near the cathode.

As some readers might recall, such dipole layers, rather than the pure accumulation layers previously described, have actually been observed in potential-probing experiments.[6,18,43] There are perfectly comprehensible reasons that account for this preference. In order to be able to perform such probing experiments, the crystal has to be sufficiently thick. Therefore, for the available doping levels, the nl product almost inevitably becomes larger than 10^{12}; which, of course, is the condition precipitating formation of mature dipole domains.

These potential-probing experiments may have resulted in the widespread misapprehension that this mature dipole-domain mode is the one in which actual microwave oscillators operate. However, this is only true to a very limited extent, the dipole mode frequently being undesirable for this application. It is, indeed, true that this mode has been of tremendous use in those scientific experiments that have clarified the basic mechanism of the Gunn effect, but these investigations were, for the most part, performed at relatively low frequencies. In realistic microwave devices, this mode has, in fact, several drawbacks. For one, the high electric fields that can build up inside the dipole domains can lead to avalanche multiplication,[18] which is noisy and may lead to an electrical breakdown of the entire device. For another, the wave shape associated with the fully matured dipole mode is not particularly suitable for microwave oscillators. Finally, in the interest of low heat dissipation, it is generally desirable to work at doping levels as low as possible, and certainly lower than those that are required for the formation of mature dipole domains. As a result, most actual non-LSA microwave oscillators employ devices that are much closer to the pure accumulation mode than they are to the mature dipole mode.

The space-charge layer propagation in the mature dipole mode is considerably slower and much more uniform than the pure accumulation model. This uniformity indicates that during a large portion of each oscillating cycle the current density is essentially constant.

Quenched space-charge instabilities in resonant circuits

The quenched dipole mode. Thus far, we have assumed that voltage across the device is constant with time. A question arises concerning the fate of space-charge instabilities within a crystal that is inserted into an oscillating, high-Q resonant circuit, particularly if this circuit causes the voltage across the crystal to dip periodically below the threshold voltage. When this occurs, an exciting phenomenon takes place, which is clearly illustrated in Fig. 10.

Let us first consider the case of the mature dipole domain.[43-45] As indicated in Fig. 9, most of the voltage across such a crystal is dropped across the high-field domain itself. Hence, as the circuit reduces the overall bias voltage, this voltage reduction decreases the thickness of the high-field domain. Ultimately, at some particular voltage, the accumulation and depletion layers will simply neutralize each other, and the dipole domain will vanish. Upon closer inspection, this quenching begins slightly above the point at which the bias field drops to that value for which the velocity on the positive mobility branch of the velocity-field characteristic is equal to the velocity in the velocity valley along the high-field (negative mobility) branch. Once the field drops below its quenching value, the entire dipole domain disappears; and when the voltage across the device recovers, a new domain becomes nucleated immediately after the bias field regains the threshold.

In such a circuit, the oscillations will take place at the resonant frequency of the circuit itself, rather than at the transit-time frequency. If this resonant frequency is substantially larger than the transit-time frequency, this implies that the domains become quenched long before they reach the anode. In that event, the remainder of the crystal merely acts as a parasitic series resistance- except during that short portion of each cycle when a new domain is being nucleated. As a result, this quenched dipole mode, although tunable, is generally a mode of low efficiency.

The quenched accumulation (LSA) mode. If one only had an accumulation layer to quench, the resulting situation would be altogether different. In this particular case, the entire crystal section between the accumulation layer and the anode would be a series resistance of negative value. This, in turn, would not dissipate, but would actually deliver oscillation energy to the circuit (except for that short portion of each cycle during which the quenching takes place). Within the limit of sufficiently high frequencies, the penetration of the accumulation layer into the crystal will be so short that essentially all the energy that is delivered to the circuit will be generated by this negative series resistance. The role of the accumulation layer itself can then be neglected.[25]

This is Copeland's LSA mode once again, and our description has now shown how it is achieved; namely, by periodic quenching of the primary accumulation layer that detaches itself from the cathode during every cycle. It can further be shown that, in order to quench a single accumulation layer, it is not necessary to reduce the field to that point at which a dipole domain would become quenched. Instead, it is sufficient to reduce the field only somewhat below the threshold field of the velocity-field characteristic.[25] This is because the field difference across a primary accumulation layer is much weaker than the field differences that build up in a mature dipole domain--making the primary accumulation layer easier to quench. Finally, if the circuit frequency is adequately higher than the transit-time frequency, the nl product need not necessarily be kept close to 10^{12} electrons per square centimeter in order to prevent the formation of dipole domains. One can (and, in fact, should) increase n beyond this value as the circuit frequency goes up. Copeland has shown[24,25] that the proper operating condition for the LSA mode requires n/f, the ratio of electron density to frequency, to be between 10^4 and 10^5.

A conclusion

After outlining the present status of negative-conductance effects in semiconductors, I would like to conclude with a few remarks concerning possible future applications of these devices. It has often been stated that the importance of bulk phenomena lies in their potential ability to replace low-power microwave tubes. I firmly disagree! I believe that they will not replace microwave tubes any faster than transistors have replaced receiving tubes — nor is replacing microwave tubes a particularly challenging objective. The true essence of these bulk-effect devices lies in their overwhelming potential for creating newer developments in the field of microwaves that ordinarily could not have been derived from tubes alone.

In the same manner that the transistor and integrated circuit created such new fields as large-scale digital computers, in *that* sense will these devices create newer microwave applications. What form this progress will take is anyone's guess. That they will arrive is a certainty.

Revised text of a paper published originally in *Festkörperprobleme*, vol. 7, published by Vieweg-Verlag, Braunschweig, Germany, 1967; and presented at the German Physical Society-IEEE Region 8 Conference on Semiconductor Device Research, Bad Nauheim, Germany, April 19, 1967.

REFERENCES

1. For a survey of the state of this development as of the fall of 1965, see the "Special Issue on Semiconductor Bulk-Effect and Transit-Time Devices," *IEEE Trans. Electron Devices*, vol. ED-13, Jan. 1966. (See also Ref. 46.)

2. Johnston, R. L., DeLoach, B. C., and Cohen, B. G., "A silicon diode microwave oscillator," *Bell System Tech. J.*, vol. 44, pp. 369-372, Feb. 1965.

3. Read, W. T., "A proposed high-frequency negative-resistance diode," *Bell System Tech. J.*, vol. 37, pp. 401–446, Mar. 1958.

4. Gunn, J. B., "Microwave oscillations of current in III-V semiconductors," *Solid-State Commun.*, vol. 1, pp. 88–91, Sept. 1963.

5. Gunn, J. B., "Instabilities of current in III-V semiconductors," *IBM J. Res. Develop.*, vol. 8, pp. 141–159, Apr. 1964.

6. Gunn, J. B., "Instabilities of current and of potential distribution in GaAs and InP," *Symp. Plasma Effects Solids*, Dunod, Paris, 1964.

7. Hines, M. E., "Noise theory for the Read type avalanche diode," *IEEE Trans. Electron Devices*, vol. ED-13, pp. 158-163, Jan. 1966.

8. Josenhans, J., "Noise spectra of Read diode and Gunn oscillators," *Proc. IEEE*, vol. 54, pp. 1478–1479, Oct. 1966.

9. Bowman, L. S., and Burrus, C. A., Jr., "Pulse-driven silicon p-n junction avalanche oscillators for the 0.9- to 20-mm band," *IEEE Trans. Electron Devices*, vol. ED-14, August 1967.

10. Ridley, B. K., and Watkins, T. B., "The possibility of negative resistance effects in semiconductors," *Proc. Phys. Soc. (London)*, vol. 78, pp. 293-304, Aug. 1961.

11. Hilsum, C., "Transferred electron amplifiers and oscillators," *Proc. IRE*, vol. 50, pp. 185–189, Feb. 1962.

12. Kroemer, H., "Theory of the Gunn effect," *Proc. IEEE*, vol. 52, p. 1736, Dec. 1964.

13. Gunn, J. B., and Elliott, B. J., "Measurement of the negative differential mobility of electrons in GaAs," *Phys. Letters*, vol. 22, pp. 369–371, Sept. 1966.

14. Chang, D. M., and Moll, J. L., "Direct observation of the drift velocity as a function of the electric field in gallium arsenide," *Appl. Phys. Letters*, vol. 9, pp. 283–285, Oct. 1966.

15. Thim, H. W., "Potential distribution and field dependence of electron velocity in bulk GaAs measured with a point contact probe," *Electron. Letters*, vol. 2, pp. 403-405, Nov. 1966.

16. Ruch, J. G., and Kino, G. S., "Measurements of the velocity-field characteristic of gallium arsenide," *Appl. Phys. Letters*, vol. 10, pp. 40–42, Jan. 1967.

17. Acket, G. A., "Determination of the negative differential mobility of n-type gallium arsenide using 8 mm microwaves," *Phys. Letters*, vol. 24A, pp. 200-202, Feb. 1967.

18. Hecks, J. S., "Some properties of the moving high-field domain in Gunn effect devices," *IEEE Trans. Electron Devices*, vol. ED-13, pp. 68–79, Jan. 1966.

19. Butcher, P. N., and Fawcett, W., "Calculation of the velocity field characteristic for gallium arsenide," *Phys. Letters*, vol. 21, pp. 489-490, June 1966.

20. Ehrenreich, H., "Band structure and electron transport of GaAs," *Phys. Rev.*, vol. 120, pp. 1951–1963, Dec. 1960.

21. Allen, J. W., Shyam, M., and Pearson, G. L., "Gunn oscillations in indium arsenide," *Appl. Phys. Letters*, vol. 9, pp. 39–41, July 1966.

22. Foyt, A. G., and McWhorter, A. L., "The Gunn effect in polar semiconductors," *IEEE Trans. Electron Devices*, vol. ED-13, pp. 79–87, Jan. 1966.

23. Ludwig, G. W., Halsted, R. E., and Aven, M. S., "Current saturation and instability in CdTe and ZnSe," *IEEE Trans. Electron Devices*, vol. ED-13, pp. 671–672, Aug./Sept. 1966.

24. Copeland, J. A., "A new mode of operation for bulk negative-resistance oscillators," *Proc. IEEE*, vol. 54, pp. 1479–1480, Oct. 1966.

25. Copeland, J. A., "LSA oscillator diode theory," *J. Appl. Phys.*, vol. 38, pp. 3096–3101, July 1967.

26. Shuskus, A. J., and Shaw, M. P., "Current instabilities in gallium arsenide," *Proc. IEEE*, vol. 53, pp. 1804–1805, Nov. 1965.

27. Shaw, M. P., and Shuskus, A. J., "Current instability above the Gunn threshold," *Proc. IEEE*, vol. 54, pp. 1580–1581, Nov. 1966.

28. Copeland, J. A., "CW operation of LSA oscillator diodes—44 to 88 GHz," *Bell Syst. Tech. J.*, vol. 46, pp. 284–287, Jan. 1967.

29. Copeland, J. A., and Spiwak, R. R., "LSA operation of bulk n-GaAs diodes," presented at the 1967 Internat'l Solid-State Circuit Conf., Philadelphia, Pa.

30. For example, Kennedy K., "Negative conductance in bulk gallium arsenide of high frequencies," Thesis, Cornell University, Sept. 1966.

31. Ridley, B. K., "Specific negative resistance in solids," *Proc. Phys. Soc. (London)*, vol. 82, pp. 954-966, Dec. 1963.

32. Kroemer, H., "Nonlinear space-charge domain dynamics in a semiconductor with negative differential mobility," *IEEE Trans. Electron Devices*, vol. ED-13, pp. 27–40, Jan. 1966.

33. McCumber, D. E., and Chynoweth, A. G., "Theory of negative-conductance amplification and of Gunn instabilities in 'two valley' semiconductors," *IEEE Trans. Electron Devices*, vol. ED-13, pp. 4–21, Jan. 1966.

34. Shockley, W., "Negative resistance arising from transit time in semiconductor diodes," *Bell System Tech. J.*, vol. 33, pp. 799-826, July 1954.

35. Thim, H. W., Barber, M. R., Hakki, B. W., Knight, S., and Uenohara, M., "Microwave amplification in a dc-biased bulk semiconductor," *Appl. Phys. Letters*, vol. 7, pp. 167–168, Sept. 1965.

36. Hakki, B. W., and Knight, S., "Microwave phenomena in bulk GaAs," *IEEE Trans. Electron Devices*, vol. ED-13, pp. 94–105, Jan. 1966.

37. Thim, H. W., and Barber, M. R., "Microwave amplification in a GaAs bulk semiconductor," *IEEE Trans. Electron Devices*, vol. ED-13, pp. 110–114, Jan. 1966.

38. Foyt, A. G., and Quist, T. M., "Bulk GaAs microwave amplifiers," *IEEE Trans. Electron Devices*, vol. ED-13, pp. 199-200, Jan. 1966.

39. Hakki, B. W., and Beccone, J. P., "Microwave negative conductance of bulk GaAs," *Proc. IEEE*, vol. 54 pp. 916-917, June 1966.

40. McWhorter, A. L., and Foyt, A. G.," Bulk GaAs negative conductance amplifiers," *Appl. Phys. Letters*, vol. 9, pp. 300-302, Oct. 1966.

41. Thim, H. W., "Temperature effects in bulk GaAs amplifiers," *IEEE Trans. Electron Devices*, vol. ED-14, pp. 59-62, Feb. 1967.

42. Kroemer, H., "Detailed theory of the negative conductance of bulk negative mobility amplifiers, in the limit of zero ion density," to be published. (This paper also contains extensive references to other theoretical works on subcritically doped devices.)

43. Gunn, J. B., "Properties of a free, steadily traveling electrical domain in GaAs," *IBM J. Res. Develop.*, vol. 10, pp. 300–309, July 1966.

44. Carroll, J. E., "Oscillations covering 4 Gc/s to 31 Gc/s from a single Gunn Diode," *Electron Letters*, vol. 2, p. 141, Apr. 1966.

45. Carroll, J. E., "Resonant-circuit operation of Gunn diodes: a self-pumped parametric oscillator," *Electron. Letters*, vol. 2, pp. 215–216, June 1966.

46. For a state-of-the-art survey as of the spring of 1967, see "Second Special Issue on Semiconductor Bulk-Effect and Transit-Time Devices," *IEEE Trans. Electron Devices*, vol. ED-14, Sept. 1967.

1844

PROCEEDINGS OF THE IEEE, NOVEMBER 1970

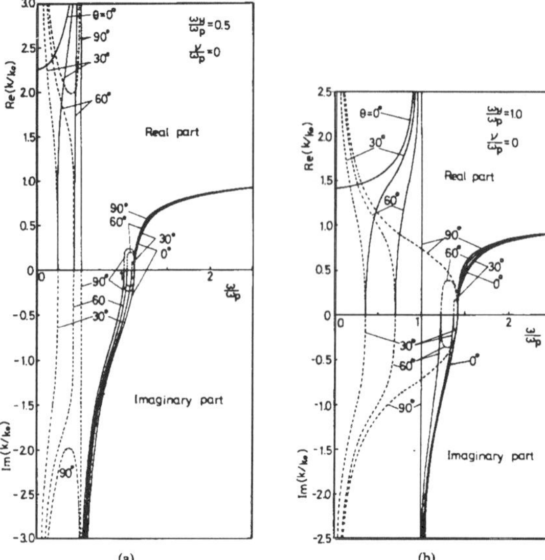

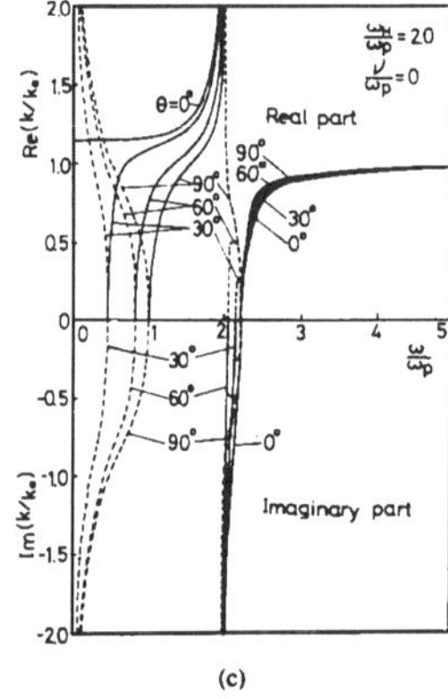

Fig. 1. Theoretical propagation constant along an infinitely long, very thin conducting wire in a magnetoplasma for various angles of inclination of the dc magnetic field. Solid curves represent the propagating mode or cutoff mode; dashed curves represent the complex mode.

It is found that the wave is always a propagating mode in the frequency region of $\omega > \omega_u = \sqrt{\omega_p^2 + \omega_H^2}$ (where ω_u is upper hybrid resonance frequency) and its dispersion curve varies very little by the inclination θ; it is also found that the cutoff frequency is given by the upper hybrid resonance frequency independently of the inclination. The complex wave exists in the frequency region of $\omega < \omega_u$ for all the angles of inclination except $\theta = 0°$. It is also seen from Fig. 1 that the dispersion curves for the propagating modes and the complex modes are considerably sensitive to the inclination in the frequency region of $\omega < \omega_u$.

The effect of the finiteness of the collision frequency on the propagation constant is described by T. Ishizone, *et al.*[4]

T. Ishizone
S. Adachi
Y. Mushiake
Faculty of Eng.
Tohoku University
Sendai 980, Japan

[4] T. Ishizone, S. Adachi, and Y. Mushiake, "Measurement of the phase constant along a conducting wire in a magnetoplasma," this issue, pp. 1852–1854.

Generalized Proof of Shockley's Positive Conductance Theorem

Abstract—Shockley's theorem states that a semiconductor with a negative differential mobility and a well-behaved cathode contact, has a positive differential conductance. The proof of this theorem is generalized to arbitrary impurity distributions and geometries.

In 1954, Shockley showed[1] that the static differential *conductance* at the terminals of a piece of semiconductor with negative differential *mobility* would be positive if the nature of the cathode contact is such that the electric field there is maintained near zero. In his proof Shockley assumed that the ionized impurity density is constant, and that the geometry is one-dimensional. We wish to show that Shockley's theorem holds for arbitrary impurity distributions and geometries as long as the local mobility remains isotropic.

We first wish to retain temporarily the assumption of a one-dimensional geometry.

The electric field inside such a semiconductor obeys Poisson's equation (using obvious sign conventions)

$$\varepsilon \frac{\partial}{\partial x} F(x, j) = \frac{j}{v(F)} - qN(x) \tag{1}$$

where j is the electrical current density, $v(F)$ the electron drift velocity (like Shockley, we neglect diffusion), and $N(x)$ the ionized donor density. If $F(0, j) = 0$, then near $x = 0$ the first term on the right side of (1) becomes very much larger than the second term, and the solution of (1) for small x becomes

$$F(x, j) = \sqrt{\frac{2jx}{\varepsilon \mu}}, \tag{2}$$

where μ is the low-field mobility.

Assume now that the current density through the structure is increased by an infinitesimal amount, from j_0 to

$$j = j_0 + \Delta j. \tag{3}$$

This leads to an infinitesimal change of the electric field, from F_0 to $F_0 + \Delta F$. For small x, ΔF follows from (2):

$$\Delta F(x, j_0) = \sqrt{\frac{x}{2j_0 \varepsilon \mu}} \cdot \Delta j > 0. \tag{4}$$

The voltage increment associated with the current increment is given by

$$\Delta U = \int_0^L \Delta F \, dx. \tag{5}$$

A negative static differential conductance is equivalent to $\Delta U < 0$. Because of (4), ΔU can become negative only if the curve of $\Delta F(x)$ crosses the x axis, from above! But at any point where $\Delta F = 0$, the slope of the $\Delta F(x)$ curve is, after (1), given by

$$\frac{d}{dx} \Delta F(x) = \frac{1}{\varepsilon} \Delta \frac{j}{v(F)} = \frac{1}{\varepsilon} \frac{\Delta j}{v(F_0)} > 0, \tag{6}$$

that is, ΔF can cross the x axis only from below. Thus a static negative differential conductance cannot occur under the assumed conditions, irrespective of $N(x)$.

The conclusion remains valid if $F(0, j) \neq 0$, as long as the cathode boundary conditions are "well-behaved," that is,

$$\frac{\partial}{\partial j} F(0, j) > 0. \tag{7}$$

Manuscript received June 1, 1970.
[1] W. Shockley, "Negative resistance arising from transit time in semiconductor diodes," *Bell Syst. Tech. J.*, vol. 33, pp. 799–826, July 1954.

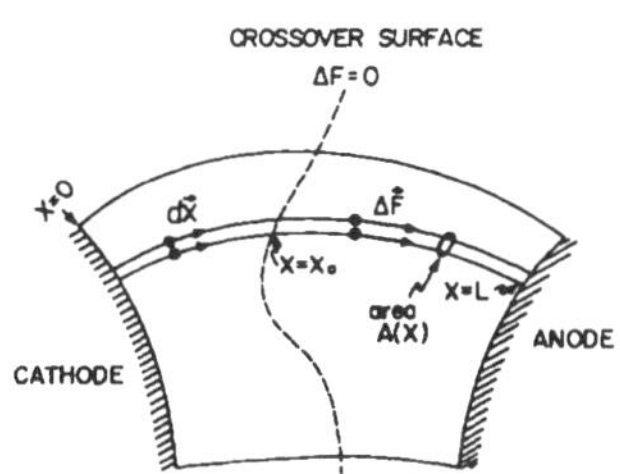

Fig. 1.   Field and coordinate lines in a structure of arbitrary geometry.

For cathode boundary conditions such that

$$\frac{\partial}{\partial j}F(0,j) < 0, \tag{8}$$

it is known[2] that Shockley's theorem need not hold. In this case the ΔF curve would start out *below* the x axis and if no crossover takes place the static differential conductance will be negative. According to (2), crossover is permitted in this case and its occurrence is necessary for a *positive* static differential conductance, although even with crossover the conductance may remain negative.

The neglect of diffusion is not believed to have any effect on the validity of the theorem, since diffusion effects play a role only in regions of low electric field, and there they do not reverse the sign of the current dependence of $F(x,j)$.

In an arbitrary geometry the current density will not be independent of position, and the field lines will in general not be parallel. It is useful to consider the field lines of the *incremental* field ΔF. If the semiconductor is isotropic, then, since the current lines cannot cross the boundaries of the semiconductor, the incremental field lines cannot cross either. It is then convenient to set up a curvilinear coordinate system inside the structure in such a way that the local x direction is everywhere parallel (or possibly anti-parallel) to the direction of the local incremental field ΔF (Fig. 1). In such a coordinate system, ΔF along any particular field line can be considered a scalar ΔF, and the incremental voltage ΔU remains as given by the scalar equation (5), if the integration is carried out along *any one* of the field lines, and if L is the length of the particular line chosen.

We will assume temporarily that as a result of the postulated increase in overall device current, the magnitude of the current density increases *everywhere* in the device. In this case (4) remains valid, and therefore ΔF must again change its sign if the static differential conductance is to be negative. This crossover must take place along every field line, so there must be a complete "crossover surface" cutting through the device, separating the cathode from the anode region. Everywhere on this surface the total field is equal to the original field, the incremental current density Δj has the direction of the original current density j_0, and the change in the *magnitude* of the current density can still be expressed by the scalar equation (3).

Now let x_0 be the crossover point for some incremental field line, that is, $\Delta F(x_0)=0$. We can again evaluate the slope of ΔF, at $x=x_0$, along this field line, from a three-dimensional generalization of Poisson's equation (6). To this end we consider not a single field line, but an infinitesimal "bundle" of adjacent field lines. Let $A(x)$ be the (infinitesimal) cross-sectional area of this bundle, which in general will vary along the length of the bundle. Then we can write the three-dimensional equivalent of the one-dimensional Poisson equation (6), at $x=x_0$, in the form

$$\frac{\partial}{\partial x}\left[A(x)\cdot\Delta F(x)\right]\bigg|_{x_0} = A(x_0)\cdot\frac{\partial}{\partial x}\Delta F(x)\bigg|_{x_0} = A(x_0)\cdot\frac{1}{\varepsilon}\frac{\Delta j}{v(F_0)} > 0. \tag{9}$$

All other terms vanish. Clearly, $A(x_0)$ cancels out, and all our earlier conclusions remain valid.

However, in a three-dimensional geometry there is no a priori assurance that Δj will be positive everywhere when the total current increases. The only statement that can be made is the following. On *any* surface that separates cathode and anode, the current density through this surface cannot decrease *everywhere*, but it must increase over at least part of this surface. Now one such surface is the cathode surface itself. From the two facts that it is an equipotential surface, and that everywhere

$$\text{curl } \Delta F = 0 \tag{10}$$

it can be shown that $\Delta F(0)$ must have the same sign across the entire cathode surface. And since it must be positive over at least part of the cathode, it must be positive over the entire cathode. Thus the conclusion remains valid that ΔF must change its sign along every field line for negative conductance to occur. Consider now the crossover surface. Over at least part of this surface $\Delta j > 0$, and at least over those parts of the surface the required crossover cannot in fact take place. Therefore, the static differential conductance cannot be negative for well-behaved boundary conditions, regardless of both doping distribution and device geometry.

A negative *static* (not merely time-average) differential conductance is occasionally observed in experimental Gunn effect devices. Our proof dispenses with any notion that such behavior can be explained by doping nonuniformities or geometry effects, rather than imperfect cathode boundary conditions.

HERBERT KROEMER
Dep. of Elec. Eng.
University of Colorado
Boulder, Colo. 80302

[2] H. Kroemer, "The Gunn effect under imperfect cathode boundary conditions," *IEEE Trans. Electron Devices*, vol. ED-15, pp. 819–837, November 1968.

Manuscript received June 18, 1970.

Reprinted from

H. Kroemer, ``Hot-Electron Relaxation Effects in Devices,''
Solid-State Electron., Vol. 21(1), pp. 61-67, 1978.

Solid-St. Electronics, 1978, Vol. 21, pp. 61–67. Pergamon Press. Printed in Great Britain

HOT-ELECTRON RELAXATION EFFECTS
IN DEVICES†

HERBERT KROEMER

Department of Electrical Engineering and Computer Science, University of California, Santa Barbara, CA 93106,
U.S.A.

Abstract—Whenever the electric field in a device varies sufficiently rapidly with distance and/or time, significant deviations of the electron drift velocity from the static velocity-field characteristic occur. These relaxation effects manifest themselves not only in the high-frequency properties, but also in the static device properties of devices such as transferred-electron devices and field-effect transistors. Important quasi-static effects are drift velocity overshoot in FET's and relaxation effects in the bistable switching in TE devices. The ultimate speed limitations of TE devices are governed by relaxation effects. Transit-time modes are superior to the LSA mode, barrier-type contacts to n^+-on-n contacts, and InP to GaAs.

INTRODUCTION

In his introductory paper to this conference, Cyril Hilsum[1] pointed out that much of the original motivation for hot-electron research was a device motivation, a search for a negative differential mobility at high electric fields. We all know how successful this search and its subsequent device utilization was. Since then, hot-electron physics has become a lively field of research in its own right, going significantly beyond specific device motivations. At the same time, the development of negative mobility hot-electron devices has posed new problems in the physics of hot electrons, problems that go beyond the current mainstream of non-device-oriented hot-electron physics. It is with some of these that this paper is concerned.

There are two ways in which device-oriented hot-electron physics differs from general hot-electron physics: A much greater emphasis on situations in which the electric field exhibits large and fast transients, and on situations in which the field is grossly non-uniform. In either case one obtains a number of effects that go beyond the description of the device behavior in terms of a local quasi-static velocity-field characteristic $v(E)$. By "local" and "quasi-static" we mean the drift velocity one would obtain in a large sample if the local electric field were in fact uniform through the sample, and time-independent.

Such an assumption is clearly an oversimplification. For example, the negative differential mobility in such semiconductors as GaAs and InP requires that the electrons first acquire a kinetic energy at least equal to the energy of the satellite valleys in the band structures of these compounds. Even at fields significantly above the quasi-static threshold field E_{th}, this requires a finite time, and the actual drift velocity during fast transients is not the quasi-static one. This leads to frequency limitations in any devices utilizing the negative differential mobility. In addition to requiring time, the acceleration also requires distance, implying that the actual drift velocity

in non-uniform fields is not the one corresponding to the local field. In devices with sufficiently small dimensions this, too, will affect the device behavior, including even the low-frequency behavior for which the time limitations are inconsequential.

Up to a point, these relaxation effects can be neglected. Much of the qualitative and semi-quantitative understanding of the role of hot electrons in devices does not involve these effects, and they were indeed largely ignored until roughly 1968. But since that time, relaxation effects have become increasingly important, both for the quantitative optimization of devices such as transferred-electron (Gunn-effect) devices and field-effect transistors, and even more for understanding the fundamental limitations imposed upon those devices by the relaxation effects.

Although far from a mature field, great progress has been made in incorporating hot-electron relaxation effects into the theory of those devices in which hot electrons are important. Still, except by those actually working on these effects, their importance has drawn far less attention than it deserves. The majority of hot-electron device physics papers even after 1968 has continued to discuss the electron dynamics in the devices in terms of a local quasi-static $v(E)$ characteristic. To the extent that such papers had as their goals a general qualitative understanding, they have contributed little that was not already known by the end of 1968. To the extent that their objective was a quantitative calculation of details, they contributed little that can be treated quantitatively by the approximations used. The present review will have served its purpose if it succeeds in drawing a greater attention to the crucial importance of relaxation effects.

2. SPATIAL RELAXATION EFFECTS

2.1 *Velocity overshoot*

Consider an electron in the central valley of a semiconductor with a band structure such as that of GaAs or InP. Assume an electron energy equal to the energy ϵ_s of the satellite valleys that are responsible for the negative differential mobility in those substances. If the central

---

†Supported by ONR.

valley is assumed parabolic with an effective mass m^*, the velocity of this electron is then

$$v_0 = \sqrt{2\epsilon_s/m^*} = 5.93 \times 10^7 \frac{\text{cm}}{\text{sec}} \times \sqrt{\frac{\epsilon_s}{1\,\text{eV}} \times \frac{m_0}{m^*}}. \quad (1)$$

Table 1 gives numerical values for GaAs and InP. For InP we also give values corrected for the strong non-parabolicity of that substance; the details of the corrections are described in Appendix A. As Table 1 shows, the velocities v_0 exceed by a factor of about 5 the quasi-static threshold velocities v_{th} of the $v(E)$ characteristics, which are of the order $2.2–2.5 \times 10^7$ cm/sec.

The reason the $v(E)$ characteristic stays so far below the actual microscopic peak velocities is, of course, that under conditions of a uniform static field only a small fraction of the electrons have a velocity near v_0 at any instant of time, anywhere in the crystal. The averaging over all electrons reduces the velocity drastically.

The important point is now that even the average drift velocity can exceed the quasi-static $v(E)$ value drastically and approach the value v_0, whenever the velocity of different electrons is highly correlated in either space or time. This will be the case whenever the electric field either undergoes sufficiently large and fast transients, or exhibits a sufficiently large spatial non-uniformity. We start our discussion with the spatial effects.

Consider, first, the extreme limit of an electric field $E(x)$ that varies abruptly with position (that is, over a distance short compared to a mean-free-path) from a near-zero value ($E \ll E_{th}$) to a value $E \gg E_{th}$. Under these circumstances there is a high probability that any electron entering the high-field region will be accelerated to the velocity v_0 without suffering any collisions. If we neglect collisions altogether, the electrons will reach this velocity after the distance

$$L = \frac{\epsilon_s}{qE}. \quad (2)$$

If we also neglect any electrons streaming backwards across the plane $x = L$, the velocity of all electrons for $0 < x \le L$ will be the same,

$$v(x) = v_0\sqrt{x/L}. \quad (3)$$

This will then also be the drift velocity at the position x, implying a huge overshoot over the peak velocity of the $v(E)$ characteristic.

In reality, neither collisions nor electron backstreaming can be neglected. As a result, the velocity overshoot over v_{th} stays below v_0. The problem was first studied quantitatively by Ruch[5], for GaAs. His main result is shown in Fig. 1(a). For a field of 10 kV/cm ($\approx 3E_{th}$) he estimates an overshoot of more than a factor of 2 over v_{th}, and of a factor between 3 and 4 over the $v(E)$ value corresponding to $E = 10$ kV/cm. More recently, these calculations have been extended by Maloney and Frey[6] to higher fields and to InP, as shown in Fig. 1(b). The results speak for themselves.

The assumption of a perfectly abrupt field change is of course an oversimplification. The actual rate of change of the field is always finite. One can expect significant overshoot effects whenever the field increases from a value $E \ll E_{th}$ by an amount exceeding E_{th}, over a distance shorter than that value of L that corresponds formally to the threshold field itself,

$$L_0 = \frac{\epsilon_s}{qE_{th}}, \quad (4)$$

that is, when

$$\frac{dE}{dx} > \frac{E_{th}}{L_0} = \frac{E_{th}^2}{q\epsilon_s}. \quad (5)$$

Table 1 gives numerical values for L_0 for GaAs and InP. Clearly, drastic overshoot effects must be expected in FET's with sub-micron gate lengths. In fact, it was the role of velocity overshoot in FET's that provided the basic motivation for the work both of Ruch and of Maloney and Frey. This role is a very beneficial one: Both the low-frequency transconductance and the ultimate upper frequency limit of FET's increase directly proportionally with the speed of the electrons. Velocity overshoot effects go a long way towards explaining why short-gate FET's made from GaAs or InP are as good as they are.

2.2 Spatial relaxation effects in transferred-electron devices

For transferred-electron (TE) devices one might at first glance expect that velocity overshoot plays a role only in the very thin devices that are required for millimeter-wave frequencies of 100 GHz or so. However, this argument is deceptive.

In most transferred-electron devices (but not in FET's) the field geometry is essentially a one-dimensional one.

Table 1. Various parameters pertaining to hot-electron relaxation effects, for GaAs and InP. For those parameters of InP for which two values are given, the upper value is obtained ignoring non-parabolicity, the lower is corrected for non-parabolicity according to Appendix A. The satellite valley energies and effective masses for GaAs are taken from Ref. [2], those for InP from Refs. [3] and [4]

	ϵ_s (eV)	m^*/m	E_{th} (kV/cm)	v_0 (10^8 cm/sec)	L_0 (μm)	n_0 (10^{15} cm^{-3})	τ_0 (psec)
GaAs	0.296	0.067	3.2	1.25	0.92	0.24	1.48
InP	$\approx$0.53	0.078	$\approx$10	$\begin{cases}1.55\\1.13\end{cases}$	0.53	1.3	$\begin{cases}0.69\\0.75\end{cases}$

63

(a)

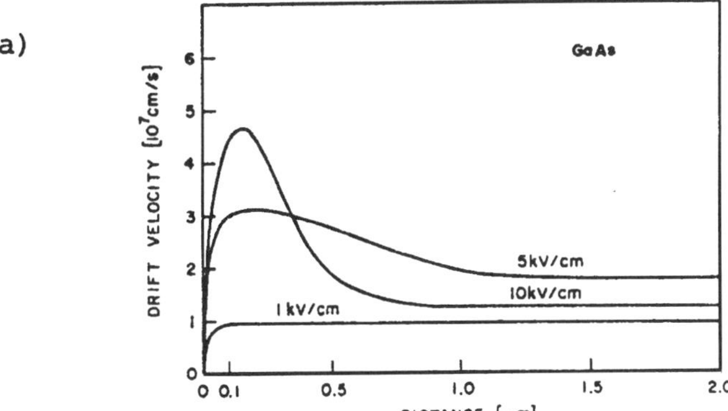

(b)

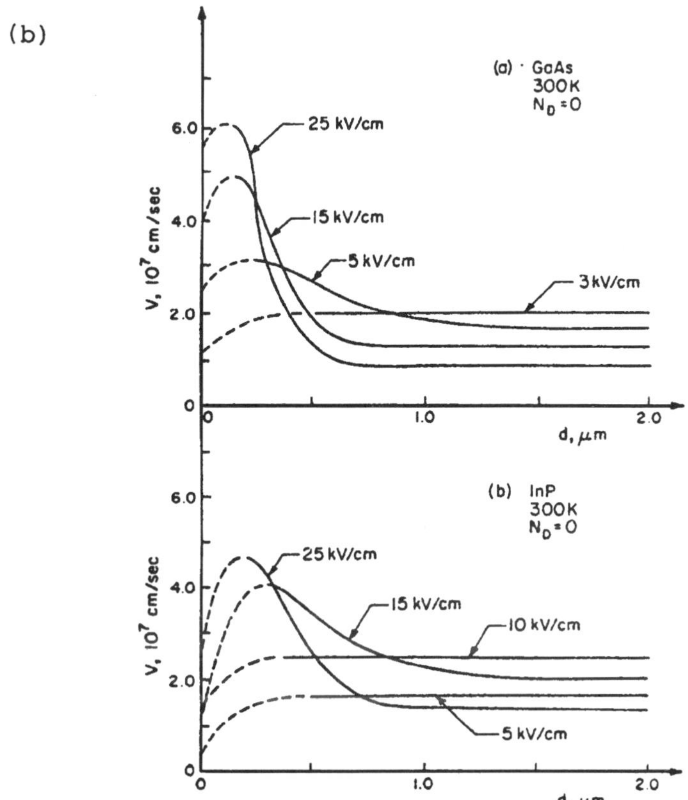

Fig. 1. Drift velocity vs position in GaAs and InP for electrons released at $x = 0$ into various uniform fields: (a) GaAs, from Ruch[5]; (b) GaAs and InP, from Maloney and Frey[6]. All results are for a lattice temperature of 300K.

In this case the condition (5) can be translated into a space-charge density condition corresponding to an excess of electrons over donors,

$$n - N_D > n_0, \qquad (6)$$

where

$$n_0 = \frac{\epsilon E_{th}^2}{\epsilon_s} \approx 6.91 \times 10^{12} \text{ cm}^{-3} \times \left(\frac{E_{th}}{1\,\text{kV/cm}}\right)^2 \times \frac{1\,\text{eV}}{\epsilon_s}. \qquad (7)$$

(Note that ϵ is the dielectric permittivity of the semiconductor, $\epsilon \approx 12.5\,\epsilon_0$, not an energy.) Table 1 again gives numerical values. These values are remarkably low, significantly below the doping levels actually employed in TE devices, and below the excess electron densities found in the diffusion tails of the $n^+ - on - n$ cathode contacts often employed in TE devices. This means that the cathode boundary conditions cannot be treated quantitatively without considering relaxation effects, even in relatively low-frequency devices in which the overall device thicknesses are much larger than L_0.

One phenomenon occurring in some quite thick TE devices is a bistable switching capability, between two stable states, one combining low voltage with high current, the other, high voltage with low current[7]. Such a behavior violates Shockley's positive conductance theorem[8], which states that in a device with a low cathode field—which these devices presumably have—the static current cannot decrease with increasing static voltage, even in the presence of a negative differential mobility. Furthermore, it has been shown[9] that this theorem is true regardless of the geometry of the device, its doping distribution, etc., so long as the local current density can be written

$$j = qnv(E), \qquad (8)$$

where n is the local electron density, E the local electric field, and $v(E)$ a fixed isotropic function of E only. Clearly, the empirical violation of the theorem must be due to this assumption being invalid. The conventional explanation of the effect is that (8) must be augmented by a diffusion term[10]. There can be no doubt that a diffusion term of suitable form could indeed lead to the observed effects—if it were present and if other deviations from (8) were small by comparison. But a concentration gradient of sufficient magnitude to cause a significant diffusion current is invariably associated with a high excess electron density on one side, that is, with a high space-charge density. But as we have seen, this implies a high rate of change of the field and hence strong relaxation effects with their drastic deviations from the $v(E)$ characteristic. It appears quite likely that these big variations play a much larger role than the comparatively weak diffusion effects. Even if the latter should be dominant, the *quantitative* form of the diffusion term is bound to be influenced just as much by the relaxation effects as the $v(E)$ characteristic. Unfortunately, the problem appears not to have been analyzed beyond a remark made by Källbäck[11] that relaxation effects should be included in the analysis of the switching effects between the two bistable states. Even Källbäck apparently does not recognize the importance of spatial relaxation effects for the stable states themselves. Whatever the ultimate outcome, it can be practically guaranteed that non-relaxation analyses could be quantitatively valid only by a lucky coincidence. In view of the practical interest in the bistable switching effects, an analysis including relaxation effects is clearly called for.

3. SPEED LIMITATIONS IN TRANSFERRED-ELECTRON DEVICES

3.1 *The basic time constant*

It was pointed out by Rees already in 1969[12] that the dominant speed limitation of hot-electron effects in GaAs is the rate with which electrons can gain or lose energy in the central (Γ) valley of the energy-band structure, rather than the intervalley scattering rate, which is much faster. Those estimates of Rees were done before it was realized that the lowest satellite valleys in GaAs are L-valleys rather than X-valleys[2]. But based on the evidence in InP, the Γ–L scattering is, if anything, even faster than the Γ–X scattering[13], and as a result, Rees' conclusion not only remains valid, but applies to InP as well.

The central valley dynamics itself is governed by two time constants:

(a) The energy relaxation time τ_ϵ due to collisions. For electrons with energies near ϵ_s, τ_ϵ is of the order 6×10^{-12} sec for GaAs and 3×10^{-12} sec for InP[14].

(b) The acceleration-deceleration time τ_{ad} required for an electron to gain or lose the energy ϵ_s,

$$\tau_{ad} = \frac{1}{qE} \sqrt{2m^* \epsilon_s} = \tau_0 E_{th}/E, \qquad (9)$$

where

$$\tau_0 = \frac{1}{qE_{th}} \sqrt{2m^* \epsilon_s} \approx 0.337 \text{ psec}$$
$$\times \frac{1 \text{ kV/cm}}{E_{th}} \times \sqrt{\frac{m^*}{m_0} \times \frac{\epsilon_s}{1 \text{ eV}}}. \qquad (10)$$

Table 1 gives values for τ_0.

Note that the values for τ_0 and, hence, those for τ_{ad} are appreciably shorter than those for τ_ϵ. Because collision processes and inertial energy changes act in parallel rather than in series, the shorter of the two time constants, τ_0, governs the speed with which the electron distribution in k-space adjusts itself to any change in electric field. We elaborate by considering an electron that has just been scattered out of a satellite valley into the central valley, where it has a velocity of magnitude v_0. Depending upon the direction of this velocity, the electron will either get accelerated or decelerated by this field. If accelerated, the electron will get scattered back into the satellite valley almost instantaneously (i.e. in a time appreciably less than τ_{ad}). If decelerated, at most the time $2\tau_{ad}$ will elapse before the electron has been re-accelerated to the energy ϵ_s. On the average, a time somewhat less than τ_{ad} will have elapsed before the electron is returned to the satellite valley, and this time must be viewed as the characteristic time constant with which the electron distribution adjusts itself to any changes in the field, for fields appreciably above E_{th}.

Philosophically, this is an interesting result: Solid-state devices have reached the point that their ultimate speed limitations are back to the same physical principles as in electron tubes: Inertial time-of-flight effects. What is different are the numbers: Both the dimensions and the voltages in solid-state devices are much smaller.

With the acceleration-deceleration time decreasing with increasing field strength, hot-electron device speeds increase with increasing field. Put differently: The higher the desired device speed, the higher the field required to achieve it. In the case of transferred-electron devices, this implies that the effective threshold field for high-frequency operation is higher than E_{th}, increasing with increasing frequency. Rees has calculated, already in 1969, for GaAs, both an effective threshold field E_0 and the field E_m for maximum (net) negative mobility, as functions of the desired frequency, as well as the maximum net negative mobility μ_m itself. His results are shown in Fig. 2. If the electric field inside the device were uniform, the field E_m would be the optimum bias field. For fields below E_m the relaxation effects become too slow; for fields above E_m the slope of the static $v(E)$ characteristic becomes too low. The value E_m is a compromise between those two different effects. The increase of E_m with frequency explains a phenomenon quite familiar to those working with actual millimeter-wave transferred-electron devices[15]: A given device, operated at different frequencies within its bandwidth (usually implying different circuits) generally requires, at the higher frequencies, a bias field that is significantly higher than the optimum bias field at the lower frequencies. In fact, its low-frequency performance at the high-frequency bias field may be below its high-frequency performance at that field, even though the *optimum* low-frequency performance will follow the μ_m curve and will therefore always be better than the optimum high-frequency performance.

As we pointed out, Rees' calculations were for GaAs. Table 1 shows that the dominant time constant τ_0 for InP is appreciably shorter than for GaAs. Everything else being equal, InP should, therefore, be about twice as fast as GaAs. Note that this prediction is a result solely of the 3-times higher threshold field of InP. This higher threshold field is the biggest single drawback of InP for low-frequency (<20 GHz) devices, because such devices are invariably thick (>10 μm) and the high required fields imply a high threshold dissipation, making InP useless for X-band TE devices. But then the dominant interest

in TE devices is at frequencies above X-band anyway, and our argument shows that this low-frequency drawback is really a blessing in disguise.

3.2 *The LSA mode*

Relaxation effects have a particularly devastating impact on the LSA mode[16] which, during the late 1960s and early 1970s had been widely believed to be the most promising approach towards higher frequencies. As the reader may recall, in the LSA mode the device is *much* longer than one transit-time length, *and* the space charge injected by the cathode during every oscillation cycle is quenched later in the cycle by dropping the field below threshold, after the charge has traveled only a small fraction of the device length. It is essential that the quenching is complete. An imcompletely quenched space charge would simply re-grow as the field returns above the threshold field, and the device would abruptly switch to a transit-time operation at a lower frequency. Now it is clear from our earlier discussion that the need for a low field for quenching conflicts with the need for a high field for highest speed. This in itself suggests that the LSA mode might in fact not be the most desirable mode for highest-frequency operation. But this is not all: The deathblow was dealt to the LSA mode in 1973 by Jones and Rees[17]. They discovered in computer simulations that, even after the space charge had decayed, a gradient in the electron temperature would remain at the location of the space charge, for a much longer time. This gradient introduces a non-uniformity into the electron drift velocity, which is capable of re-nucleating the space-charge layer upon return of the field to above-threshold values. Jones and Rees estimate that true LSA operation becomes impossible in GaAs not far above 20 GHz. In InP this frequency limit is likely to be somewhat higher, but in neither of the two materials does it approach the frequencies that can be obtained in transit-time operation.

Our description is at variance with numerous papers in the literature in which LSA operation was claimed at frequencies above 20 GHz. The evidence is invariably indirect, usually based on no more than a frequency larger than the rule-of-thumb transit-time value $f[\mathrm{Hz}] = 10^7/L[\mathrm{cm}]$, combined with good circuit tunability. We are inclined to agree with Jones' and Rees' evaluation that in all these observations a transit-time mode was present in which the accumulation layers are not quenched but re-cycle with a speed appreciably higher than 10^7 cm/sec, a phenomenon the possibility of which is quite familiar to those having performed extensive computer simulations of accumulation layers[18, 19], but which is unfortunately not widely appreciated.

3.3 *The Jones–Rees effect*

The critique of the LSA mode leads naturally to the discussion of relaxation-effect speed limitations in transit-time TE devices. In such devices, traveling accumulation layers are an essential feature of the device operation. The accompanying traveling *spatial* non-uniformity of the electric field interacts with the time relaxation effect in a very surprising but highly beneficial way,

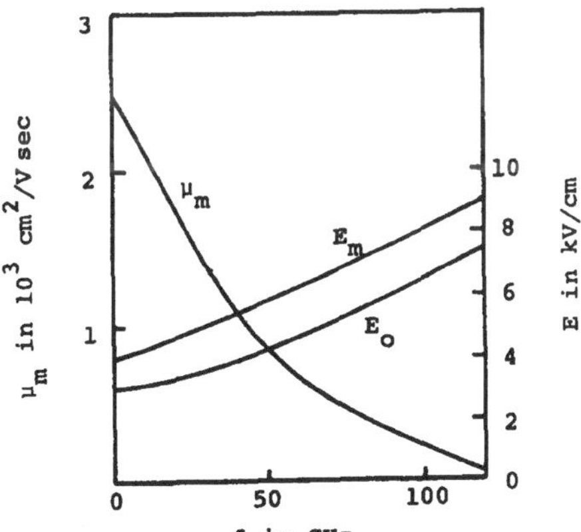

Fig. 2. The frequency dependence of the small-signal negative differential mobility (NDM) behavior of hot electrons in GaAs, after Rees[12]. E_0 is the threshold field for the onset of the NDM, E_m the field at which the NDM reaches its maximum, and μ_m the maximum NDM value reached at that field.

which I would like to call the Jones–Rees effect, in honor of its 1972 discoverers who first described the effect—together with a wealth of other insights—in a series of three papers[17, 19, 20] which collectively represent probably the most important account ever written on the internal space-charge dynamics in TE devices. Particularly the 1973 paper[20] is required reading for an in-depth understanding of these devices.

Consider Fig. 3, showing both the electron concentration and the field distribution in the vicinity of a traveling accumulation layer. As Fig. 3(a) shows, much of the electron accumulation consists of satellite valley electrons. Follow once again an electron that has just been scattered out of the satellite valley into the central valley, where it finds itself with a high velocity near v_0, a velocity much larger than the accumulation layer speed. What matters now is the direction of this velocity, more exactly, the component of this velocity in the direction of the field and of the accumulation layer motion. If the electron moves *with* the accumulation layer, it will overtake the latter, enter a region of much higher electric field, which quickly accelerates the electron to energies much higher than ϵ_s. Such an electron is very quickly scattered back into the satellite valleys, at a position further downstream in the device than the position where the electron left the satellite valleys. At least part of the accumulation layer propagation is due to this sequence. But the really interesting events happen to those electrons whose velocity, after being scattered into the central valley, is opposite to that of the accumulation layer. They, too, will quickly leave the accumulation layer, being *decelerated* in the process. But in doing so they will find themselves in a region in which the electric field is lower than E_{th}, insufficient to accelerate the electrons back to the energy ϵ_s, and cooling most of them all the way down to the lattice temperature in a time roughly equal to τ_0. In effect the traveling accumulation layer segregates the electrons into two classes: Those whose direction of motion is such that they will pick up energy from the field, end up automatically where the device needs high-energy electrons. Those whose directions of motion leads to a loss of energy end up where low-energy electrons are needed.

We saw earlier that it is not possible to quench the accumulation layer as rapidly as appeared desirable. The Jones–Rees effect now shows that it would not even be desirable to quench the layer if it could be done: The traveling accumulation layer actually aids high-frequency operation by its electron-segregation action.

3.4 *Contact effects*

We are left with rate effects at the two contacts. As the accumulation layer reaches the anode region it dumps its high-energy electrons into the presumably heavily doped, low-field anode body. In effect, those high-energy electrons, which are no longer needed, are "thrown away".

The situation is quite different at the cathode. This is where the needed hot electrons must first be generated. Presumably, the faster this is done, the better. This calls for contacts other than n^+-on-n contacts, for which the field increases only gradually with distance, and hence with time in the frame of reference of the moving electrons. What is desirable is a cathode contact that *somehow* incorporates a potential drop of height $\approx \epsilon_s$, which accelerates the electrons rapidly to the satellite valley energy. Exactly by what means such a drop is created is quite immaterial, but one obvious possibility is the potential drop at a reverse-biased Schottky barrier. Indeed, it has been observed that GaAs TE devices with metal alloy cathode contacts tend to operate to higher frequencies than devices with n^+-on-n regrown contacts[15, 21].

Ideally, the postulated cathode barrier should have a height $\Delta\epsilon$ that depends on the downstream field, in such a way that for $E < E_{th}$ the barrier height is less than ϵ_s, while for $E > E_{th}$, $\Delta\epsilon > \epsilon_s$. In this way, the cathode would automatically inject the electrons into those states that they would eventually reach in the bulk field anyway, without wasting either time or space. In fact, by combining such a hypothetical cathode with a device length short compared to the transit-time length, one would no longer be restricted by relaxation effects within the active layer, simply disposing of the electrons after one transit time by dumping them into the anode. Obviously, this is a design philosophy diametrically opposite to the LSA approach; whether it is a promising one remains to be seen: The physics of hot-electron relaxation effects is clearly not at its end yet.

Fig. 3. Electron concentrations (a) and electric field (b) around an accumulation layer. n is the total electron concentration, n_s the concentration of satellite valley electrons.

4. CONCLUSIONS

I have tried to show that relaxation effects play a large role in devices containing hot electrons, particularly in devices whose operation depends on such electrons. Often the relaxation effects limit the device performance; sometimes they help it. In either case, the effects must be understood to optimize the devices, and by understanding them it is often possible to design around their limitations and, sometimes, to convert a nuisance in one application into an asset in another.

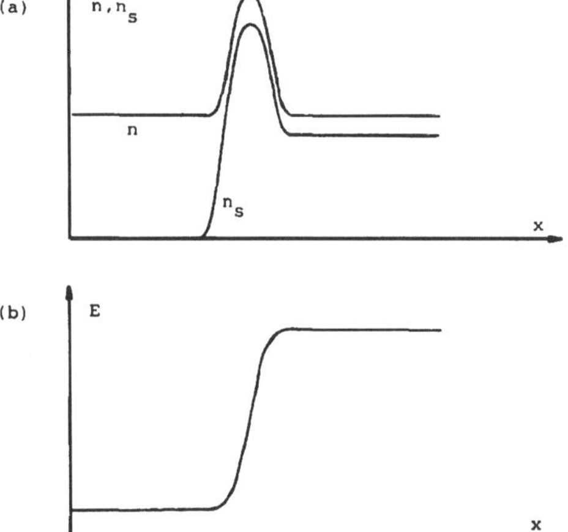

Hot-electron relaxation effects in devices

Acknowledgements—My own appreciation of hot-electron relaxation effects has benefited greatly by uncounted discussions, over several years, with many individuals at Varian Associates, particularly Bob Goldwasser, Steve Long, Tom Ruttan and Karl Varian. Their work on millimeter-wave TE devices provided a wealth of inputs from the world of real devices, constantly challenging my own understanding and providing the humbling corrective that theorists seem to need to bring them back down to earth periodically.

REFERENCES

1. C. Hilsum, *Solid-St. Electron.* **20**, 5 (1977).
2. D. E. Aspnes, *Phys. Rev. B* **14**, 5331 (1976).
3. L. Eaves, *et al., J. Phys. C* **4**, L42 (1971).
4. L. James, *J. Appl. Phys.* **44**, 2746 (1973).
5. J. Ruch, *IEEE Trans. Electron Dev.* **ED-19**, 652 (1972).
6. T. J. Maloney and J. Frey, *J. Appl. Phys.* **48**, 781 (1977).
7. H. Thim, *Proc. IEEE* **59**, 1285 (1971), and *Electron. Lett.* **7**, 246 (1971).
8. W. Shockley, *Bell Syst. Tech. J.* **33**, 799 (1954).
9. H. Kroemer, *Proc. IEEE* **58**, 1844 (1970).
10. For the two most recent papers on this topic giving extensive earlier references, see P. Jøndrup, P. Jeppesen and B. Jeppson, *IEEE Trans. Electron Dev.* **ED-23**, 1028 (1976), and H. Hasuo, *et al.*, **ED-23**, 1063 (1976).
11. B. Källbäck, *Solid-St. Electron.* **18**, 257 (1975).
12. H. D. Rees, *IBM J. Res. Develop.* **13**, 537 (1969).
13. W. Fawcett and D. C. Herbert, *Electron. Lett.* **9**, 308 (1973), and *J. Phys. C* **7**, 1641 (1974).
14. G. H. Glover, *J. Appl. Phys.* **44**, 1295 (1973).
15. T. Ruttan and K. Varian, Personal communications.
16. J. Copeland, *J. Appl. Phys.* **38**, 3096 (1967).
17. D. Jones and H. D. Rees, *Electron Lett.* **8**, 363 (1972).
18. H. Kroemer, *IEEE Trans. Electron Dev.* **ED-13**, 27 (1966).
19. D. Jones and H. D. Rees, *Electron. Lett.* **8**, 566 (1972).
20. D. Jones and H. D. Rees, *J. Phys. C* **6**, 1781 (1973).
21. R. Goldwasser and S. Long, Personal communications.

APPENDIX A

Corrections for non-parabolicity in InP

At sufficiently high energies the effective masses in all semiconductors become negative. The simplest $\epsilon(k)$ relation capable of describing this behavior is

$$\epsilon(k) = \frac{\hbar^2 k^2}{2m^*}\left[1 - \frac{1}{6}\left(\frac{k}{k_i}\right)^2\right],$$

where k_i is the wave number corresponding to the inflection point, related to the energy ϵ_i of the inflection point via

$$k_i = \sqrt{\frac{6}{5}} \times \sqrt{2m^* \epsilon_i}. \tag{1A}$$

For GaAs $\epsilon_i \gg \epsilon_t$ and non-parabolicity can be neglected, but for InP $\epsilon_i \simeq \epsilon_t$ [4]. Setting $\epsilon_i = \epsilon_t$ one obtains for v_0, instead of (1),

$$v_0 = 2\hbar k_i / 3m^* = \sqrt{8/15} \times \sqrt{2\epsilon_t / m^*},$$

and for τ_0, instead of (10),

$$\tau_0 = \frac{\hbar k_i}{qE_{th}} = \sqrt{\frac{6}{5}} \times \frac{1}{qE_{th}} \sqrt{2m^* \epsilon_t}. \tag{10A}$$

Reprinted from

H. Kroemer, ``Polar-on-Nonpolar Epitaxy,''
J. Cryst. Growth, Vol. 81, pp. 193-204, 1987.

Journal of Crystal Growth 81 (1987) 193–204
North-Holland, Amsterdam

POLAR-ON-NONPOLAR EPITAXY

Herbert KROEMER

Department of Electrical and Computer Engineering, University of California, Santa Barbara, California 93106, USA

One of the most fundamental problems that must be solved if device-quality GaAs is to be grown on Si substrates is that of suppressing antiphase disorder. Recent experimental evidence shows that such disorder can be suppressed not only on the (211) orientation, but also on (100), contrary to earlier theoretical expectations. A detailed discussion is given of the mechanism by which this suppression takes place, through a combination of slight misorientation and a high-temperature surface anneal, which lead to the pairing of all Si surface steps into a particular kind of double-height steps. A recent model by Aspnes and Ihm explains the energetic preference for this kind of step by postulating a drastic reconstruction of the atomic configuration at the step edge through the formation of a π-bonded chain running along the step. Another unexpected puzzle is posed by the recent observation that on a given Si(100) surface antiphase disorder-free growth with both possible Ga–As sublattice allocations can be achieved, depending on initial nucleation conditions. A new detailed nucleation model is proposed that explains these observations, by drawing heavily on earlier considerations of Harrison et al. concerning the electrostatics of a polar–nonpolar interface.

1. Introduction

When a polar (compound) semiconductor like, say, GaAs, is grown on a nonpolar (elemental) substrate of similar crystal structure, like Ge or Si, at least three new problems arise that are not present in the more conventional heteroepitaxial growth of one III/V compound upon another: (a) the problem of antiphase disorder on the compound side of the interface; (b) the problem of lack of electrical neutrality at the interface; (c) the problem of cross-doping. Although the existence of these problems has been recognized for some time, they have assumed central importance with the recent progress in the epitaxial growth of GaAs on Si substrates, a development of potentially very large practical significance.

Although the polar-on-nonpolar problems are by no means the only problems that must be solved if high-quality GaAs-on-Si epitaxy is to be achieved – the misfit dislocation problem is at least as severe – a solution of these problems is a *necessary* step without which this goal could not be achieved.

The present paper reviews the progress that has been made recently both in the understanding of the problems of polar-on-nonpolar epitaxy, and in their solution. As a result of this progress, it is probably safe to say that the principal problem in GaAs-on-Si materials quality is no longer the polar-on-nonpolar problem, but the misfit dislocation problem, a discussion of which lies outside the scope of this paper. The emphasis of the paper will be on the antiphase disorder and the interface neutrality problem, both because they are the most fundamental ones from a purely scientific point of view, and because the writer's own work has been concentrating on these. The cross-doping problem is not totally ignored; it is strongly interrelated with the the interface neutrality problem, and some comments about it will be made in that context.

2. Suppression of antiphase disorder

2.1. The problem

The diamond structure in which Si and Ge crystallize consists of two interpenetrating face-centered cubic sublattices. The two sublattices differ from each other only in the spatial orientation of the four tetrahedral bonds that connect each atom to its four nearest neighbors (which are on the other sublattice). For example, in fig. 1 the atoms with the bond orientations indicated as "A"

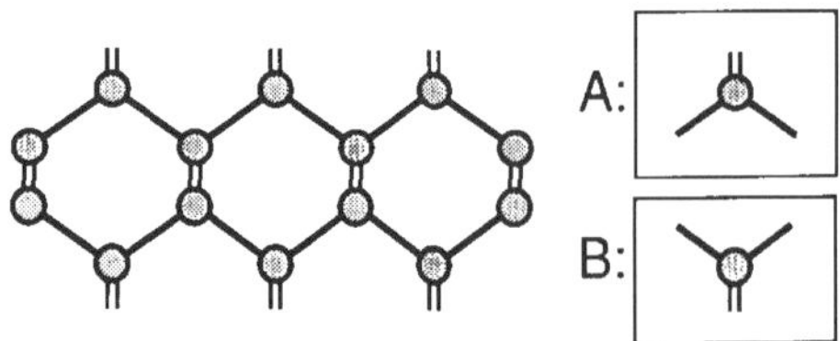

Fig. 1. Two sublattices in a Si crystal, distinguished only by bond orientation in space.

and "B" belong to different sublattices. There is no distinction between the two sublattices otherwise; both are occupied by the same atomic species.

In the zincblende structure, in which GaAs crystallizes, the two sublattices are occupied by different atoms, in the case of GaAs one by Ga atoms, the other by As atoms. In a crystal without antiphase disorder the sublattice allocation is the same throughout the crystal. But if this allocation changes somewhere inside the crystal (fig. 2), the interface between domains with opposite sublattice allocation forms a two-dimensional structural defect called an *antiphase boundary* (APB). The domains themselves are called *antiphase domains* (APDs).

Such APBs can be expected to form when GaAs is grown on Si or Ge, especially on a (100)-oriented substrate, the most widely used crystallographic orientation for MBE and MOCVD growth. Inasmuch as As forms strong bonds with Si, whereas Ga does not, the first atomic layer bonding to the Si substrate should be expected to be an As layer. Now, any *real*(100)

surface will *always* exhibit steps. At any step only one atomic layer high (or an odd number of layers high) the sublattice site allocation of Ga and As on opposite sides of the step is interchanged (fig. 3), and an APB results.

The APBs are structural defects, and we have little reason to expect that they might turn out to be the first benign defects in the history of semiconductor technology. Antiphase boundaries in GaAs contain Ga–Ga and As–As bonds. Such bonds represent electrically charged defects: A comparison of the number of bonding orbitals with the number of valence electrons available to fill them shows that Ga–Ga bonds act as acceptors, and As–As bonds as donors, with effective charges $\pm q/2$ per bond. In general, an APB will contain roughly equal numbers of both charges, thus acting as an extremely highly compensated doping sheet with very little *net* doping. The situation is least bad for an APB that follows exactly anly {110} plane, as in fig. 2: In that case Ga–Ga and As–As bonds will alternate within each crystallographic unit cell, leading to perfect *local* charge compensation. But for the deviations from this idealized arrangement, the lack of exact *local* charge balance will lead to potential fluctuations that will affect the electronic properties. Inasmuch as the initial Si surface steps are not likely to have the exact orientation within the surface plane that would lead to comparatively benign perfect-{110} APBs, the APBs actually resulting from the

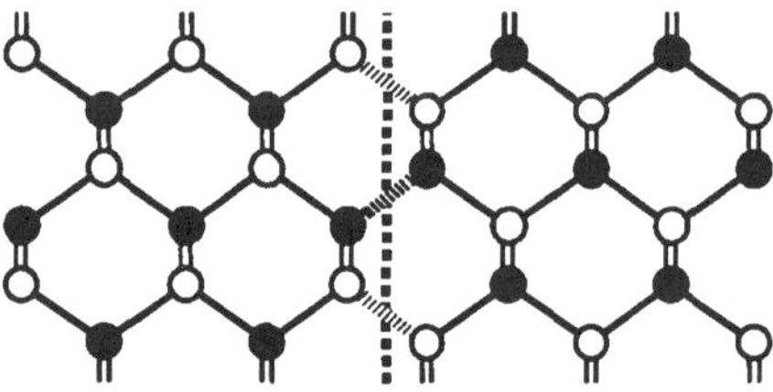

Fig. 2. Antiphase boundary (APB) formation in the zincblende structure, containing (in the case of GaAs) both Ga–Ga and As–As bonds. The configuration shown is the simplest possible one, a perfectly (110)-oriented APB, with alternating Ga–Ga and As–As bonds.

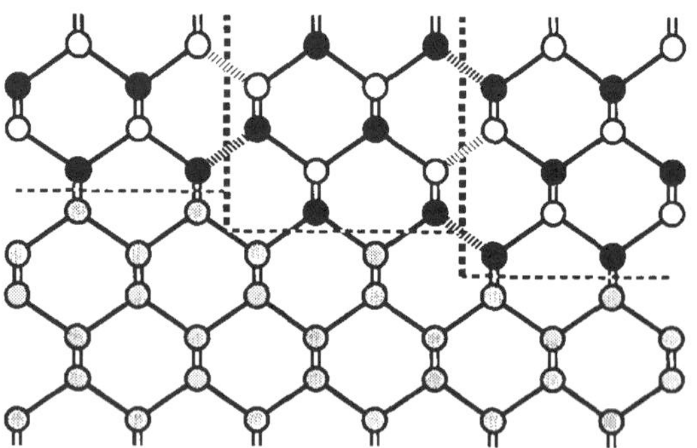

Fig. 3. Mechanism of APB formation during polar-on-nonpolar growth due to the presence of single-height steps on the substrate surface.

nucleation on a real surface must be expected to exhibit local charge fluctuations with large amplitude, and hence be harmful.

There are at least two approaches towards the essentially complete avoidance of APBs: One involves a switch to a different crystallographic orientation on which APBs do not form in principle, even in the presence of steps, like the (211) orientation employed by us [1–4] and described below. The other is to *somehow* enforce a perfect doubling of the height of *all* surface steps, an approach that has proven tenable since early 1985, contrary to all prior expectations.

2.2. The (211) solution

In our own work we have used the (211) orientation for the growth [1–4], rather than relying on perfect step doubling on the (100) surface. On a (211) surface the atomic sites of the two sublattices have a different number of back bonds to the Si substrate (fig. 4): One of the sublattices has two back bonds, the other has only one, *and this difference remains even in the presence of steps.* As a result, the two sublattice sites are no longer energetically and chemically equivalent, and elementary bonding energy considerations suggest that the As atoms, which have a much stronger tendency than Ga atoms to bond to Si, will seek out the more strongly binding doubly back-bonded

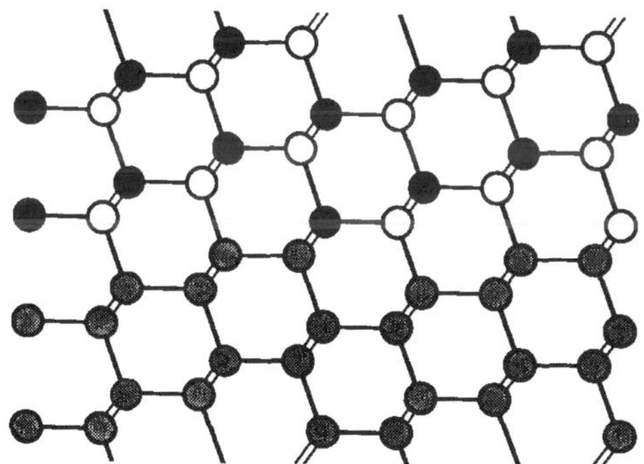

Fig. 4. A (211)-oriented polar-on-nonpolar interface. On such a surface, APBs do not form even in the presence of steps, because the two sublattices differ in the way they are back-bonded to the substrate, leading to sublattice control by chemical bonding preferences.

sites, displacing Ga atoms to the single back-bonded sites. Using crystallographic etching techniques, we were able to demonstrate that the GaAs layers are indeed free of APBs, and that the sublattice allocation is as stated above [1–4].

2.3. Step doubling on (100) surfaces

Most investigators working on GaAs-on-Si growth have preferred to continue to work with the conventional (100) orientation, or with wafers deliberately misoriented from the (100) orientation by a few degrees, relying on step doubling for the suppression of APBs.

When a step on a Si(100) surface is an even number of atomic layers high, the two sublattices on the GaAs side are in registry again, and an APB will not occur at this step. Unfortunately, it is well established experimentally that for "as-polished" exactly (100)-oriented Si surfaces the most common step height is one atomic layer [5,6] and there is in fact ample evidence [7] that not only the growth of GaAs on Si, but the growth of other III/V compounds such as GaP, on exactly (100)-oriented Si or Ge substrates *usually* exhibits copious APBs.

It was first indicated by Henzler and Clabes [5] that on misoriented Si(100) surfaces there is tendency towards step doubling with increasing annealing temperature. This was subsequently followed up in careful detail by Kaplan [6], who reported that on Si surfaces tilted by a few degrees from the (100) plane towards the (011) plane, *most* steps are two atoms high. Inasmuch as the step density on deliberately misoriented surfaces is much higher than for accurately oriented (100) surfaces, a certain amount of step doubling is to be expected, and the step doubling might be extensive if there is a simple energetic preference for double steps over single steps. However, unless the number of remaining single-height steps is drastically reduced, such tilting would not aid in the drastic suppression of APBs. In any event, it is hard to see how APBs could be avoided *completely* over the entire area of an entire wafer: In order to achieve APB-free growth, it is necessary that *all* steps be two atoms high, not just the majority of steps. At first glance, such a proposition appears hopeless. Yet it has become clear since early 1985 that such a perfect step doubling can indeed be achieved, leading to perfectly APB-free epitaxial growth of GaAs on Si(100):

(a) Recently, Fischer et al. [8,9] have reported growth on deliberately misoriented substrates, which does indeed appear to be free of APBs, judging from the anisotropic etching patterns of device structures on the epitaxial layers. Anisotropic etching is one of the simplest and most powerful techniques to test for APDs. We have had an opportunity to investigate one of the layers grown by this group, using our own etch pit technique [1–3], and we confirm the absence of APDs in that layer.

(b) Similarly convincing evidence of APD-free growth, based on an anisotropy of the RHEED patterns that was uniform over the entire wafer area, was presented by Nishi et al. [10]. The Si wafers in that work were not deliberately misoriented, but probably had a small amount of accidental misorientation. Similar results had been reported earlier by the same group for MOCVD-grown GaAs on Si [11]. RHEED evidence similar to that of Nishi et al., but less direct, had been earlier presented by Wang [12], and by hindsight it appears likely that Wang also had achieved APB-free growth.

(c) Perhaps the most convincing direct evidence for perfect step doubling already on the pre-growth Si(100) surface is contained in the stunning recent work by Sakamoto and Hashiguchi [13], who showed that a nominally (100)-oriented Si surface would go from a singly-stepped surface to a doubly-stepped surface during a prolonged high-temperature anneal (20 min at 1000°C), with *all* step terraces belonging to the same sublattice!

2.4. Step doubling mechanism

The empirical observation that a Si(100) surface transforms itself into a single-domain surface upon annealing raises the question: How is the possible? A little reflection shows that *perfect* long-range step doubling can *never* be explained by the elementary proposition that double steps are simply energetically preferred over single steps, *without any additional assumptions*. If that were the whole

story, each single-height step on the original surface would pair at random either with the step to its right or the step to its left. In such a model there is a 50% probability that the step to the left first pairs up with the next step even further to the left, and a 50% probability that the step to the right first pairs up with the next step even further to the right, leading to a $(50\%)^2 = 25\%$ probability that *neither* of these nearest-neighbor steps are available for pairing up first with the step of interest, leaving the latter unpaired. In such a case an APB would form on the average at every fourth initial step, and a misorientation would increase the density of APBs rather than decrease it, despite the energetic preference for double steps.

In order to explain the long-range suppression of APBs, the left/right randomness of the step pairing must somehow be eliminated. In the presence of double-height steps the surface layers on all terraces belong to the same Si sublattice; hence there must clearly be a preference for one of the two Si sublattices over the other. Inasmuch as the two different kind of sublattice planes differ from each other only by a 90° rotation in space, the preference mechanism can only reside inside the atomic arrangement at the edges.

At this point it is important to realize that the dangling bond configuration at a step edge depends not only on the direction of the step but also on which of the two sublattices forms the top of the step. The difference between the two different sublattices is most pronounced for steps that run along one of the two $\langle 011 \rangle$ directions within the (100) plane, precisely the kinds of steps generated by a tilt about one of those directions. It was pointed out already by Kaplan [6] that there are then two different kinds of terrace-and-edge combinations possible: (a) "type-A terraces" (my terminology), on the surface of which the dangling bonds of the Si atoms point parallel to the step edge, and (b) "type-B" terraces whose dangling bonds point perpendicularly to the step edge (fig. 5). Suppose now that there is – for whatever reasons – a strong energetic preference for one of the two bond configurations, say, for that of a type-A step. It was recently pointed out by this writer [14] that Kaplan's LEED data strongly suggested just such a preference [15], and that under

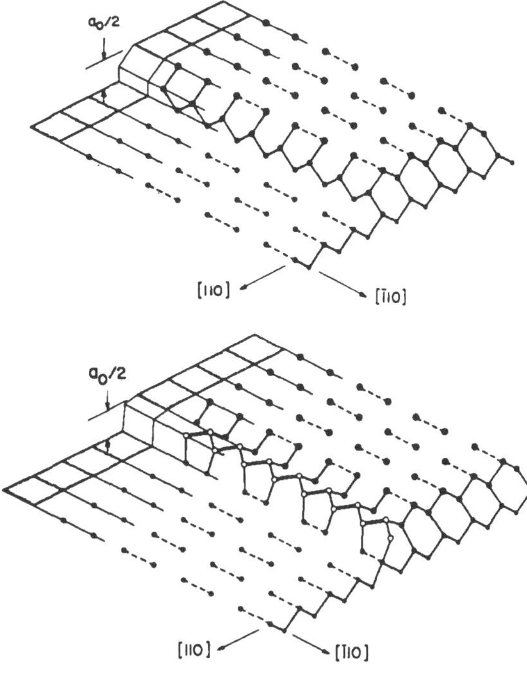

Fig. 5. Two kinds of bond configurations at [011]-oriented single-height atomic steps. For what we call a *type-A step*, the dangling bonds run parallel to the step, for the *type-B step* they run perpendicular.

Fig. 6. Atomic reconstruction proposed by Aspnes and Ihm [16] for the type-A double step, explaining the energetic preference for double-height type-A steps over all other kinds of steps. Top: unreconstructed edge. Bottom: a π-bonded atomic chain is formed along the step edges.

such conditions, and at sufficiently high temperatures, atoms from type-B edges would diffuse towards the type-A edges, until the former had simply disappeared by forming double-height steps that are bounded by edges of type A. The result would be a perfectly doubly-stepped Si surface, with all terraces belonging to sublattice A, and leading to GaAs growth free of APBs.

The only difficulty with this hypothesis in its original form is that it is hard to see how a sufficiently large energy difference could arise between the two kinds of step edges. Single-bond arguments fail to predict any energy difference, and it is clear that a significant energetic difference requires some sort of reconstruction, but the writer was unable to give a specific reconstruction model. This difficulty has recently been overcome by Aspnes and Ihm [16], who have pointed out that at a type-A *double*-height step the atomic configuration can lower its energy significantly, by about 40 meV per edge atom, through a drastic reconstruction during which π-bonded chains are formed, similar to the π-bonded chains that are believed to be present in the 2×2 reconstruction of the Si(100) surface (fig. 6). The Aspnes–Ihm model differs from the earlier model of ref. [14] in that the energetic preference is not one for type-A single steps over type-B single steps, but one for double-height type-A steps only, over *all* other kind of steps, including single-height type-A steps. But the ultimate consequences regarding APB-free growth are of course the same under both models.

It thus appears that the Aspnes–Ihm model is the best model proposed so far to offer an explanation of why single-domain growth of GaAs on Si can be achieved.

2.5. The role of temperature and misorientation

An essential ingredient of the step-doubling hypothesis is a surface temperature sufficiently high for a sufficiently long time to permit diffusion of Si atoms from the energetically unfavorable steps to the favorable ones. In Kaplan's LEED work, these conditions were clearly met: the Si surfaces had been heated to 1100°C, and because of the deliberate large misorientation the distance between steps was small. A study of the recent papers reporting reasonably convincing evidence of a single-domain surface indicates that in all cases the pre-growth Si surface was subjected to a high-temperature "heat-cleaning" step of one kind or another, although usually at a lower temperature and/or for less time than in the case of Sakamoto and Hashiguchi [13]. Although the utility of high-temperature heat treatment for surface cleaning purposes is well established, our model suggests that a second and possibly more important function is to permit the step doubling to take place. In fact, to achieve such a step doubling it may be necessary to perform the high temperature treatment even on perfectly clean surfaces.

Our model suggests further, and the experience of Nishi et al. [10] and of Akiyama et al. [17] confirm it, that a major deliberate misorientation is not really necessary. Its principal benefit would be to decrease the distance between surface steps and thereby to decrease the time and/or temperature required to achieve the desired step doubling. But it would appear that, given a sufficiently high annealing temperature for a sufficiently long time, even a small accidental misorientation always present might be sufficient to achieve the desired goal.

Another important question concerns the *direction* rather than the magnitude of the surface tilt. In our above discussion we had assumed, for simplicity, that the misorientation of the surface away from the exact [100] orientation corresponds to a rotation about one of the two $\langle 011 \rangle$ directions, thus leading to steps that can line up parallel to that direction. It is this assumption that led to two different kinds of steps with different kinds of dangling bond configurations. The situation would be quite different for a rotation about either the [010] or the [001] direction. In that case both kinds of steps have dangling bonds whose projections upon the [100] plane runs at the same angle relative to the edge direction (differing only in the sign of that angle), and which are energetically equivalent by symmetry. For such a tilt there is no mechanism enforcing a coherent step doubling with no remaining single steps, and it has in fact been reported [17,18] that GaAs growth on such surfaces leads to copious APBs.

In practice, the exact direction of tilt will rarely agree with an ideal exact $\langle 011 \rangle$ rotation, but will be somewhere in between a favorable $\langle 011 \rangle$ and the unfavorable [010] or [001] orientation, either

accidentally or deliberately. The steps created by such a tilt must necessarily contain sections that cannot benefit from the energy lowering due to the Aspnes–Ihm mechanism, or whatever other mechanism might be present. In such cases one should expect that the step geometry that forms on the surface would run in a zigzag direction, with the longer portions being of the energetically favored type-A double-step kind, the shorter kind of type-B double steps or possibly type-A and type-B single steps in close proximity (fig. 7). If the stabilization energy of the type-A double steps is sufficiently large, APBs would still be suppressed even for very large deviation of the tilt axis from a $\langle 011 \rangle$ axis, except possibly for small regions between any pair of single steps. Empirically, this is apparently what happens:

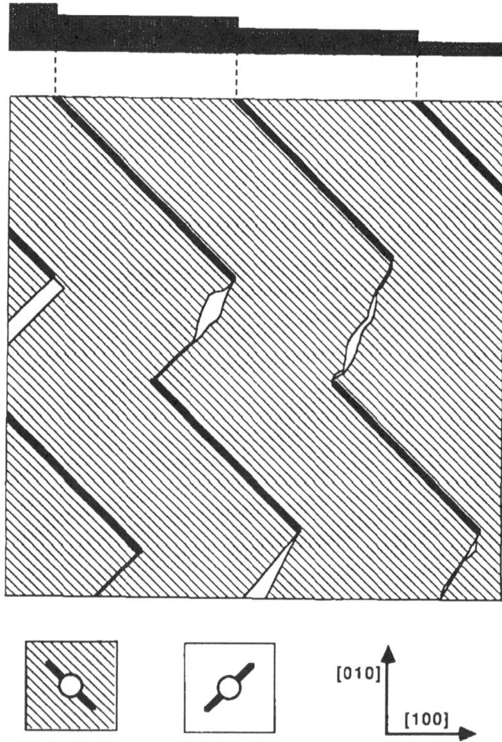

Fig. 7. Zigzag edge model for surfaces with a tilt axis deviating from $\langle 011 \rangle$. Sections of type-A double steps alternate with shorter sections that might be type-B double steps or regularly or irregularly shaped pairs of single steps.

Akiyama et al. [17] have performed GaAs-on-Si growth on lens-shaped surfaces, which present a continuum of both direction and magnitude of tilt, and they have reported that APBs occur only in a narrow band of tilt directions near a perfect [001] and [010] tilt.

A final point that requires discussion concerns the need for a high anneal temperature before nucleation. If a doubly-stepped surface is energetically preferred, then it should be possible to create such a surface by means other than a high-temperature anneal, for example by a sufficiently weak chemical etch, which might attack the Si surface only at the weak steps, but leaves the stronger steps alone, until the entire surface consists only of the more stable steps. In fact, the literature on GaAs-on-Si growth contains many reports of seemingly APB-free growth following heat treatments much gentler than those reported by Sakamoto et al. [13], suggesting that a certain amount of step doubling might already take place during the chemical polish and/or etch treatment currently employed. This is clearly a fertile field for future experimentation.

3. Interface atomic structure and neutrality

In our original discussion of the initial nucleation of GaAs on a Si(100) surface we made the simplifying assumption that the last Si plane was an unbroken plane, in which case chemical bonding arguments led to the conclusion that the first atomic plane on the GaAs side should be an unbroken As plane, as inside bulk GaAs. However, it was pointed out by Harrison, Kraut, Waldrop, and Grand (HKWG) [19] already in 1978 in the context of a GaAs/Ge interface that an atomic configuration composed of unbroken bulk planes at a polar–nonpolar (100) interface is energetically highly unfavorable [4]. This can easily be understood by arguments more adapted to our needs than those given by HKWG, as follows. Recall that Ga–Ga and As–As bonds are changed defects, each carrying a charge $\pm q/2$. The same argument applies to Ga–Si and As–Si bonds, except that the defect change is only half as large, $\pm q/4$ per bond. If the first atomic plane adjacent

to the Si substrate were a perfect As plane, the two back bonds per As atom would imply a donor-like defect charge of $+q/2$ per atom, or a charge density of $+q/a^2$, where a is the lattice constant. This is a very large charge (about 3×10^{14} donors/cm^2). If not neutralized, it would support an electric field of about 4×10^7 V/cm inside the growing GaAs layer! On a macroscopic scale, this charge would of course be neutralized by mobile electrons in the conduction band. However, the neutralizing charge would extend over an appreciable distance into the semiconductor, and the field seen by the atoms at the interface itself would be almost undiminished.

As HKWG point out, such a large field would lead to massive atomic re-arrangements during the high-temperature growth itself, attempting to neutralize the interface charge. One possibility – not considered by HKWG – would be the formation of a very large concentration of negatively charged antisite defects (Ga atoms on As sites) on the GaAs side. Harrison et al. themselves propose that the re-arrangement is one of the GaAs/Si interface itself, in such a way that a significant fraction of the Si atoms is removed from the top Si layer and replaced by Ga atoms, whose back bonds to the Si substrate have the opposite charge imbalance and hence neutralize the As–Si bond charge [20]. Electrical neutrality would be reached when the number of Ga–Si bonds created in this breakup of the last Si plane equals the number of As–Si bonds, and the authors propose that the reconstruction proceeds close to this stage.

Harrison et al. consider two limiting cases of idealized atomic arrangements, both of which would restore a perfectly neutral interface, shown in fig. 8. In the first of these, all Si atoms broken out from the top Si plane (referred to as plane No. 0 in what follows) are removed from the vicinity of the interface, to the surface of the growing epilayer. Interface neutrality is then achieved when one-half of the Si atoms are replaced by either Ga or As atoms. In the second arrangement, all broken-out Si atoms remain in plane 1, the plane directly atop the original Si surface. In this case, neutrality is achieved already when one-quarter of the Si atoms are broken out. The authors point out that the first of these arrangements, while free

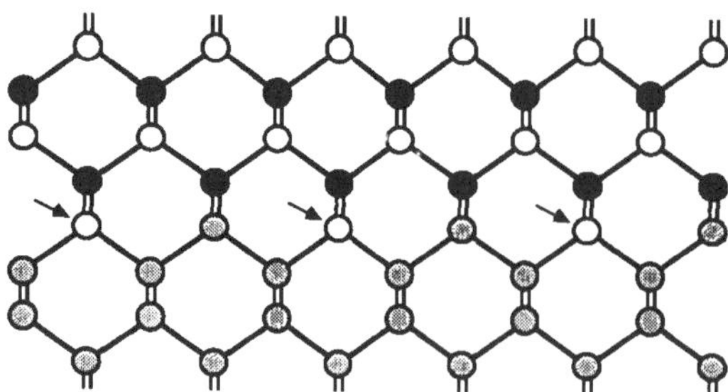

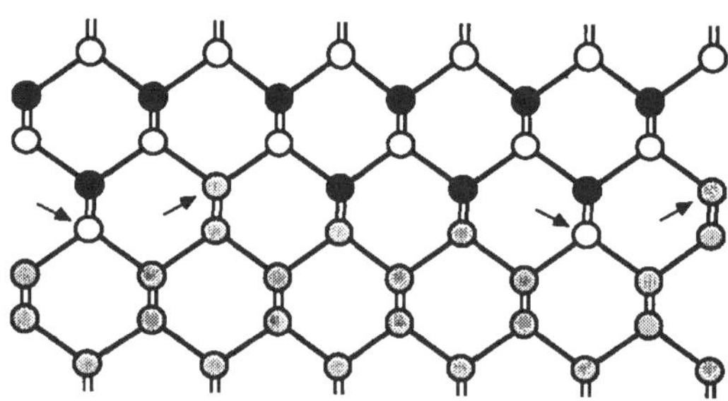

Fig. 8. Two models of atomic re-arrangement proposed by Harrison et al. [19] to achieve electrical neutrality at the interface, both involving the removal of Si atoms from the top Si layer. Top: single transition layer model, requiring removal of one-half the Si atoms, without their re-incorporation into the GaAs near the interface. Bottom: double transition layer model, requiring the removal of one-quarter of the Si atoms, and their re-incorporation into the first atomic layer on the GaAs side; this configuration has the lower energy.

One would certainly expect that the As-first hypothesis remains valid under nucleation conditions under which the Si is first exposed to an As flux, permitting the formation of copious As–Si bonds, before turning on the Ga flux. But the outcome is far less clear if the initial exposure of the Si surface is to Ga rather than As, as advocated by us for the (211) orientation. The question clearly calls for an experimental answer.

The question is readily tested by the etch pattern geometry generated by anisotropic etched on deliberately misoriented (100) surfaces. The orientation of the etch pits relative to the rotation axis depends on the sublattice allocation. Such tests have been performed by Fischer et al. [8,9], using two different nucleation conditions, by depositing either an As prelayer or a Ga prelayer before the actual growth. They found that APB-free growth could be achieved in both cases, but the sublattice ordering depended on the nature of the prelayer. Evidently a sublattice switch does take place! In their most recent work [22], Fischer et al. show that the sublattice allocation also changes with nucleation temperature, under what are implied to be otherwise unchanged conditions: For nucleation at low temperatures of 450–500°C they find APB-free growth with one particular sublattice ordering. There follows a temperature range (of unstated width) inside which copious APBs are observed. Above a certain (unstated) temperature, APB-free growth is again achieved, but with a sublattice ordering opposite to that at low temperatures.

The authors interpret their observations in terms of a model in which the first atomic plane following an unbroken Si surface is either an unbroken As plane or an unbroken Ga plane, depending on nucleation conditions. They suggest that the switch from As to Ga for the first layer is simply a consequence of the loss of As by evaporation at higher temperatures.

We would not wish to rule out an unbroken As plane model for the As-dominated case, despite the electrostatic argument that speak against it. But the formation of a simple unbroken Ga plane bonded to an unbroken Si plane as a result of As loss by evaporation is extremely unlikely, on the purely chemical grounds of the very different

of a net electric charge, still carries a residual electric dipole, whereas the second arrangement is free of both, and hence represents a state of lower overall electrostatic energy.

At this point an interesting question arises. If the HKWG re-arrangement towards an essentially neutral interface should indeed go to completion, or at least near-completion, then there would be no longer any energetic preference for this first plane after the original Si surface to be an As plane, and hence not for a specific sublattice allocation for Ga and As on the GaAs side! Which of the two sublattices would be which would then be decided by the kinetics of the nucleation process.

strength of Ga–Si and As–Si bonds. In our own work on the growth of GaP on Si we found that during the thermal decomposition and desorption of a GaP film the last Ga would evaporate long before the last phosphorous [23]. One would expect the same to be true for GaAs on Si, and recent work by Bringans et al. [24] strongly supports this expectation. Hence the formation of an unbroken Ga plane bonded to an unbroken Si surface is extremely unlikely so long as there is any As present at all.

The observations of Fischer et al. [8,9,22] clearly call for a different explanation. In the next section we propose a mechanism for the Ga-dominated nucleation mode that leads to the observed final result, but from diametrically opposite initial assumptions.

4. Proposed nucleation model

4.1. Ga-dominated nucleation: the As–Si site exchange postulate

We make the following two initial postulates:
(a) Arriving Ga atoms bond to Si atoms *only* when at least one Ga–As bond can be formed along with every Ga–Si bond. The idea behind this postulate is that the 1/4 electron excess of the Ga–As bond is transferred to the Ga–Si bond, where it helps forming the latter bond.
(b) Even the formation of As–Si bonds is facilitated if at the same time Ga–Si bonds are formed, to take up the electron excess of the As–Si bond.

These postulates lead to the idea that the initial nucleation of GaAs on Si does not simply take the form of either As or – much less likely – Ga first bonding to the original Si surface, especially not in the presence of a sufficiently large non-bonded and hence mobile Ga concentration on the Si surface. Instead, we therefore make the *central postulate* that:
(c) the initially arriving As atoms will undergo an exchange reaction during which a Si atom from the last Si plane (plane 0) exchanges sites with an arriving As atom, with two Ga atoms simultaneously bonding to both the As atom and two adjacent Si atoms of the original surface, as shown in fig. 9.

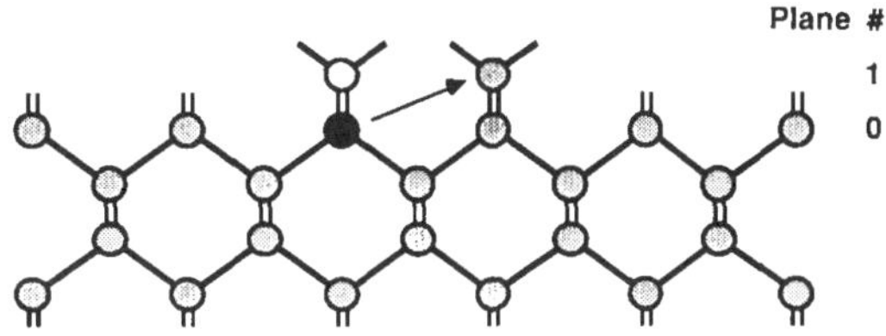

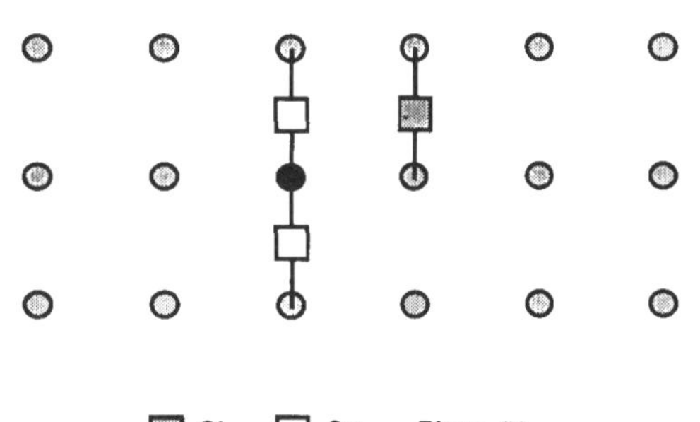

Fig. 9. Proposed first stage of the nucleation of GaAs on Si under Ga-rich conditions: an incoming As atom interchanges sites with a Si atom in plane 0 (the original top Si plane), simultaneously bonding two "waiting" Ga atoms. The ejected Si atom is placed on an adjacent site in plane 1. Top: [011] view; bottom: [100] (i.e. downward) view.

Note that in this initial nucleation step two As–Si bonds and two Ga–Si bonds are formed; hence the nucleus is an electrically neutral object in the sense of the work of HKWG. The number of Si–Si bonds does not change during the site exchange.

If we make the plausible assumption that the Si atom ejected from plane 0 is simply placed on one of the adjacent sites in plane 1, the site in plane 2 that has back bonds to both this Si atom and to the adjacent Ga atom, will form a natural bonding site for another As atom, and an As–Ga pair will connect between the remaining bond of the Si atom in plane 1 and a Si atom in plane 0, as shown in fig. 10. During this second nucleation step again two As–Si and two Ga–Si bonds are formed.

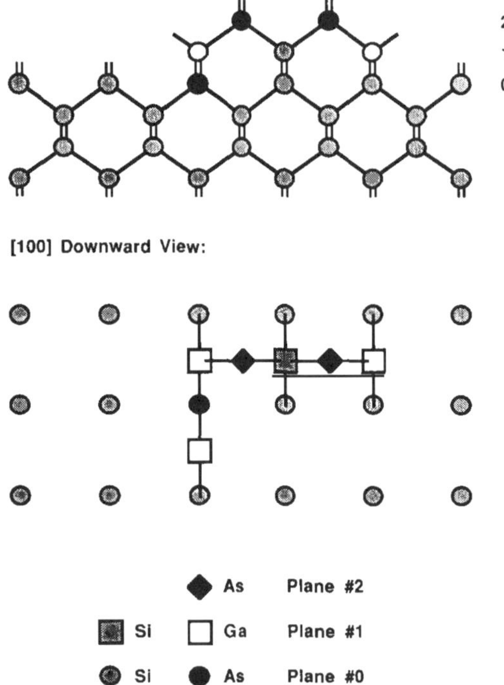

Fig. 10. Proposed second stage of the nucleation of GaAs on Si under Ga-rich conditions: two As atoms bond to the Si atom in plane 1, and a Ga atom closes the bonding loop of the outer of the two As atoms to the Si surface.

It is evident that this nucleation process automatically leads to a placement of all Ga atoms into plane 1 and of all As atoms into planes 0 and 2, the opposite choice from what would be present if the As–Si site exchange did not take place, but exactly the order observed by Fischer et al. under nucleation conditions of sufficiently high temperature and Ga flux. There can be little doubt that any subsequent lateral spreading of each nucleus thus formed will retain this sublattice order. This spreading can proceed either by adding As–Ga pairs, or by adding two As atoms and performing an As–Si site exchange. Because the two processes lead to opposite bond charges, a combination of both may be expected to take place, to aid in the establishment of electrical neutrality, but it is doubtful that perfect neutralization will take place.

We note however that, if perfect neutralization

did take place, the resulting composition of the different planes adjacent to the interface would be exactly what is demanded by the HKWG 2-layer reconstruction model.

4.2. The As-dominated case

The experiments of Fischer et al. [8,9,22] make it clear that under lower-temperature As-dominated conditions the As–Si site exchange postulated for Ga-dominated nucleation does not occur. The reason for this is probably one or both of the following: either the presence of an excess of Ga may be necessary to drive the exchange, or the temperature is too low for the exchange reaction to overcome an energetic reaction barrier that is very likely present, for example the As_2 dissociation barrier. Quite possibly both may play a role.

Nor is it clear whether or not the HKWG interface reconstruction process takes place at the lower temperatures and, if not, exactly how electrical neutrality is subsequently established – if indeed it is. Quite possibly the HKWG mechanism occurs even then, but with the Si atoms ejected from plane 0 now replaced by Ga rather than As atoms, because of the prior formation of a tightly-bonded partial As coverage in plane 1. An alternate possibility would be that the top Si plane remains intact, but is neutralized by the formation of a very high concentration of Ga-on-As site antisite defects, as mentioned earlier. Or maybe the interface remains highly charged, being neutralized only by mobile electrons in the conduction band. This is evidently another fertile field for future research.

4.3. Residual defects

The high-temperature Ga-dominated nucleation model presented above is an idealization. In practice, one must expect numerous defects to occur, especially the following three:

(a) Occasionally, As atoms may end up in plane 1, by bonding to Si plane 0 without undergoing the site exchange reaction. So long as these atoms remain in the minority, they would simply appear as local antisite point defects (double donors) on what is otherwise a Ga sublattice, without causing actual finite-size antiphase domains.

(b) As mentioned already, in those areas of the interface where the growth proceeds by lateral spreading from the initial nuclei, the neutralization of As–Si bonds by Ga–Si bonds formed may be incomplete, leading to a net doping of the interface, which may be either donor-like or acceptor-like, depending on the exact details of the process.

(c) Finally, some of the Si atoms removed from plane 0 might not be incorporated into plane 1, but are taken up by the growing GaAs bulk instead. On purely thermodynamic grounds one should expect at least the GaAs layers closest to the interface to be Si doped to the thermodynamic solubility limit.

Additional Si atoms may accumulate on the GaAs surface, from where they are gradually incorporated into the growing GaAs as bulk dopant.

Because of these various kinds of charged defects, the GaAs-on-Si(100) interface will almost certainly be one with a residual interface charge sufficiently large that it cannot be ignored for device purposes, combined with a heavy Si doping of at least the near-interface region of the crystal. The extent to which these two effects will take place will depend strongly on details of the exact growth procedure.

Up to a point, these defects are largely inconsequential, so long as they remain confined to within a few atomic monolayers of the original interface, and do not have a deleterious effect on the quality of subsequent layers. The defect structure near the interface is likely to be dominated by the very large density of misfit dislocations there, compared to which the other defects are a comparatively minor disturbance. Largely because of the misfit dislocations, the GaAs/Si interface itself is not likely to be usable as a part of the "intrinsic" device for most devices under current consideration, and its short-range properties are therefore not of primary concern. Apart from the dislocations, the defect making itself felt farthest from the interface is probably Si uptake by the growing GaAs, and its suppression probably deserves the highest priority after the suppression of the propagation of misfit dislocations.

5. Conclusions

With the 1985 emergence of convincing evidence that APBs in GaAs-on-Si (100) growth can be suppressed – and the understanding of the suppression mechanism presented here – the most urgent next problem becomes that of the suppression of the propagation of misfit dislocation, especially for minority carrier device applications such as lasers. For heterostructure FETs, the interface charge and the related cross-doping problem may be of similar importance. With both problems, the long-term issue is not just to make "good" devices, but to do so without having to resort to thick buffer layers.

The achievement of excellent APB suppression on the (100) orientation makes the switch to (211), long advocated by us [1–4], a less urgent one. However, the (211) orientation not only remains a perfectly viable option, but we continue, in fact, to believe that the *long-term* potential of the (211) orientation is still the better one for many devices. Most of the nucleation and interface charge problems discussed in sections 2–4 of this paper pertained almost exclusively to the problems of the (100) orientation, and are absent on the (211) orientation. Not only is there no problem about the sublattice allocations during nucleation, there also is no natural interface charge: on a perfect (211) surface, the numbers of As–Si and Ga–Si bonds are exactly the same (see fig. 4), leading naturally to an electrically neutral interface. The lack of a charge implies the lack of an electric field driving any atomic re-arrangement; hence any takeup of Si by the growing GaAs should also be much weaker. This has indeed been observed: In their work on the MBE growth of GaP on Si, Wright et al. [2] found that the uptake of Si by a (211) layer was much less than that by a (100) layer grown side by side. All of these could be sizeable advantages of the (211) orientation over the (100) orientation for applications in which the near-interface quality, and especially a low near interface impurity uptake of the GaAs are important.

Probably the biggest unknown in the (100) versus (211) competition is the behavior of misfit dislocations on the two orientations. Nothing is known yet about differences between the two orientations in this regard, but the differences could be severe, and they could decide the issue.

Acknowledgements

The writer gratefully acknowledges contributions from many individuals. Dr. B.A. Joyce (Philips) was a discussion partner during the early phases of this work. Dr. W.I. Wang (IBM) was the first to try to persuade me that APB-free growth could indeed be achieved on mis-oriented substrates, followed by Professor H. Morkoc and Dr. R.J. Fischer·(University of Illinois). Drs. M. Akiyama and S. Nishi (Oki Electric Co.) and especially Dr. T. Sakamoto (ETL-Tsukuba) completed that persuasion. To Professor Morkoc and Dr. Fischer additional thanks are due for intense discussions, for providing a wafer to test for freedom from APBs, and for making valuable information available prior to publication. Dr. D.E. Aspnes (Bellcore) resolved many puzzles with a preprint of his work. My former co-worker, Dr. P.N. Uppal, now at Martin Marietta, participated in many discussions; he also performed the tests for APBs on the Illinois wafer. Dr. E.A. Kraut (Rockwell) served as a patient sounding board throughout this work. Last but not least, much gratitude is due to the US Army Research Office for supporting this work.

References

[1] S.L. Wright, M. Inada and H. Kroemer, J. Vacuum Sci. Technol. 21 (1982) 534.
[2] S.L. Wright, H. Kroemer and M. Inada, J. Appl. Phys. 55 (1984) 2916.
[3] P.N. Uppal and H. Kroemer, J. Appl. Phys. 58 (1985) 2195.
[4] For a review, see also: H. Kroemer, Surface Sci. 123 (1983) 543.
[5] M. Henzler and J. Clabes, in: Proc. 2nd Intern. Conf. on Solid Surfaces, Kyoto, 1974 [Japan. J. Appl. Phys. Suppl. 2, Part 2 (1974) 389].
[6] R. Kaplan, Surface Sci. 93 (1980) 145.
[7] For extensive references see Uppal and Kroemer [3].
[8] R.J. Fischer, N.C. Chand, W.F. Kopp, H. Morkoc, L.P. Erickson and R. Youngman, Appl. Phys. Letters 47 (1985) 397.
[9] R.J. Fischer, N. Chand, W.F. Kopp, C.-K. Peng, H. Morkoc, K.R. Gleason and D. Scheitlin, IEEE Trans. Electron Devices ED-33 (1986) 206.
[10] S. Nishi, H. Inomata, M. Akiyama and K. Kaminishi, Japan. J. Appl. Phys. 24 (1985) L391.
[11] M. Akiyama, Y. Kawarada and K. Kaminishi, Japan. J. Appl. Phys. 23 (1984) L843.
[12] W.I. Wang, Appl. Phys. Letters 44 (1984) 1149.
[13] T. Sakamoto and G. Hashiguchi, Japan. J. Appl. Phys. 25 (1986) L57.
[14] H. Kroemer, in: Heteroepitaxy on Si Technology, Materials Research Society Proc. 1986 Spring Meeting, in the press.
[15] In ref. [6], Kaplan himself writes: "On relatively low step density, i.e. accurately cut (100) crystals, regions terminating on the different sublattices occur with equal probability. This should be true also of high step density vicinal surfaces, *unless the surface energy is highly sensitive to the dangling bond configuration relative to the steps*" [emphasis mine]. Kaplan did not follow up this remark any further; his data show that type-A terraces *must* be present, but he fits his LEED data to a model that assumes both kinds of terraces to be present, without stating whether the data could be fitted just as well or better by an "A-only" model. In ref. [14] the present writer argues that one could have concluded already from Kaplan's data that there must be a strong energetic preference for type-A steps present.
[16] D.E. Aspnes and J. Ihm, to be published.
[17] M. Akiyama, K. Kawarada, S. Nishi and K. Kaminishi, 1986 Spring Meeting, Materials Research Society.
[18] S. Sakai, T. Soga, M. Takeyasu and M. Umeno, 1986 Spring Meeting, Materials Research Society.
[19] W.A. Harrison, E.A. Kraut, J.R. Waldrop and R.W. Grant, Phys. Rev. B18 (1978) 4402.
[20] Our discussion here and in the rest of this section goes beyond HKWG in specifically assuming that the first atomic plane above the original Si surface is, at least initially, an As plane. The arguments in Harrison et al., apart from discussing GaAs-on-Ge rather than GaAs-on-Si, apply independently of which of the two sublattices on the GaAs side is which, so long as the two different kinds of atoms occur in alternating planes.
[21] R.J. Fischer, W.T. Masselink, J. Klem, T. Henderson, T.C. McGlinn, M.V. Klein, H. Morkoc, J. Mazur and J. Washburn, J. Appl. Phys. 45 (1985) 374.
[22] R. Fischer, H. Morkoc, C. Choi, N. Otsuka, M. Longerbone and L.P. Erickson, to be published.
[23] S.L. Wright, PhD Dissertation, University of California, Santa Barbara, CA (1982), unpublished.
[24] R.D. Bringans, R.I.G. Uhrberg, M.A. Olmstead, R.Z. Bachrach and J.E. Northrup, 1986 Spring Meeting, Materials Research Society.

Reprinted with permission from

T.-Y. Liu, P. M. Petroff, and H. Kroemer, ``Luminescence of GaAs/(Al,Ga)As superlattices grown on Si substrates, containing a high density of threading dislocations: Strong effect of the superlattice period,'' J. Appl. Phys., Vol. 64(12), pp. 6810-6814, 1988.

Luminescence of GaAs/(Al,Ga)As superlattices grown on Si substrates, containing a high density of threading dislocations: Strong effect of the superlattice period

T. Y. Liu, P. M. Petroff, and H. Kroemer

Department of Electrical and Computer Engineering and Materials Department, University of California at Santa Barbara, Santa Barbara, California 93106

(Received 26 April 1988; accepted for publication 19 August 1988)

In heavily dislocated material (dislocation density of mid-10^7 cm^{-2}), the low-temperature cathodoluminescence intensity of a sufficiently short-period superlattice comes remarkably close to that of comparison structures grown on low-dislocation material. We attribute this effect to a redistribution of the recombination-active sites along the dislocation cores taking place in the material. The driving force of such redistribution may be electrical or chemical in nature.

INTRODUCTION

The presence of dislocations in semiconductors is detrimental to material and device properties. They are, for example, responsible for a reduction of minority-carrier lifetimes and an increase in leakage currents of minority-carrier devices. Perhaps most important, dislocations severely limit the operating lifetime of solid-state lasers.[1]

Dislocations can originate from a variety of sources. Some of the highest dislocation densities arise when an epitaxial layer is grown on a lattice-mismatched substrate. It is generally found that, depending on the amount of mismatch, when the epitaxial layer thickness exceeds a certain critical value, misfit dislocations are formed to accommodate the stress in the epitaxial overlayer. Such misfit dislocations do not remain confined to the interfacial plane between the two materials, but many of them propagate into the epitaxial layer, forming a network of *threading dislocations*.

Because of the increasing interest in growing GaAs on Si substrates for integrating optical and electronics devices on the same substrate (a system with a lattice mismatch of $\approx 4\%$), it is important to understand much better the effect of dislocations on the electronic properties of the material. The best material to date still has threading dislocation densities in the range between 10^6 cm^{-2} and 10^7 cm^{-2} or more, as established by transmission electron microscopy. Nevertheless, cw room-temperature *quantum-well* lasers have been demonstrated in such material.[2] On the other hand, "ordinary" double-heterostructure lasers have not been made to operate cw in such material. We therefore believe that superlattices or quantum wells might play a central role in modifying the electronic and optical properties of dislocations.

There are other observations pointing in the same direction. For example, in our own extensive earlier work on the photoluminescence properties of GaSb/AlSb multiple-quantum-well superlattices grown on grossly lattice-mismatched GaAs substrates,[3] we had found excellent photoluminescence properties, initially suggesting that the dislocations had been suppressed by the first-grown portion of the superlattice, adjacent to the GaAs substrate. However, transmission electron microscope (TEM) studies subsequently showed that no such suppression had taken place,

and that the threading dislocation density was in fact very high, about 10^8 cm^{-2}.[4]

To understand the effects of dislocations in lattice-mismatched semiconductor epitaxial layers, and to see if there exist any effects associated with a superlattice on the electronic and optical properties of dislocations, we have performed a series of experiments designed to look at the materials quality of heavily dislocated epitaxial layers of GaAs grown by molecular-beam epitaxy (MBE) on Si substrates, each layer containing several superlattice "stacks" with different superlattice periods. We used the intensity of low-temperature cathodoluminescence (CL) intensity for quantifying the overall materials quality. Spectrally resolved CL images at high magnification were used to monitor the uniformity of the materials. Conventional cross-section TEM was used to directly obtain the threading dislocation densities in the epitaxial layer structures.

EXPERIMENT

The silicon wafers used in this study were (211) oriented, with no deliberate misorientation. This choice of substrate orientation was based on our previous work of growing GaP[5] and GaAs[6] on silicon. The wafers were chemically degreased and rinsed in running deionized water, after which they were oxidized in concentrated sulphuric acid at 180 °C for 15–30 min. The oxide was then removed again by dipping the wafers into 5%–10% dilute hydrofluoric acid. The samples were next reoxidized with a solution of 1 NH$_4$OH : 1 H$_2$O$_2$(30%) : 10 H$_2$O, by volume. The samples were finally rinsed under running deionized water for 7 min and spun dry before loading into the load lock of a Varian-360 MBE system. The thin silicon oxide thus formed was desorbed in the growth chamber at 800 °C under a Ga beam.[7]

A strained-layer superlattice (SLS) buffer layer was first grown on the silicon substrate. The SLS consisted of a 10-period 50 Å GaAs + 50 Å In$_{0.25}$Ga$_{0.75}$As superlattice, grown at 505 °C. Such a superlattice buffer has been shown to drastically improve the growth front morphology. It *may* also be effective in reducing the threading dislocation density to the mid-10^7-cm^{-2} range.[8,9] However, the actual efficiency of a SLS in reducing dislocations at such a high level is not clear. In particular, it is not clear to what extent this

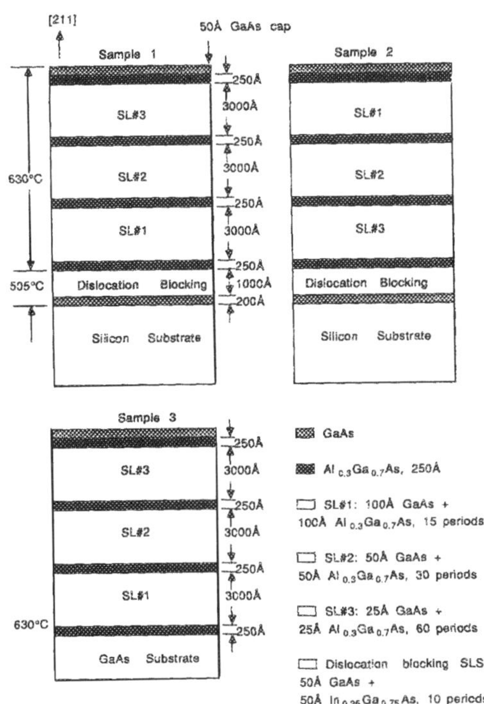

FIG. 1. Samples used to study the effect of the superlattice period on dislocation properties. To facilitate comparison, the three superlattice packets contain the same amount of GaAs and $Al_{0.3}Ga_{0.7}As$ in each packet.

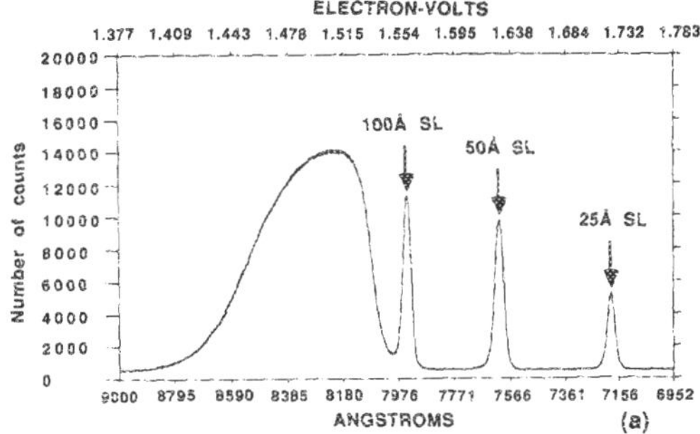

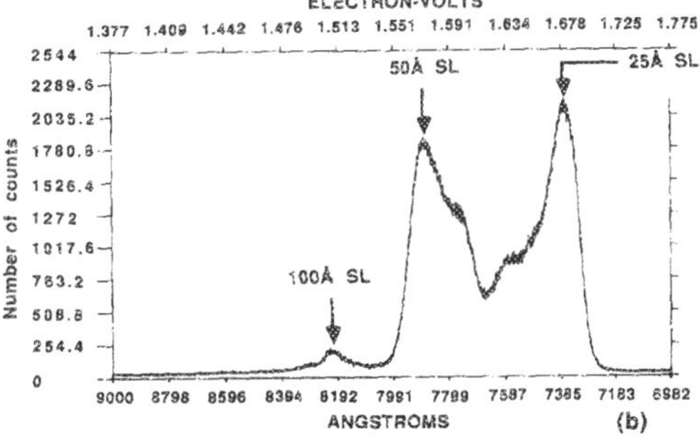

FIG. 2. Typical low-temperature cathodoluminescence (CL) spectra. (a) Spectrum recorded from the low-dislocation-density reference sample No. 3, grown on a GaAs substrate. The decrease in the luminescence intensity towards shorter SL period is a characteristic behavior of an increase in interface recombination velocity. (b) A typical low-temperature CL spectrum from sample No. 1. A trend opposite to sample No. 3 is evident.

reduction would have taken place even for a non-SLS buffer layer of the same thickness.

Four samples were grown. In sample No. 1 (Fig. 1), the structure, grown at 630 °C, consisted of three superlattice (SL) "packets," labeled SL Nos. 1, 2, and 3, separated from each other by 250 Å of $Al_{0.3}Ga_{0.7}As$ barriers. To facilitate comparison, each of the three SL packets contained the *same amount* of GaAs and $Al_{0.3}Ga_{0.7}As$. In SL No. 1, the 3000-Å packet was partitioned into15 periods of 100 Å of GaAs and 100 Å of $Al_{0.3}Ga_{0.7}As$. In SL No. 2, the same amount was partitioned into 30 periods of 50 Å of GaAs and 50 Å of $Al_{0.3}Ga_{0.7}As$. Finally, in SL No. 3, 60 periods of 25 Å of GaAs and 25 Å of $Al_{0.3}Ga_{0.7}As$ constituted the 3000-Å layer.

To account for a possible attenuation of the dislocation density with increasing distance from the interface, sample No. 2 was prepared (Fig. 1). It had the same growth parameters and buffer layers as sample No. 1, the only difference being that the stacking sequence of the three SL packets was reversed.

A control sample, No. 3, was also prepared on a GaAs (211B) substrate. All the growth parameters were kept the same as in other growths, and the same stacking sequence of the SL packets as in sample No. 1 was employed. Only the SLS buffer was omitted from this control sample (Fig. 1).

Finally, an "ordinary" double hererostructure sample (No. 4, not shown in Fig. 1) was also prepared, which con-

sisted of a thick 5000-Å GaAs region sandwiched between two 2500-Å $Al_{0.3}Ga_{0.7}As$ barrier regions.

Low-temperature CL measurements were used to assess the quality of the different SL packets. They were carried out in a modified JEOL STEM model 200 equipped with a liquid-helium stage. A relatively low-energy (150-keV) electron beam was used, to avoid formation of radiation-induced damage to the material during the observation. No correction to the system transfer function was applied to the collected data.

To take into account any complications due to the fact that photons from the three SL packets have different escape depths, the cathodoluminescence was also monitored from the substrate side of the sample. To this end the silicon substrates of both samples No. 1 and No. 2 were thinned down and dimpled to within tens of microns away from the epitaxial layers. A hot aqueous solution of 20% potassium hydroxide at 120 °C was used as a highly selective etch to remove the dimpled silicon region until the epitaxial GaAs layer was exposed. Cathodoluminescence measurements were done on both the epitaxial sides and the substrate sides of both samples.

Figure 2(a) shows a typical spectrum from the control sample. One can clearly identify the three peaks corresponding to the 100-, 50-, and the 25-Å SLs. As can be seen from Fig. 2(a), there exists a trend in which the CL intensity *decreases* as the SL period decreases. This is probably due to an increase in the recombination at the increased number of interfaces or—more likely—inside the (Al,Ga)As barriers, as the SL period decreases.[10,11]

Figure 2(b) shows a spectrum taken from the epitaxial side of sample No. 1, with the CL peaks from the three SLs labeled. The luminescence data from the substrate side were essentially identical to those from the epitaxial side, indicating that the intensity trends seen cannot be due to differences in photon escape. Figure 3(a) illustrates the *normalized* CL *peak* intensities of sample No. 1 as a function of SL period. The normalization is done by simply taking the ratio of the heights of the line peaks in sample No. 2 to those in the control sample No. 3. No attempt was made to integrate over the broadened lines. The error bars indicate the spread of data due to (a) spatial variations of relative intensity on the same side and (b) differences in relative intensity on different sides of the same sample. The trend, evident in Figs. 2 and 3(a), is just the opposite of what one can observe in the control sample. The 25 Å + 25 Å SL recovers to about 20% of the intensity of the corresponding SL grown on a GaAs substrate, despite the fact that cross-sectional TEM data indicate that all three SL packets have dislocation densities in the mid-10^7-cm^{-2} range. An additional data point has been appended to Fig. 3, at 5000 Å, corresponding to the intensity from the "ordinary" double heterostructure of sample No. 4.

Clearly, in sample No. 1 the luminescence intensity *increases* very strongly as the well width decreases. One might be tempted to explain such an observation in terms of a reduction of the dislocation density away from the GaAs/silicon interface, even though TEM measurements (on other samples; we do not have TEM data on sample No. 1 itself)

show only a very slight decrease in dislocation density, totally insufficient to explain the extremely strong luminescence trend. Such an explanation is fully ruled out by the data of sample No. 2, for which the normalized CL intensity from sample No. 2 is shown in Fig. 3(b). One can see the same basic trend as in sample No. 1, in which the narrow-well ($<$50-Å) SL packets show a much higher CL intensity than the wider-well ($\geqslant$100-Å) ones, up to 20%–25% of the control sample. Considering that in sample No. 2 the narrow-well sample is adjacent to the Si substrate, this clearly rules out any *major* effect due to a spatial dislocation density variation. In fact, the fairly strong CL intensity of the 25-Å well packets is essentially the same in both samples, despite their locations at opposite ends of the stacks.

Still, there are clearly differences between the two samples, the most notable being that in sample No. 2 the 50-Å wells have a 5 times higher CL intensity than in sample No. 1 (higher even than the 25-Å wells), even though the 50-Å wells are at the same distance from the interface in both samples. The 100-Å wells show a similar enhancement in sample No. 2. Evidently, there are sample-to-sample variations, the origins of which are not understood. However, these variations do not in any way weaken the basic conclusion drawn from our work, that there is a drastic overall increase in luminescence output, by about two orders of

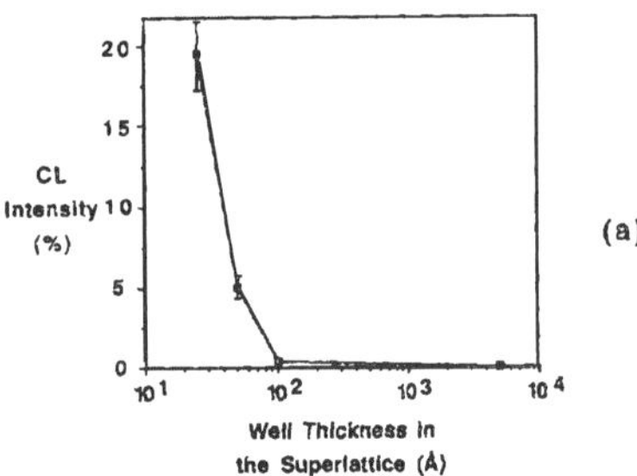

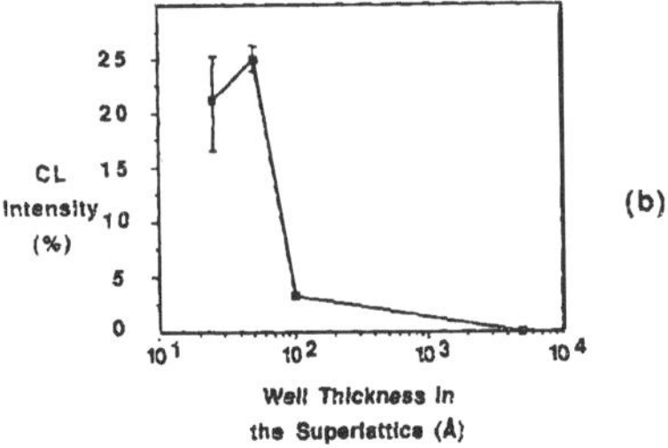

FIG. 3. Normalized low-temperature CL intensity, relative to the intensity of the reference sample No. 3. (a) Sample No. 1. (b) Sample No. 2.

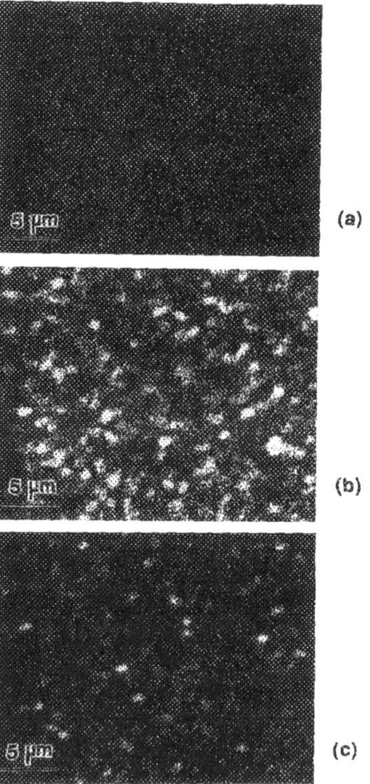

FIG. 4. Spectrally resolved images of SL's grown on silicon substrates. (a) Image taken at a wavelength that reveals the nonuniformities of the 100-Å SL. (b) The same location, imaged for the 50-Å SL. (c) The same location, imaged for the 25-Å SL.

magnitude, in the narrow-well ($\leqslant$50-Å) structures, compared to the wide-well ($\gtrsim$100-Å) structure.

As shown in Fig. 2(b), the spectra of the dislocated samples exhibit complex line-broadening effects. At least part of this broadening is evidently the result of the presence of strong spatial inhomogeneities, which were clearly visible in spectrally resolved CL images, as in Fig. 4. The images shown were formed at exactly the same location, but using different wavelengths to image the different SL packets. One can readily see an increase in nonuniformity as the SL period decreases. This is what one would expect if the inhomogeneities were simply due to spatially varying well thicknesses. The origin of these variations is not clear, nor is it clear to what extent they might be related to the differences between samples Nos. 1 and 2. In fact, it is not even certain that these inhomogeneities have anything at all to do with the growth on a lattice-mismatched substrate: some inhomogeneities have been observed even in the quantum-well luminescence in low-dislocation structures grown on GaAs substrates.[12]

DISCUSSION

The nonradiative recombination effects associated with dislocations have been attributed to several different possible causes. The most likely possibility is that the recombination centers are associated with deep levels due to unreconstructed dangling bonds along the dislocation cores. It was shown by Kimerling and Patel[13] that most of the dangling bonds at the dislocation cores in silicon are reconstructed or rearranged so that only a small fraction of the sites are electrically active. The authors estimated a 2.5% site occupation in the Shockley[14] dangling-bond structure or a 1% site occupation in the Hirsch[15] dissociated dislocation model. Defect-site spacings of the order of 200 Å along the dislocation core were measured. In zinc-blende semiconductors, the situation is anticipated to be similar. Kimerling and Patel proposed that the electrically active sites are located at the dislocation *kink sites* that are associated with dislocation motions from one Peierls valley to the other in the same glide plane. These kinks are known to be highly mobile along the dislocation core. Their motion is easily thermally activated at the growth temperature.

If one accepts the kink model of recombination at dislocations, the simplest explanation of our observations would be that in sufficiently narrow-well superlattices the overwhelming majority of dislocation kinks are *somehow* expelled from the GaAs portion of the SL structures into the (Al,Ga)As barriers. Conceivably, the strong electric fields that are present at heterointerfaces, or even chemical potential gradient, might play a role. Recombination-enhanced kink migration due to minority-carrier generation during the CL observation is unlikely, since no change in the luminescence yield of the material has been observed as a function of time. Whatever the mechanism, any spatial expulsion mechanism would clearly run much closer to completion in narrow-well structures than in wide-well ones. Further experiments are needed to clarify the exact mechanism. Several such experiments—suggested by several possible mechanisms—are currently in preparation.

One alternate to dislocation kinks as the recombination centers would be the *Cottrell atmosphere* of native point defects surrounding the dislocation cores. There are a variety of ways in which native point defects can be created around the dislocation cores. For example, when a dislocation jog moves from one glide plane to an adjacent one, a row of vacancies or interstitials results, depending on the direction of motion. At the typical growth temperature of the GaAs epitaxial layer (580–630 °C), dislocations are highly mobile. Depending on the nature of the point defects, the latter may be singly or multiply charged, or remain neutral. Such defects can induce deep trap states in the gap region and are thus electrically active.

A less likely possibility than either dislocation kinks or dislocation-associated native defects would be chemical impurities that are trapped along the dislocation cores. Impurities have been observed to diffuse along dislocations (by pipe diffusion) in semiconductors. However, we do not believe that this is a likely mechanism: Many different kinds of devices have operated successfully in GaAs/(Al,Ga)As heterostructures grown on Si substrates, and if extensive pipe diffusion were a problem, this would almost certainly have been noticed, particularly when one considers that such material typically contains dislocation density of the order of 10^8 cm^{-2}.[16]

The various driving forces discussed above for the spatial expulsion of dislocation kinks from the quantum wells could also be invoked for a point defect or an impurity model of the recombination at dislocations.

Finally, an alternate to the *spatial* expulsion of the energy states responsible for the recombination activity might be an *energetic* expulsion of the associated energy levels from the forbidden energy gap into the conduction band and/or the valence band, thus rendering the recombination centers ineffective. It has been shown[17] that, theoretically at least, a deep trap in the bulk can become a shallow energy state at a heterointerface and *vice versa*. The effect of an interface is essentially to shift the s-like energy level and to split the three p-like levels, with the p-like state that points in the direction perpendicular to the interface being affected most. However, we find this—or any other—energetic expulsion model least likely: Deep levels tend to be highly localized, extending over only a few atomic distances. Hence one would expect an efficient energetic expulsion only for quantum wells that are only a few atoms wide. It is hard to see how a two-orders-of-magnitude increase in CL intensity could result already for wells that are still 50 Å wide.

In summary, we have shown that in a sufficiently short-period superlattice in a heavily dislocated material, strong interaction exists between the recombination centers and the superlattices so that the effective recombination rate is highly reduced. The interaction is a strong function of the superlattice period. The origin of such an interaction is likely to be a spatial expulsion of dislocation-related recombination centers from the well portion of the superlattices into the barriers, driven by either electrical or chemical forces. These results strongly suggest that minority-carrier devices in GaAs on silicon substrates should exhibit a superior performance if the active part of the devices is composed of a narrow GaAs quantum well.

ACKNOWLEDGMENT

We wish to express our appreciation to the Army Research Office for supporting this work.

[1] P. M. Petroff, in *Semiconductors and Insulators*, edited by F. C. Brow and N. Itoh (Gordon and Breach, New York, 1983), Vol. 5, p. 307.

[2] H. Z. Chen, A. Ghaffari, H. Wang, H. Morkoç, and A. Yariv, Appl. Phys. Lett. **51**, 1320 (1987).

[3] G. Griffiths, K. Mohammed, S. Subbanna, H. Kroemer, and J. L. Merz, Appl. Phys. Lett. **43**, 1059 (1983).

[4] J. S. Ahearn and P. Uppal (personal communication).

[5] S. L. Wright, M. Inada, and H. Kroemer, J. Vac. Sci. Technol. **21**, 534 (1982).

[6] P. Uppal and H. Kroemer, J. Vac. Sci. Technol. B **4**, 641 (1986).

[7] S. L. Wright and H. Kroemer, Appl. Phys. Lett. **36**, 210 (1980).

[8] Z. L. Weber, E. R. Weber, J. Washburn, T. Y. Liu, and H. Kroemer, in *Heteroepitaxy on Silicon II*, Vol. 91 of *Materials Research Society Symposium Proceedings*, edited by J. C. C. Fan, J. M. Phillips, and B. Y. Tsaur (Material Research Society, Pittsburgh, 1987), p. 91.

[9] J. S. Ahearn, P. Uppal, T. Y. Liu, and H. Kroemer, J. Vac. Sci. Technol. B **5**, 1156 (1987).

[10] G. Duggan, H. I. Ralph, and R. J. Elliott, Solid State Commun. **56**, 17 (1985).

[11] B. Sermage, M. F. Pereira, F. Alexandre, J. Beerens, R. Azoulay, C. Tallot, A. M. Jeanlans, and D. Mechenin, J. Phys. (Paris) Colloq. **48**, C5-135 (1987).

[12] P. M. Petroff, R. C. Miller, A. C. Gossard, and W. Wiegmann, Appl. Phys. Lett. **44**, 217 (1984).

[13] L. C. Kimerling and J. R. Patel, VLSI Electron. **12**, 223 (1985).

[14] W. Shockley, Phys. Rev. **91**, 228 (1953).

[15] P. B. Hirsch, J. Phys. (Paris) Colloq. **40**, C6-27 (1979).

[16] M. M. Al-Jassim, A. E. Blakeslee, K. M. Jones, and S. E. Asher, Inst. Phys. Conf. Ser. No. **87**, 99 (1987).

[17] R. E. Allen, J. P. Buisson, and J. D. Dow, Appl. Phys. Lett. **39**, 975 (1981).

Reprinted from

H. Kroemer, T.-Y. Liu, and P. M. Petroff, ``GaAs on Si, and Related
Systems: Problems and Prospects,'' J. Cryst. Growth, Vol. 95,
pp. 96-102, 1989.

96

Journal of Crystal Growth 95 (1989) 96–102
North-Holland, Amsterdam

GaAs ON Si AND RELATED SYSTEMS: PROBLEMS AND PROSPECTS

Herbert KROEMER, Tak-Yu LIU and Pierre M. PETROFF

Department of Electrical and Computer Engineering, and Department of Materials, University of California, Santa Barbara, California 93106, USA

The dominant problem in the epitaxial growth of GaAs and other III–V compounds on silicon is the problem of threading dislocations caused by the large lattice mismatch of the compound semiconductors relative to the Si substrate. Efforts to suppress these dislocations to levels as low as are routinely achieved in epitaxy on lattice-matched substrates, have fallen far short of the goal, and there are strong theoretical arguments against this possibility. However, promising devices are being achieved despite the high dislocation densities, even demanding minority carrier devices such as quantum well lasers. Dislocations threading through narrow quantum wells are evidently far less deleterious than bulk dislocations. The prospects for VLSI HBT circuits are also promising.

1. Introduction

Very impressive progress continues to be made in the performance of GaAs/(Al,Ga)As and other compound semiconductor devices grown on Si substrates, by MBE or OMVPE. Probably nothing illustrates this better than the achievement, by several groups, of cw lasers with non-negligible operating lifetimes [1,2]. Considering that cw lasers are more demanding of crystal quality than any other device, this progress is certainly gratifying. But serious questions remain. Put bluntly, we do not really understand why the material works as well as it does, and in order to reach the limits of its capability, such an understanding will almost certainly be necessary.

There are (at least) two quite different kinds of problems:

(a) The most urgent problem is the misfit threading dislocation problem, caused by the 4% lattice mismatch between Si and GaAs. In the best material grown to date, the dislocation densities are still above 10^6 cm^{-2}, more often in the 10^7 cm^{-2} range. In bulk, such material would be essentially useless. Nevertheless, it is in such material that room-temperature cw lasers with non-negligible operating lifetimes have been achieved.

(b) A second problem is the site allocation problem, that is, the problem of which of the two fcc sublattices of the Si crystal becomes the Ga sublattice, and which becomes the As sublattice [3,4], and how confusion in this site allocation is avoided.

The present paper concentrates on the dislocation problem. But we continue to be puzzled by the site allocation problem. In the early days of GaAs-on-Si technology, it was believed by some – including one of the present writers [5] – that growth on the favored (100) orientation was bound to lead to heavy antiphase disorder. Fortunately, this problem was easily overcome by simple misorienting the Si substrate slightly [6], but it is still not clear why this simple recipe works!. One explanation was that the substrate preparation led to a Si surface in which all terraces belonged to the same sublattice [3,7], but recent data have clearly shown that not even that is necessary [8–10]. The mystery is further deepened by the observation that both of the two possible sublattice allocations are achievable without antiphase disorder, depending on the nucleation conditions [6,8,9,11].

However, even if the site allocation problem

remains a fascinating scientific puzzle, it does not appear to pose a serious problem from the device performance point of view. For this reason, the remainder of this paper concentrates on the dislocation problem.

2. The threading dislocation problem

The lattice constant of GaAs is about 4% larger than that of Si. In epilayers of the kind of thickness required for almost all devices, the mismatch leads to the formation of misfit dislocation at or near the interface, running (ideally) parallel to the interface. A 4% misfit requires a dislocation roughly every 25 atomic rows. The relevant Burgers vector have the $\langle 110 \rangle$ directions, with a magnitude $a/\sqrt{2} \sim 4$ Å, where a is the lattice constant. That is, there must be two orthogonal dislocation networks with a spacing between dislocations in each network of roughly $S = 100$ Å. For a variety of reasons, these dislocations do not stay confined near the vicinity of the interface, but bend upwards into the substrate, where they form threading dislocations (recall that a dislocation cannot end inside the crystal, but only on its surface). If each dislocation remained confined for a length L, the density D of (primary) threading dislocations would be

$$D = 4/SL, \tag{1}$$

where the factor 4 arises from the fact that there are two orthogonal misfit dislocation networks, and each dislocation has two ends. It is not well understood what controls the confinement length, and a discussion of this topic is outside the scope of this paper, but empirically, typical confinement lengths are less than 1 μm. Assuming, for simplicity, $L = 1$ μm, yields $D = 4 \times 10^{10}$ cm^{-2}, a huge density. To achieve a primary threading dislocation density comparable to bulk material, say, 10^4 cm^{-2}, would require confinement lengths of the order 400 cm! Evidently, in order to achieve dislocation densities sufficiently low to make the material useable for devices, the overwhelming majority of the primary dislocation must be annihilated by recombination of pairs of threading dislocations.

To a considerable extent, such annihilation takes place naturally as the epitaxial layer grows thicker. Whenever the surface ends of two threading dislocations with the same Burgers vector approach each other, they may recombine, and that leads to a rapid thinning-out of the dislocations as the epitaxial layer grows thicker. If it were feasible to grow sufficiently thick epitaxial layers, very low dislocation densities would presumably result [12]. Unfortunately, very thick buffer layers are ruled out in the case of GaAs-on-Si by the large difference in thermal expansion coefficient between Si and GaAs, which caused a large tensile strain to be built into the epilayer during cooling from the growth temperature. As a result, GaAs epilayers thicker than about 4 μm tend to exhibit massive cracking. The question then is what dislocation densities can be achieved within about 1–2 μm from the Si interface, leaving about 2–3 μm for the device itself. In the absence of any specific dislocation suppression schemes, but under otherwise "good" growth conditions, one find dislocation densities as low as 10^8 cm^{-2} at such distances, about four orders of magnitude higher than routine values for GaAs-on-GaAs growth.

3. Dislocation reduction schemes

There have been a large number of attempt in recent years to do reduce the threading dislocation densities, by various kinds of buffer layers, especially strained-layer superlattice buffers [13]. Although drastic improvements are often claimed, in the last analysis all these efforts have fallen far short of bridging a gap some four orders of magnitude wide, and they are more remarkable for what they have *failed* to achieve than for what they have achieved: Improvements in dislocation density by a factor of two, compared to doing nothing at all, while hailed as improvements, are simply uninteresting, and even the best results have given not much more than a factor of ten in improvement, leading to dislocation densities around 10^7 cm^{-2}, still several orders of magnitude too high to meet the goal of "bulk-quality" material. This failure of the SLSL buffer layer approach came as a surprise to many investigators (including this writer, who had attempted to use this approach), because SLSL layers had been remarkably effective in suppression of threading dislocations at lower dislocation densities. El Masry et al. [14] have recently proposed that dislocation tangling at high dislocation densities places a natural upper limit to the dislocation densities that can be suppressed by SLSL buffer layers. As a result of statistical fluctuations, these authors found that although there tend to occur regions several μm in diameter that tend to be dislocation-free, the dislocation density outside these islands tends to more than make up for the reduction, leading to undiminshed average dislocation densities. We ourselves have performed work along the same lines as El Masry et al., with essentially the same results [15].

The best (or better: least-bad) results have been achieved by a technique that comes closest to doing nothing spectacular at all: A simple high-temperature anneal of the epilayer [16]; which seems to yield values around 10^7 cm^{-2} reliably. What appears to be the best (believable) result reported in the literature was achieved by a brute-force extension of the thermal anneal technique: Using a sequence of 13 in-situ anneal steps at 800°C alternating with OMVPE growth, Itoh et al. [17] were able to reduce the dislocation density to between 2×10^6 and 5×10^6 cm^{-2}.

Barring any major unexpected breakthroughs, it is hard to see how any of these approaches can lead to material of "bulk-quality". Perhaps the only major hope that is left is to work deliberately with small islands of material. Initial results in this direction, in the (Ga,In)As/GaAs system [18] are promising, but it remains to be seen how successful this approach will ultimately be.

4. Model for annihilation kinetics of threading dislocations

In order to understand better the fundamental reasons for the failure to achieve the desired dislocation suppression, it is instructive to model this problem mathematically.

Let $D(x)$ be the areal density (number per unit area) of threading dislocations. The rate at which dislocations disappear by recombination will then be proportional to the square of their concentration, as for other binary recombination laws,

$$\mathrm{d}D/\mathrm{d}x = -\lambda D^2, \tag{2}$$

where λ is an unknown proportionality factor of the dimension of a length.

Integration of (2) yields

$$D(x) = \frac{D(0)}{1 + D(0)\lambda x} \rightarrow \frac{1}{\lambda x}. \tag{3}$$

where $D(0)$ is the primary threading dislocation density, and where the asymptotic limit refers to the situation sufficiently far from the interface that the majority of dislocations has recombined. Empirically, one finds between 10^7 and 10^8 dislocations per cm^2 at a distance of about 1 μm, implying a value of λ between 10^{-3} and 10^{-4} cm.

The asymptotic behavior in (3) has a number of important consequences:

(a) The first of these is that the asymptotic density at a given distance is independent of the initial density. This means that any reduction in the primary threading dislocation density, by whatever means, will not lead to a proportional reduction in the asymptotic density, unless the *primary* dislocation density is reduced all the way to the desired final value! It also explains, for example, why the quality of, say, GaSb/AlSb structures grown on Si is no poorer than that of GaAs/AlAs structures, despite a much larger lattice mismatch.

(b) The second consequence is that the decay of the dislocation density is not exponential, but only inverse linear. What this means that any initial reduction of the dislocation density by a large factor, achieved by the incorporation of a suitable buffer layer, will not lead to a further reduction by the same factor by simple doubling the buffer layer.

To achieve the desired reduction in dislocation density by the desired factor of 10^3 or more clearly then calls for an increase in the capture crosssection and hence an increase in the characteristic length λ by the same factor. It is hard to see how this can be achieved.

5. Dislocations in quantum wells

As stated above, the best material to date still has threading dislocation densities in the range between 10^6 and $10^7 \mathrm{cm}^{-2}$, or more, as established by transmission electron microscopy. Nevertheless, room-temperature cw quantum well lasers have been demonstrated in such material. On the other hand, no room-temperature cw operation has been reported for "ordinary" double heterostructure lasers made from such material. This suggests that superlattices or quantum wells might drastically modify the electronic properties of dislocations in such a way as to make them far less deleterious.

There are other observations pointing in the same direction. For example, in our own extensive earlier work on the photoluminescence properties of GaSb/AlSb multi-quantum well superlattices grown on grossly lattice-mismatched GaAs substrates [19], we had found excellent photoluminescence properties, initially suggesting that the dislocations had been suppressed by the first-grown portion of the superlattice, adjacent to the GaAs substrate. However, TEM studies subsequently showed that no such suppression had taken place [20], and that the threading dislocation density was in fact very high, about $10^8 \mathrm{cm}^{-2}$.

Prompted by such considerations, we have performed a series of experiments designed to look at the luminescence properties of heavily dislocated epitaxial layers of GaAs grown by MBE on Si substrates, each layer containing three superlattice "packets" with different superlattice periods [21].

Three samples were grown by MBE, all at 630°C, all on (211)-oriented substrates. Each contained three superlattice (SL) "packets", labeled SL#1, #2 and #3, separated from each other by 250 Å of $Al_{0.3}Ga_{0.7}As$ barriers. Each of the three SL packets contained the same amount of GaAs and $Al_{0.3}Ga_{0.7}As$. In SL#1, the 3000 Å packet was partitioned into 15 periods of 100 Å of GaAs and 100 Å of $Al_{0.3}Ga_{0.7}As$. In SL#2, the same amount was partitioned into 30 periods of 50 Å + 50 Å. Finally, SL#3 consisted of 60 periods of 25 Å + 25 Å. Samples #1 and #2 were both grown on Si substrates; the only (intentional) difference being the order of the three superlattice

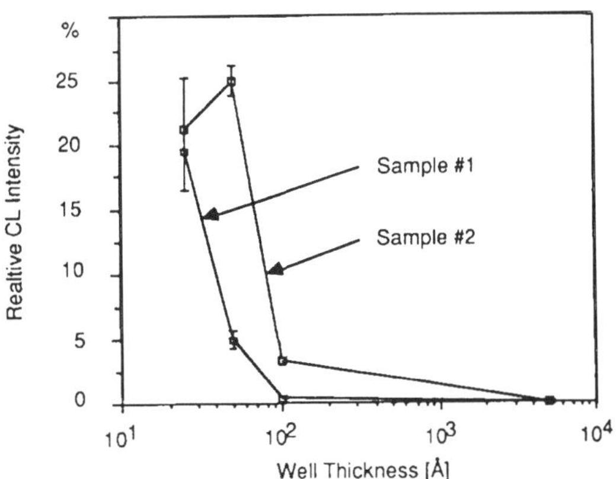

Fig. 1. Cathodoluminescence intensity of GaAs/(Al,Ga)As superlattices with various well widths, grown on Si substrates, relative to identical superlattices grown on GaAs substrates.

stacks: In sample #1, the 100 Å + 100 Å SL was closest to the substrate and the 25 Å + 25 Å SL was farthest away; in sample #2 the stacking order was reversed. Sample #3 was a control sample, with the same stacking order and growth parameters as sample #1, but on a GaAs rather than a Si substrate. An "ordinary" double heterostructure sample was also prepared, consisting of a thick 5000 Å GaAs region sandwiched between two 2500 Å $Al_{0.3}Ga_{0.7}As$ barrier regions, on a Si substrate.

Low-temperature cathodoluminescence (CL) measurements were used to assess the quality of the different SL packets. To eliminate any uncertainties due to the fact that photons from the three SL packets have different escape depths, the cathodoluminescence was also monitored from the substrate side at least of samples #1 and #2, after removing the Si substrates with a KOH etch until the epitaxial GaAs layer was exposed.

Fig. 1 illustrates the *normalized* CL peak intensities of sample #1 and #2 as a function of SL period. The normalization is done by simply taking the ratio of the *heights* of the line peaks in samples #1 and #2 to those in the control sample #3. No attempt was made to integrate over the broadened lines. The error bars indicate the spread of data due to (a) spatial variations of relative intensity on the same side, and (b) differences in relative intensity on different sides of the same sample. The trend speaks for itself: Although the luminescence of the 100 Å + 100 Å SL remains poor, that of the 25 Å + 25 Å SL has recovered to about 20% of the intensity of the corresponding SL grown on a GaAs substrate, despite the fact that cross-sectional TEM data indicate that all three SL packets have dislocation densities in the mid-10^7 cm^{-2} range, not decreasing significantly from the bottom SL packet to the top packet. The additional data point at 5000 Å corresponds to the intensity from the "ordinary" double hetero-structure sample, showing the very poor luminescence for that sample.

Clearly, the luminescence intensity increases very strongly as the well width decreases, for essentially fixed dislocation densities, indicating a drastic reduction in the undesirable recombination efficiency of the dislocations in narrow quantum wells.

The most likely explanation is that the re-combination centers are associated with deep levels due to unreconstructed dangling bonds along the dislocation cores. It was shown by Kimerling and Patel [22] that in silicon only a small fraction of the sites at the dislocation cores are electrically active. Defect site spacings of the order of 200 Å along the dislocation core were estimated. In zincblende semiconductors, the situation is expected to be similar. Kimerling and Patel proposed that the electrically active sites are located at the dislocation kink sites that are associated with dislocation motion within their glide plane. These kinks are known to be highly mobile along the dislocation core, and their motion is easily thermally activated at the growth temperature.

If one accepts the kink model of recombination at dislocations, the simplest explanation of our observations would be that in sufficiently narrow-well superlattices the overwhelming majority of dislocation kinks are somehow expelled from the GaAs portion of the SL structures into the (Al,Ga)As barriers. Conceivably, the strong electric fields that are present at hetero-interfaces, or chemical potential gradients, might play a role. For a discussion of alternate possible mechanisms, considered less likely, the reader is referred to our original paper [21].

Whatever the exact mechanism, we believe that our our results explain the surprisingly good performance of GaAs/(Al,Ga)As quantum well lasers grown on Si substrates, as opposed to "ordinary" double heterostructure lasers. Going beyond that, our results strongly suggest that minority carrier devices in GaAs on silicon substrates should quite generally exhibit a performance approaching that of devices grown on GaAs substrates *whenever* the active part of the devices is can be constructed from narrow GaAs quantum wells, or narrow-well superlattices.

6. Dislocations in heterostructure bipolar transistors

The situation in heterostructure bipolar transistors (HBTs) is quite different than in quantum well lasers, because the incorporation of superlattices in the base region is not practical. However, we come to a similarly positive assessment, for quite different reasons. In fact, HBTs with remarkably good properties made from (Al,Ga)As/GaAs on Si substrates have already been demonstrated [23], despite the fact that the epi-layers of GaAs on Si typically probably had threading dislocation densities exceeding 10^7 cm^{-2}.

The central issue in an HBT is the competition between minority carrier capture by a dislocation and capture by the collector depletion layer: The worst a dislocation can do in the base of an HBT is to act as a perfect sink for minority carriers. But if the nearest dislocation is farther away from an injected carrier than the collector depletion layer, that carrier is more likely to be collected by the collector than to be captured by a dislocation. In many HBTs there is a built-in field driving the carriers towards the collector [24]; this would reduce the capture by dislocations further. Even neglecting such fields, but assuming that dislocation lines run perpendicular to the base plane, one estimates a capture cross section of about πw^2, where w is the base width. If D is the dislocation density, then, for uniform carrier injection by the emitter, one estimates that the fraction

$$f_r = \pi D w^2 \tag{4}$$

of the injected carriers is captured by dislocations. Assuming $D = 10^7$ cm^{-2} (as a typical value for GaAs on Si) and $w = 10^{-5}$ cm (as a typical value for HBTs), one predicts a surprisingly low capture fraction of about 0.3%, essentially negligible – at least for discrete transistors with a large enough area $A(A \gg 1/D)$ that there are several dislocations present per device.

The argument becomes more complicated for VLSI circuits with very small devices. Emitter areas A as small as $(1 \ \mu\text{m})^2 = 10^{-8}$ cm^2 can be anticipated. For dislocation densities $D \leq 10^7$ cm^{-2}, over 90% of the transistors would have no dislocations at all, however, the remaining transistors would be influenced proportionately more strongly: Transistors threaded by a single dislocations would lose 3% of the injected carriers rather than only 0.3%. This is still an essentially negligible loss, but in a sufficiently large VLSI circuit, there will always be some transistors threaded by several dislocations, with proportionately larger losses. For discrete devices, a very small fraction of transistors with unacceptably high losses would represent an inconsequentially small reduction of the manufacturing yield, but in a VLSI circuit they would jeopardize the entire circuit.

To the first order, one might expect that the threading dislocations are randomly ($=$ Poisson) distributed, implying a probability

$$P(N) = \frac{\langle N \rangle^N}{N!} e^{-\langle N \rangle} \tag{5}$$

that a transistor will be threaded by N dislocations, if $\langle N \rangle$ is the average number of dislocations per transistor. Suppose that we have $\langle N \rangle = 0.1$, and that the circuit becomes inoperative when a single transistor is threaded by four dislocations (a current loss $> 10\%$). From (4), for $\langle N \rangle = AD = 0.1$, we find $P(4) \sim 4 \times 10^{-6}$. Evidently, under our assumptions, serious yield problems would arise only for circuits with more than about 10^5 transistors. Inasmuch as circuits of this size should be anticipated, the problem is not negligible. But is is also evident that even a small reduction in dislocation density, or an increased tolerance of the individual device to dislocations, would all but alleviate the remaining threat.

Acknowledgements

This work was supported by the US Army Research Office.

References

[1] D.W. Nam, N. Holonyak, K.C. Hsieh, R.W. Kaliski, J.W. Lee, H. Shichijo, J.E. Epler, R.D. Burnham and T.L. Paoli, Appl. Phys. Letters 51 (1987) 39.

[2] H.Z. Chen, A. Ghaffari, H. Wang, H. Morkoç and A. Yariv, Appl. Phys. Letters 51 (1987) 1320.

[3] H. Kroemer, J. Crystal Growth 81 (1987) 193.

[4] H. Kroemer, in: Proc. 14th Intern. Symp. on GaAs and Related Compounds, Heraklion, Crete, 1987, Inst. Phys. Conf. Ser. 91, Eds. A. Christou and H.S. Rupprecht (Inst. Phys., London–Bristol, 1988) p. 21.

[5] For a 1986 review of this topic, see H. Kroemer, in: Heteroepitaxy on Silicon, Eds. J.C.C. Fan and J.M. Poate, Materials Research Society Symposia Proceedings 67 (Mater. Res. Soc., Pittsburgh, PA, 1986) p. 3.

[6] R.J. Fischer, N.C. Chand, W.F. Kopp, H. Morkoç, L.P. Erickson and R. Youngman, Appl. Phys. Letters 47 (1985) 397;
see also R.J. Fischer, H. Morkoç, D.A. Neumann, N. Otsuka, M. Longerbone and L.P. Erickson, J. Appl. Phys. 60 (1986) 1640.

[7] D.E. Aspnes and J. Ihm, Phys. Rev. Letters 57 (1986) 3054.

[8] P.R. Pukite and P.I. Cohen, J. Crystal Growth 81 (1987) 214.

[9] P.R. Pukite and P.I. Cohen, Appl. Phys. Letters 50 (1987) 1739.

[10] K. Kawabe and T. Ueda, Japan. J. Appl. Phys. 26 (1987) L944.

[11] K. Kawabe, T. Ueda and H. Takasugi, Japan. J. Appl. Phys. 26 (1987) L114.

[12] See, for example, G.H. Olsen, J. Crystal Growth 31 (1975) 223.

[13] The literature on this topic is extensive; see, for example, the numerous papers in the two volumes Heteroepitaxy on Silicon, of the Materials Research Society Proceedings: Vol. 67, Eds. J.C.C. Fan and J.M. Poate (1986) and Vol. 91, Eds. J.C.C. Fan, J.M. Phillips and B.-Y. Tsaur (1987).

[14] N. El-Masry, J.C.L. Tarn, T.P. Humphreys, N. Hamaguchi, N.H. Karam and S.M. Bedair, Appl. Phys. Letters 51 (1987) 1608.

[15] T.Y. Liu, H. Kroemer and Z. Liliental-Weber, unpublished.

[16] J.W. Lee, H. Shichijo, H.L. Tsai and R.J. Matyi, Appl. Phys. Letters 50 (1987) 31.

[17] Y. Itoh, T. Nishioka, A. Yamamoto and M. Yamaguchi, Appl. Phys. Letters 52 (1988) 1617.

[18] E.A. Fitzgerald, P.D. Kirchner, R. Proano, .D. Petit, J.M. Woodall and D.G. Ast, Appl. Phys. Letters 52 (988) 1496.

[19] G. Griffiths, K. Mohammed, S. Subbana, H. Kroemer and J.L. Merz, Appl. Phys. Letters 43 (1983) 1059.

[20] J.S. Ahearn and P. Uppal, personal communication.

[21] T.Y. Liu, P.M. Petroff and H. Kroemer, to be published.

[22] L.C. Kimerling and J.R. Patel, in: VLSI Electronics, Vol. 12, Ed. N.G. Einspruch (Academic Press, Orlando, FL, 1985) p. 223.

[23] R. Fischer, J. Klem, C.K. Peng, J.S. Gedymin and H. Morkoç, IEEE Electron Device Letters EDL-7 (1986) 112.

[24] See, for example, H. Kroemer, J. Vacuum Sci. Technol. B1 (1983) 126.

Reprinted from

H. Kroemer, C. Nguyen, and E. L. Hu, ``Electronic Interactions at Superconductor-Semiconductor Interfaces,''
Solid State Electron., Vol. 37(4-6), pp. 1021-1025, 1994.

Pergamon

Solid-State Electronics Vol. 37, Nos 4–6, pp. 1021–1025, 1994
Copyright © 1994 Elsevier Science Ltd
Printed in Great Britain. All rights reserved
0038-1101/94 $6.00 + 0.00

ELECTRONIC INTERACTIONS AT SUPERCONDUCTOR–SEMICONDUCTOR INTERFACES

HERBERT KROEMER, CHANH NGUYEN and EVELYN L. HU

Department of Electrical and Computer Engineering, University of California, Santa Barbara,
CA 93106, U.S.A.

Abstract—Two current flow mechanisms across a superconductor–semiconductor–superconductor double heterostructure are discussed: the conventional proximity effect, and Andreev reflections. The emphasis is on Nb–InAs–Nb structures, with the InAs being in the form of a quantum well with AlSb barriers, for which current flow by multiple Andreev reflections can lead to an enhancement of the zero bias conductance by a large factor. For sufficiently short inter-electrode spacings, the multiple Andreev reflections can lead to a true supercurrent flow.

1. INTRODUCTION

When a superconductor and a semiconductor are brought together into atomically intimate contact, with an interface that is free from intervening oxides and/or contaminants, and which does not form an electron-blocking Schottky barrier, the electrons in the two materials can interact with each other in ways that can drastically alter the current flow through what may be called "Super–semi–super double heterostructures". An example of a particularly suitable structure for the observation of such interaction effects is shown in Fig. 1[1–3]. It consists of a thin layer of InAs, in which the electrons are confined at top and bottom by AlSb barriers, forming a two-dimensional electron gas. This gas is then contacted by superconducting Nb electrodes.

One of the reasons for the use of InAs is that the Fermi level at metal–InAs contacts tends to be pinned inside the InAs conduction band, thus leading to an absence of Schottky barriers impeding the flow of electrons. As a result, such structures behave like pure resistors above the critical temperature of the Nb electrodes (9.2 K), and any new effects due to super–semi interactions are especially pronounced, unencumbered by non-superconducting complications. The reason for singling out a quantum well over a bulk structure is to achieve high electron concentrations by modulation doping while retaining high mobilities[4], and to suppress mobility reductions due to surface scattering, a problem especially severe with InAs, because of the absence of surface band bending. Typical sample parameters are: Well width of 15 nm, a channel length ranging from sub-μm dimensions to several μm, and an electron sheet concentration of several-times $10^{12}\,\mathrm{cm}^{-2}$.

The super–semi interaction effects in such structures are pronounced. Figure 2 shows the 4.2 K differential conductance of a structure as in Fig. 1, as a function of bias voltage[3]. The device shows a very narrow conductance spike around zero bias, inside which the conductance is enhanced by a factor 7 relative to the conductance just above the critical temperature of Nb (9.2 K). With increasing bias the conductance decreases, but shows a rich structure up to bias voltages equivalent to the superconducting gap of Nb ($\approx 3.2\,\mathrm{mV}$). These phenomena disappear when the Nb electrodes "go normal".

The structure and the behavior shown are by no means the only manifestation of super–semi interactions, nor are advanced quantum well structures necessary for all such observations. A variety of interaction phenomena have been observed in a variety of structures, employing a variety of semiconductors, including GaAs[5], (Ga,In)As[6], and Si[7]. Complete references can be found in the papers cited.

2. PROXIMITY EFFECT, WEAK LINKS AND JOSEPHSON FETs

There are two distinct basic forms of super-semi interactions: the well-known *Proximity Effect*, and the less-well-known, but perhaps more important *Andreev Reflections*.

In the conventional proximity effect, the Cooper pairs that are the carriers of supercurrent inside the superconductor, can tunnel into the a normal conductor, causing induced superconductivity there, falling off exponentially with distance, with a characteristic length called the *coherence length*. If the separation between the superconducting electrodes is sufficiently small—typically of sub-μm dimensions—this can lead to what is called a *weak link*, a structure capable of carrying a true resistance-less supercurrent through the semiconductor.

In 1980, Clark *et al.*[8] drew attention to the promise of semiconductors rather than conventional metals as the non-superconductor in proximity effect studies. They proposed a Hybrid Josephson FET (=JOFET), basically a weak link the critical current of which can be modulated, leading to a current–voltage characteristic resembling that of a field

1022   HERBERT KROEMER *et al.*

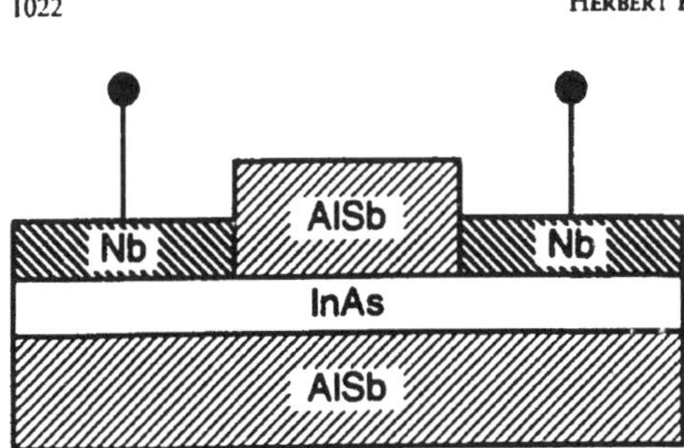

Fig. 1. Schematic InAs–AlSb quantum well structure with superconducting Nb electrodes, for the investigation of electron–electron interaction effects across a super–semi interface.

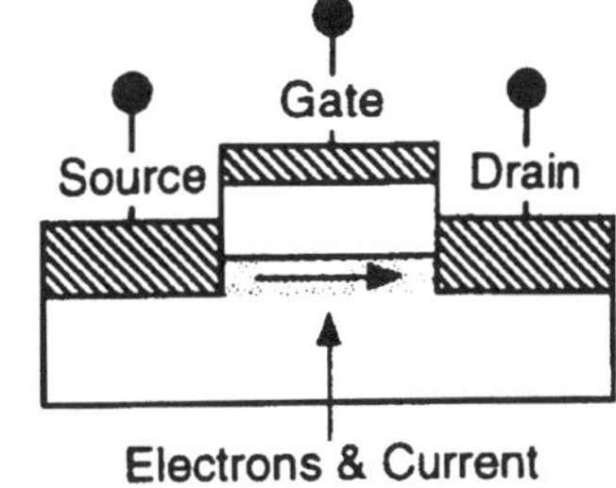

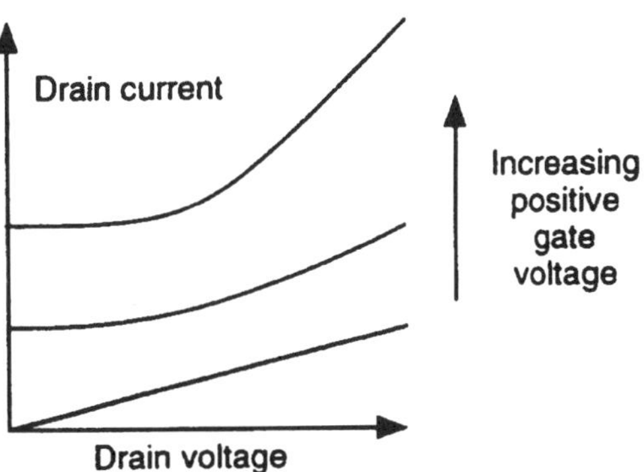

Fig. 3. Schematic JOFET structure and its I–V characteristics.

effect transistor, except for a very different voltage scale, in the low-mV range, and of course a very different physics. The central idea was that the critical current that can be passed through a weak link employing the proximity effect depends strongly on the superconductive coherence length inside the semiconductor, which in turn depends on the electron concentration in the semiconductor, which can be modulated with a gate electrode. The overall result would be a current–voltage characteristic as shown schematically in Fig. 3.

What distinguishes JOFETs from conventional FETs are not only the much lower voltage (and current) scales, but the existence of a true zero-resistance on-state. This makes a JOFET a device of potential interest as a current routing switch in superconducting networks. There is some doubt as to whether such JOFETs would ultimately be useful as

amplifiers or logic gates: the gate voltage swings required for current modulation tend to be larger than the drain voltage swings obtainable from the current modulation.

Clark *et al.* pointed out that InAs appeared to be the ideal semiconductor for such studies, not only because of the absence of Schottky barriers at metal-to-InAs interfaces, but also because of its unusually high electron mobilities, which in turn reflect the low effective mass of electrons in InAs. Because of this low effective mass, heavily n-type doped InAs has a Fermi velocity approaching that of many true metals, and as a result, InAs in contact with a superconductor behaves more like a high-mobility metal than like a semiconductor. In particular, large coherence lengths should be achievable.

Weak links and JOFETs employing a Nb–InAs–Nb structure were subsequently demonstrated, by Takayanagi *et al.*[9,10], followed by others. However, the current–voltage characteristics of those early structures were relatively poor, and JOFETs with much better characteristics were obtained in GaAs and even Si[7], despite the theoretical superiority of InAs. Perhaps the most interesting of those early JOFET structures was that of Ivanov *et al.*[5], which appears to have been the first to employ a quantum well channel [GaAs–(Al,Ga)As] in a weak link or JOFET, demonstrating the superiority of such a design.

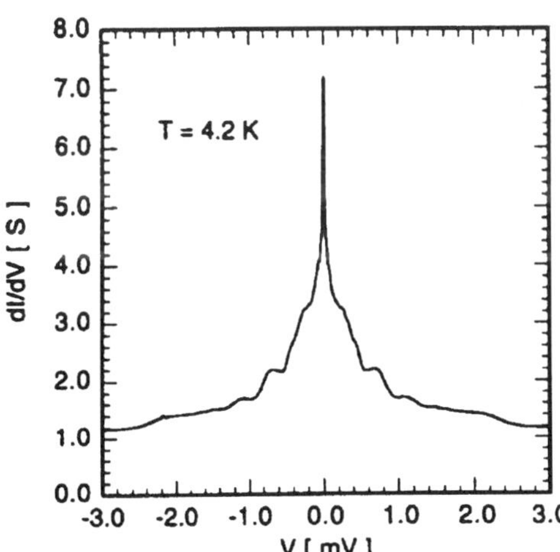

Fig. 2. Very strong enhancement at zero bias of the differential conductance of a recent InAs–AlSb quantum well structure with Nb electrodes[3], of the type shown in Fig. 1, with a 1 μm electrode separation. The rich structure shown on the flanks of the central identifies the conductance peak as due to multiple Andreev reflections (see text).

Following the development of a technology for high-quality InAs–AlSb quantum wells during the 1980s, we ourselves turned to the problem of InAs weak links, and the balance of this paper deals with that work. In 1990 we were able to demonstrate weak links showing unprecedentedly high critical current densities above 2×10^5 A/cm^2, for a remarkable large inter-electrode spacings of 0.6 μm[1]. We were naturally interpreting these results as caused by the conventional proximity effect. More recent observations challenge this interpretation, and suggest a different superconductivity mechanism in terms of multiple Andreev Reflections, our next topic.

3. ANDREEV REFLECTIONS

Consider a semi–super interface between a degenerately doped semiconductor and a superconductor, with a band diagram as shown in Fig. 4. On the superconductor side, a superconducting energy gap has opened up. If now a small bias voltage V is applied, as shown, the existence of the gap then prevents a *single* electron at the Fermi level of the semiconductor from entering the superconductor. This argument suggests that, in the absence of the proximity effect, the onset of superconductivity in the metal thus actually *increases* the electrical resistance to current flow across the interface, due to this gap formation. However, even a single electron may pair up with a second electron at the bias energy qV *below* the Fermi level, forming a Cooper pair, which *can* enter the superconductor, causing a doubling of the current compared to that in the absence of superconductivity, rather than the reduction that would occur in the absence of this pair formation. As the electron below the Fermi level is removed from the semiconductor, it leaves behind a hole below the surface of the Fermi sea. The generally accepted jargon associated with this phenomenon is to say that the incident electron is *reflected as a hole*, a kind of reflection process called an *Andreev reflection*, honoring the originator of the concept[11]. The *Andreev hole* left behind, being a "bubble" under the surface of the

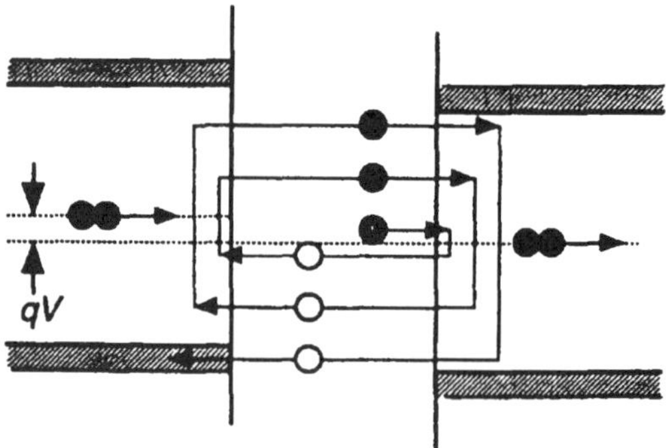

Fig. 5. Multiple Andreev reflections (AR) alternating between the super–semi interfaces at opposite ends of the semiconductor region.

Fermi sea in the conduction band, must not be confused with a valence band hole.

In a semiconductor with a large mean free path for the electrons ($\approx 3 \mu$m in our structures), the Andreev hole left behind at the interface has a large mean free path itself, roughly equal to that of the electrons, and theory shows that the hole travels back into the semiconductor along a trajectory that is essentially the time reversal of the trajectory of the original incident electron. If its mean free path is sufficiently large, the hole will eventually reach the negative superconducting electrode. If the bias across the structure is sufficiently small, the energy of the hole is still within the superconducting gap on that side. Such a hole cannot enter the superconductor, but it can be annihilated by breaking up a Cooper pair inside the adjacent superconductor: one of the electrons of the pair annihilates the hole, the other electron takes up the annihilation energy, and is injected into the semiconductor as a ballistic electron above the Fermi level, at an energy above that of the initial electron. This process, illustrated in Fig. 5, can evidently be repeated, until either an electron or a hole has been "pumped up" to an energy outside the superconducting gap, on one of the two sides of the structure. If all reflections of electrons and holes were Andreev reflections rather than "ordinary" reflections, the result would be an enhancement of the conductivity by a factor equal to the number of ballistic round trips before escape or before collision events randomize either the electron or the hole flow in this chain reaction. As a rule, the conductance enhancement in past structures has been much smaller, presumably due to a low AR probability, itself caused by strong normal reflections due to residual potential barriers at the interfaces.

One of the "fingerprints" of multiple ARs is a rich "sub-harmonic gap structure" in the conductance-vs-voltage characteristic, with steps occurring at voltages equal to the integer fractions of the superconducting gap voltage[12–14]. A discussion of this structure lies outside the scope of the present paper,

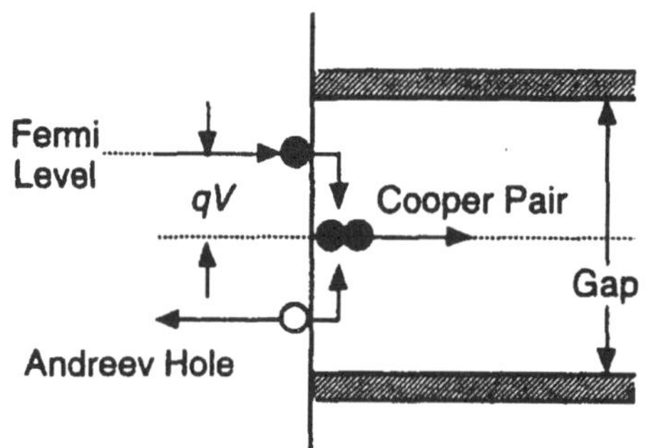

Fig. 4. Andreev reflection (AR) of an electron at a biased super–semi interface.

but the occurrence of such a structure is evident in the characteristics of Fig. 2, thus clearly indicating the multiple-AR origin of the conductance peak. What is new compared to earlier data reported in the literature is the huge enhancement in the differential conductance, by a factor of 7 in the example of Fig. 2. The behavior appear superficially as if the proximity effect were present. However, we will show below that contact resistance measurements rule out such an explanation.

4. ANDREEV-REFLECTION-INDUCED SUPERCONDUCTIVITY

The large conductance enhancement suggests that it might be instructive to carry the above multiple-AR argument to its extreme limit, the case of zero applied bias, and assuming that *all* reflection events at the super–semi interfaces are AR events, and that *no* scattering of any kind *inside* the semiconductor randomizes the electron and hole velocities. In this case, a given AR "chain" would go on forever. During each electron–hole round trip, one Cooper pair is annihilated at one of the electrodes, and re-constituted at the other electrode on the opposite side, leading to the net transfer of one Cooper pair per round trip. Given an initial net current, this current would persist, just as in the proximity effect, but by an altogether different mechanism.

These are extreme assumptions, especially the assumption of a 100% AR probability, yet the final conclusion appears to be correct. The quantum mechanics of this hypothetical multiple-AR mechanism has recently been analyzed in detail by Schüssler and Kümmel (SK)[15], using a model assuming the existence of a *definite fixed phase difference* between the pair potentials in the two superconducting electrodes, and neglecting scattering in the semiconductor channel, but not assuming a 100% AR probability. The authors showed that under their conditions the multiple Andreev Reflections of phase-conjugate ballistic quasi-particles (i.e. electrons and Andreev holes) form indeed a very effective mechanism for Cooper pair transfer between the electrodes, capable of carrying a much higher zero-resistance current densities than the conventional proximity effect.

We believe that the narrow central conductance spike shown in Fig. 2, with the up-to-sevenfold enhancement of the differential conductance, is a *precursor* of the true Andreev-caused supercurrent postulated above, and analyzed by SK. We have to call it a precursor, because our data indicate a still-finite conductance, occurring over a narrow but nonzero voltage range ($\approx 50\,\mu$V). Presumably, the finite height and width of the central conductance spike is the result of residual scattering events present in the relatively long (1 μm) InAs-AlSb QW channel, eventually randomizing the quasiparticle velocities. Furthermore, we believe that the true superconducting limit can indeed be achieved in Nb–InAs–Nb

quantum well structures with a shorter inter-electrode spacing.

In their work, SK *assume* that there is a fixed phase relation between the pair wave functions in the two superconducting electrodes, and analyze the consequences. They do not address the question of how such a phase relation, and with it any supercurrent, might be maintained in the presence of scattering in the semiconductor channel. In the absence of such scattering, the assumption of a fixed phase relation between the pair wave functions in the two superconducting electrodes is entirely self-consistent. On the other hand, in the presence of sufficiently strong scattering, as in the case of a sufficiently wide inter-electrode spacing, any current not driven by an external voltage must eventually decay. This raises the question as to the nature of the transition to the SK superconducting limit, as the scattering in the semiconductor channel is reduced, by reducing the temperature and/or the inter-electrode spacing: will the zero-bias resistance of the overall structure drop towards zero continuously, without ever reaching the true superconducting limit? Or will collective effects cause a "condensation" of the Andreev pairs into a new correlated many-body state, in which the dephasing effects of scattering are quenched, similar to the way the BCS transition quenches the ordinary resistivity in a BCS superconductor?

We believe that the latter is indeed the case, and that our earlier observation of very large weak link current densities in structures with 0.6 μm electrode spacing was indeed a manifestation of such a mechanism. To pursue this idea further, we have utilized laser holography to prepare what is essentially a grating of ≈ 300 parallel Nb lines making periodic contact to an InAs quantum well with AlSb barriers, with a 1 μm period and a $\approx 0.4\,\mu$m spacing between the Nb lines. The rest of the technology was basically the same as in the structure whose data were shown in Fig. 2. In the direction perpendicular to the grating lines, the structure acts basically as a series-connection of 300 diodes of the type shown in Fig. 1. At 4.2 K, this structure showed a characteristic qualitatively similar to that of Fig. 2, with the "Andreev fingerprint" of sub-gap harmonics, but with a ≈ 300-fold enhanced voltage scale. More importantly, the conductance enhancement was by a factor 75, presumably as a result of the shorter inter-electrode spacing. With decreasing temperature, the zero-bias resistance dropped further, reaching an immeasurably low value between 3.9 and 3.8 K.

This grating structure was still being evaluated at the time of the deadline for this manuscript; up-to-date results will be presented at the conference.

5. THE CONTACT RESISTANCE PROBLEM

Our interpretation of the conductance enhancement in terms of multiple ARs rather than as a precursor of the ordinary proximity effect is sup-

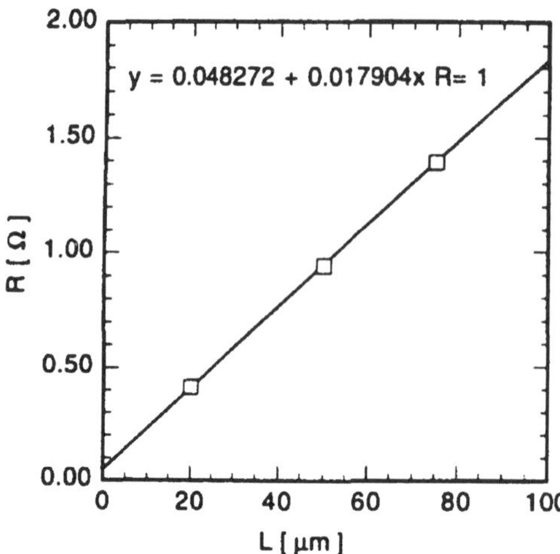

Fig. 6. Zero-bias differential resistance $R = dV/dI$ at 4.2 K of a set of Nb–InAs(QW)–Nb structures with different inter-electrode spacings, plotted as function of spacing. The straight line is a linear fit through the data, including a point at $L = 200\,\mu$m, not shown on the plot. The intercept value, representing twice the line contact resistance, is positive.

ported by measurements of the specific contact resistance at the Nb–InAs interface, using the conventional *transmission line method* widely used in semiconductor technology[16]. The latter consists of measuring the set of voltage drops across a monolithic array of metal contacts to a thin semiconductor layer, with various lithographic inter-contact spacings L, and fitting the measured voltages and their current derivatives to an expression of the form:

$$\frac{dV}{dI} = 2\frac{dV_c}{dI} + \rho_s \cdot \frac{L}{w}. \tag{1}$$

Here I is the current through the array, w is the width of the array, and ρ_s is the ordinary sheet resistance of the semiconductor layer in the limit that L and w are large compared to the electron mean free path. In eqn (1), the length-proportional term represents the "ordinary" path resistance of a semiconductor path of length L, and $2dV_c/dI$ represents the effects of whatever additional voltage drops are present at or near the two contacts. The latter include the true contact resistances at the two interfaces, plus any deviations from *bulk* behavior inside the semiconductor near the electrodes, for example due any proximity effect. If the latter is present, a d.c. current across the super-semi interface would be carried entirely by

Cooper pairs. We would then expect the true contact resistance associated with the Nb–InAs interface to be zero, and the resistive portion of the semiconductor path to be shortened below the lithographic length, leading to a *negative* value of the intercept voltage $2V_c$ and to a negative *apparent* contact resistance, represented by the leading term in eqn (1). Our measurements, shown in Fig. 6, indicate that the apparent contact resistance remains positive, thus ruling out the proximity effect as an explanation of the zero-bias conductance spike.

Acknowledgements—This work was supported in part by the Office of Naval Research and in part by the National Science Foundation, the latter through the NSF Science and Technology Center for Quantized Electronic Structures, grant no. DMR 91-20007, as well as through the NSF Materials Research Laboratory Program, Award no. DMR 912-3048. One of us (C.N.) wishes to acknowledge the financial support from the UCSB Vice Chancellor's Fellowship for Advanced Research on Quantized Structures.

REFERENCES

1. C. Nguyen, J. Werking, H. Kroemer and E. L. Hu, *Appl. Phys. Lett.* **57**, 87 (1990).
2. C. Nguyen, H. Kroemer and E. L. Hu, *Phys. Rev. Lett.* **69**, 2847 (1992).
3. C. Nguyen, H. Kroemer and E. L. Hu, to be published.
4. C. Nguyen, B. Brar, C. B. Bolognesi, J. J. Pekarik, H. Kroemer and J. H. English, *J. Electron. Mater.* **22**, 255 (1993).
5. Z. Ivanov, T. Claeson and T. Andersson, *Japan. J. appl. Phys.* **26(Suppl. 3)**, DP31 (1987). (Proc. 18th Int. Conf. Low Temperature Physics, Kyoto, 1987).
6. A. Kastalsky, A. W. Kleinasser, L. H. Greene, R. Bhat, F. P. Milliken and J. P. Harbison, *Phys. Rev. Lett.* **67**, 1326 (1991).
7. T. Nishino, M. Hatano, H. Hasegawa, F. Murai, T. Kure, A. Hiraiwa, K. Yagi and U. Kawabe, *IEEE Electron. Device. Lett.* **10**, 61 (1989).
8. T. D. Clark, R. J. Prance and A. D. C. Grassie *J. appl. Phys.* **51**, 2736 (1980).
9. H. Takayanagi and T. Kawakami, *Phys. Rev. Lett.* **54**, 2449 (1985).
10. H. Takayanagi and T. Kawakami, *Proc. Int. Electron Devices Meeting*, p. 98 (1985).
11. A. F. Andreev, *Sov. Phys. JEPT* **19**, 1228 (1964).
12. T. M. Klapwijk, G. E. Blonder and M. Tinkham, *Physica B + C* **109 & 110**, 1657 (1982).
13. M. Octavio, M. Tinkham. G. E. Blonder and T. M. Klapwijk, *Phys. Rev. B* **27**, 6739 (1983).
14. K. Flensberg, J. B. Hansen and M. Octavio, *Phys. Rev. B* **38**, 8707 (1988).
15. U. Schüssler and R. Kümmel, *Phys. Rev. B* **47**, 2754 (1993).
16. R. E. Williams, *Gallium Arsenide Processing Techniques*. Artech House, Dedham, Mass. (1984).

``I did not intend to invent compact disc players.''

Herbert Kroemer

Reprinted with permission from

H. Kroemer, ``Superconductor-Semiconductor Devices,'' NATO Adv. Res.
Workshop Future Trends in Microelectronics: Reflections on the Road to
Nanotechnology, Ile de Bendor, France, S. Luryi, J. Xu, and A. Zaslavsky,
Eds., NATO ASI Series; Series E: Applied Sciences, Vol. 323, Kluwer
Academic Publishers, pp. 237-250, 1996.

With kind permission of Springer Science and Business Media.

SUPERCONDUCTOR-SEMICONDUCTOR DEVICES

HERBERT KROEMER

*ECE Department, University of California
Santa Barbara, CA 93106, USA*

1. Introduction

1.1 THE PREMISE

It has long been recognized that electronic devices operating at reduced temperatures—including both semiconductor and superconductor devices—can often offer much higher performance (by several criteria) than room-temperature devices. But the need for cooling has greatly retarded their use, and there exists an almost-universal persistent belief that low-temperature devices just don't have a chance to find significant practical applications.

My presentation is based on the premise that this belief is a myth, and that the future of electronics is likely to draw increasingly, within the next decade or two, on low-temperature devices, at least in applications such as high-performance workstations and scientific and medical instrumentation, where increasing performance requirements can justify the additional cost of the cryogenics, which is itself decreasing

However, the performance-to-cost relation is by no means the only issue: No matter how favorable that relation is, no system engineer is going to fool around in a "real" commercial system with cryogenics under conditions that resemble those of a research laboratory. What is absolutely essential is "user-friendly" cryogenics! The enabling technology for the widespread actual use of cryogenic electronics is likely to be the increasing availability of small self-contained closed-cycle refrigerators. The development of the latter (mainly Stirling-cycle machines), originally driven by IR detector technology, has more recently found increasing use in high-T_c superconductor applications. It is rapidly approaching the point that we may begin to view such a refrigerator as just another module inside a piece of electronic equipment, somewhat analogous to, say, a fancy high-voltage power supply.

237

S. Luryi et al. (eds), Future Trends in Microelectronics, 237-250.
© *1996 Kluwer Academic Publishers. Printed in the Nettherlands.*

238

Suppose I offered you a self-contained box, about 2-3 liters in volume, drawing less than 100 Watts, and I would provide inside this box a volume of about $100 cm^3$ inside which I guarantee a temperature T, of say, 77K, with a cooling capacity of, say 3-4 Watts. Given a reasonable cost, such a box would evidently meet our demand for user-friendly cryogenics. The above specifications are not fictitious, they are those of actual hardware about to go into production, interestingly by a company whose business is in the field of high-T_c superconductors, and which has found it necessary to provide integrated system solution to its customers, solutions that include a user-transparent cryogenics package (Superconductor Technologies, Santa Barbara, CA).

The principal bottleneck to their more widespread use is their cost, but this is likely to follow the classical pattern of dramatic cost reduction in the wake of building up mass production. Furthermore, the specifications are likely to improve with time, including rapid progress to lower temperatures with time, at least to about 20K, the practical limit of the Stirling cycle, with slower progress below that.

1.2. SUPERCONDUCTOR-SEMICONDUCTOR DEVICES

1.2.1. *Hybrids With Buffer Layers*

The devices that very likely will emerge in the wake of this development will not only be supercohducting devices using high-T_c superconductors, and conventional devices such as FET's explicitly designed to operate at low temperatures, but also integrated super-semi hybrids. The first class likely to emerge are high-T_c superconductors integrated on-chip with semiconductor devices, like a superconducting SQUID integrated with GaAs or InAs electronics. Because of processing compatibility limitations, such devices require a buffer layer between the two kinds of materials, a technology in which much progress has been made recently [1, 2]. But, being devices operating at temperatures within easy range of the Stirling cycle, such devices should emerge relatively soon.

1.2.2. *Monolithic Integration without Interface Barrier*

As "practical" temperatures get pushed lower, we will also see devices in which a low-T_c superconductor, such as Nb, has been integrated with a semiconductor, such as InAs, without an intervening layer, in such a way that the electrons can cross the interface while retaining the phase information that is the essence of superconductivity, thereby inducing superconductivity in the semiconductor. New kinds of Josephson devices based on this principle are rapidly emerging, offering advantages over more conventional Josephson devices. In fact, much of my presentation—all of Section 2—will deal with this particular combination, as a look far ahead at a branch of low-temperature transport physics that is likely to become important over the long term.

For the near-term future (< 10 years), the need for operating temperatures below the Stirling-cooler range (< 20K) implies more elaborate cryogenic techniques, and these devices may, for some time, remain restricted to two kinds of applications environments: (a) Environment where cryogenic temperatures are available in any event, and where cryogenic electronics can be piggy-backed on the existing cryogenics with minimal additional cost. (b) Large-scale "ultimate-performance" computer mainframes the cost of a helium liquefier would represent only a small fraction of the cost of the overall machine.

1.3. ON NOT REPEATING THE PAST

Anybody invoking this last scenario as a realistic one for the future must address him- or herself to the fact that a huge effort of precisely this kind was undertaken by IBM during the 70-s, only to be abandoned in 1983. The failure of this project had a terribly discouraging effect on the whole field of low-temperature electronics, and anybody re-considering this approach is in danger of running afoul of Santayana's famous dictum that "those who do not remember the past are condemned to repeat it."

It has been argued persuasively by Likharev [3] that this failure was due, not to the need for liquid-helium temperatures, but to two quite unrelated reasons: (a) The use of a unsuitable non-refractory metallurgy based on lead as a superconductor, which was not sufficiently stable under thermal cycling. The resulting reliability problems would have been avoided by using niobium as a superconductor. (b) The use of a logic principle, employing voltage-state logic, that was basically too imitative of semiconductor logic, and which had inherent power dissipation limits that negated much of the speed advantage of Josephson junctions. As Likharev points out, a much more suitable form of superconducting logic would be one that is based on the unique property of super-conductors that magnetic flux in superconducting loops is quantized, and which shuffles single flux quanta rather than shuffling voltage states. Likharev's own presentation at this workshop reviews the present state

It would constitute a major breakthrough for superconductor-semiconductor devices if a high-temperature superconductor could be found that is technologically compatible with existing semiconductors, especially III-V semiconductors. As it stands now, all the high-T_c superconductors are oxides that must either be deposited, or require a post-deposit anneal, in a high-temperature oxidizing atmosphere that will simply destroy any of the semiconductors it is in atomic contact with, thereby elimi-nating barrier-free structures. Current research on high-T_c superconductors stresses the achievement of higher critical temperature, rather than elimination of the need for a high-temperature oxidizing environment. From the point of view of super-semi devices, the achievement of semiconductor-compatible materials would be a far more valuable goal, even if it meant a drastic reduction in critical temperature, say, to 40K.

240

2. Semiconductor-Coupled Superconducting Weak Links

2.1. INTRODUCTION

As the title of my presentation indicates, its objective is restricted to low-temperature devices in which superconductors and semiconductors are monolithically integrated into a common device, ignoring both "pure" superconductor devices—such as Josephson tunnel junctions—that do not involve a semiconductor, and pure semiconductor devices that just happen to be specifically designed for low-temperature use. In fact, my presentation concentrates on what I consider the potentially most interesting form of monolithic superconductor-semiconductor integration, namely, *semiconductor-coupled superconducting weak links*. Much of contents of this section is based on a recent longer introductory review of this topic by Professor Hu and myself [4], where the interested reader may find additional details and additional references. An earlier elementary introductions is found in [5].

The term *weak links* refers to superconducting devices in which two superconducting "banks" are coupled through another *conducting* medium, as opposed to Josephson *tunnel* junctions, in which the current flow is by Cooper pair tunneling through an *insulating* barrier. In the case of interest here, the conducting medium is a semiconductor rather than a metal. More specifically, it is a narrow (~15nm) InAs quantum well with AlSb barriers, forming a short ($<1\mu$m) conducting link between two Nb superconducting banks, schematically shown in Figure 1. For reasons I will discuss below, this combination has emerged as a particularly promising one.

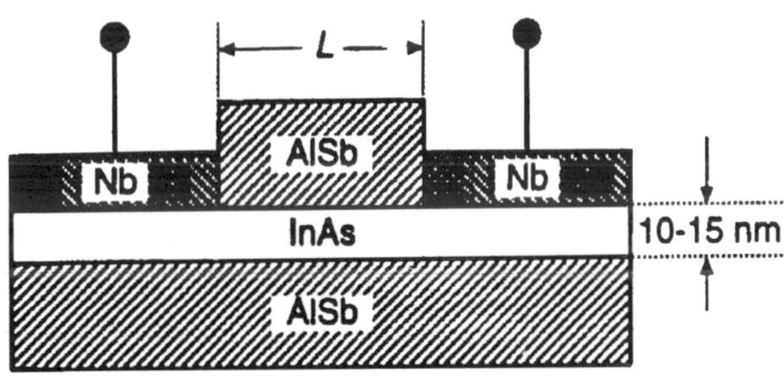

Figure 1. Semiconductor-coupled superconducting weak link based on an InAs-AlSb quantum well forming a conducting link between two superconducting Nb electrodes.

Like Josephson tunnel junctions, weak links exhibit a pronounced *Josephson effect*, manifested by a current-voltage characteristic as in Figure 2, which shows data from a semiconductor-coupled weak link of the kind shown in Figure 1. The characteristic feature of the Josephson effect is the existence of a current range inside which a resistance-less *supercurrent* can flow between the two superconducting banks, up to a certain critical current I_c. Only when this current is exceeded does a voltage appear between the superconducting terminals.

Compared to tunnel junctions, weak links have a much larger inter-electrode separation between the two superconducting banks, which leads two potential major advantages: (a) much lower capacitances, an important consideration for the use of these devices as high-speed devices (b) a much smaller sensitivity of the characteristics to variations in the electrode separation.

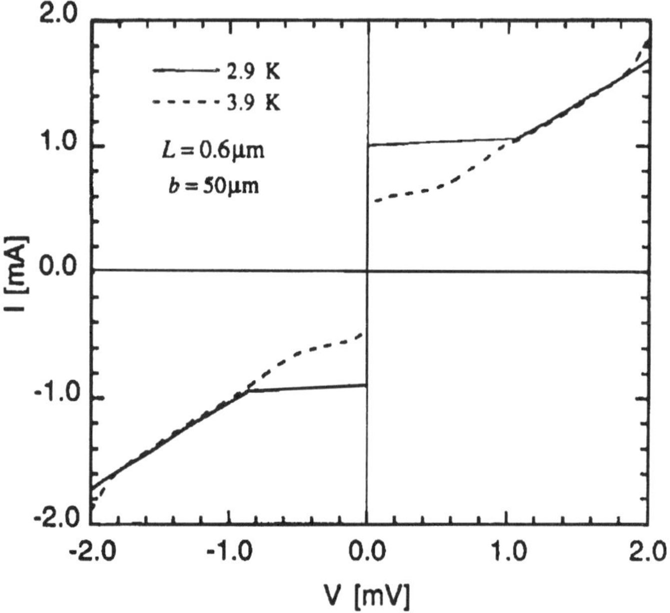

Figure 2. Josephson-type *I-V* characteristics of a device as shown in Figure 1, with 0.6µm electrode separation, at two temperatures [6].

I will not address myself here to the actual *applications* of semiconductor-coupled weak links. In principle, weak links are candidates for all applications for which Josephson tunnel junctions are candidates, with the advantage of higher potential speed, and a technology that lends itself naturally to integration with semiconductor circuitry, including monolithic integration in which the latter operates at the temperature of the weak link itself. One specific application is of course in computers based on the Josephson effect, a topic where I gladly defer to Likharev's presentation at this workshop. However, as I have stated in my earlier presentation at this workshop, the principal applications of any sufficiently new technology tend to be applications *created* by the new technology—which at this time must be left open to speculation.

242

2.2.   BASIC WEAK LINK PHYSICS: A TUTORIAL

2.2.1.   *Current-Phase and Phase-Voltage Relations*

An understanding of weak links requires at least a rudimentary understanding of the basic physics underlying Josephson junctions in general, and weak links in particular. I summarize her the basic facts, without justifications or derivations, for which I must refer to relevant texts (see, for example, refs. [7-10]).

The Pair Wave Function and its Phase.   The essence of superconductivity is the existence of a common pair wave function for the Cooper pairs in the superconductor, which may be written

$$\psi(\mathbf{r}) = |\psi(\mathbf{r})| \cdot \exp\left[i\theta(\mathbf{r})\right]. \tag{1}$$

Here, the magnitude $|\psi(\mathbf{r})|$ of the pair wave function is related to the local Cooper pair density $n(\mathbf{r})$ via

$$|\psi(\mathbf{r})|^2 = n(\mathbf{r}), \tag{2}$$

and $\theta(\mathbf{r})$ is a phase. The key point is that this phase is coherent over macroscopic distances, and, in the absence of a current, it is the same throughout the entire superconductor.

Supercurrent as a Function of the Phase Difference.   In a weak link, two superconductors are coupled through another conducting medium, through which electrons can pass in such a way that the phase of the electrons is preserved in the process. If the phases of the two superconductors are the same, there will be zero net current, but if there is a phase *difference* between the two superconductors, a resistanceless Josephson supercurrent can flow from one superconductor to the other, the magnitude of which is a function of the phase difference $\theta_2 - \theta_1$. For Josephson *tunnel* junctions the functional relationship is simply sinusoidal,

$$I = I_c \cdot \sin\left(\theta_2 - \theta_1\right). \tag{3}$$

Here, with the ordering of the two phases as given, a positive current designates a flow of Cooper pairs from bank #2 to bank #1. Because the pairs carry a negative charge $-2e$, the *electrical* current is in the opposite direction, from bank #1 to bank #2. In weak links, more complicated relations may occur, but $I(\theta_2 - \theta_1)$ is always an odd function of the phase difference, and inasmuch as phase differences have a physical meaning only modulo 2π, the $I(\theta_2 - \theta_1)$ relation is necessarily a periodic one, with a period 2π.

A.C. Josephson Effect. In the absence of a bias voltage between the two superconducting banks, whatever phase difference $\theta_2 - \theta_1$ may be present, will not change with time, hence the current will continue to flow—which is why it is called a supercurrent. If an external bias voltage is present, the difference becomes time-dependent according to the simple law

$$\frac{d}{dt}(\theta_2 - \theta_1) = \frac{2e}{\hbar} \cdot (V_2 - V_1). \tag{4}$$

The supercurrent-vs.-phase relation $I(\theta_2 - \theta_1)$ remains valid in the presence of such a voltage, but the supercurrent now oscillates about zero, with the Josephson frequency

$$v_J = \frac{2e}{h} \cdot (V_2 - V_1), \tag{5}$$

where $h = 2\pi\hbar$ is Planck's constant.

2.3. ANDREEV REFLECTIONS

2.3.1. *Semiconductor-Coupled Weak Links as "Clean" Weak Links*

The weak-link physics of Sec. 2.2 holds independently of the nature of the mechanism that preserves the phase of the electrons. In the semiconductor-coupled weak links discussed here, the mean free path of the electrons tends to be larger than the inter-electrode separation, in which case the mechanism for the phase transfer tends to be dominated by the phase-coherent flow of *ballistic* electrons between the banks. In the jargon of superconductivity, such weak links are called "clean" weak links, in contrast to the more common "dirty" weak links extensively studied in the past, in which the electron transport is diffusive. Unfortunately, much of the literature on weak links, including Likharev's classical review of weak links [11], is still dominated by considerations of dirty weak links.

The mean free path that matters for the phase transfer is not the elastic mean free path that determines the low-field mobility, but the *inelastic* mean free path that is responsible for any de-phasing of the electron waves, and which is typically much longer than the elastic mean free path. For example, in impurity scattering the phase of the scattered wave is coherent with the phase of the incident wave, and while such scattering may create a chaotic wave front, this does not constitute phase-incoherence

244

in the sense of weak link theory: there is still a fixed phase relation between any two points in the wave field.

Given an inelastic mean free path much longer that the inter-electrode spacing, the dominant phase-altering process for the electrons becomes the scattering, not inside the semiconductor, but at the semiconductor-superconductor interface, between the electrons in the semiconductor and those in the superconductor. Now, electron-electron scattering is normally a phase-destroying process. However, at a super-semi interface, at sufficiently low temperatures, the only electrons available for participation in scattering on the superconductor side are the Cooper pairs. But, as we saw earlier, the Cooper pairs all have the same well-defined phase. As a result, the scattering interaction of electrons in the semiconductor with electrons in the superconductor becomes itself a phase-coherent process. It is universally referred to as Andreev scattering or, more commonly, as *Andreev reflections* (AR's), in honor of the man who discovered the possibility of such a process in 1964 [12].

Although postulated over thirty years ago, Andreev reflections have received major attention only during the last few years, when it became clear that their understanding is central to the understanding of clean-limit weak links. As a result of this belated recognition, they have not yet found their way into current textbooks on superconductivity. Even the 1979 weak-ink review by Likharev, written just before clean weak links became technologically realizable, mentions Andreev reflections only in passing. In fact, on page 132 of his paper [11], Likharev explicitly lists a number of experimental observations that are not consistent with the then-existing theoretical understanding, all of which find their explanation via Andreev scattering. I therefore provide here the necessary background on this topic.

2.3.2. *Andreev Reflections: Basic Concept*

The basic idea behind Andreev reflections is simple. Consider an interface between a degenerately doped semiconductor and a superconductor. As shown in Figure 3a, a superconducting energy gap has opened up on the superconductor side. If now an electron with an energy $\mathcal{E}$ *above* the Fermi level (but still inside this gap) is incident on the interface from the semiconductor side, the absence of single-particle states within the gap prevents that electron from entering the superconductor as a *single* electron, and one might expect this electron to be reflected, and the electrical resistance to current flow across the interface actually to increase at the onset of superconductivity in the metal.

However, the electron may pair up with a second electron at the same energy $\mathcal{E}$ *below* the Fermi level, forming a Cooper pair, which *can* enter the superconductor, causing a doubling of the current compared to that in the absence of superconductivity, rather than a suppression. The electron removed from the semiconductor below the Fermi level leaves behind a hole in the Fermi sea. The generally accepted jargon associated with this phenomenon is to say that the incident electron is *reflected as a hole*.

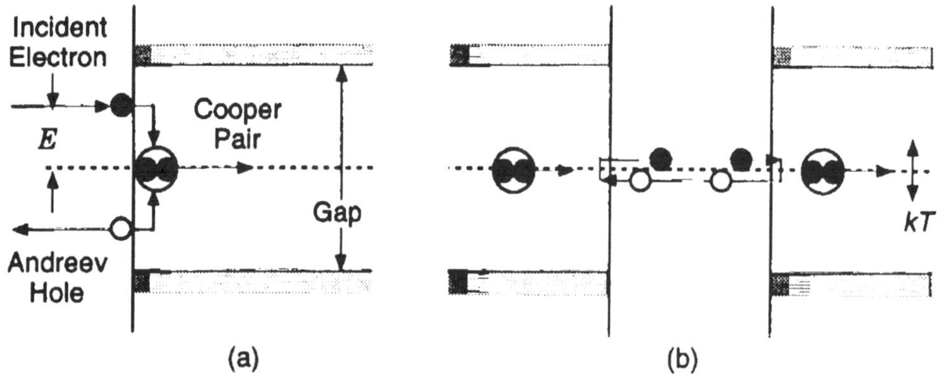

Figure 3. Andreev reflections. (a): Basic concept. (b): Persistent current flow by multiple Andreev reflections.

In a semiconductor with a large mean free path for the electrons, the Andreev hole left behind has a large mean free path itself, roughly equal to that of the original electron, and theory shows that the hole travels back into the semiconductor along a trajectory that essentially re-traces the trajectory of the original incident electron. If its mean free path is sufficiently large, the hole will eventually reach the opposite superconducting electrode. In the absence of any bias across the structure, the energy of the hole is still within the superconducting gap on that side. Such a hole cannot enter the superconductor, but it can be annihilated by breaking up a Cooper pair inside the other superconductor: One of the electrons of the pair annihilates the hole, the other electron takes up the annihilation energy, and is injected into the semiconductor as a ballistic electron *above* the Fermi level, at an energy exactly equal to that of the initial electron. This process, illustrated in Figure 3b, can evidently be repeated: The Andreev reflections act as what I like to call a "Cooper pair pump," annihilating Cooper pairs on one side, and re-creating them on the other. If *all* reflections of electrons and holes were Andreev reflections rather than "ordinary" reflections, and if there were no other kinds of scattering processes, the result would be a persistent current.

However, perturbations are always present, and we would expect simply an enhancement of the conductivity by a factor equal to the number of ballistic traverses before an unfavorable reflection or collision event randomize either the electron or the hole flow in this chain reaction. Also, even if no unfavorable reflection and collision events took place, the diagrams in Figure 3b show only half the story: For each state with a given direction of arrows there exists another state with all current flows reversed. If both states of such a pair are occupied, their currents will cancel. To understand how a supercurrent can arise, we must go beyond the pure ballistic particle

246

picture of Figure 3b, and must take into account the wave properties of the unpaired electrons and holes, and of the Cooper pairs [13].

2.3.3. *Andreev Supercurrents*

Waves have phase, and even in the absence of any scattering events, the simple current-carrying state illustrated schematically by Figure 3b is a quantum-mechanically allowed stationary state only if the round-trip phase shift along the electron-hole loop is an integer multiple of 2π,

$$\Delta\phi_{RT} = 2\pi n. \tag{6}$$

For every value of n, there will actually be two states, corresponding to opposite directions of the flow arrows in Figure 3b.

Up to a point, the above is exactly the same condition as for the bound states in an "ordinary" one-dimensional semiconductor quantum well. These, too, are states for which the round-trip phase changes are the different multiples of 2π. In fact, with regard to the spatial confinement of the *unpaired* electrons and holes inside the semiconductor portion of the structure, the stationary states may indeed be viewed as a new kind of bound states [13], the difference being that the "heterojunction" barriers are now formed, not by the conventional energy gap of another semiconductor, but by the superconducting energy gap of the two superconducting electrodes.

However, there are two decisive differences. The first is that for an AR state confined by superconducting energy gap barriers, the phase on one of the two traverses is carried by an electron, on the other traverse by a hole. This means that these kinds of bound states actually carry a current across the semiconductor, in contrast to the current-less conventional bound states in a conventional quantum well. The two states belonging to a given n belong to opposite directions of that current flow.

A second difference is the following. As in a conventional quantum well with barriers of finite height, the round-trip phase shift contains a contribution from the reflections at the two superconductor barriers. In a semiconductor quantum well, these contributions simply represent the finite penetration of the wave function into the barrier, and they are responsible for lowering the bound state energies with decreasing barrier height.

But in the case of Andreev reflections there is an *additional* phase shift at each bank, equal in magnitude to the phase of Cooper pair wave function in that bank, but with a sign depending on whether an electron or a hole is reflected: When an electron is reflected at a superconductor with phase θ, the wave function of the hole resulting from the reflection acquires an additional phase shift by $-\theta$. This can be readily understood by realizing that the Andreev reflection of an incident electron creates an additional

Cooper pair with phase θ. The phase shift $-\theta$ of the reflected hole simply compensates for the phase of the new Cooper pair.

Conversely, if a hole is reflected, the wave function of the resulting electron acquires the phase $+\theta$, with a similar interpretation. What matters for the Andreev bound states is of course the *net* round-trip phase shift. If the two superconducting banks have the same phase, the phase shifts by $\pm\theta$ at the two banks cancel, but if there is a phase difference between the two banks, it will make a contribution

$$\Delta\phi = \pm\left(\theta_2 - \theta_1\right) \tag{7}$$

to the round-trip phase shift, with the following sign rule: If, in Figure 3b, the left-hand bank is bank #1, the minus-sign applies, otherwise the plus sign.

In order to retain the round-trip condition (6) in the presence of the phase shift contribution $\Delta\phi$, the latter must be compensated for by an opposite change in the phase shift contribution associated with the ballistic flight through the semiconductor itself. But this leads to a change of the energy of the Andreev bound states: A positive contribution to the round-trip phase shift requires a lowering of the ballistic phase contribution, and hence a lowering of the bound-state energy, while a negative contribution raises the latter. Because of the sign difference in (7), in the presence of a nonzero phase difference $\theta_2 - \theta_1$, the energies of the bound states will depend on the direction of current flow in each state, in such a way that the states with a current flow in the direction proper for a Josephson supercurrent will have a lower energy and hence a higher thermal occupation probability, than those with a current flow in the opposite direction. Hence, in this case there will be a thermodynamically stable net current flow, even in the presence of scattering events.

Recall finally that a time-independent phase difference corresponds to zero bias voltage. Hence the stable current is a true zero-resistance supercurrent, with a certain maximum value, the critical current, for some particular value of the phase difference $\theta_2 - \theta_1$.

When the current through the device exceeds the critical current, a bias voltage develops across the semiconductor, leading to the bending-over of the *I-V* characteristic seen in Figure 2. This dissipative regime contains itself a rich variety of physical phenomena, the discussion of which would again go beyond the scope of this paper; the interested reader is referred to the literature, probably starting with a few existing elementary review papers [4-6], which contain extensive references to key original papers, including specifically to papers on the detailed theory for the various phenomena.

12

certificate that says, in effect: "This individual has proven that he/she is capable to perform independent high-quality engineering or scientific work, has adaptability to a wide range of needs, and the ability to make, within a broad strategic context, the decisions about how to conduct that work." This is far more useful than, say: "This individual has spent over four years studying the low-temperature optical absorption of sowhatnium, has honed the technique involved to perfection, and knows more about this specific topic than anyone else in the world."

4. References

1. Lepselter, M. (1974) Integrated Circuits—The New Steel, *IEDM Digest*.

2. Kroemer, H. (1982) Heterostructure Bipolar Transistors and Integrated Circuits, *Proc. IEEE* **70**, 13-25.

3. Kroemer, H. (1963) A Proposed Class of Heterojunction Lasers, *Proc. IEEE* **51**, 1782-1783.

4. Kroemer, H. (1967) Solid State Radiation Emitters, *U.S. Patent* 3,309,553.

However, work on true (three-terminal) JOFET's forms only a small fraction of the overall recent work on semiconductor-coupled weak links that was stimulated by the original JOFET proposal. None of the JOFETs actually reported to-date have shown the kind of performance that offers promise for practical applications. One of their most severe problems is that the obtainable drain-to-source voltage swings are typically much less (<< 1mV) than the gate voltage swings (>> 1mV) required to achieve significant drain current changes. These devices therefore have painfully low voltage gains, which appear to be inherent in their physics. Even the recent NTT devices just barely achieve a voltage gain of unity under optimal loading conditions. It remains to be seen whether or not future developments will overcome this problem. As it stands now, a more likely application of JOFETs is as current-routing switches in superconducting networks, drawing on the fact that a JOFET is an FET with a true zero-resistance on-state, something no pure semiconductor device can offer.

2.4.2 *Multi-Gap Grating Structures*

We ourselves have found it useful to go beyond a single-gap device geometry of Figure 1, and to study series-connected periodic arrays, prepared by laser holography, involving a large number (≥ 300) of gaps, shown schematically in Figure 4.

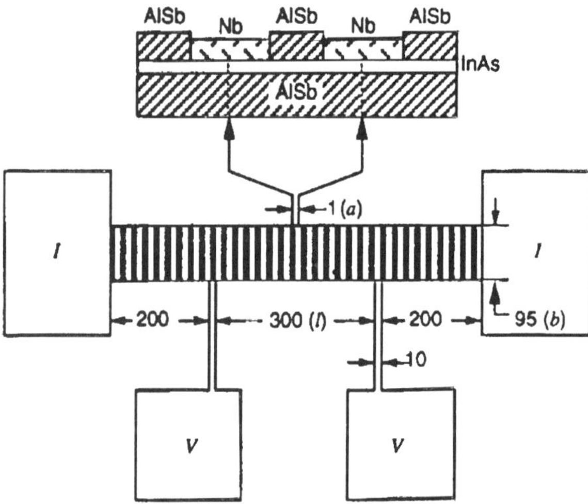

Figure 4. Overall layout (bottom) of Nb grating structure, along with (top) a schematic cross-section through a pair of Nb lines separated by a narrow stripe of InAs-AlSb quantum well. All dimensions are in μm.

We consider such grating structures particularly promising for future applications, and extensive studies of such structures are currently underway, to be reported in due course. Initial results are found in [6] and [4].

250

References

1. Chang, L. D., Tseng, M. Z., Fork, D. K., Young, K. H., and Hu, E. L. (1992) Epitaxial MgO buffer layer for $YBa_2Cu_3O_{7-x}$ thin films on GaAs, *Appl. Phys. Lett.* **60**, 1753-1755.

2. Tseng, M. Z., Jiang, W. N., and Hu, E. L. (1994) Measurements and analysis of Hall effect of a two dimensional electron gas in the close proximity of a superconducting $YBa_2Cu_3O_{7-x}$ film, *J. Appl. Phys.* **76**, 3562-3565.

3. Likharev, K. K. and Semenov, V. K. (1991) RSFQ Logic/Memory Family: A new Josephson-Junction Technology for Sub-Terahertz-Clock-Frequency Digital System, *IEEE Trans. Appl. Supercond.* **1**, 3-28.

4. Kroemer, H. and Hu, E. (1996) "Semiconducting and Superconducting Physics and Devices in the InAs/AlSb Materials System," in *Nanotechnology*, G. Timp, Ed., New York, AIP Press. In the press.

5. Kroemer, H., Nguyen, C., and Hu, E. L. (1994) Electronic Interactions at Superconductor-Semiconductor Interfaces, *Solid-State Electron.* **37**, 1021-1025. (Proc. MSS-6, Garmisch-Partenkirchen, Germany, Aug. 1993).

6. Kroemer, H., Nguyen, C., Hu, E. L., Yuh, E. L., Thomas, M., and Wong, K. C. (1994) Quasiparticle transport and induced superconductivity in InAs-AlSb quantum wells with Nb electrodes, *Physica B* **203**, 298-306. (Proc. NATO Advanced Research Workshop on Mesoscopic Superconductivity, Karlsruhe, 1994).

7. Feynman, R. P., Leighton, R. B., and Sands, M. (1965) *The Feynman Lectures on Physics; Vol. 3: Quantum Mechanics*, Addison-Wesley, Reading. See Sec. 21-9.

8. Kittel, C. (1986) *Introduction to Solid State Physics*, Wiley, New York.

9. Tinkham, M. (1975) *Introduction to Superconductivity*, McGraw-Hill, New York.

10. de Gennes, P. G. (1966) *Superconductivity of Metals and Alloys*, Benjamin, New York.

11. Likharev, K. K. (1979) Superconducting weak links, *Revs. Mod. Phys.* **51**, 101-158.

12. Andreev, A. F. (1964) The thermal conductivity of the intermediate state in superconductors, *Sov. Phys. JETP* **19**, 1228-1231.

13. van Houten, H. and Beenakker, C. W. J. (1991) Andreev reflection and the Josephson effect in a quantum point contact, *Physica B* **175**, 187-197.

14. Mead, C. A. and Spitzer, W. G. (1964) Fermi Level Position at Metal-Semiconductor Interfaces, *Phys. Rev.* **134**, 713-716.

15. Nakagawa, A., Kroemer, H., and English, J. H. (1989) Electrical properties and band offsets of InAs/AlSb *n-N* isotype heterojunctions grown on GaAs, *Appl. Phys. Lett.* **54**, 1893-1895.

16. Silver, A. H., Chase, A. B., McColl, M., and Millea, M. F. (1978) Superconductor-Semiconductor Device Research, *Future Trends in Superconductive Electronics*, Charlottesville, VA, J. B. S. Deaver, C. M. Falco, H. H. Harris, and S. A. Wolf, Eds., Am. Inst. Phys. Conf. Ser., vol. 44, Am. Inst. Physics, pp. 364-379.

17. Clark, T. D., Prance, R. J., and Grassie, A. D. C. (1980) Feasibility of hybrid Josephson field effect transistors, *J. Appl. Phys.* **51**, 2736-2743.

18. Takayanagi, H., Akazaki, T., Nitta, J., and Enoki, T. (1995) Superconducting Three-Terminal Devices Using an InAs-Based Two-Dimensional Electron Gas, *Jpn. J. Appl. Phys.* **34**, 1391-1395.

19. Akazaki, T., Nitta, J., and Takayanagi, H. (1995) Superconducting Junctions using a 2DEG in a Strained InAs Quantum Well Inserted into an InAlAs/InGaAs MD Structure, *IEEE Trans. Applied Supercond.* **5**, 2887-2891.

Reprinted from

P. M Petroff, K. Ensslin, M. S. Miller, S. A. Chalmers, H. Weman, J. L. Merz, H. Kroemer, and A. C. Gossard, ``Novel Approaches in 2 and 3 Dimensional Confinement Structures: Processing and Properties,'' Superlattices and Microstructures, Vol. 8(1), pp. 35-39, 1990.

Superlattices and Microstructures, Vol. 8, No. 1, 1990

NOVEL APPROACHES IN 2 AND 3 DIMENSIONAL CONFINEMENT STRUCTURES: PROCESSING AND PROPERTIES.

P. M.Petroff, K.Ensslin, M. Miller, S. Chalmers, H. Weman, J. Merz, H. Kroemer and A. C. Gossard

Materials Department and Electrical and Computer Department, University of California. Santa Barbara. CA. 93106

(Received 30 July 1990)

In this paper we review two novel types of quantum structures. The first, aimed at producing during growth quantum wire superlattices relies on the deposition of tilted superlattices. Some of the difficulties associated with the growth of tilted superlattices and the novel serpentine superlattice have been discussed and solutions proposed. The second type of quantum structures aimed at producing zero dimensional confinement structures relies on the formation of an antidot lattice. The transport properties of antidot lattices with various periodicities are presented.

1) Introduction

The present interest in nanostructures and quantum structures properties has led to the development of sophisticated and novel processing and testing procedures. Depending on wether mesoscopic or true quantum structures are desired, the processing requirements differ drastically. Indeed, the mesoscopic regime requires only devices with dimensions smaller than the coherence length of the carriers. This regime for high quality III–V compounds semiconductors requires devices with sizes larger than a few 1000Å. Processing such devices is rather easily done with standard lithography techniques (electrons, X-rays and UV lithography). To exhibit easily detectable quantum confinement effects, two and three dimensional carrier confinement structures must have sizes below 500Å. Confinement effects have been demonstrated [1–8] in structures with larger dimensions, however the confining potentials are always small and the properties are detectable only at low temperatures. The proper choice of semiconductors will relax the dimensional requirements [8] but the available systems are few. In all cases, the lithography techniques have been stretched to their limits and because lithography methods are used, the quantum structures density is small.

In this paper, we report on recent progress achieved in the processing of quantum wire superlattices using the tilted superlattice (TSL) structures and a novel type of superlattice, the "Serpentine superlattice". The second half of the paper describes recent advances in antidots structures processing using the focused ion beam. The optical properties of quantum wire superlattices and the transport properties of antidot structures are presented.

2) Advances In Direct Growth of Quantum Wire Superlattices

The process relies on the fabrication of Tilted Superlattices (TSL) which offer the possibility of tilting the superlattice periodicity axis at any angle with the substrate [9,10]. The quantum wire superlattice consists of a thin TSL layer sandwiched by two layers of wider band gap material.

The TSL is fabricated by alternate deposition of fractional monolayers of two III–V compounds on a vicinaly oriented substrate. First demonstrated for the GaAs-AlGaAs system, TSL structures have also been demonstrated for the GaSb-AlGaSb system [11].

The method allows for TSL whose periodicity is function of the substrate misorientation angle and of the TSL tilt angle ß, with respect to the terraces normal. The TSL periodicity for a given substrate misorientation angle, α, can be continuously tuned by changing the tilt parameter p. The TSL period is given by:

$$T = \frac{pd}{\left[\tan^2\alpha + (1-p)^2\right]^{\frac{1}{2}}}$$

The fraction of monolayers for the two semiconductors are m and n and the tilt parameter is p=m+n. The step height d, for GaAs is 2.83Å.

The 3 difficulties associated in the TSL deposition are:

a) the requirement of a periodic step array over the entire substrate, during the deposition of the TSL as well as the deposition of the buffer layer and the cladding layers required for the fabrication of a quantum wire superlattice.

b) the requirements of a uniform tilt angle of the TSL over the entire wafer.

c) the necessity of maintaining sharp interfaces between the quantum wires and the cladding layers.

We examine subsequently the recent progress made in solving these 3 problems.

A) Step Ordering on a Vicinal Semiconductor Surface

The vicinal surface as delivered by the manufacturer has a mean misorientation α, which does not correspond to the presence of a periodic step array on the surface. A Gaussian distribution of terraces with a mean dimension l=d/tgα is present on the surface. Fortunately, for the GaAs [12] and GaSb [11] {100} substrates, nature provides us with a self correcting process which allows us to obtain a periodic step lattice out of a gaussian distribution of steps around a mean misorientation α. If a potential barrier to

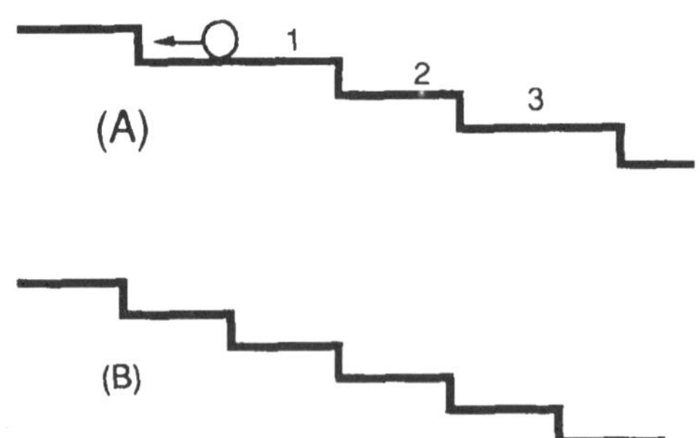

Figure 1: Schematic of two misoriented surfaces with a random step array (A) and a periodic step array (B).

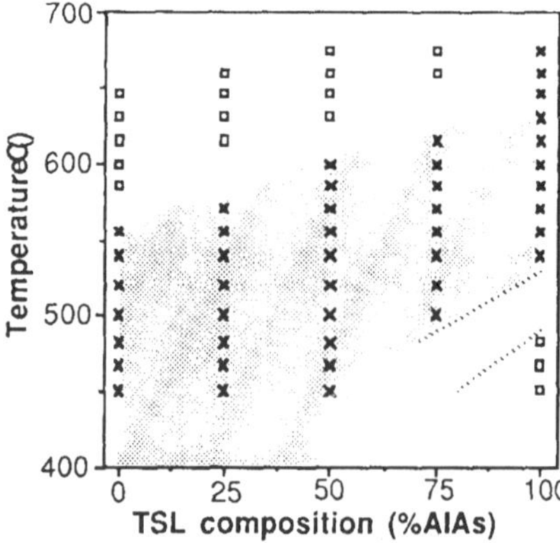

Figure 2: Phase diagram of TSL growth surface morphology as a function of substrate temperature and AlAs composition obtained from the RHEED data analysis. MEE deposition. Data points represent smooth growth. Lightly shaded area represents regions where growth is mostly smooth but appreciable island nucleation is taking place on the terraces. Heavily shaded area represents rough growth [14].

atomic motion prevents atoms from going down from one terrace to the other, e.g. from 1 to 2 or 2 to 3 (Figure 1), before it is incorporated at the step as part of the growing layer, an equalization of the terrace length takes place providing that a layer growth regime is established. This effect was demonstrated analytically and by Monte Carlo simulations [13]. The existence of a potential barrier to atom motion from one terrace to the other is found to be necessary to the self correcting process. The origin of this potential barrier is not clear, however one might speculate that the bond breaking mechanism is more difficult if hybridized bonds are formed at step edges and more bonds have to be broken if the atoms goes from one terrace to another. Intuitively, the short terrace, e.g. 2 in Figure 1, will grow laterally faster than the larger adjacent one, e.g. 3 in Figure 1, since the number of atoms impinging on terrace 3 is larger.

The preparation of the vicinal surface is done by observing during growth of a buffer layer, the double peak structure of the specular beam in the RHEED pattern when the incident electron beam is orthogonal to the step edges. The full width at half maximum is directly correlated to the step periodicity and the distance between these peaks is related to the vicinal surface misorientation. The proper conditions [14] for producing a periodic array of steps will depend on the surface composition, the growth temperature T and wether growth is taking place in the molecular beam epitaxy (MBE) or the migration enhanced epitaxy (MEE) mode [15].

For the $Al_xGa_{1-x}As$ system grown in the MEE mode, a phase diagram has been established experimentally [14] as a function of x and T. As shown in Figure 2 for a As/Ga flux ratio of 6, there is no ideal deposition condition that will preserve steps for a GaAs-AlAs surface during MEE deposition of a TSL. However, a GaAs-$Al_xGa_{1-x}As$ TSL can be grown while preserving a good step structure at a temperature T>600°C for x<0.5.

The preservation of a step lattice at lower temperatures T<500°C for x>0.75 is not presently understood.

B) Tilt Angle Uniformity Issue

The ultra fast variation of the tilt angle with the tilt parameter has serious consequences on the uniformity of the characteristics of the TSL grown on a wafer scale. As seen in Figure 3, small (1 or 2%) variations of the flux incident on the wafer will induce large changes (20–30°) in the TSL orientation. This type of variations is expected

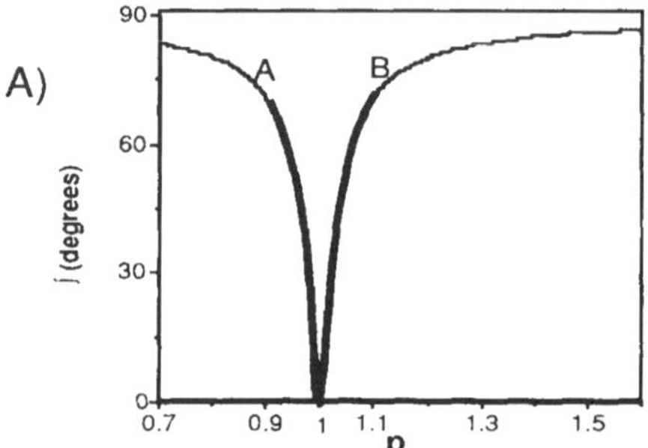

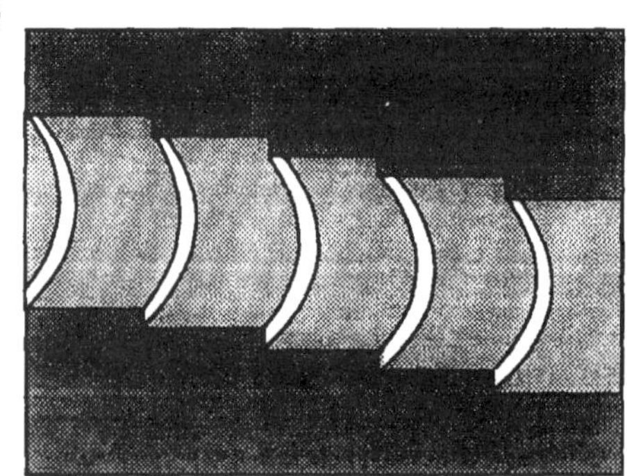

Figure 3: (A) TSL tilt angle versus tilt parameter for a 2° vicinal surface. (B) Serpentine Tilted superlattice schematic. The GaAs region are shown as clear regions while the shaded regions represent the AlGaAs. The curvature of the GaAs quantum wells is determined by continuously changing the variations of p with time from a value smaller than 1 to a value larger than 1.

Superlattices and Microstructures, Vol. 8, No. 1, 1990

even with wafer rotation during growth. Figure 3A shows for a 2° vicinal surface the computed variations of the tilt angle with the tilt parameter [10].

A solution to this problem has recently been proposed [16]. It involves insuring that everywhere on the wafer, there will be a region of material with the proper periodicity. This is obtained by growing a so called "Serpentine superlattice" (SSL).

For example, by imposing a linear time dependence of the tilt parameter p between A and B (Figure 3A) during the growth of the SSL, one can insure that there will always be on the wafer a region for which p=1. A schematic of the resulting SSL is shown in Figure 3B. This type of superlattice structure warrants that a uniform quantum wire superlattice is formed everywhere on the wafer. Two dimensional carrier confinement is achieved in the regions of the SSL with the smallest radius of curvature [16,17]. A three dimensional QWW superlattice is formed by continuously varying the p value alternatively between 2 values which span the p=1 value. The resulting SSL has an S shape. By changing the rate of change with time during growth, the curvature of the SSL quantum well can be varied at will.

Because of the variable width in the SSL's quantum wells, the SSL yields a quantum wire like confinement in the region of largest curvature. The photoluminescence spectrum of a "C" shaped serpentine superlattice deposited on a 2° vicinal substrate is shown in Figure 4. The main peak is attributed to recombination in parts of the SSL which are essentially a random alloy and shows no polarization dependence when a polarized filter is placed in between the sample and the monochrometer. The shoulder, which shows a pronounced polarization dependence, is tentatively assigned to recombination from a quantum wire state. The PL intensity is at a maximum, in the curve labeled 90°, when the polarizer passes light perpendicular to the quantum wire. A weak polarization dependence can also be seen in the background at lower energies. These polarization effects have been seen on 1° and 2° substrates, but are much weaker or absent on 0.5° and 4° substrates. A fuller interpretation of these spectra is still being developed.

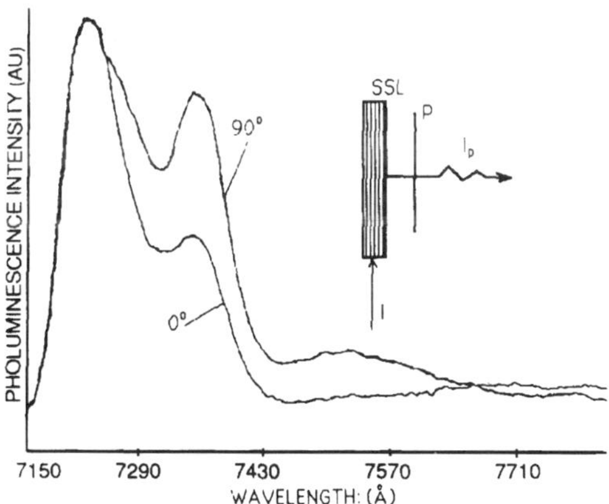

Figure 4: PL spectrum of a "C" shaped serpentine superlattice deposited on a 2° vicinal surface. With the polarizer P set at 90° the electric vector is perpendicular to the wires. The wires are running vertically in the sample and are excited end on [16].

C) The Interface Sharpness in the TSL Structures

One of the more serious issue remaining in the TSL or SSL structures is that of interface sharpness. Measurements have indicated that the interface sharpness of 4 to 5 monolayers are usually present in the TSL. The interface sharpness is controlled by the diffusion kinetics of atoms at the surface and by the equilibrium shape of steps during growth. The presence of kinks at the step edges which is required to insure a layer growth regime is one of the essential component affecting the interface sharpness. Since there are no experimental data on this problem, a modeling approach has been adopted.

An attempt at understanding the partitioning of Al and Ga at the surface is achieved by using a stochastic kinetics simulation method [18] aimed at reproducing the Al-Ga segregation which has been recently observed [19] during MEE growth of the coherent tilted superlattice (CTSL).

The CTSL is realized through the self organization and phase segregation which take place when Al and Ga atoms are simultaneously arriving on a vicinal surface. The modeling reproduces well the observed self organization and phase segregation providing that interactions of the Al and Ga atoms with their nearest neighbors and next nearest neighbors are made anisotropic [18].

Figure 5 shows results corresponding to the model yielding the sharpest interface for 2 temperatures. There is obviously an optimal temperature for growing this interface. A similar approach is under investigation for obtaining conditions which will yield the sharpest interfaces during growth of the normal TSL.

3) Antidot Lattices and Properties

One and zero dimensional transport structures have been realized through a variety of technologies. These structures showed novel quantized conductance and ballistic transport effects. Studies of the electronic properties of quantum dots have been realized by far infrared spectroscopy [3], resonant tunneling [4] and capacitance techniques [5]. Transport through an array of quantum dots (QD) is a challenging problem because of difficulties associated with size irregularities in the QDs and the height of the barrier between dots. Rather, by studying transport through an antidot array [20], some of these difficulties are minimized.

The antidot is produced by damaging locally the two dimensional electron gas (2 DEG) of a modulation doped structure. With this approach, the depletion layer induced by the damage traverses the 2 DEG region and this will produce a much higher confinement potential barrier than the standard surface depletion layer approach. By changing the periodicity in the antidot array, it is possible to investigate the transition between a periodic lattice of scattering centers to an array of localized states which arise from overlap of the antidots. Through the persistent photoconductivity effect or by application of a gate bias over the antidot lattice, the carrier density N_s of the 2 DEG and therefore the effective size of the antidots can be changed. This allows tuning the mobility in the antidot lattice which acts as an array of scattering centers [20].

The 2 DEG is produced by modulation doping an MBE grown GaAs-AlGaAs heterostructure. The 2 DEG interface is located deep (3000Å) below the sample surface. Two samples have been investigated: for one, the carrier density is low, 3.10^{11} cm^{-2} and the mobility high,

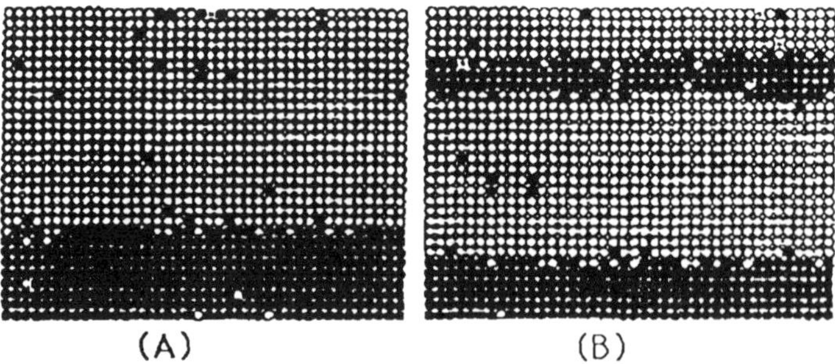

(A)   (B)

Figure 5: Surface image of the Al-Ga (ratio of Al to Ga atoms is 0.33) deposition on a 2° vicinal surface. Full circles denote Al atoms and open ones the Ga atoms. A) After complete deposition of a monolayer (1200 atoms are deposited) at the optimal temperature (T=600°C). B) After complete deposition of a monolayer at lower temperature (kT=1/7). The [110] step edge is located at the bottom of each figure (After reference 18).

$3.10^5 \text{cm}^2/\text{V.s}$; for the other the carrier density is 6.10^{11}cm^{-2} and the mobility at 4.2°K: $10^5 \text{ cm}^2/\text{V.s}$.

The antidot is produced by damaging the 2 DEG with a well focused Ga^+ beam of a focused ion beam implanter operated at 150KV accelerating voltage. The resulting damage is partly annealed at 700°C for 1 minute in a rapid thermal annealer with a forming gas atmosphere. The position of the 2 DEG beyond the projected range (500-600Å) of the Ga^+ beam insures that the defect annealing in the 2 DEG region will be optimized since at this depth only channeled ions [21] will produce elemental defects in low concentration. Another advantage of using the channeled ions resides in the preservation of the FIB probe size (800Å). A dose of 21 ions per antidots has been used. The periodicity for the antidots lattice was varied from 500nm to 110nm.

A mesa structure with a Hall geometry which comprises both an antidot lattice and adjacent to it a non implanted 2 DEG is defined by chemical etching and photolithography. The magnetoresistance of these structures is investigated at 4°K in the range of fields 0<B<8T.

Figure 6 presents, for the low mobility sample, the longitudinal magnetoresistance for an antidot lattice with a periodicity of 500nm. The series of curves correspond to various gate bias voltages of the antidot lattice. For low magnetic fields, ρ_{xx} first drops and passes through a minimum and then increases continuously. For increasing gate bias, the carrier density and the mobility increase. An evaluation of the minima in ρ_{xx} for fields B> 1T indicate a periodicity in 1/B. Thus the magnetoresistance fluctuations correspond to the well known Shubnikov de Haas (SdH) oscillations. The carrier density for the various gate bias are deduced from the Hall measurements as well as the SdH minima positions. Similar results are obtained for antidot structures in the high mobility sample after an illumination pulse.

The minimum in ρ_{xx} at low magnetic field is assumed to correspond to electron delocalization when the diameter of the cyclotron orbits is smaller than the spacing between two antidots. The antidot size is controlled by the size of its associated depletion layer which is assumed to be a

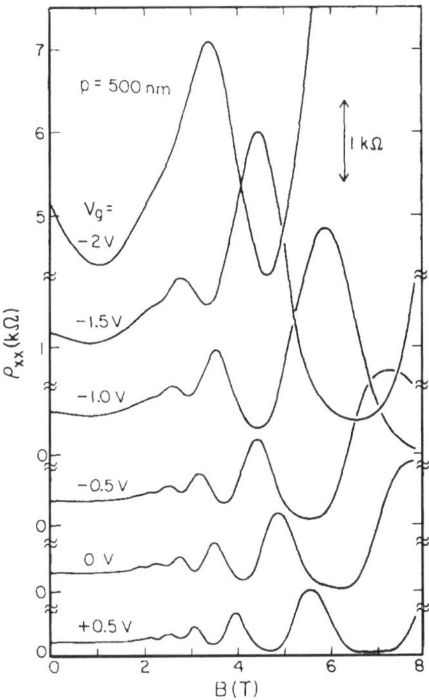

Figure 6: Longitudinal magnetoresistance versus magnetic field at 4°K for an antidot lattice with a periodicity p=500nm. The gate bias voltages are also indicated.

cylinder. An upper limit for the depletion layer length d_{depl} of the carriers is given by $d_{depl.} = p/2\text{-}R_c$. The diameter of the cyclotron orbit, $2R_c$ corresponding to the minimum in ρ_{xx} is given by $2R_c=(2\pi N_s)^{1/2}h/\pi e B_m$. B_m is the field corresponding to the minimum in the magnetoresistance.

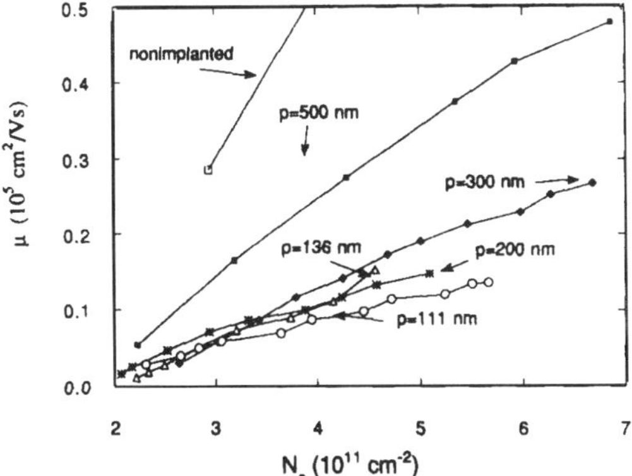

Figure 7: Mobility versus carrier density for the antidot lattice in the low mobility sample. The carrier density was changed by applying a gate bias.

A striking consequence of the effective screening of the antidot potential by the 2 DEG is the depletion layer dependence upon the antidot periodicity. We have observed that the smaller periodicities lead to a more effective screening and therefore smaller depletion layer. The smallest depletion layer dimension was about 500Å for a periodicity of 2000Å. This leads to the possibility of producing antidot lattices with extremely small periodicities providing the focused ion beam probe size can be reduced below its present value (800Å).

The mobility in these antidot lattices varies greatly with their periodicity. As shown in Figure 7, for the same value of N_s, the antidot lattice acts as a scattering center array. This was clearly revealed by the absence of mobility changes (N_s=constant) due to illumination of the antidot lattice in the low mobility sample.

4) Conclusions

In this paper we have reviewed two novel types of quantum structures. The first aimed at producing during growth quantum wire superlattices relies on the deposition of tilted superlattices. Some of the difficulties associated with the growth of tilted superlattices have been discussed and solutions proposed. The serpentine superlattice structure which offers the possibility of producing quantum wire structures has been presented.

The second type of quantum structures aimed at producing zero dimensional confinement structures relies on the formation of an antidot lattice. The transport properties of antidot lattices with various periodicities have been discussed. A delocalization regime in the magnetoresistance corresponding to a transition to a classical 2 DEG transport has been observed as a function of the applied magnetic field.

Acknowledgements—It is a pleasure to thank Y. T. Lu and H. Metiu for the modeling of the crystal growth and J. English for his valuable help with the MBE growth. We acknowledge financial support by the AFOSR and QUEST, a National Science Foundation Science and Technology center.

References:

[1] B.J.vanWees, H.VanHouten, C.W.Beenakker, J.G.Williamson, L.P.Kouwenhoven, D.van der Marel, and C.T.Foxton, Physical Review Letters 60, 848 (1988).

[2] T.J.Thornton, M.Pepper, H.Ahmed, D.Andrews, and G.J.Davies, Physical Review Letters 56, 1198, (1986).

[3] Ch.Sikorki and U.Merkt, Physical Review Letters 62, 2164 (1989).

[4] M.A.Reed, J.N.Randall, R.J.Aggarwal, R.J.Matyi, T.M.Moore, and A.E.Wetsel, Physical Review 60, 535 (1988).

[5] T.P.Smith III, K.Y.Lee, C.M.Knoedler, J.M.Hong, and D.P.Kern, Physical Review B38, 2172 (1988).

[6] T.Hiramoto, K.Hirakawa and T.Ikoma, Journal of Vacuum Science and Technology B6, 1014 (1988).

[7] Y.Hirayama, T.Saku, and Y.Horikoshi, Physical Review B39, 5535 (1989).

[8] W.Hansen, M.Horst, J.P.Kotthaus, U.Merkt, Ch.Sikorski and K.Ploog, Physical Review Letters 58, 2586 (1987).

[9] J.M.Gaines, P.M.Petroff, H.Kroemer, R.J.Simes, R.S.Geels and J.H.English, Journal of Vacuum Science and Technology B6, 1378 (1988).

[10] P.M.Petroff, J.M.Gaines, M.Tsuchiya, R.Simes, L.A.Coldren, H.Kroemer, J.H.English, and A.C.Gossard, Journal of Crystal Growth 95, 260 (1989).

[11] S.A.Chalmers, A.C.Gossard and H.Kroemer, Applied Physical Letters (submitted 1990).

[12] S.A.Chalmers, A.C.Gossard, P.M.Petroff, J.Gaines, and H.Kroemer, Journal of Vacuum Science and Technology B7,1357 (1990).

[13] H-G.Gossmann, S.W.Siden, and L.C.Feldman, Journal of Applied Physics 67, 745 (1990).

[14] S.A.Chalmers, A.C.Gossard, P.M.Petroff, and H.Kroemer, Journal of Vacuum Science and Technology B8, 431 (1990).

[15] Y.Horikoshi and M.Kawashima, Journal of Crystal Growth 95, 17 (1989).

[16] M.L.Miller, H.Wehman, L.Somoska, C.Prior, H.Kroemer, and P.M.Petroff, International Conference of Physics Semiconductors Proceedings (Submitted Thessaloniki 1990).

[17] H.Wehman, M.Miller, J.Merz and P.M.Petroff, Materials Research Society Proceedings (submitted 1990).

[18] Y.T.Lu, P.M.Petroff and H.Metiu, Applied Physical Letters (submitted).

[19] M.Tsuchiya, P.M.Petroff, and L.A.Coldren, Applied Physical Letters 54, 1690 (1989).

[20] K.Ensslin and P.M.Petroff, Physical Review B41, 12307 (1990).

[21] F.Laruelle, A.Bagchi, M.Tsuchiya, J.Merz and P.M.Petroff, Applied Physical Letters 56, 1561 (1990).

``When you look at the history of technology, you see that the principal applications do not evolve Incrementally, but are created by the technology. Until you come up with such applications, you cannot judge how promising the technology is. It is utterly foolish to ask immediately what a new technology is good for.''

Herbert Kroemer

Reprinted with permission from

H. Kroemer,``Heterostructures Tomorrow: From Physics to Moore's Law,'' Inst. Phys. Conf. Ser., Vol. 166, pp. 1-11, 1999.

Inst. Phys. Conf. Ser. No 166: Chapter 1
Paper presented at 26th Int. Symp. Compound Semiconductors, Berlin, Germany, 22–26 August 1999
© *2000 IOP Publishing Ltd*

Heterostructures Tomorrow:
From Physics to Moore's Law

Herbert Kroemer
ECE Department, University of California
Santa Barbara, CA, USA 9310*

Abstract

Research on heterostructures with below-2D dimensionality is predicted to be the most challenging research area in heterostructures for the next few decades. The central problem will be the suppression of statistical fluctuations in size, shape, and placement of these structures. Large progress may be expected, but specific results are almost impossible to predict.

1) Introduction

Heterostructures used to *mean* compound semiconductors. Today, compound semiconductors *means* heterostructures [1]. Even studies of bulk properties nowadays are ultimately undertaken because the materials and properties studied are important for heterostructures (HSs). Anyone having any doubt about this is invited to consider what would be left of compound semiconductor research and technology in the absence of HSs—and whether this symposium would even exist. In fact, HSs are assuming an increasing role even for Si devices.

Nothing illustrates the importance of HSs better than the award of the 1998 Nobel Prize in Physics for the discovery and understanding of the Fractional Quantum Hall Effect (FQHE) in a HS-confined quantized quasi-2-dimensional electron gas (2DEG), following the 1982 prize for the discovery of the "ordinary" (= integer) Quantum Hall Effect (first seen in the 2DEG at a Si/SiO$_2$ interface in what were basically MOSFET structures).

The FQHE also illustrates another point that is central to my presentation: The futility of making long-term technological predictions. The FQHE discovery was completely unexpected, and the history of semiconductor technology is in fact littered with unexpected discoveries. As a result, much of the history of long-term technology forecasts has been a history of failures—the longer the forecast period, the larger the failure [2], because the larger will then be the impact of new discoveries that could not be taken into account at the time the predictions were made. I believe that this unpredictability is a characteristic of all really big research breakthroughs. In fact, I shall refrain from making predictions of

---

* Electronic mail: kroemer@ece.ucsb.edu

2

specific research results; all I shall attempt is to predict research *directions* for, say, the next decade or so, not their results.

Some people might argue that my critique of predictability may apply to future research breakthroughs, but that it should be possible to predict the applications of already ongoing research. I believe that this is a fallacy, too, and I will, in fact devote the last section of my presentation to this point.

For now, let us take today's 2DEG as our point of departure for looking at "tomorrow." The quantized 2DEG was a first case of a successful structure with *reduced dimensionality*. Much of today's quantized-HS research is concerned with reducing the dimensionality of the electron system further, from 2-D quantum wells (with one direction of quantization) to 1-D quantum wires (two directions of quantization), and ultimately to 0-D quantum dots, with all three directions quantized. This trend brings altogether new promises—and problems. Although non-quantized HS devices have by no means disappeared (for example, they form the basis of HBTs), I shall concentrate here on these newer quantized structures, with emphasis on the most extreme case, 0-D quantum dots. This is where I see the most challenging research problems for the "tomorrow" in the title of my contribution.

2) The 2-D Electron Gas—A Review

A 2DEG is more than just a thin 3-D electron gas; the term refers to a gas in a narrow HS potential well for which the electron motion perpendicular to the plane of the well has become quantized to the point that there is no longer any transverse motion across the well.

For simplicity, consider a quantum well of width w with infinitely high walls. Its n-th transverse state (ignoring the unquantized motion *along* the well) has the confinement energy

$$\mathcal{E}_n = \frac{n^2 \hbar^2 \pi^2}{2m^* w^2}, \tag{1}$$

where m^* is the electron effective mass. Assuming, for example, $w = 10\text{nm}$ and $m^* = 0.1 m_e$, we obtain an energy level separation between the two lowest states of $3\mathcal{E}_1 = 113\text{meV}$, which is large compared to kT even at room temperature, and much more so at lower temperatures. Actual quantum wells do not have infinitely high walls, and hence have lower energy level separations. But energy level separations large compared to kT are readily achievable, for sufficiently narrow wells even at room temperature.

At sufficiently low electron concentrations, only the lowest subband will then be occupied. According to the laws of quantum mechanics, any current *across* the well requires a superposition of at least two different transverse quantum states with different confinement energy, and if only the lowest state is occupied, all transverse motion is suppressed, even though the wave function still extends over the full width of the well. In this way we obtain a quasi-two-dimensional electron gas, the "quasi" alluding to the fact that the wave functions are still in 3-D; only the *current* is restricted to the plane of the well, without any lateral excursions even within the well.

The high mobilities in such wells that formed the basis for the QHE—fractional or otherwise—are achieved by applying the trick of modulation doping, a core ingredient of today's HS technology. It is this modulation-doped 2DEG that has dominated the physics research on heterostructures during the last two decades. As we shall see shortly, the transverse quantization is also a key ingredient on today's quantum well lasers.

3) Quantum Wires: Ultra-High Mobilities?

The first impetus for a further reduction in dimensionality was probably Sakaki's work in which it was pointed out that a 1-D electron gas could, in principle, exhibit vastly larger mobilities than even a modulation-doped 2DEG [3]. The central idea was the following. Mobilities are limited by the rate with which electrons can lose their momentum in the current flow direction by scattering events. In a 2DEG, just as in 3-D, this loss of forward velocity can be accomplished by a series of small-angle scattering events within the plane of the quantum well; there is no particular scattering bottleneck.

The transition from 2-D to 1-D is a much more drastic step in the basic physics than the step from 3-D to 2-D. In a 1DEG, there are no longer any energetically accessible states with a sideways momentum, and the scattering physics changes dramatically. The mobility-limiting scattering process is now pure backward scattering, in which the electron must completely reverse its direction of motion in a single scattering event. For quantum wires with the relatively high electron concentrations that would be of practical interest, these scattering events involve a relatively large change in the wave vector k of the electron. But such processes tend to be relatively inefficient, hence the prediction of huge mobilities, which would be of obvious device interest—if the quantum wires were otherwise ideal.

Unfortunately, a new scattering process now rears its ugly head: Quantum wires have a large surface-to-volume ratio, and interface roughness scattering (IFRS) at the seemingly inevitable atomic-scale irregularities in the electron confinement potential now tends to limit the mobility. In principle, IFRS plays already a role in narrow quantum wells. But it is possible to reduce the atomic roughness at *planar* hetero-interface to a very low level: Atomically flat quantum wells, grown over islands with a useful area, are in principle achievable, and have in fact been achieved, at least for selected crystallographic orientations.

But the problem of reducing irregularities at the hetero-interfaces of a wire structure is vastly more difficult: Here, imperfections in lithographic dimensions inevitably enter the problem. With the help of several clever techniques, impressive progress has been made, but Sakaki's original goal has remained elusive.

The reason I bring all of this up here is to illustrate a key point that will dominate much of the rest of my presentation:

4

With the transition from 2-D to 1-D and below, the interface irregularity problem becomes THE fundamental obstacle to the realization of the theoretical promise of structures with reduced dimensionality.

As we shall see, this is true not only for electron transport structures, but for photonic structures as well.

4) Going to the Limit: Quantum Dots

As I said earlier, much of today's research is concerned with reducing the dimensionality of the electron system further, from 2-D quantum wells to 1-D quantum wires, and ultimately to 0-D quantum dots, all with dimensions sufficiently small to lead to quantum effects in all confinement dimensions. These developments are what I think of foremost when I think of "heterostructures tomorrow."

What are the motivations for this trend, other than pure curiosity? One of the key points is the dimensionality-dependence of energy level distribution.

4.1) Density of States Distributions

The simplest representation of the static energy level distribution is in terms of the *density of states* (DOS) of the structure, defined as the number of states per unit energy interval, as a function of the energy at which a narrow energy interval is centered. This density of states is fundamentally different for different dimensionality of the electron system, as illustrated in Fig. 1.

It is a standard textbook fact that for a 3-dimensional gas of free electrons, the DOS increases monotonically and continuously with energy; more specifically, it is proportional to the square root of the electron kinetic energy:

$$D_3\left(\varepsilon\right) = \frac{1}{2\pi^2} \cdot \left(\frac{2m^*}{\hbar^2}\right)^{3/2} \cdot \sqrt{\varepsilon}. \tag{2}$$

Here m^* is again the effective mass of the electrons, and the energy is measured from the bottom of the conduction band (for holes downward from the top of the valence band).

For a 2DEG, the lowest subband has a constant energy density,

$$D_2\left(\varepsilon\right) = \frac{m^*}{\pi\hbar^2}, \tag{3}$$

provided the total electron energy ε exceeds the energy ε_1 is of the lowest transverse bound state in the well. Higher subbands obey a law just like (3), but with a different starting energy, leading to an overall staircase distribution as shown. So long as the Fermi level stays below the bottom of the second subband, we have a quasi-2DEG, the case of greatest fundamental interest (at least to us). Note the discontinuous jump from zero to a constant value, especially of the lowest subband, in contrast to the continuous increase in D in the 3-D case. As we shall see, this jump has important beneficial consequences for QW lasers.

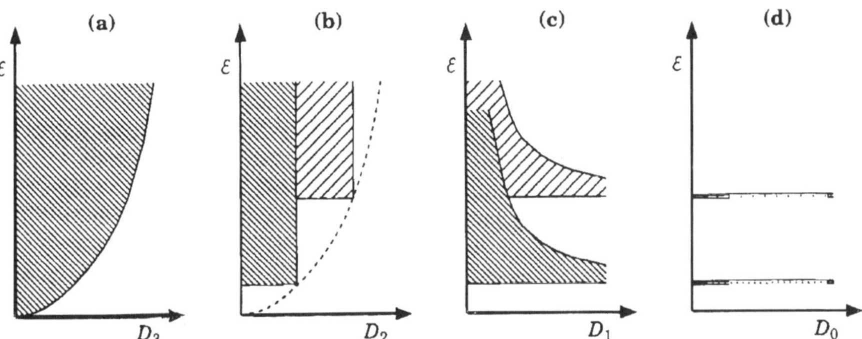

Fig. 1. Density of states for simple systems with dimensionality decreasing from 3-D (a) to 0-D (d). For (b) through (d) only the lowest two subbands are shown.

The really big transition in the physics occurs again when we go from a 2-D gas to a 1-D quantum wire, with *two* quantized dimensions. In the ideal "textbook" limit, the density of states now jumps to infinity at the bottom of each 1-D subband, falling off with increasing energy according to the *inverse*-square-root law for the n-th subband

$$D_{1,n}(\mathcal{E}) = \frac{1}{\pi} \cdot \left(\frac{2m^*}{\hbar^2} \right)^{1/2} \cdot \frac{1}{\sqrt{\mathcal{E} - \mathcal{E}_n}}, \tag{4}$$

where $\mathcal{E}_n$ is the energy of the bottom of the subband. Note that, although D_1 goes to infinity, the square-root singularity remains integrable.

The ultimate limit of quantization is the zero-dimensional quantum dot, in which all three dimensions are quantized, leading to density of states distribution in the form of a string of delta functions, at the energy levels of the bound states of the confining box.

4.2) Laser Implications

Nowhere is the density-of-states distribution more important than in semiconductor lasers. Consider first the 3-D case. If we plot, not the density of *all* states, but only that of *occupied* states, we obtain a distribution with a peak somewhere *above* the minimum allowed energy. The electrons with an energy in the vicinity of that peak (and the holes around an equivalent energy peak) dominate the stimulated emission; there simply are not enough electrons at lower energy to provide enough gain. But the energy of that peak depends both on the amount of population inversion, and—worse—on the temperature. This implies an

6

undesirably large temperature dependence of the laser wavelength (and other laser properties).

In a true 2DEG, because of the abrupt jump of the density of states from zero to a finite value, the maximum density of *occupied* states always coincides with the bottom of the lowest 2-D band, regardless of everything else. While this does not eliminate undesirable temperature dependences, it greatly reduces them. In fact, essentially all of today's practical semiconductor laser diodes are based on quantum wells, for this reason as well as others (of lesser interest to us here).

The situation evidently becomes even better for the sharply peaked distribution of the density of states in quantum wires (Fig. 1c), and the delta function-like distribution for quantum dots (Fig. 1d) would evidently be ideal.

The trouble with this idea is that, for a useful dot laser, we need a very large number of participating dots, just like a conventional laser needs a large number of participating atoms or molecules. But while *atoms* of a given species are naturally identical, technology-dependent *dots* inevitably exhibit statistical fluctuations in their size and shape, with the result that the sharp delta-function distribution of Fig. 1d becomes strongly broadened, each dot contributing a slice of the overall broadened distribution. (Similar problems arise already in arrays of quantum wires.)

At any given laser wavelength only a fraction of the dots can participate, the rest "just sit there." This is not a total disaster: If the electron-hole pairs in the non-participating dots would not undergo spontaneous emission or radiationless recombination, their presence would be of limited consequence. Unfortunately, significant inefficiencies are all but inevitable. Despite the latter, very impressive results with such lasers have been reported —including at this symposium—especially for multimode power lasers driven sufficiently hard to swamp the loss processes. But there can be no doubt that a more perfect control over the dot size and shape remains a central research goal—probably *the* central research goal— the solution of which is essential if the full potential of the reduced dimensionality is to be realized. As a glance at the program of this symposium shows, this is a key research topic already, and I predict that it will remain so for the "tomorrow" in the title of my presentation.

4.3) The Size/Shape/Placement Problem

Better control over dot size (and shape) will almost certainly require a regular dot placement into a periodic lattice. The approach of random nucleation of "self-organizing" dots on an unstructured substrate is not likely to lead to the size (and shape) uniformity ultimately required.

This need for controlled placement will become particularly important if we want to make individual electrical contact to each dot in future electronic circuits, for which we will need a technology that automatically creates a regular array of dots, initially a periodic lattice of dots at pre-determined locations, later extending to non-periodic controlled arrays.

One approach towards this goal would be "smart substrates", with lithographically pre-defined dot positions. However, even a pre-determination of the dot positions does not necessarily lead to a uniform dot size when the dots become small. Consider a scenario in which, during MBE growth, all the atoms landing in a lithographically defined target area coalesce into a single dot at some pre-defined location within the target. If the atoms arrive in an uncorrelated beam, the number per target will be Poisson-distributed, that is, for dots containing an average of N atoms (or formula units, like Ga-As), the resulting dot sizes will fluctuate with a standard deviation of $\sqrt{N}$. For example, dots with an average size of 1000 atoms will fluctuate by about 3%, which for many applications will be too much.

To achieve the desired sub-Poisson "squeezed" distributions calls for more sophisticated approaches than the "hit-and-stick" technique of MBE growth *on top* of a pre-existing lithographic pattern. I suspect that chemical vapor phase epitaxy offers a better chance of uniformity, but my own favored approach would be to first grow an *unpatterned continuous* layer of atomically-controlled thickness, and to pattern this layer *afterwards* into dots of lithographically-defined size and shape. This, of course, calls for the development of suitable nanoscale lithography, but this is a something we will need anyway.

5) Coupled Quantum Dots as Future Electronic Circuits

An altogether new field opens up when we consider quantum dots that are weakly coupled, for example by tunneling between adjacent dots. Controlling the tunneling by suitable gate electrodes offers a new kind of active electronic circuits that operate on the single-electron level. This is already an active research field [4], and I predict that it will become one of the most important areas within tomorrow's heterostructure research.

5.1) Charge Quantization and Single-Electron Devices

Charge is rigidly quantized. Yet, because the charge quantum is so small, all of today's devices treat the electronic charge as a continuous fluid. If anything, charge quantization is considered a nuisance, because it causes shot noise. Yet, from the fundamental perspective, charge quantization could be the ultimate digital property in future nanoscale circuits.

Consider a small heterostructure capacitor with a "plate" area L^2, and a plate separation L. Ignoring edge effects, this capacitor has a capacitance $C \approx \varepsilon L$, where ε is the permittivity of the "dielectric." To change the number of electrons on the plates by one electron, requires a voltage change

$$\Delta V = q\varepsilon / L. \tag{5}$$

For a "nano-capacitor" with, say, $L = 10$nm and $\varepsilon = 10\varepsilon_0$, we obtain $\Delta V = 181$ mV, a value sufficiently large to permit, at least in principle, the control of the number of electrons on the single-electron level, with a

8

tolerable noise margin. Conversely, the voltage appearing across the capacitor will be a measure of the *exact* number of electrons on the plates.

An especially interesting phenomenon takes place when, in a string of weakly coupled identical capacitors an applied voltage is exactly halfway between two values corresponding to two exact integer numbers of electrons. The charge on the plates can then fluctuate, which implies a relatively free current flow along the string at this voltage, but not at other voltages, a phenomenon referred to as *Coulomb Blockade*, and playing a central role in ideas for future nanoscale circuits.

All of this requires—of course—that the capacitors in such a circuit have a sufficiently well-controlled value. Thus we are back to the problem of precise size control. But to achieve tight control over a capacitance is probably less difficult than to achieve similarly precise control over more size- and shape-dependent properties. All that is needed is control over the capacitor *area*; the exact plate *shape* does not matter, in contrast to the energy levels inside quantum dots, which are significantly shape-dependent. Nor do we have to worry about interface roughness scattering: Capacitance is a purely static property. The real problem is that of statistical spatial charge fluctuations of the semiconductor background [4], but this appears solvable. It is considerations such as these that make me expect such devices to have a promising future.

Because of the minimal currents flowing, and the presumably small voltages involved, single-electron circuits would meet another desirable goal: A low dissipation.

5.2) Stacked Quantum Dots as Modulated Quantum Wires.

One potential form of coupled quantum dots that intrigues me personally is that of a vertical periodic stack of identical dots, coupled to each other. In effect, such a stack would represent a true one-dimensional conductor, but with the addition of a periodic potential along the conductor. As a result, the 1-D band structure would break up into alternating allowed and bands and forbidden gaps, a situation similar to what we all have seen in our textbooks, but technologically undoable until now. The transport properties in such a mini-band structure could be extremely interesting, including such possibilities as Bloch oscillations and negative differential mobilities at high fields.

6) Will Quantum Devices Extend Moore's Law?

Much of the semiconductor community has developed an obsession with Moore's Law, that is, the observation that, for more than three decades, the dimensions of devices have shrunk exponentially, thereby making possible chips with an exponentially increasing number of devices. This process must of course saturate *somewhere*, and at least for the last ten years, there have been regular predictions that this saturation is imminent—only to be regularly surpassed by actual continued progress. As a result one now sometimes sees Moore's law viewed as if it were a law of nature, which somehow *must* continue to be valid until we reach atomic device dimensions, and our job is simply to find out what makes it continue to be valid.

This is of course nonsense, but it has had a pernicious effect on the non-Si community: the over-selling quantized HSs as the magic ingredient that takes over when CMOS "runs out of technology." The argument runs roughly as follows. If continued, the reduced device scales *must* eventually lead to structures that obey quantized transport laws rather than the drift-and-diffusion laws of mainstream Si devices. Now HS-based quantum devices are of course devices of this kind. By a leap of faith, much of the current research on HS-based quantum devices is therefore sometimes justified as research towards this post-Si era of devices. I wished it were so!

I am not going to claim that this hope is without *some* rational basis. But if we are serious about it, we must pay more attention to another problem: Moore's law is ultimately a law of the exponential increase with time in the number of devices processed *per processing step*. The reduction of device dimensions was *necessary* to make this possible, but it was not by itself sufficient. Yet almost all current work on quantum devices is using *serial* manufacturing, one device at a time. That is fine for physics research. But if HS-based quantum devices with nanoscale dimensions are to fulfill the post-Si promises often made for them, we must pay more attention to the development of massively parallel assembly techniques that go far beyond "Y2K" Si technology. In fact, a look at present-day one-at-a-time quantum devices shows that the overall device dimensions are actually larger than the overall dimensions of present-day CMOS ICs, not to mention readily foreseeable CMOS dimensions a few years down the road.

I do not say that it cannot be done; I am simply trying to draw attention to this need, and I *do* predict that work along this direction will play an increasing role.

In fact, I consider the situation far from hopeless, and I cannot resist the temptation to speculate which directions this research might take. The Silicon Road Map seems to place its bets on two technologies: (a) Going to even shorter wavelengths in optical projection lithography [5], and (b) a form of e-beam projection lithography know as SCALPEL (= Scattering with Angular Limitation Projection Electron-beam Lithography) [6], which is a parallel rather than serial process I suspect that going to shorter-wavelength UV is not going to go far enough to carry us to the dimensions of ultimate interest. SCALPEL looks more promising, but my personal suspicion is something altogether different: Abandoning *projection* lithography altogether and going to nano-scale *contact* printing techniques that have recently shown remarkable promise [7].

7) So what will be the Applications?

Let me now return to a claim that I made in the *Introduction*, about the futility of predicting applications. Whenever we work on exotic new-physics device structures with an uncertain future, we are almost

10

invariably pressured into stating their practical applications—inevitably meaning applications within a few years. This is a worldwide problem, and it has been getting out of hand in recent years. The honest answer would be to reply that we don't know, and that it is in fact *one* of the objectives of the research itself to look into what applications that research might have.

But the temptation is to justify our research by making speculative promises—which may or may not be realistic, and which at best will be realized only after a much longer time than the questioners expect. Such promises tend to diminish our credibility and may ultimately backfire in the form of diminished research support. We must fight this nonsense on a sounder basis. Ultimately, the justification of more open-ended research lies in its historical record, expressed by what I have once called the *Lemma of New Technology* [8]:

The principal applications of any sufficiently new and innovative technology always have been—and will continue to be—applications *created* by that technology.

But this means that is fundamentally wrong to evaluate the promise of newly emerging dramatically different technologies by asking what their applications might be. Worse, an insistence on such "visible" applications tends to suppress rather than advance progress. The future will belong to those who do not restrict themselves in this narrow-minded way!

One manifestation of this shortsightedness that we must fight particularly, is the cry for "more relevance" in university research, a cry that might be safely translated into a call for less open-ended research. But university research plays a central role in the education of the top technological leaders of the next generation. Restricting that research to what outsiders consider relevant simply deprives our students of acquiring an education preparing them for the future—to the detriment of society as well.

References

[1] For a recent review of driving forces behind these developments, see H. Kroemer, "Band Offsets and Chemical Bonding: The Basis for Heterostructure Applications," *Physica Scripta*, vol. T68, pp. 10-16, 1996.

[2] An elaboration on this point can be found in: H. Kroemer, "Devices for the Future: A Peek into the Next Century," *Int. Conf. on Solid State Devices and Materials*, Yokohama, Japan, 1994, pp. 397-399 (Extended Abstracts).

[3] H. Sakaki, "Scattering suppression and high-mobility effect of size-quantized electrons in ultrafine semiconductor wire structures," *Jpn. J. Appl. Phys.*, vol. 19, pp. L735-L738, 1980.

[4] For a recent review, see K. K. Likharev, "Physics and Possible Applications of Single-Electron Devices," *Future Electron Dev. J.*, vol. 6, Supplement 1, pp. 5-14, 1995.

[5] For a recent review, see N. Harned, "Ultralight lithography," *IEEE Spectrum*, vol. 36, pp. 35-40, 1999.

[6] See, for example, L. R. Harriott, "A new role for e-beam: Electron projection," *IEEE Spectrum*, vol. 36, pp. 41-45, 1999.

[7] S. Y. Chou, "Sub-10 nm imprint lithography and applications," *J. Vac. Sci. Technol. B*, vol. 15, pp. 2897-2904, 1997.

[8] H. Kroemer, "All that Glitters isn't Silicon — or 'Steel and Aluminum Re-Visited'," *NATO Adv. Res. Workshop "Future Trends in Microelectronics: Reflections on the Road to Nanotechnology"*, Ile de Bendor, France, 1995, S. Luryi, J. Xu, and A. Zaslavsky, Eds., NATO ASI Series; Series E: Applied Sciences, vol. 323, Kluwer Academic Publishers, pp. 1-12.

Nano-whatever: Do we *really* know where we are heading?

` `We *really* don't know where we are heading''.

I believe that this is an extraordinary important field and one of the things, I believe is important, that we do not restrict our interest in this work to applications and that we do not let this be driven by perceived applications. If you look at one of the things, I keep repeating is Kroemer's lemma of new technology.

The principal applications of any sophisticated, new and innovative technology have always been and will continue to be applications created by that technology, rather than being pre-existing applications, where the new technology simply provided improvements.

I think this is exactly going to be true in Nanotechnology. I think, we must follow the opportunities that the technology offers us and then see what applications might spread out.

Herbert Kroemer

Reprinted from

H. Kroemer, ` `Speculations about Future Directions,''
J. Cryst. Growth, Vol. 251, pp. 17-22, 2003.

Available online at www.sciencedirect.com

SCIENCE @ DIRECT®

Journal of Crystal Growth 251 (2003) 17–22

JOURNAL OF **CRYSTAL GROWTH**

www.elsevier.com/locate/jcrysgro

Speculations about future directions

Herbert Kroemer[a,b,*]

[a] *ECE Department, University of California (UCSB), Santa Barbara, CA 93106, USA*
[b] *Materials Department, University of California (UCSB), Santa Barbara, CA 93106, USA*

Abstract

Although MBE technology is over a quarter-century old, and has been outstandingly successful in the growth of semiconductor heterostructures, it has a large reserve of as-yet unexplored capabilities left, many of which are likely to play a role in the future evolution of MBE. Developments that can be anticipated are the additions of OMVPE techniques to MBE, for example, for gas etching and surface cleanup. A central problem will be finding MBE-compatible ways to achieve lateral pattern control down to the nanometer scale. Nanoimprint techniques are a good candidate for that. Self-assembled quantum dots will probably give way to lithographically defined quantum dots with much better control over size and placement. Heterostructures of materials other than semiconductors will be increasingly explored, like magnetic and superconducting structures, and may be even organics.
© 2002 Elsevier Science B.V. All rights reserved.

PACS: 81.15.Hi; 81.16.−c; 85.30.−z; 85.35.−p

Keywords: A1. Nanostructures; A3. Molecular beam epitaxy; B3. Heterojunction semiconductor devices

1. Introduction

Speculations about the future of technology are a hazardous business. Much of the history of long-term technology forecasts has been a history of failures, so I undertake the theme of my title with some trepidation. My only consolation is that I am sufficiently old that it is unlikely that I can be called to account for those of my speculations that will turn out to be wrong. But then, maybe some of them will turn out to be right.

Let me start out by telling you what I do *not* intend to talk about. I will not talk about the growth of MBE as a production technology. Perhaps more importantly—and maybe more surprisingly—I will say almost nothing about the application of MBE to specific individual devices.

One of my reasons for the second restraint is that others, more involved than myself, will present much of the future of specific devices at this conference anyway. But my reasoning goes beyond that—which leads me right to the heart of my intended topic. A study of the history of technology presents staggering evidence for what I have called, on other occasions, *Kroemer's Lemma of New Technology:*

The principal applications of any sufficiently new and innovative technology have always been—and will continue to be—applications *created* by that technology. [1]

*Corresponding author. ECE Department, University of California (UCSB), Santa Barbara, CA 93106, USA. Tel.: +1-805-8933078; fax: +1-805-8937990.

E-mail address: kroemer@ece.ucsb.edu (H. Kroemer).

0022-0248/03/$ - see front matter © 2002 Elsevier Science B.V. All rights reserved.
PII: S 0 0 2 2 - 0 2 4 8 (0 2) 0 2 1 9 9 - 1

Now MBE is hardly a new technology anymore, but I am convinced that it has huge reserves left in itself, and that its future probably contains much more than we can currently predict.

But if this is so, then we should not judge the future of MBE technology from the perspective of already-recognized applications, be they heterostructure lasers, HBTs, or what-have-you. Work on those amounts to simply doing something better than we can do it already, but does not represent new applications yet to be *created* by MBE.

I am the first to admit that this is a very speculative proposition, which will not be universally welcome, for two reasons: (a) Applications that will be *generated* by future MBE technology can, by their very nature, not be readily predicted; so I evidently talk about something that is anathema to any control-centered industrial manager. (b) Many of you—perhaps most—are doing MBE in an environment where you do not have the "luxury" of doing MBE in a context of open-ended research, but are compelled to work on very specific applications. But this does not in any way diminish the usefulness—even for those of you—of remaining aware of unanticipated things to come. And there are probably a few members of funding agencies in the audience, who should perhaps be reminded that the current obsession with so-called *strategic research* is little more than a fancy-sounding justification for the discouragement of open-ended research, even though the latter has historically been the ultimate source of most long-term progress. Nobody has said that better than Mermin in his delightful put-down:

> I am awaiting the day when people remember the fact that discovery does not work by deciding what you want and then discovering it. [2]

2. What is MBE?—a broad generic view

My approach to the future of MBE calls for a rather broad generic view of MBE that may go beyond present-day realized capabilities, and before turning to specifics, let me explain this

point of view. A good point of departure is to compare MBE with OMVPE. To me, the two have more similarities than differences. Both produce carefully controlled high-quality epilayers from a stream of incident atoms or molecules. The principal applications of both are in growing heterostructure; "plain" non-hetero layers hardly deserve the elaborate equipment. Finally, both are ultimately technologies for chemical reaction synthesis, with the difference that in MBE the reactions take place only on the growth surface itself, while in OMVPE reactions in the gas phase play an important role. All differences arise from a difference in the mean free path of the molecules on their way from some source to the growth surface. In MBE, this path is large compared to the distance traveled, in OMVPE is short. This central difference gives each of the two techniques certain advantages and disadvantages relative to its competitor, and I believe the future development of both techniques will include attempts to minimize the disadvantages by incorporating some aspects of the "other" technique.

Hence I anticipate future MBE equipment that contains, within the same envelope, an OMVPE capability. In fact, combining the two technologies—albeit not in an integrated piece of equipment—is already being practiced: Some of my colleagues at UCSB working on the new nitrides have found it useful to grow structures where an OMVPE nucleation and template growth is followed by an MBE growth. I expect that we will see more of this kind of hybrid growth, using whichever of the two technologies is better for whichever part of the overall structure.

Finally, some indium compounds grow well under In-stabilized or even In-rich conditions, bordering on a new form of beam-fed liquid-phase epitaxy.

3. The lateral resolution problem

MBE has been spectacularly successful in the degree of control and design freedom on the "vertical" scale along the growth direction, down to individual atomic monolayer control, but it lacks—in common with other crystal growth

H. Kroemer / Journal of Crystal Growth 251 (2003) 17–22

technologies—any significant lateral pattern control *within* those beautiful monolayer planes, especially on the sub-micron scale. These limitations have always been present; they will simply become more severe as we wish to grow increasingly sophisticated structures, especially if the sophistication calls for smaller lateral dimensions, which is likely to be the case. Hence, this can be readily predicted to be one of the dominant developments of the future.

There are two separate aspects involved in this: Multi-step growth, and high-resolution lithography. Let me start with the former.

3.1. Multi-step interrupted-growth techniques

At present, we are still relying almost exclusively on post-growth conventional photolithography. Worse, we are relying on what I would like to call *single-shot* growth followed by lithography-based processing. By the latter I mean a *single* MBE growth sequence—no matter how complicated the internal layer structure—followed by one or more processing steps. What we really need is the capability to have multiple growth sequences separated by processing steps that take place outside the MBE chamber. In Si technology, this capability is routinely present; we would benefit from it, too.

There has recently been some progress in this direction, often referred to as MBE "regrowth" techniques, where a second MBE growth follows some ex situ processing after a first growth. I predict that research in this direction will be one of the important research topics in the years to come.

The problem with all such regrowth techniques is the introduction of interface contamination and defects at the restart interface, especially if ex situ chemical processing has taken place. Simply stopping GaAs MBE growth and exposing the surface to air, without doing anything else, introduces interface defect concentrations (in this case acceptors) exceeding $2 \times 10^{11}\,\mathrm{cm}^{-2}$, with much higher concentrations on processed surfaces. There are of course applications where such defect concentrations are acceptable, for example, when the doping levels on both sides of the interface are

sufficiently high to swamp the interface defects. But we do not want to be restricted to such cases; we want to be able to have "invisible" stop-and-restart interfaces, say, inside a laser structure.

Protecting a GaAs surface during ex situ exposure with a film of As, a technique used successfully for GaAs surface studies, is not the answer for processing, because it protects only those parts of the exposed surface that are left alone during; it does nothing for a surface exposed during the processing, for example, by etching.

More research on those interruption-induced defects is called for, along with the development of in situ cleaning techniques within the vacuum envelope. I doubt that "energetic" techniques, such as ordinary sputtering, e-beam bombardment, or ion-assisted etching will be a fully satisfactory answer: These techniques create damage; and while this damage may be acceptable in many structures, it will be unacceptable in others, and if we do not wish to limit ourselves in what we can do, we need some less-energetic techniques, presumably purely chemical or photo-chemical ones.

This is an area where OMVPE has an advantage. Thermal gas etching is a standard part of OMVPE, and I anticipate that it will become more widely accepted in MBE, too. This will obviously not be done inside the UHV MBE growth chamber itself, but in an interlocked chamber for gas processing. Once we have "lost our innocence" by taking this step, I would not be surprised if we equip the gas chamber with a separate OMVPE-like growth capability of its own. In fact, we might wish to mix MBE with OMVPE even for the growth itself, as is already done in some nitride technology.

Once we have "benign" surface cleaning techniques, we will also increasingly employ *pre-growth* patterning technologies, including the patterned deposition of non-volatile metal precursors. Re-awakening the old vapor–liquid–solid growth technique on a nanoscale appears a possibility.

3.2. Beyond optical lithography?

When discussing structuring on a nanometer scale, people often propose the use of AFM or

20   *H. Kroemer / Journal of Crystal Growth 251 (2003) 17–22*

STM tips as quasi-lithographic tools for achieving the desired resolution (and precise placement). The trouble with such probe methods is that they are *serial*, that is, one object at a time. This is fine for building physics research structures that require just a handful of devices, but it is far too slow for structures on the level of complexity as today's integrated circuits. The same comment remains largely true even for serial electron beam writing, except for relatively simple structures. Those who dream about extending Moore's Law by such serial techniques might do well to recognize that anything with less than 10^9 devices per chip is just not interesting as a competition to CMOS, and then do their own throughput calculations—along with what this means for the equipment amortization cost per chip.

For demanding applications involving a large production volume of structures with high complexity, a *parallel* assembly technique is absolutely required. In the last analysis, Moore's Law is simply a statement about the triumph of parallel assembly via optical lithography with finer and finer resolution, which ultimately required shorter and shorter wavelengths. Current trends in mainstream IC technology are toward extreme-ultraviolet (EUV) lithography—at an astronomical equipment cost.

Such costs may be economically acceptable in the IC industry with its huge production volume, but many of the applications of MBE are not of this kind. There is no doubt in my mind that we *do* wish to participate in the push towards nanoscale dimensions. But in this case we should expand our lithography horizons beyond optical lithography, and I do not think that X-ray projection lithography is the answer. There has recently been a rapidly increasing interest in going back some 550 years to Gutenberg's printing press, but on the nanometer scale (see, for example, Ref. [3]). While I am not persuaded that nanoimprinting will take over from EUV lithography in silicon IC technology, I believe that we should consider it as a natural partner of MBE. I can visualize the printing, not only of masks, but of growth precursors that subsequently react with incoming molecular (or atomic) beams.

4. On self-assembled quantum dots

One way around the lithography resolution problem is to work with nanoscale self-assembled quantum dots that form under certain conditions during MBE growth. I am very impressed by what has been achieved with this technology, for example, in the low-threshold laser field; I refer the reader to the numerous papers at this conference for details. But I believe that the approach of using spontaneously nucleated dots is ultimately too limited, or—to put it positively—is only the proverbial tip of the iceberg of what *might* be achievable. If and when we achieve dots with a much better uniformity and with a tightly controlled placement, this will open up a much wider range of capabilities. Many of my QD friends consider 10% (linear) size fluctuations as excellent uniformity; but that means a 30% volume fluctuation, and 20% fluctuations in the quantum energies. There certainly are applications for which this is sufficient, like LEDs and low-threshold lasers without tight spectral constraints, and possibly other devices, especially if the dots need not be electrically contacted individually. But I am convinced that the true potential of QDs will require us to do much better. In order to achieve the kind of size uniformity that will ultimately be required, controlled placement will almost certainly be necessary—which calls for some sort of the *pre-growth* lithography to which I alluded earlier (maybe imprint-based). The sooner we start moving in that direction, the better it will be. And of course, we need not only dots, many applications will require interconnect lines. This calls not only for a nanoscale line technology, it also calls for a predictable placement of the things to be connected.

Ultimately, we will almost certainly want to go to much smaller dots (and narrower lines) than what we are exploring today. At that point, statistical Poisson fluctuations will seriously enter the picture. Any technique that relies on simply collecting the atoms impinging over a certain target area will suffer from these. For example, a Poisson distribution with an average of 1000 atoms will have a standard deviation of ± 33 atoms, or about 3%. At that point we will need

H. Kroemer / Journal of Crystal Growth 251 (2003) 17–22

growth techniques that are more deterministic in assembling the correct number of atoms.

One obvious way of minimizing Poisson fluctuations would be to first grow extended layers of controlled thickness (something we know how to do very well) and then create the dots (and lines) by "cookie cutter" lithography with nanometer resolution. Furthermore, that continuous layer need not be the final material itself; it could be a precursor, for example, an In film, to be reacted with As after patterning. Regardless of details, nanometer lithography will again be required.

I will probably be told by some that there are no applications for dots this small. That would almost certainly be true *if* we restricted ourselves to the kind of applications that dominate the MBE usage of today. But remember what I said in the *Introduction* about new technology *creating* its own applications. I am convinced that this would be true again here.

5. Beyond "classical" semiconductors

MBE, as a crystal growth technique, started with III–V compounds, especially GaAs and (Al, Ga)As, and that continues to be its mainstream, although by now all III–V compounds have been grown for one purpose or other, almost invariably in the form of heterostructures. The "hottest" III–V materials are of course the nitrides. There has also been significant work on II–VI compounds, but work on other materials is only now becoming a major part of MBE research and technology.

The most active emerging class of new MBE-grown materials is that of magnetic materials, especially magnetic semiconductors. Much of that work sails under the flag of *spintronics*. Inasmuch as there are numerous papers on this topic at this conference, I will simply refer readers to those papers, and only express my expectation that these, and other magnetic materials will be an increasingly important application of MBE technology. What MBE technology, with its tightly controlled and highly instrumented growth procedures, can bring to bear on such materials is not simply an ability to grow thin films with—

maybe—better quality. Instead, the emphasis will be on heterostructure involving layers of different materials; including heterostructures with non-magnetic semiconductors. This is in fact a trend we see already.

I am not persuaded that all the applications that have been predicted for spintronics are realistic. But this skepticism should under no circumstances be interpreted as a criticism of the research itself. I am guided here by my own *Lemma of Technological Innovation*, stated earlier, which suggests that this particular new technology, too, will *create* its own applications, which may or may not have anything to do with the predictions made today.

Another class of materials that I believe will play an important role in the future of MBE technology is that of high-T_c superconductors, including superconductor–semiconductor hybrids. Much of my own research during the last 12 years has been on such hybrids, more specifically on so-called *superconductive weak links* in which an MBE-grown heavily modulation-doped InAs quantum well (with AlSb barriers) acts as a coupling link between two superconductor bodies (Nb) deposited on the InAs by ordinary sputtering [4]. The quality of the super-semiinterface has emerged to be crucial, and it would probably be beneficial if the superconductor, too, could be grown by MBE. Given this background, I hope I can be forgiven for saying a few words about the combination of MBE and superconductors.

Some of the high-T_c cuprates, especially YBCO ($=\mathrm{YBa_2Cu_3O_{7-x}}$), have been prepared by MBE, but the cuprates are poor candidates for super–semi hybrids: They require deposition (or a post-deposition anneal) in a strongly oxidizing environment at high temperatures, a deadly combination for any classical semiconductor. Nor has the inverse approach of growing the semiconductor on top of the cuprate superconductor been more successful: The semiconductor tends to reduce the superconductor, which destroys the super-conductivity. Perhaps the most interesting (and challenging) superconductor candidate for MBE growth is the new intermetallic (non-oxide) superconductor magnesium diboride ($\mathrm{MgB_2}$), with its remarkably high critical temperature (for a non-oxide) of 39 K. Being non-oxidic, it might be compatible

22 *H. Kroemer / Journal of Crystal Growth 251 (2003) 17–22*

with semiconductors for future super–semi hybrids, an old favorite topic of mine. In fact, initial reports on the MBE growth of MgB_2, at remarkably low growth temperatures ($\leqslant 320°C$) on various substrates, including specifically Si (1 1 1), look promising [5]. Given the low growth temperatures, growth on III–V compounds might be possible, including specifically on InAs, the ideal coupling medium for semiconductor-coupled superconductive weak links.

Finally, I would not be surprised if MBE were applied to organic materials. The driving force to do so would be the tightly controlled and highly instrumented growth procedures of MBE, which might offer capabilities beyond those of classical organic chemistry.

References

[1] H. Kroemer, Rev. Mod. Phys. 73 (2001) 783–793.
[2] D. Mermin, Phys. Today. 52 (1999) 11–13.
[3] C. Kim, M. Shtein, S.R. Forrest, Appl. Phys. Lett. 80 (2002) 4051 (this paper contains extensive references to earlier work. See also a series of conference papers in J. Vac. Sci. Technol. B 19 (2001) 2707).
[4] M. Thomas, H.-R. Blank, K.C. Wong, H. Kroemer, E. Hu, Phys. Rev. B. 58 (1998) 11676.
[5] K. Ueda, M. Naito, Appl. Phys. Lett. 79 (2001) 2046.